AF509466

Advances in High-Performance
Non-Ferrous Materials

Advances in High-Performance Non-Ferrous Materials

Editors

Hailiang Yu
Zhilin Liu
Xiaohui Cui

MDPI • Basel • Beijing • Wuhan • Barcelona • Belgrade • Manchester • Tokyo • Cluj • Tianjin

Editors
Hailiang Yu
Central South University
China

Zhilin Liu
Central South University
China

Xiaohui Cui
Xiaohui Cui
China

Editorial Office
MDPI
St. Alban-Anlage 66
4052 Basel, Switzerland

This is a reprint of articles from the Special Issue published online in the open access journal *Materials* (ISSN 1996-1944) (available at: https://www.mdpi.com/journal/materials/special_issues/ferrous_mater).

For citation purposes, cite each article independently as indicated on the article page online and as indicated below:

LastName, A.A.; LastName, B.B.; LastName, C.C. Article Title. *Journal Name* **Year**, *Volume Number*, Page Range.

ISBN 978-3-0365-6652-8 (Hbk)
ISBN 978-3-0365-6653-5 (PDF)

Contents

About the Editors

Hailiang Yu

Hailiang Yu, Professor of Central South University, PhD supervisor, National Overseas High-Level Talent Young Talent, has been engaged in non-ferrous metal forming and processing. He has published 220 papers in mainstream academic journals in this field and has published two monographs, two edited books, and two chapters of English monographs with Science Press, Tsinghua University Press, and Elsevier. He is currently a member of the Board of Directors of the Nonferrous Metals Society of China, the Plasticity Engineering Branch of the Chinese Mechanical Engineering Society, the Composites Committee of the Nonferrous Metals Society of China, the Materials Science and Engineering Academic Committee of the Nonferrous Metals Society of China, the Metallurgical Equipment Branch of the Chinese Society for Metals, the CAE Simulation Committee of the Chinese Society for Simulation, and a member of the Expert Committee of the China Nonferrous Metals Industry Technology Innovation Strategic Alliance. He is a member of Metall. Mater. Trans. A Key Reader, editorial board of *Chinese Mechanical Engineering*, *Journal of Plasticity Engineering*, *Ferrous Metall.* and *Sci. Rep.*, young communication expert of Engineering, young editorial board of *Chinese Journal of Nonferrous Metals* (Chinese and English), and secretary of J. Cent. South Uni. He is the Secretary General of the Youth Editorial Board.

Zhilin Liu

Zhilin Liu has been a professor of both Mechanical Engineering and Materials Science and Engineering at the Light Alloys Research Institute, Central South University since 2021 and is a supervisor for PhD and MPhil students. His courses are "Metal Solidification Theories and Casting Facilities", "Mechanical Engineering Project (I)", and "Introduction to Materials and Manufacturing Engineering" for undergraduate students in Mechanical Engineering. His research mainly focuses on (i) Solidification and Grain Refinement of Cast Metals/Alloys, (ii) Nanomechanics and Micromechanics of Advanced Materials, (iii) Crystallography of Phase Transformation, and (iv) the Ultrasound-assisted casting of large-scale ingots. He received some prestigious awards and honors, including a Marie Curie fellowship (EU), Talent Attraction fellowship (Spain), UQI scholarship (Australia), UQ top-up scholarship, ICM honorable mention paper (USA), CSC scholarship, First prize of China Nonferrous Metal Industry Science and Technology Award, etc.

Xiaohui Cui

Xiaohui Cui has been an Associate Professor at the Light Alloy Research Institute of Central South University. His courses include "Sheet Metal Forming Process and Equipment" and "Heat Treatment Process and Equipment" in mechanical engineering. His research interests include electromagnetic pulse-forming technology, theory and equipment for lightweight materials. The focus of his research is on precision forming and properties of complex thin-walled components of aluminium alloys, magnesium alloys and titanium alloys. He carried out extensive work on reducing residual stress and the spring back of thin-walled parts using electromagnetic force, realizing large size part deformation with incremental loading, and changing the performance of parts through the induction of current and magnetic pressure.

Editorial

Advances in High-Performance Non-Ferrous Materials

Hailiang Yu [1,2,*], Zhilin Liu [1,2] and Xiaohui Cui [1,2]

[1] Light Alloy Research Institute, College of Mechanical and Electrical Engineering, Central South University, Changsha 410083, China
[2] State Key Laboratory of High Performance Complex Manufacturing, Central South University, Changsha 410083, China
*   Correspondence: yuhailiang@csu.edu.cn

Citation: Yu, H.; Liu, Z.; Cui, X. Advances in High-Performance Non-Ferrous Materials. *Materials* **2023**, *16*, 1186. https://doi.org/10.3390/ma16031186

Received: 10 January 2023
Revised: 17 January 2023
Accepted: 19 January 2023
Published: 30 January 2023

Yang et al. [1] investigated both the tensile behavior and failure mechanism of the neutron-irradiated LT21 Al alloy. The significant variation of the tensile properties was due to the neutron transmutation-produced precipitates (i.e., Si and Mg_2Si). Using electron backscatter diffraction, transmission electron microscopy, and X-ray diffraction, Li et al. [2] explored the effects of room temperature rolling and cryogenic rolling on the microstructure, textures, and mechanical properties of the Cu–Ti–Cr–Mg alloy. Material strengthening and microstructural evolution are closely related to the dynamic recovery and dynamic recrystallization that were impeded by cryogenic temperature. Meanwhile, the cryogenic temperature facilitates the production of dislocations and nano-twins, further enhancing the corresponding mechanical properties. Zhang et al. [3] studied the influence of cryogenic treatments on the microstructure and mechanical characteristics of Al–SiC composites. The cryogenic treatment increased the dislocation density and micro defects near the boundaries. More nucleation sites for precipitation during aging could be provided to increase the hardness and yield strength. Wang et al. [4] unraveled the microstructure–mechanical properties relationship of the $W–Al_2O_3$ composite that was prepared *via* induction sintering and rolling processes. Shuai et al.'s work revealed the alloying elements' (i.e., La and Y) effects on the microstructure and mechanical properties of cast Al–Si–Cu alloys [5]. The addition of La and Y had an evident effect on refining and/or modifying α-Al grains, the eutectic Si phase, and $β-Al_5FeSi$ intermetallics. Huang et al. [6] elucidated the precipitation behavior of the $T_B(Al_7Cu_4Li)$ and $S(Al_2CuMg)$ phases of the 2195 Al–Cu–Li alloy at different homogenization temperatures.

Mao et al. [7] and Gao et al. [8] studied the bonding strength of Al-matrix laminated composites subjected to liquid/solid casting and roll bonding, respectively. Zhao et al. [9] analyzed the mechanical parameters of asymmetrical rolling. Ma et al. [10] investigated the influence of CeO_2 particle size on the microstructure, synthesis mechanism, and refining performance of an Al–Ti–C Alloy. Yu et al. [11] explored the correlation between the grain refiner and fracture toughness in a 7050 Al alloy ingot and plate. Moreover, the tensile deformation and fracture behavior of a Ti–5Al–5Mo–5V–1Cr–1Fe alloy were studied in situ by Pan et al. [12]. The effect of the element cobalt on the magnetic properties of $Ni_{50-x}Co_xMn_{35.5}In_{14.5}$ annealed ribbons was studied by Dubiel et al. [13]. Du et al. [14] investigated the deformation behavior and properties of a 7075 Al alloy under electromagnetic hot forming. Xu et al. [15] proposed an efficient approach to the five-axis flank milling of 7075 Al alloy spiral bevel gears. Zhang et al. [16] performed a correlation analysis of the microstructure and tribological properties of an *in situ* VC_p reinforced iron-based composite. Wang et al. [17] conducted systematic experimental work regarding the high temperature oxidation behavior of an equimolar Cr–Mn–Fe–Co high-entropy alloy. Li et al. [18] revealed the microstructural evolution of lamellar-structured high-purity Ni through cold rolling and cryo-rolling. Finally, Manu et al. [19] reviewed the roles of titanium in cast Cu alloys. Their review work summarizes various casting techniques to fabricate bronze alloys, mainly focusing on the microstructures, tensile properties, and tribological characteristics of Cu–Sn and Cu–Sn–Ti alloys [19]. All these theoretical and experimental works aforementioned in this SI will shed new light on fabricating the non-ferrous materials with high performances.

Conflicts of Interest: The authors declare no conflict of interest.

References

1. Yang, W.; Qian, J.; Zhong, W.; Ning, G.; Peng, S.; Yang, W. Tensile Property of Irradiated LT21 Aluminum Alloy Sampled from Decommissioned Irradiation Channel of Heavy Water Research Reactor. *Materials* **2023**, *16*, 544. [CrossRef] [PubMed]
2. Li, R.; Xiao, Z.; Li, Z.; Meng, X.; Wang, X. Work Hardening Behavior and Microstructure Evolution of a Cu-Ti-Cr-Mg Alloy during Room Temperature and Cryogenic Rolling. *Materials* **2023**, *16*, 424. [CrossRef] [PubMed]
3. Zhang, M.; Pan, R.; Liu, B.; Gu, K.; Weng, Z.; Cui, C.; Wang, J. The Influence of Cryogenic Treatment on the Microstructure and Mechanical Characteristics of Aluminum Silicon Carbide Matrix Composites. *Materials* **2023**, *16*, 396. [CrossRef] [PubMed]
4. Wang, C.; Dong, X.; Liu, Y.; Wei, S.; Pan, K.; Zhang, C.; Xiong, M.; Mao, F.; Jiang, T.; Yu, H.; et al. Microstructure and Mechanical Properties of W-Al$_2$O$_3$ Alloy Plates Prepared by a Wet Chemical Method and Rolling Process. *Materials* **2022**, *15*, 7910. [CrossRef]
5. Shuai, L.; Zou, X.; Rao, Y.; Lu, X.; Yan, H. Synergistic Effects of La and Y on the Microstructure and Mechanical Properties of Cast Al-Si-Cu Alloys. *Materials* **2022**, *15*, 7283. [CrossRef] [PubMed]
6. Huang, H.; Xiong, W.; Jiang, Z.; Zhang, J. A Quasi In-Situ Study on the Microstructural Evolution of 2195 Al-Cu-Li Alloy during Homogenization. *Materials* **2022**, *15*, 6573. [CrossRef] [PubMed]
7. Mao, F.; Zhang, P.; Wei, S.; Chen, C.; Zhang, G.; Xiong, M.; Wang, T.; Guo, J.; Wang, C. Interface Microstructure and Mechanical Properties of Al/Steel Bimetallic Composites Fabricated by Liquid-Solid Casting with Rare Earth Eu Additions. *Materials* **2022**, *15*, 6507. [CrossRef] [PubMed]
8. Gao, H.; Gu, H.; Wang, S.; Xuan, Y.; Yu, H. Effect of Annealing Temperature on the Interfacial Microstructure and Bonding Strength of Cu/Al Clad Sheets with a Stainless Steel Interlayer. *Materials* **2022**, *15*, 2119. [CrossRef] [PubMed]
9. Zhao, Q.; Liu, X.; Sun, X. Analysis of Mechanical Parameters of Asymmetrical Rolling Dealing with Three Region Percentages in Deformation Zones. *Materials* **2022**, *15*, 1219. [CrossRef] [PubMed]
10. Ma, Y.; Chen, T.; Gou, L.; Ding, W. Effect of CeO$_2$ Size on Microstructure, Synthesis Mechanism and Refining Performance of Al-Ti-C Alloy. *Materials* **2021**, *14*, 6739. [CrossRef] [PubMed]
11. Yu, F.; Wang, X.; Huang, T.; Chao, D. Effect of Grain Refiner on Fracture Toughness of 7050 Ingot and Plate. *Materials* **2021**, *14*, 6705. [CrossRef] [PubMed]
12. Pan, S.; Fu, M.; Liu, H.; Chen, Y.; Yi, D. In Situ Observation of the Tensile Deformation and Fracture Behavior of Ti–5Al–5Mo–5V–1Cr–1Fe Alloy with Different Microstructures. *Materials* **2021**, *14*, 5794. [CrossRef] [PubMed]
13. Dubiel, Ł.; Wal, A.; Stefaniuk, I.; Żywczak, A.; Potera, P.; Maziarz, W. Cobalt Content Effect on the Magnetic Properties of Ni$_{50-x}$Co$_x$Mn$_{35.5}$In$_{14.5}$ Annealed Ribbons. *Materials* **2021**, *14*, 5497. [CrossRef] [PubMed]
14. Du, Z.; Deng, Z.; Cui, X.; Xiao, A. Deformation Behavior and Properties of 7075 Aluminum Alloy under Electromagnetic Hot Forming. *Materials* **2021**, *14*, 4954. [CrossRef] [PubMed]
15. Xu, H.; Zhou, Y.; He, Y.; Tang, J. An Efficient Approach to the Five-Axis Flank Milling of Non-Ferrous Spiral Bevel Gears. *Materials* **2021**, *14*, 4848. [CrossRef] [PubMed]
16. Zhang, Y.; Lai, R.; Chen, Q.; Liu, Z.; Li, R.; Chen, J.; Chen, P. The Correlation Analysis of Microstructure and Tribological Characteristics of In Situ VCp Reinforced Iron-Based Composite. *Materials* **2021**, *14*, 4343. [CrossRef] [PubMed]
17. Wang, L.; Zeng, Q.; Xie, Z.; Zhang, Y.; Gao, H. High Temperature Oxidation Behavior of an Equimolar Cr-Mn-Fe-Co High-Entropy Alloy. *Materials* **2021**, *14*, 4259. [CrossRef] [PubMed]
18. Li, Z.; Wu, Y.; Xie, Z.; Kong, C.; Yu, H. Grain Growth Mechanism of Lamellar-Structure High-Purity Nickel via Cold Rolling and Cryorolling during Annealing. *Materials* **2021**, *14*, 4025. [CrossRef] [PubMed]
19. Manu, K.; Jezierski, J.; Ganesh, M.R.S.; Shankar, K.V.; Narayanan, S.A. Titanium in Cast Cu-Sn Alloys—A Review. *Materials* **2021**, *14*, 4587. [CrossRef] [PubMed]

Article

Tensile Property of Irradiated LT21 Aluminum Alloy Sampled from Decommissioned Irradiation Channel of Heavy Water Research Reactor

Wanhuan Yang, Jin Qian *, Weihua Zhong *, Guangsheng Ning, Shunmi Peng and Wen Yang

Reactor Engineering Technology Research Institute, China Institute of Atomic Energy, Beijing 102413, China
* Correspondence: qianjin@ciae.ac.cn (J.Q.); zhongwh@ciae.ac.cn (W.Z.)

Abstract: LT21 a type of aluminum alloy used for the irradiation channel of the first heavy water research reactor (HWRR) in China. Studying the mechanical property of irradiated LT21 aluminum under actual service conditions is essential for evaluating its application property. In this paper, tensile specimens of irradiated LT21 were manufactured from the decommissioned irradiation channel of an HWRR; then, tensile tests were carried out, and then the fracture surfaces were observed. The effect of neutron irradiation on tensile behavior and the failure mechanism was analyzed by comparing the result of irradiated and unirradiated LT21 specimens. The results show that, with the thermal neutron flux increasing to 2.38×10^{22} n/cm^2, the YS gradually increased from the initial 158 MPa to 251 MPa, the UTS increased from 262 MPa to 321 MPa, and the elongation decreased from 28.8% to about 14.3%; the brittle fracture of the LT21 specimen appeared after irradiation, and the proportion of brittle fracture increased as the neutron fluence increased; the nanophase structures, with a size of less than 50 nm, were precipitated in the LT21 aluminum alloy after neutron irradiation. Transmutation Si is presumed to be the main cause of the radiation effect mechanism of LT21.

Keywords: aluminum alloy; neutron irradiation; tensile property; irradiation embrittlement; transmutation Si

Citation: Yang, W.; Qian, J.; Zhong, W.; Ning, G.; Peng, S.; Yang, W. Tensile Property of Irradiated LT21 Aluminum Alloy Sampled from Decommissioned Irradiation Channel of Heavy Water Research Reactor. *Materials* **2023**, *16*, 544. https://doi.org/10.3390/ma16020544

Academic Editor: Hideki Hosoda

Received: 14 November 2022
Revised: 24 December 2022
Accepted: 28 December 2022
Published: 5 January 2023

1. Introduction

Aluminum and its alloy have a low melting point. Thus, they have better resistance to the irradiation effect than other metals at low operating temperatures [1–3]. Aluminum is widely used in structural components of research reactors, including the High Flux Isotope Reactor (HFIR), the Savannah River Laboratory (SRL), and the Oak Ridge Research Reactor (ORRR) [4–7]. Munitz et al. [8] studied the effects of neutron irradiation on the mechanical properties and fracture morphology of Al-6063 using scanning electron microscopy (SEM), transmission electron microscopy (TEM), and the tensile tests method. Their results show that with the increase in neutron fluence, the uniform elongation and ultimate tensile strength increase, the intergranular fracture area increases, in addition to the fracture mechanism changes from ductile transgranular shear fracture to the combination of transgranular shear and intergranular dimple fracture. Gussev et al. [9] studied the effect of neutron irradiation on the mechanical properties and fracture behavior of Al-6061 alloy manufactured using ultrasonic additive. Their results show that the specimens exhibit noticeable hardening, and the ductility of the specimens decreases after radiation. The fracture mechanism of the specimen is mainly a ductile-type fracture with many dimples on the fracture surface.

LT21 is an Aluminum alloy used in the irradiation channel of the first HWRR in China. It is significant to study the mechanical properties of irradiated aluminum alloy under experimental conditions to evaluate its application performance [10–12]. In this paper, the tensile specimens of LT21 were manufactured using an irradiated channel decommissioned by the HWRR. By comparing the experimental results of irradiated and non-irradiated LT21 specimens, the effect of neutron irradiation on the tensile properties of LT21 aluminum alloy was discussed, and the failure mechanism of LT21 aluminum alloy before and after irradiation was analyzed.

2. Material and Experiment

2.1. Material

The test materials are irradiated and unirradiated LT21 aluminum from the decommissioned irradiation channel of HWRR. The dimension of the irradiation channel is 53 (outside diameter) $\times$ 1.5 (thickness) mm^2. The main chemical compositions of LT21 aluminum alloy are as shown in Table 1. Figure 1 shows the metallograph of LT21 aluminum alloy. The average grain size is about 108 μm according to the direct measurement result of metallograph, and the primary phase is Al with BCC structure. The rod-like precipitates, which are composed of Mg and Si, are observed either in the Al matrix or at the grain boundary; the length is about 2.8~3.5 μm, and the width is about 1 μm; the coarse rod-shaped precipitates can be found in some place of matrix, with the length at about 10 μm.

Table 1. Chemical composition of A508-3 steel in wt%.

Element	Mg	Si	Fe	Cu	Al
Content	0.96	0.76	0.2	<0.01	Bal.

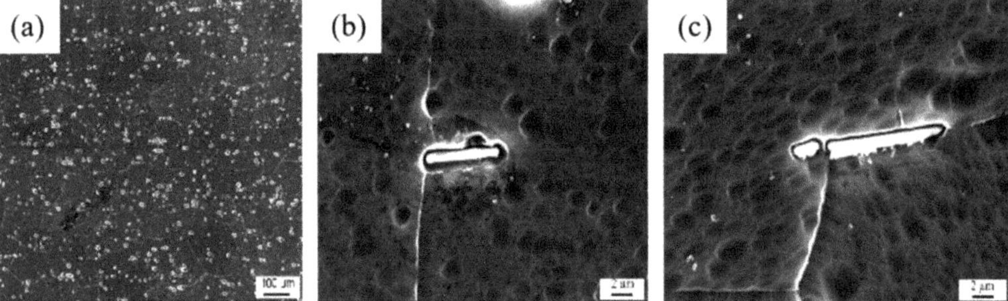

Figure 1. Metallograph of LT21 aluminum alloy: (**a**) Al matrix; (**b**) precipitate in the matrix; (**c**) precipitate at the grain boundary.

The irradiated specimens were sampled from 3 irradiation channels with different service times. The operation parameters are as follows: effective full power day is 801–3020 days, the fluence range of thermal neutron (<0.625 eV) is 5.4×10^{20}~2.38×10^{22} n/cm^2, while the fluence range of fast neutron (>0.1 MeV) is 4.57×10^{17}~2.37×10^{19} n/cm^2. The thermal to fast neutrons ratio (TFR) is about 1000. The operation temperature is about 70 °C (343 K).

2.2. Experiment

Tensile specimens were machined by using a vertical machining center in the hot cell. The dimension of the specimen and sampling schematic is shown in Figure 2. As shown in Figure 3, tensile tests were carried out on a Zwick/Roell Z100 testing machine in a hot cell at ambient temperature, with a test speed of 0.00025 s^{-1}. The extensometer used for tensile tests is Epsilon SN E91695, with an accuracy level of 0.1, and the gage length is 25 mm.

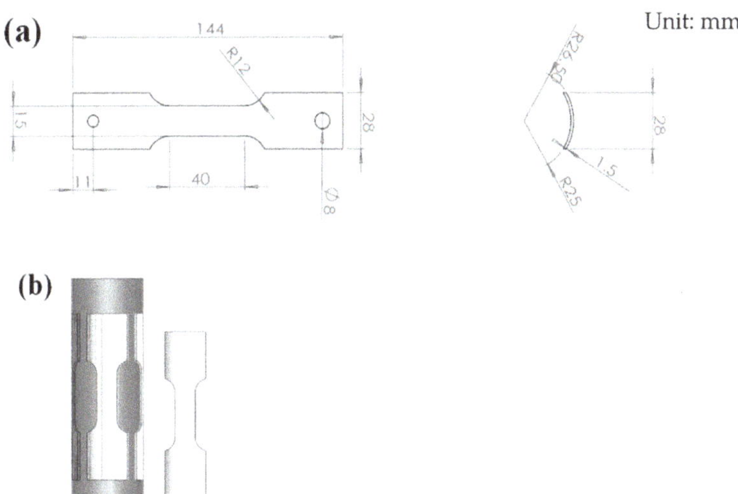

Figure 2. Tensile specimen size and sampling schematic diagram of the irradiation channel tube: (a) tensile specimen size; (b) sampling schematic diagram.

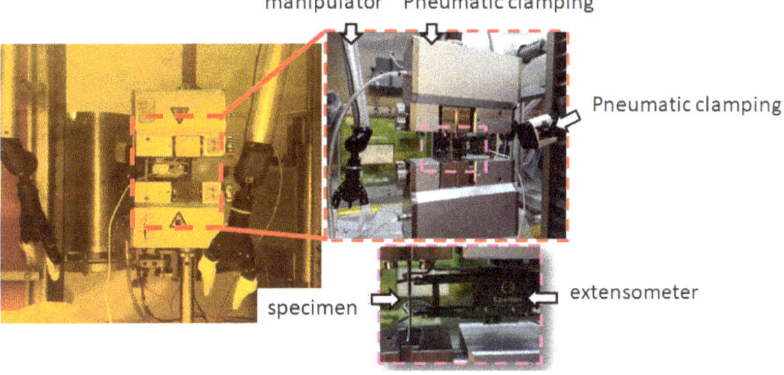

Figure 3. Tensile test machine in hot cell.

After the test, the fracture surfaces of both irradiated and unirradiated specimens were observed by KYKY-EM6900 SEM in the hot cell. The microstructures of both irradiated and unirradiated specimens were observed by JOEL 2100F TEM. The TEM sample preparation process is as follows: the 10×10 mm^2 sized sample was cut from both irradiated and unirradiated LT21 aluminum alloy tube; then, mechanical pre-thinning was carried out in the hot cell until the thickness of the sample became 100~150 nm, and finally, the $\Phi 3$ mm TEM samples were prepared by punch and Twin-jet electropolishing successively. The TEM was operated at 200 kV accelerating voltage. And the microstructure of the LT21 aluminum alloy matrix was observed by bright field image. Under the condition of matrix deviation diffraction, the second-phase particles were imaged.

3. Results

3.1. Tensile Test Results

Figure 4 shows the stress–strain curve of the LT21 aluminum tensile specimen with different irradiation fluence. There are three deformation stages that exist in both irradiated and unirradiated specimens: elastic deformation, plastic deformation, and instability deformation [4]. As the irradiation fluence increased, the strength of LT21 increased while the total elongation rate decreased.

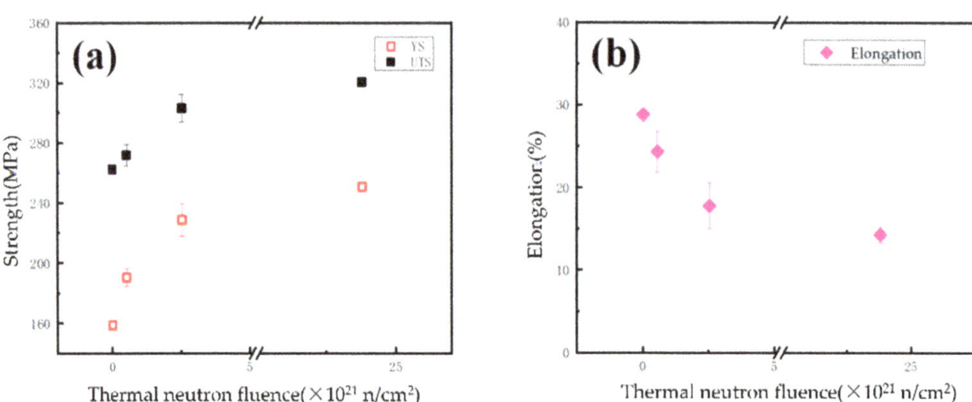

Figure 4. Stress–strain curve of the tensile specimen with different irradiation fluence.

Figure 5 shows the relationship between the tensile properties of LT21 and the thermal neutron fluence. The tensile properties change exponentially with the fluence increase. As the thermal neutron fluence increased up to 2.38×10^{22} n/cm^2, the yield stress (YS) increased gradually from the initial 158 MPa to 251 MPa, and the ultimate tensile stress (UTS) increased from 262 MPa to 321 MPa, while the elongation decreased from 28.8% to about 14.3%. The above test data indicated that, to the LT21, both the YS and UTS increased, while elongation decreased after neutron irradiation. Additionally, this phenomenon obviously proved that irradiation hardening and embrittlement occurred.

Figure 5. The relationship between the tensile properties of LT21 and the thermal neutron fluence: (**a**) strength vs thermal neutron fluence; (**b**) elongation vs thermal neutron fluence.

3.2. Fracture Surface Morphology Observation

Figure 6 shows the overall fracture morphology of tensile specimens with different irradiation fluence. The fracture characteristics of the edge and center are consistent with each other. The fracture surface of the unirradiated specimen shows dimple fracture characteristics, while the irradiated ones show quasi-cleavage characteristics.

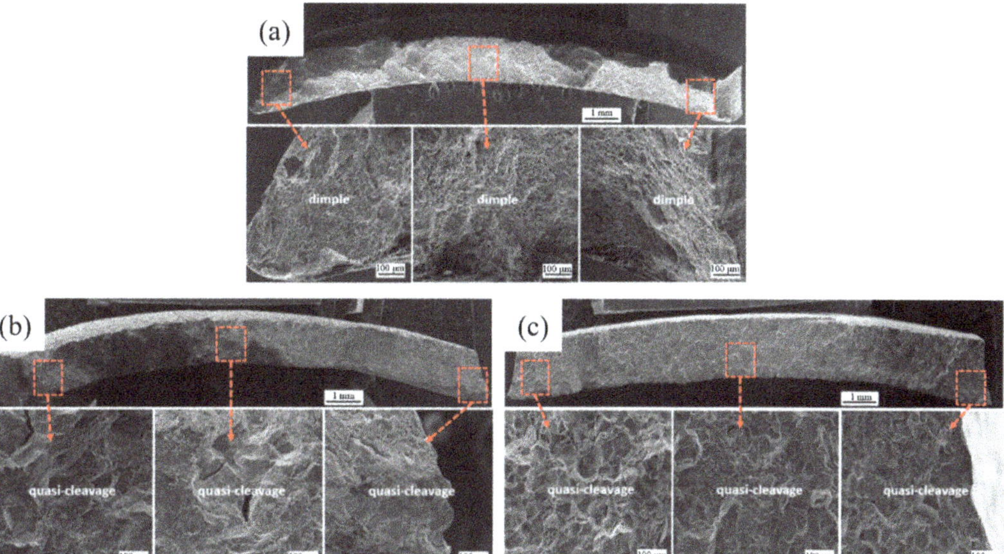

Figure 6. Fracture surface morphology of tensile specimen with different irradiation fluence: (**a**) unirradiated specimen; (**b**) irradiation fluence of 0.54×10^{21} n/cm^2; (**c**) irradiation fluence of 2.38×10^{22} n/cm^2.

Figure 7 shows the local morphology of the fracture surface. For the unirradiated specimen, the fracture surface morphology is mainly in dimple character, while for the low-fluence of the 0.54×10^{21} n/cm^2 specimen, there are intergranular morphologies mixed with dimples, and it shows a quasi-cleavage character. The fracture surface of the high-fluence tensile specimen with the thermal fluence of 2.38×10^{22} n/cm^2 is similar to that of 0.54×10^{21} n/cm^2, which is in quasi-cleavage morphology [13], but the proportion of intergranular cleavage was more than that of lower irradiation fluence. This phenomenon indicates that, after irradiation, the brittle fracture of the LT21 specimen appeared, and the proportion of brittle fracture increased as the fluence increased. Therefore, it can be concluded that LT21 aluminum alloy was embrittled after neutron irradiation; however, the toughness remained on the specimen with a thermal neutron fluence of 2.38×10^{22} n/cm^2.

Figure 8 shows the microstructure of unirradiated LT21 aluminum alloy. The second phase particles were existed in the matrix, which are elongated and rounded. The rod-like second phase particles are about 500 nm in diameter and more than 1300 nm in length. EDS analysis showed that the second phase mainly contained Mg and Si elements.

Figure 9 shows the microstructure of LT21 aluminum alloy after neutron irradiation. The original round particles, which are similar in shape and size to those of unirradiated LT21, can still be seen in the irradiated LT21 aluminum alloy. A certain amount of nanophase structures—nano particles (NP) for which the size is less than 50 nm—were precipitated in the matrix after neutron irradiation. Additionally, as the neutron fluence increases, the density of the nanometer-precipitated phase increases. This phenomenon is similar to other irradiated aluminum alloys [4,14]. These nanophases were produced due to the (n, γ) reaction between the aluminum and thermal neutron, resulting in the transmutation of Al into Si through the (n, γ) reactions. The solid transmutation product Si can have a strong effect on radiation damage structure. The transmutation Si can be precipitated either by elemental precipitation or by forming Mg$_2$Si until the Mg in solid solution is completely consumed in the alloy [4].

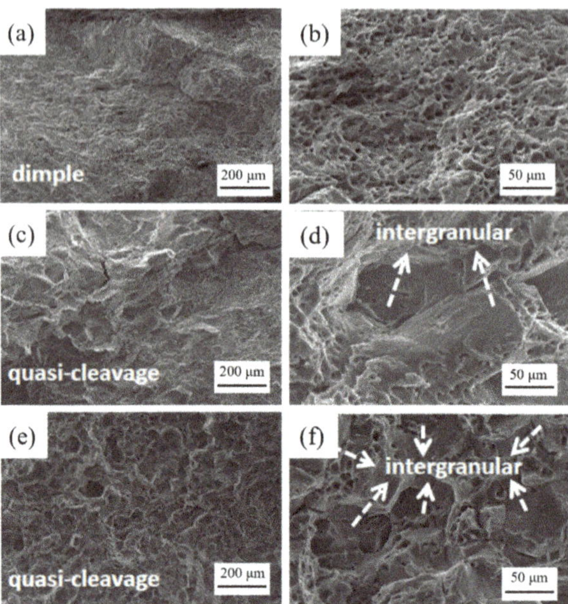

Figure 7. Local characteristics of tensile fracture of samples with different irradiation fluence: (**a**,**b**) unirradiated specimen; (**c**,**d**) thermal neutron fluence of 0.54×10^{21} n/cm^2; (**e**,**f**) thermal neutron fluence of 2.38×10^{22} n/cm^2.

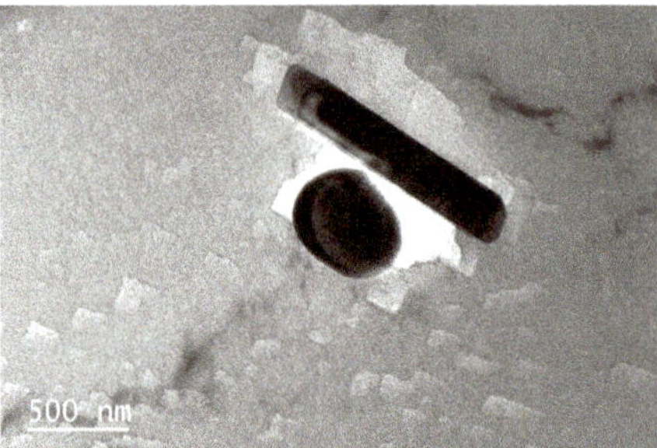

Figure 8. Microstructure characteristics of unirradiated LT21 aluminum alloy.

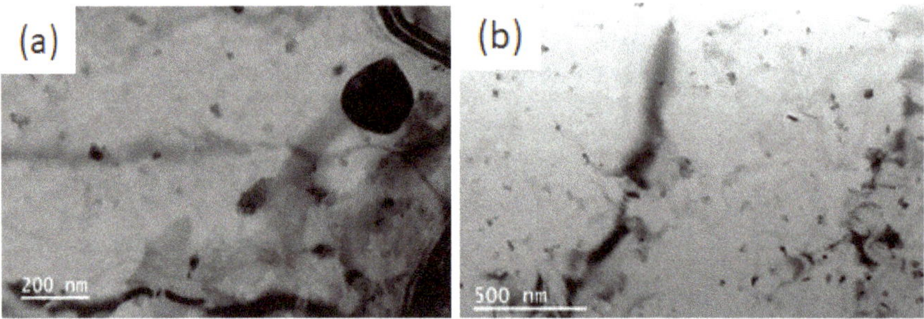

Figure 9. Microstructure characteristics of irradiated LT21: (**a**) thermal neutron fluence of 0.54×10^{21} n/cm^2; (**b**) thermal neutron fluence of 2.38×10^{22} n/cm^2.

4. Analysis

4.1. Tensile Property Evaluation

Figure 10a shows the relationship between the YS of aluminum alloy and thermal neutron fluence. The figure shows the test data of this paper and the other data of reference [14–16]. The YS of aluminum alloy increased with the thermal neutron fluence increase. However, the increased rates are different among different aluminum alloy because of their different manufacturing process and the ratio of thermal-to-fast flux (TFR). There is no obvious relationship between TFR and the variation trend of YS, which indirectly reflects that the thermal neutron irradiation dominates the irradiation hardening of aluminum alloy. The variation trend of the YS of LT21 aluminum alloy with irradiation fluence is similar to that of other aluminum alloys. The YS of LT21 at 23.8×10^{21} n/cm^2, by contrast, is lower than the other aluminum alloy. This may be related to the fact that LT21 has a TFR of 1000, which is much larger than the other aluminum alloys.

Figure 10b shows the relationship between the UTS of aluminum alloy and thermal neutron fluence. The UTS of aluminum alloy increased with the thermal neutron irradiation fluence increase. However, the increased rates are different among different aluminum alloy because of their different manufacturing process and TFR. There is no obvious relationship between TFR and the variation trend of UTS, which indirectly reflects that thermal neutron irradiation dominates the irradiation strengthening of aluminum alloy. The variation trend of the UTS of LT21 aluminum alloy with irradiation fluence is similar to that of other aluminum alloys.

Figure 10c shows the relationship between the elongation of aluminum alloy and thermal neutron fluence. The elongation of aluminum alloy decreased with the thermal neutron irradiation fluence increase. However, the increased rates are different among different aluminum alloys because of their different manufacturing process and TFR. The variation trend of the elongation of LT21 aluminum alloy with irradiation fluence is similar to that of other aluminum alloys. The elongation of LT21 at 23.8×10^{21} n/cm^2, by contrast, is larger than the other aluminum alloys. This may be related to the fact that LT21 has a TFR of 1000, which is much larger than that of other aluminum alloys, and additionally, the gage length of the specimens is different among different tests.

In summary, the variation trend of YS, UTS, and elongation of LT21 aluminum alloy with the thermal neutron fluence is very similar to that of other aluminum alloys. However, after irradiation, the strength of LT21 is lower than that of other alloys, and the elongation is higher than that of other alloys, especially at the thermal neutron fluence of 23.8×10^{21} n/cm^2. Additionally, the tensile property difference between LT21 and other Aluminum alloys after irradiation may be related to the TFR and tensile test method.

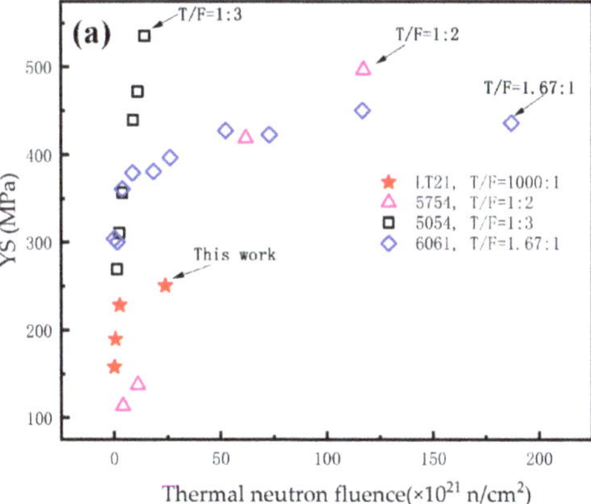

Figure 10. *Cont.*

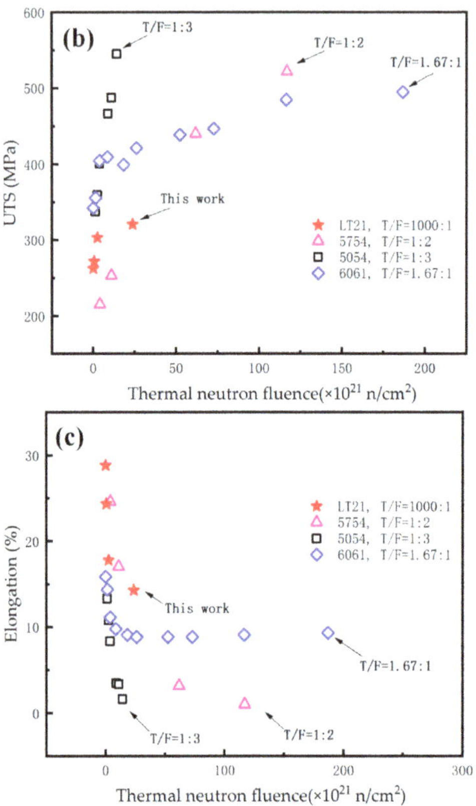

Figure 10. The relationship between (**a**) YS, (**b**) UTS, (**c**) elongation and thermal neutron fluence.

4.2. Irradiation Effect Mechanism

According to the results of mechanical analysis and microscopic analysis, it can be inferred that the effect of neutron irradiation on the tensile property of LT21 aluminum alloy is mainly due to the transmutation-produced Si by thermal neutrons. Similar phenomena have been reported by previous studies [3–5,16]. The mechanism is speculated to be as follows [3,4]. After neutron irradiation, the transmutation product Si forms in the Aluminum alloy, as the solubility of Si in the Al matrix below 373 K is negligible [4]. Hence, the transmutation-produced Si either precipitates in elemental form or forms Mg_2Si precipitates in aluminum alloys. The structure, size, and distribution of these precipitates (Si and Mg_2Si) in the microstructure determines the resulting mechanical properties of the irradiated alloys [4]. For a given volume fraction of precipitates in the microstructure, the precipitates result in higher strength, but lower ductility, and thus cause irradiation hardening. The mechanism continues until saturation in the density of precipitates is expected to occur, leading to the formation of no new Mg_2Si precipitates. With further irradiation, the hardening continues at a decreasing rate as Mg_2Si precipitates continue to grow until all the Mg is pulled out from the Al solid solution and leads to a slow irradiation hardening rate. The above analysis shows that transmutation to Si is the main cause of the irradiation effect mechanism of LT21, and this is the same as previous studies on other aluminum alloys [3–5,9,14–16].

5. Conclusions

In this paper, a tensile specimen of LT21 was prepared from the irradiation channel decommissioned by HWRR. By comparing the experimental results of irradiated and nonirradiated LT21 samples, the effect of neutron irradiation on the tensile properties of LT21 aluminum alloy was discussed. The failure mechanism of LT21 aluminum alloy before and after irradiation was analyzed. The conclusions are as follows:

1. Irradiation hardening and irradiation embrittlement occurred on the LT21 after neutron irradiation. With the thermal neutron flux increasing to 2.38×10^{22} n/cm^2, the YS gradually increased from the initial 158 MPa to 251 MPa, the UTS increased from 262 MPa to 321 MPa, and the elongation decreased from 28.8% to about 14.3%.
2. The variation trend of YS, UTS, and Elongation of LT21 with thermal neutron fluence is very similar to that of other aluminum alloys. The neutron irradiation effect on aluminum alloy mainly depends on the thermal neutron, and the specific tensile properties are different among different irradiated aluminum alloys because of their different manufacturing process and TFR, especially at high thermal neutron fluence, e.g., 2.38×10^{22} n/cm^2.
3. A brittle fracture appeared on the LT21 specimen after irradiation, and the proportion of brittle fracture increased with the neutron fluence increased. After neutron irradiation, the LT21 aluminum alloy had been embrittled, but the toughness remained at the neutron irradiation fluence of 2.38×10^{22} n/cm^2.
4. Transmutation to Si is presumed to be the main cause of the radiation effect mechanism of LT21. The nanophase structures, for which the size was less than 50 nm, were precipitated in the LT21 aluminum alloy after neutron irradiation. These nanophases were produced due to the (n, γ) reaction of the aluminum element and the thermal neutron, and can cause an increase in strength and a decrease in ductility.

Author Contributions: Conceptualization, W.Y. (Wanhuan Yang) and J.Q.; methodology, W.Z.; formal analysis, G.N.; investigation, S.P.; resources, W.Y. (Wen Yang); writing—original draft preparation, W.Y. (Wanhuan Yang); writing—review and editing, J.Q. and W.Z.; visualization, G.N.; supervision, W.Y. (Wen Yang); funding acquisition, S.P. All authors have read and agreed to the published version of the manuscript.

Funding: This work is supported by National key research and development program (2019YFB1900902).

Institutional Review Board Statement: Not applicable.

Informed Consent Statement: Not applicable.

Data Availability Statement: Data available on request from the authors.

Conflicts of Interest: The authors declare no conflict of interest.

References

1. Azevedo, C. A review on neutron-irradiation-induced hardening of metallic components. *Eng. Fail. Anal.* **2011**, *18*, 1921–1942. [CrossRef]
2. Duan, Z.; Yang, H.; Satoh, Y.; Murakami, K.; Kano, S.; Zhao, Z.; Shen, J.; Abe, H. Current status of materials development of nuclear fuel cladding tubes for light water reactors. *Nucl. Eng. Des.* **2017**, *316*, 131–150. [CrossRef]
3. Wolfer, W.G. Performance of Aluminum in Research Reactors. *Comprehens. Nucl. Mater.* **2012**, 143–157. [CrossRef]
4. Kolluri, M. Neutron Irradiation Effects in 5xxx and 6xxx Series Aluminum Alloys: A Literature Review. *Radiat. Eff. Mater.* **2016**. [CrossRef]
5. Chandler, D. High Flux Isotope Reactor (HFIR). *Encycl. Nucl. Energy* **2021**, *2021*, 64–73.
6. Ha, T.-W.; Yun, B.-J.; Jeong, J.J. Improvement of the subcooled boiling model for thermal–hydraulic system codes. *Nucl. Eng. Des.* **2020**, *364*, 110641. [CrossRef]
7. Ragan, G.; Mihalczo, J.; Robinson, R. Prompt neutron decay constants at delayed criticality for the Oak Ridge Research Reactor with 20 and 93 wt% 235U enriched fuel. *Ann. Nucl. Energy* **1997**, *24*, 21–31. [CrossRef]
8. Munitz, A.; Shtechman, A.; Cotler, C.; Talianker, M.; Dahan, S. Mechanical properties and microstructure of neutron irradiated cold worked Al-6063 alloy. *J. Nucl. Mater.* **1998**, *252*, 79–88. [CrossRef]

9. Gussev, M.; Sridharan, N.; Babu, S.; Terrani, K. Influence of neutron irradiation on Al-6061 alloy produced via ultrasonic additive manufacturing. *J. Nucl. Mater.* **2021**, *550*, 152939. [CrossRef]
10. Pornphatdetaudom, T.; Yoshida, K.; Yano, T. Recovery behavior of neutron-irradiated aluminum nitride with and without containing interstitial dislocation loops. *J. Nucl. Mater.* **2020**, *543*, 152584. [CrossRef]
11. Quaireau, S.L.; Colas, K.; Kapusta, B.; Verhaeghe, B.; Loyer-Prost, M.; Gutierrez, G.; Gosset, D.; Delpech, S. Impact of ion and neutron irradiation on the corrosion of the 6061-T6 aluminium alloy. *J. Nucl. Mater.* **2021**, *553*, 153051. [CrossRef]
12. Yano, T.; Ichikawa, K.; Akiyoshi, M.; Tachi, Y. Neutron irradiation damage in aluminum oxide and nitride ceramics up to a fluence of $4.2 \times 1026 \mathrm{n/m^2}$. *J. Nucl. Mater.* **2000**, *283–287*, 947–951. [CrossRef]
13. Chen, H.; Cheng, Y.; Lin, C.; Peng, J.; Liang, X.; Gao, J.; Liu, W.; Huang, H. In-situ TEM observation of nanobubbles evolution in helium-irradiated aluminium upon tensile stressing. *J. Nucl. Mater.* **2019**, *520*, 178–184. [CrossRef]
14. Marchbanks, M.F. *ANS Materials Databook, ORNL/M-4582, Lockheed Martin Energy Systems*; Oak Ridge Natlional Laboratory: Oak Ridge, TN, USA, August 1995.
15. Kapusta, B.; Sainte Catherine, C.; Averty, X.; Campioni, G.; Ballagny, A. Mechanical characteristics of 5754-NET-O Aluminum alloy irradiated up to high fluences: Neutron spectrum and temperature effects. In Proceedings of the Joint Meeting of the National Organization of Test, Research, and Training Reactors and the International Group on Research Reactors, Gaithersburg, MD, USA, 12–16 September 2005.
16. Luzginova, N.V.; Nolles, H.; Berg, F.V.D.; Idsert, P.V.D.; Van Der Schaaf, B. Surveillance Program Results for the High Flux Reactor Vessel Material. *ASTM Int.* **2014**, *2014*, 30–41. [CrossRef]

Article

Work Hardening Behavior and Microstructure Evolution of a Cu-Ti-Cr-Mg Alloy during Room Temperature and Cryogenic Rolling

Rong Li [1], Zhu Xiao [1,2,*], Zhou Li [1,3], Xiangpeng Meng [1,4] and Xu Wang [1]

[1] School of Materials Science and Engineering, Central South University, Changsha 410083, China
[2] Key Laboratory of Non-Ferrous Metal Materials Science and Engineering, Ministry of Education, Changsha 410083, China
[3] State Key Laboratory for Powder Metallurgy, Central South University, Changsha 410083, China
[4] Ningbo Boway Alloy Material Co., Ltd., Ningbo 315135, China
* Correspondence: xiaozhumse@163.com; Tel.: +86-139-7491-0804

Abstract: A Cu-1.79Ti-0.39Cr-0.1Mg (wt.%) alloy was prepared by a vacuum induction melting furnace in a high-purity argon atmosphere. The effects of room temperature rolling and cryogenic rolling on the microstructure, textures, and mechanical properties of the alloy were investigated by means of electron backscatter diffraction, transmission electron microscopy, and X-ray diffraction. The results show that the hardness of the cryogenically rolled alloy is 18–30 HV higher than that of the room temperature rolled alloy at any tested rolling reduction. The yield strength and tensile strength of the alloy cryogenically rolled by 90% reduction are 723 MPa and 796 MPa, respectively. With the increase of rolling reduction, the orientation density of the Cube texture decreases, while the Brass texture increases. The Brass texture is preferred especially during the cryogenic rolling, suggesting that the cross-slip is inhibited at the cryogenic temperature. The dislocation densities of Cu-Ti-Cr-Mg alloy increase significantly during the deformation, finally reaching 23.03×10^{-14} m^{-2} and 29.98×10^{-14} m^{-2} after a 90% reduction for the room temperature rolled and cryogenically rolled alloys, respectively. This difference could be attributed to the impediment effect of cryogenic temperature on dynamic recovery and dynamic recrystallization. The cryogenic temperature promotes the formation of the dislocation and the nano-twins, leading to the improvement of the mechanical properties of the alloy.

Keywords: Cu-Ti-Cr-Mg alloy; cryogenic rolling; hardness; texture; twins

Citation: Li, R.; Xiao, Z.; Li, Z.; Meng, X.; Wang, X. Work Hardening Behavior and Microstructure Evolution of a Cu-Ti-Cr-Mg Alloy during Room Temperature and Cryogenic Rolling. *Materials* **2023**, *16*, 424. https://doi.org/10.3390/ma16010424

Academic Editors: Frank Czerwinski and Dezső Beke

Received: 9 November 2022
Revised: 7 December 2022
Accepted: 29 December 2022
Published: 2 January 2023

1. Introduction

Elastic copper alloys with ultra-high strength have been widely used in electronics, aerospace, and marine engineering fields [1–3]. Copper beryllium alloy is the typical elastic alloy with a good combination of strength and conductivity, however, the high carcinogenicity of the Be limits the application of Cu-Be alloy in environment-friendly areas. Cu-Ti alloy possesses ultra-high strength, good elasticity, and moderate electrical conductivity, and has been an ideal candidate for the replacement of Cu-Be alloy [4,5].

In recent years, much work has been made to further improve the comprehensive properties of Cu-Ti alloy through either composition design or thermo-mechanic treatment. Typically, a reduction of Ti content in Cu-Ti alloy could intensively increase the conductivity of the alloy, however, this could also reduce the strength of Cu-Ti alloy [6]. Markandeya et al. reported that the addition of the Cr element could improve the yield strength and tensile strength of Cu-Ti alloy [7]. Li et al. investigated the microstructure and properties of Cu-2.7Ti-0.15Mg-0.1Ce-0.1Zr alloy and found that the addition of trace Mg element could improve the strength and conductivity of Cu-Ti alloy due to the dragging effect of Mg atoms [8]. Rouxel et al. found that the addition of trace Fe element in Cu-Ti alloy inhibited the modulation decomposition at the early stage of aging, leading to improvements in ductility and strength [9].

For metallic materials, work hardening is a common method to improve the microstructure and mechanical properties of the alloys [10,11]. Severe plastic deformation has been proven to be an ideal method to obtain the sub-micron or nanocrystalline grain structures, leading to the improvements of the room temperature strength and high-temperature ductility of copper alloys [12]. So far, several plastic deformation technologies have been developed to improve the mechanical properties of the alloys. The typical SPD methods include equal-channel angular pressing [13], groove rolling [14], accumulative roll-bonding [15], high-pressure torsion [16], slope extrusion [17], and cryogenic rolling. In most cases, it is difficult to fabricate large-sized products with uniform microstructure by equal-channel angular pressing, slope extrusion, and high-pressure torsion, and the materials prepared by accumulative roll-bonding often have low inter-facial bonding strength and poor edge quality. In recent years, cryogenic temperature rolling (CTR) has been attracted much attention and considered a promising industrial production method to produce alloy [18,19]. Cryogenic temperature rolling can inhibit dynamic recovery and dynamic recrystallization, and effectively increase the dislocation density [20], promoting the formation of an ultra-fine grain of the alloys. Especially, cryogenic temperature rolling could also promote the formation of deformation twins and stacking faults into the matrix [21], and the increased resistivity of the twin boundaries could be about an order of magnitude lower than that from the grain boundaries [22], improving the comprehensive properties of metallic materials. Shanmugasundaram et al. investigated the effect of cryorolling, low-temperature annealing, and aging treatments on the microstructure and mechanical properties of Al-Cu alloy, and the ultra-fine crystal structure and excellent properties were obtained through the cryogenic temperature rolling [23]. Yu et al. found that the asymmetric-cryorolled sheet had higher strength and better thermal stability compared to the asymmetric-rolled sheet [24]. Wang et al. prepared Cu-35Zn alloy with a good combination of strength and ductility due to the ultra-fine microstructure, high-density dislocations, and nanometer-scale deformation twins [25]. However, there was little work on the investigation of the microstructure evolution of Cu-Ti alloy during cryogenic rolling.

The typical Cu-Ti-based alloys included Cu-1.8Ti and Cu-2.4Ti alloys. In this paper, Cu-1.8Ti alloy was chosen due to its relatively low Ti content and high electrical conductivity. The addition of the Cr element could improve the comprehensive performance of Cu-Ti alloy, and the content of Cr was determined to be 0.4 wt.% according to its solid-solution limitation in the copper matrix. It had been widely reported that the addition of micro-alloying Mg element had a dragging effect on dislocation in copper alloys. As a typical trace element, its content was often chosen between 0.05–0.15 wt.% Mg, and we chose 0.1 wt.% in this study. Due to the low Ti element content, the strength of the alloy prepared by room temperature rolling could sometimes not meet the performance requirements in some specific application scenarios. Therefore, cryogenic rolling was used in this study in order to further improve the strength of the alloy. The evolution of microstructure, textures, and properties of the alloy during rolling was investigated in details, which could be expected to provide a theoretical basis for improving the comprehensive performance of Cu-Ti-based alloys.

2. Materials and Experimental Procedures

The Cu-Ti-Cr-Mg alloy was prepared by vacuum induction melting furnace in a high-purity argon atmosphere to avoid oxidation, with the raw materials of pure copper (99.9%), pure titanium (99.9%), pure chromium (99.9%) and Cu-20 wt.% Mg master alloy, and all the chemical components were placed in the crucible before melting. The measured compositions were detected by inductively coupled plasma-optical emission spectroscope (ICP-OES). The nominal and measured compositions were shown in Table 1. After removing the defects on the surface, the ingot was homogenization-treated at 880 °C for 6 h, and then hot-rolled with a reduction of 60%. The hot-rolled samples should be solid-solution treated to ensure that the solute atoms were completely dissolved into the matrix, and the hot deformation microstructure could be completely recrystallized. The solid-solution temperature and soaking time were 880 °C and 2h, respectively. Then the samples were rolled at room temperature and cryogenic temperature, respectively. In order to investigate the microstructure evolution of the alloy during rolling, the solid-solution treated samples were rolled with different reductions of 30%, 60%, 80%, and 90%, respectively. A handheld infrared thermometer was used to measure the temperature value of the alloy before each pass of rolling, keeping the cryogenic temperature at about −50 °C. After each pass of cryogenic rolling, the rolled samples were immediately immersed in liquid nitrogen and kept for at least 10 min to warrant the temperature of the samples at a low level. The total processing time of each cryogenic rolling pass (refers to the time when the samples were exposed to the room temperature environment during rolling) was less than 30 s. All rolling processes were carried out on a four-high rolling mill with $\Phi 80 - \Phi 120 \times 480$ mm/$\Phi 300 - \Phi 350 \times 450$ mm. For simplification, the room temperature rolled samples and cryogenic temperature rolled samples are abbreviated to RTR samples and CTR samples, respectively. For example, the sample with cryogenic rolling by 30% reduction is abbreviated to CTR30, as shown in Table 2.

Table 1. The nominal and measured compositions of the alloy.

Composition Element	Ti	Cr	Mg	Cu
Nominal composition(wt.%)	1.8	0.4	0.1	Bal.
Measured composition(wt.%)	1.85	0.39	0.097	Bal.

Table 2. The details of the abbreviation.

Rolling Reduction	Room Temperature	Cryogenic Temperature
30%	RTR30	CTR30
60%	RTR60	CTR60
80%	RTR80	CTR80
90%	RTR90	CTR90

The texture measurement was carried out on the D8 Discover X-ray diffractometer, with a sample size of 12 mm (RD) × 10 mm (TD). The orientation distribution function (ODF) plots ($\varphi_2 = 0°$, 45° and 60°) were calculated by the {200}, {220}, {111} pole diagrams according to the Bundle method. The hardness measurement was conducted on the HV-1000 microhardness tester, with a load of 1 kg and a dwell time of 15 s. Each sample was measured at least five times, and the average value was taken. The mechanical performance test was carried out at room temperature using an MTS-810 mechanical property tester following the GB/T 228.1-2010, with a strain rate of 0.5 min^{-1}. The dimensions of the samples tested were shown in Figure 1.

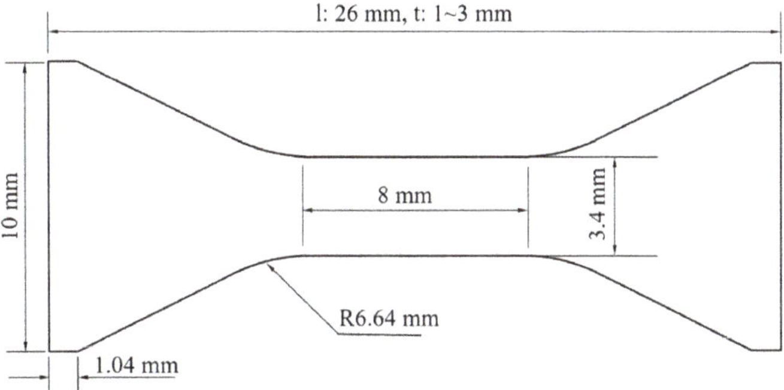

Figure 1. The dimensions of mechanical property test samples.

The microstructure in the longitudinal section was observed using Leica EC3 optical microscopy, FEI Helios Nanolab 600i scanning electron microscope with a NordlysMax^2 electron backscatter diffraction (EBSD) detector, and FEI Tecnai G^2 F20 field-emission transmission electron microscope (TEM).

Metallographic samples were prepared by mechanical polishing and then corroded in an aqueous solution of ferric chloride and hydrochloric acid(FeCl$_3$:HCl:CH$_3$CH$_2$OH = 5 g:25 mL:100 mL). EBSD was observed on the RD-ND plane (RD is the rolling direction, ND is the normal direction, and TD is the transverse direction of the samples). EBSD samples were prepared by mechanical polishing, and then electropolishing in an aqueous phosphoric acid solution of 70% volume fraction with a voltage of 2 V and a polishing time of 30 s. TEM samples were prepared by ion bean thinning.

3. Results

3.1. Microstructure of the Casted, Homogenization, Hot-Rolled, and Solid-Solution Treated Alloys

Figure 2a shows the typical metallograph of the casted Cu-Ti-Cr-Mg alloy. The sample shows the dendritic structure with a discontinuous pit-like characteristic, which resulted from the segregation of titanium. Normally, the electrode potential of titanium is lower than that of copper, so it is often corroded as an anode. Therefore, the dendritic structure under the metallograph is pitted. Figure 2b shows the metallograph of the sample after homogenization treatment, and the dendrite segregation is eliminated in the sample. After hot-rolled, the grain size of the sample is greatly reduced, while a partially deformed microstructure is observed (shown in Figure 2c). The metallograph of the solid-solution treated sample is shown in Figure 2d, presenting a homogeneous structure with a large number of annealed twins. The deformation microstructure after hot rolling treatment is basically eliminated. The inserted image on the right corner of Figure 2d shows the XRD diffraction pattern of the solid-solution treated sample. Only the characteristic diffraction peaks of copper appear in the inserted image in Figure 2d, indicating that the solid-solution treated sample is a single-phase solid-solution of copper.

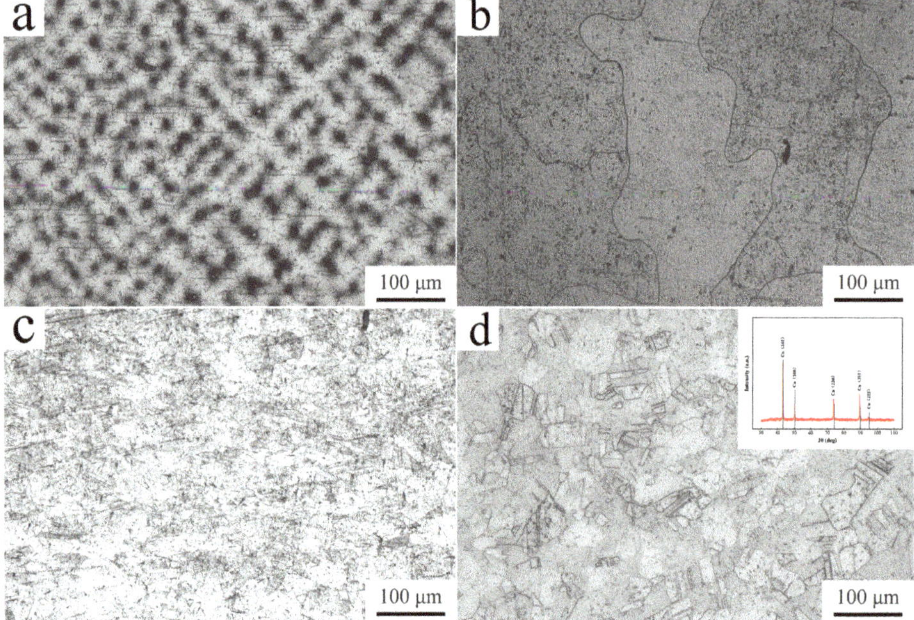

Figure 2. Metallograph of (**a**) casted, (**b**) homogenization, (**c**) hot-rolled, and (**d**) solid-solution-treated Cu-Ti-Cr-Mg alloy.

3.2. Properties of Deformed Alloy

Figure 3 shows the variations of hardness with different rolling reductions at room temperature and cryogenic temperature. The average hardness of the solid-solution treated sample is 98 HV. After a rolling reduction of 30%, the hardness of the samples increases significantly. With the further increase of rolling reduction, the rising trend of hardness begins to slow down, but it still shows an upward trend. At any given rolling reduction, the hardness of CTR samples is always larger than that of RTR samples. The results show that the hardness of the cryogenically rolled alloy is 18–30 HV higher than that of the room-temperature rolled alloy at any tested rolling reduction. Compared with room temperature rolling, Cu-Ti-Cr-Mg alloy has a higher work hardening rate at cryogenic temperature. The average hardness of the CTR90 sample is up to 231 HV, which is about 29 HV higher than that of the RTR90 sample.

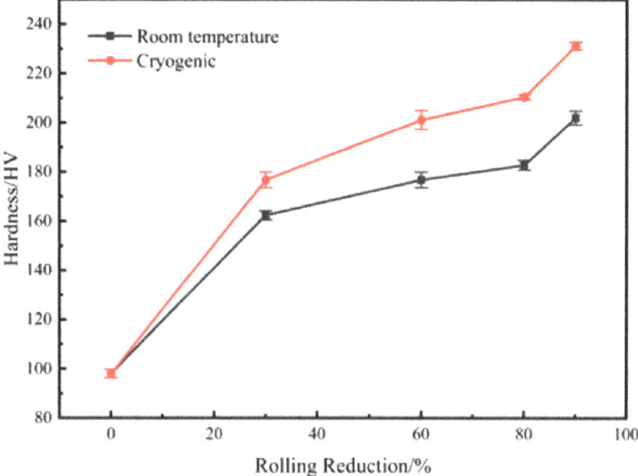

Figure 3. The variations of hardness with different rolling reductions.

The yield strength, tensile strength, and elongation of the solid-solution treated sample, RTR samples, and CTR samples with different rolling reductions are shown in Table 3. The solid-solution treated sample has a relatively low tensile strength of 319 ± 30 MPa but a high elongation of $46.8 \pm 0.3\%$. The strength of both RTR and CTR samples increases significantly with the increase of rolling reduction, but the increase in strength is accompanied by a decrease in the ductility of the alloy. Compared with RTR samples, CTR samples have higher strength at any given rolling reduction. However, the elongation of CTR samples is close to that of RTR samples. The yield strength and tensile strength of the CTR90 sample are 723 MPa and 796 MPa, increasing by about 25% and 23% compared with those of the RTR90 sample, respectively. This suggests that CTR treatment can improve the mechanical properties of Cu-Ti-Cr-Mg alloy more effectively than RTR treatment. It is worth noting that the elongations of both the RTR samples and CTR samples reach a minimum value at a rolling reduction of 80%.

Table 3. Mechanical properties of the Cu-Ti-Cr-Mg alloy after different rolling treatments.

	ST	RTR30	RTR60	RTR80	RTR90	CTR30	CTR60	CTR80	CTR90
Yield strength (MPa)	262 ± 12	411 ± 1	506 ± 2	560 ± 11	588 ± 13	474 ± 10	568 ± 13	676 ± 10	723 ± 6
Tensile strength (MPa)	319 ± 30	442 ± 2	552 ± 7	605 ± 2	634 ± 3	513 ± 1	627 ± 11	702 ± 14	796 ± 3
Elongation (%)	46.8 ± 0.3	21.6 ± 1.0	16.3 ± 0.3	14.6 ± 0.1	15.6 ± 0.5	20.3 ± 0.6	16.2 ± 1.8	14.6 ± 0.4	16.0 ± 0.2

3.3. Evolution of Microstructure during the Rolling Process

3.3.1. Metallograph Observation

Figure 4 shows the metallograph of Cu-Ti-Cr-Mg alloy with different rolling treatments. In the RTR process, very few deformation structures appear in the alloy with a rolling reduction of 30%. When the rolling reduction increases to 60%, the grains are elongated along the rolling direction, and deformation twins appear in some grains. Further increasing rolling reduction to 90%, the grains become much thinner, and the fibrous structure forms. The CTR process has a similar microstructure evolution to the RTR process. However, it is noted that a large number of deformation twins appear in the CTR30 sample.

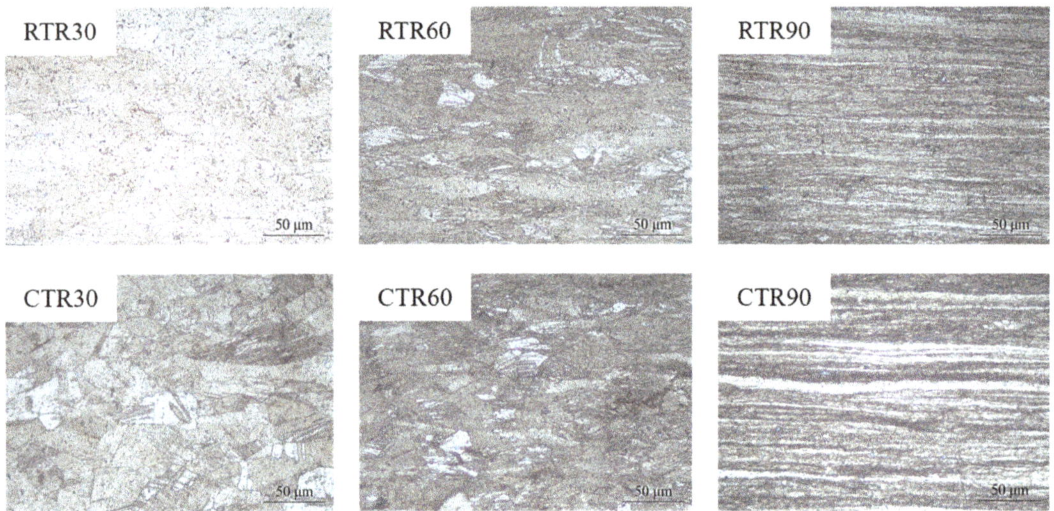

Figure 4. Metallograph of Cu-Ti-Cr-Mg alloy with different rolling conditions.

3.3.2. XRD and EBSD Analysis

Figure 5 shows the typical texture composition of the face-centered cubic metal during the rolling process and the orientation distribution function of the Cu-Ti-Cr-Mg alloy rolled at room temperature and cryogenic temperature ($\varphi_2 = 0°$, $45°$, and $65°$), respectively. The results show that the rolling reduction and rolling temperature have obvious effects on the texture evolution of the rolled samples. The orientation density of the solid-solution treated samples (0% reduction) is low, and there are a small number of Cube texture, weak Brass texture, and weak S texture. Prominent Brass texture, Goss texture, and a small number of S texture begin to appear in RTR30 and CTR30 samples. With the increase of rolling reduction, the orientation density of Brass texture, Goss texture, and S texture in the samples increases, while that of Cube texture decreases. The Cube texture is barely visible in the CTR90 sample. At any given rolling reduction, the orientation density of several major textures in the CTR samples is higher than that in the RTR samples.

EBSD analysis was used to compare the grain size and twin content of the samples rolled at different temperatures (represented by samples with a reduction of 60%).

Figure 6 shows the inverse pole figure (IPF) map of RTR and CTR samples with a rolling reduction of 60%. It is evident that grains are elongated along the rolling direction in both samples. The grains of the sample deformed at cryogenic temperature have a greater aspect ratio, indicating that the deformation of the sample at cryogenic temperature is much more serious. Moreover, there is a higher density of twins in the CTR sample (as shown in the box in Figure 6).

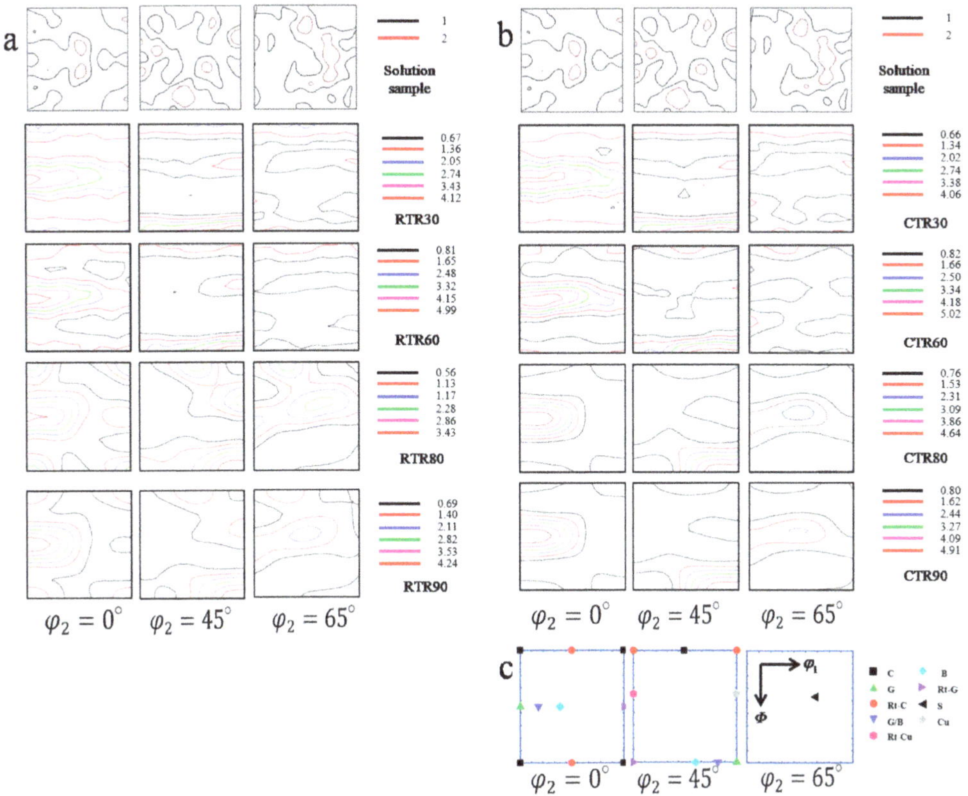

Figure 5. ODF plots of rolled samples at (**a**) room temperature and (**b**) cryogenic temperature; (**c**) schematic of the position of different texture components on the ODF section ($\varphi_2 = 0°$, $\varphi_2 = 45°$, and $\varphi_2 = 65°$).

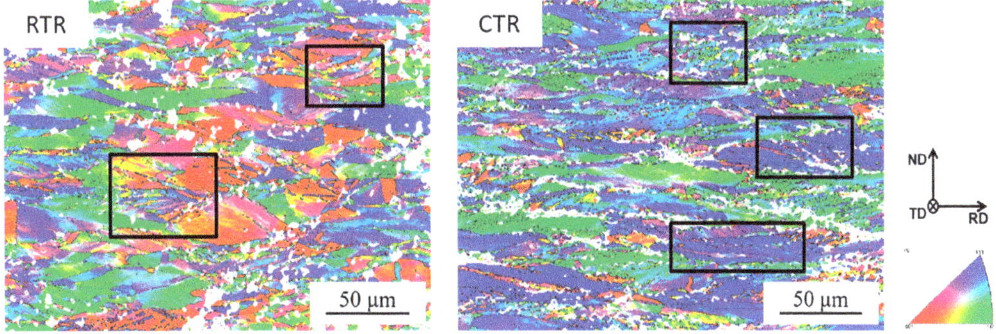

Figure 6. IPF of RTR and CTR Cu-Ti-Cr-Mg alloy with a rolling reduction of 60%.

The distribution of grain sizes of RTR and CTR samples with a rolling reduction of 60% is shown in Figure 7. After 60% deformation, the grains in both samples are broken into small sizes. The fraction of grains of the CTR sample with the sizes of 0.5 μm is up to 55%, higher than that of the RTR sample. The average grain sizes of RTR and CTR samples are 1.58 μm and 1.27 μm, respectively.

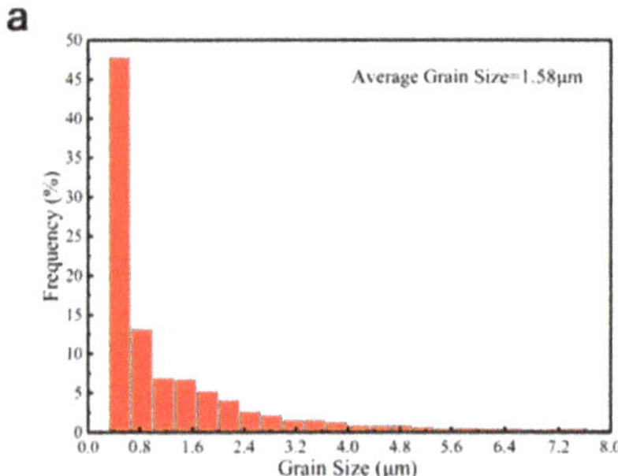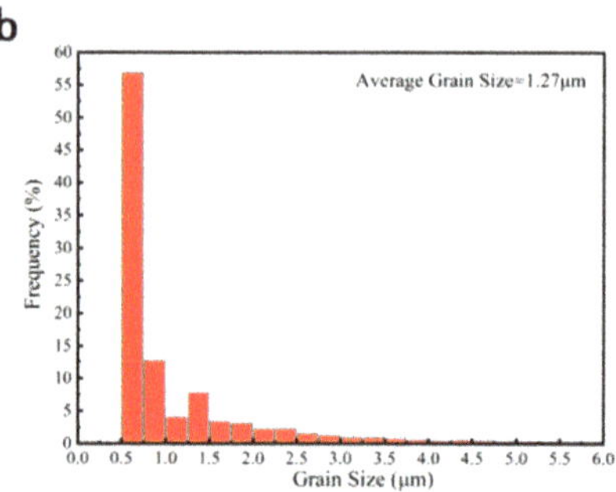

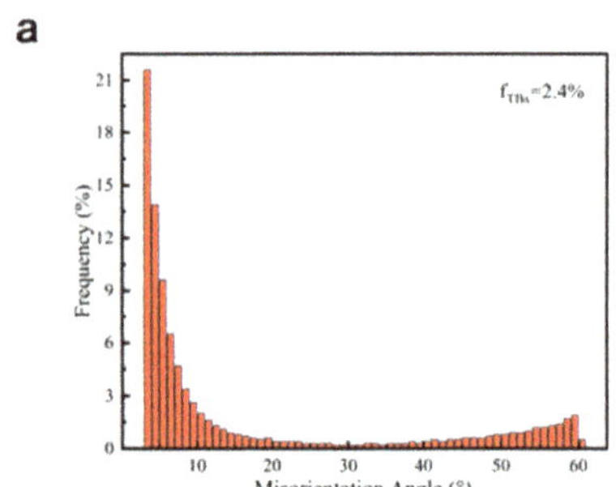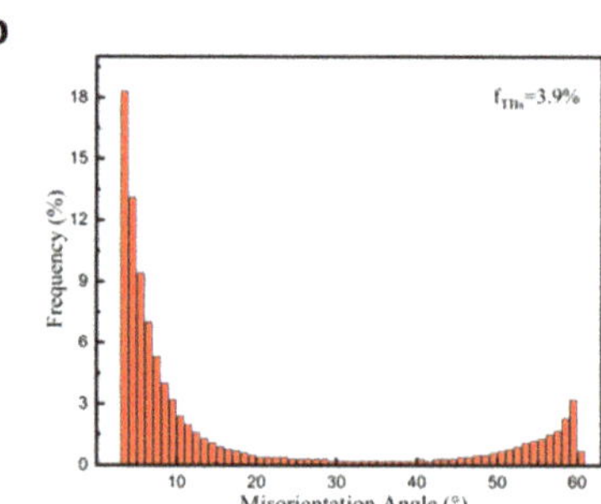

Figure 7. Grain sizes distribution of (**a**) RTR and (**b**) CTR Cu-Ti-Cr-Mg alloy with a rolling reduction of 60%.

Figure 8 shows the distribution of boundary misorientation angle of RTR and CTR samples with a rolling reduction of 60%. Two peaks are shown at a low angle and a high angle of 60°, respectively, corresponding to the broken grains with small misorientation and the deformation twins. Comparing the twin densities of RTR and CTR samples, the percentage of twin boundaries in the CTR sample (3.9%) is slightly higher than that of the RTR sample (2.4%).

Figure 8. Distribution of boundary misorientation angle of (**a**) RTR and (**b**) CTR Cu-Ti-Cr-Mg alloy with a rolling reduction of 60%.

3.3.3. TEM Observation

Figure 9 shows the microstructure evolution of Cu-Ti-Cr-Mg alloy during the room temperature rolling process. A large number of deformation twins and high-density dislocation walls (HDDWs) can be observed in the RTR30 sample (Figure 9a,b). Figure 9c is the typical high-resolution TEM image of the twin boundary in Figure 9a, and the stacking fault (SF) is observed along the {111} plane. With the increase of the rolling reduction, the density of nano-twins (twins with a thickness between 10 and 100 nm) in alloys increases, and the average size of nano-twins decreases. The dislocation tangling zones (DTZ) are observed in the samples (Figure 9e). Figure 9f is the high-resolution image of Figure 9d, showing a large number of SFs (indicated by red arrows in the figure) and nano-twins in the sample. In the RTR80 sample, the HDDW interlaces with the twins, and the twins begin to deform and distort. When the rolling reduction increases to 90%, the deformation twins are further distorted, as evidenced by the elongation of the diffraction spot in the illustration of Figure 9j. Figure 9l is the high-resolution image of Figure 9j, and there is a circle of satellite spots around each diffraction spot in the corresponding FFT figure, suggesting that the twin distortion leads to a slight change in orientation. In addition, two new twin systems intersecting the rolling direction appear in the twin crystals parallel to the rolling direction.

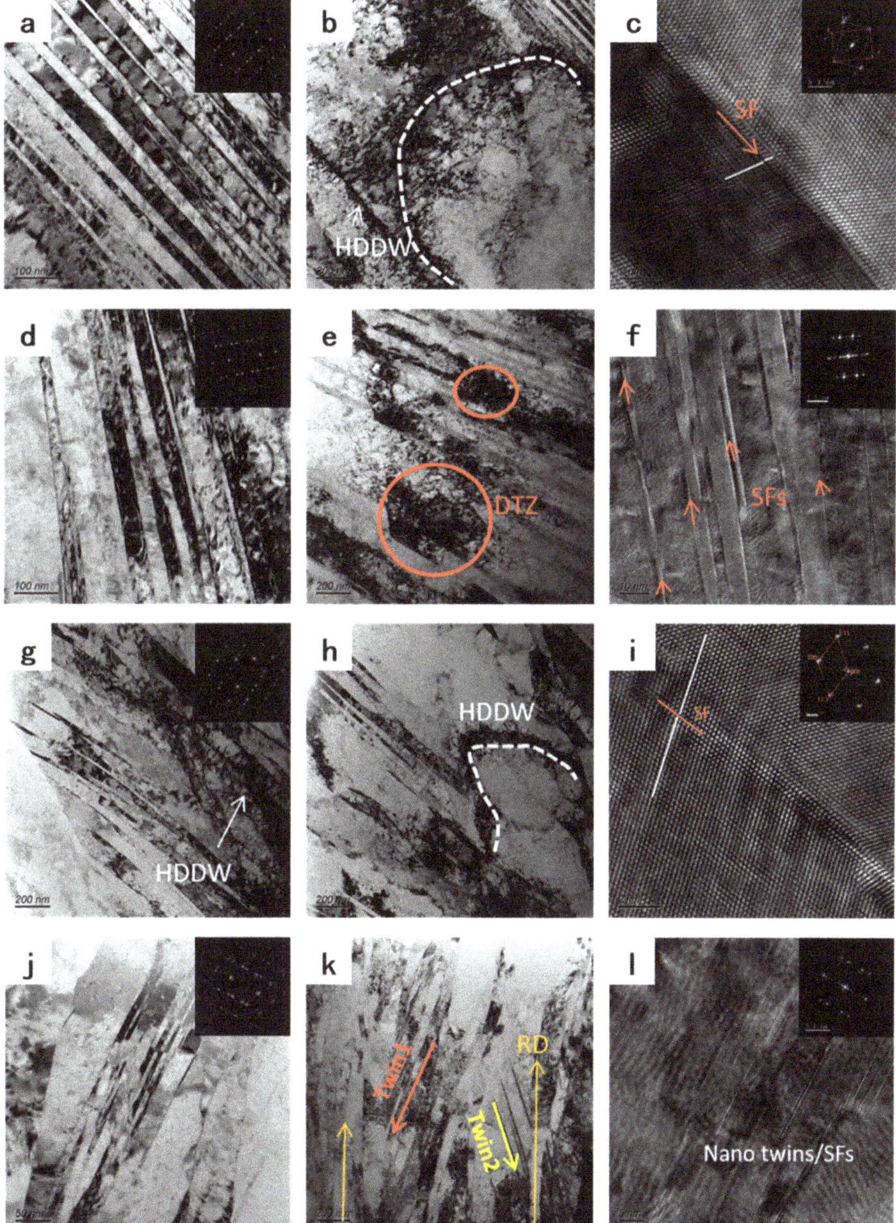

Figure 9. TEM images showing the microstructural evolution of RTR samples with different reductions: (**a–c**) RTR30, (**d–f**) RTR60, (**g–i**) RTR80, and (**j–l**) RTR90.

Figure 10 shows the microstructure evolution of Cu-Ti-Cr-Mg alloy during the CTR process. The microstructure evolution in CTR samples is similar to that in RTR samples during rolling, but the evolution process is significantly faster than that in RTR samples. The twin size in the CTR30 sample is smaller than that in the RTR30 sample. Two kinds of deformed twin systems are shown in Figure 10d,f, which can be commonly found in the

CTR60 sample. The post-activated twin divides the primary twin into smaller nanostructure. After an 80% reduction at cryogenic temperature, the twins in the sample are distorted and deformed, resulting in a large number of shear bands. In the high-resolution image (Figure 10i), many kinks between twins can be observed. Further increasing the rolling reduction to 90%, the deformation twins fracture, leading to the formation of a large number of cell block tissues in the sample, and only a small number of very small-size twins remain.

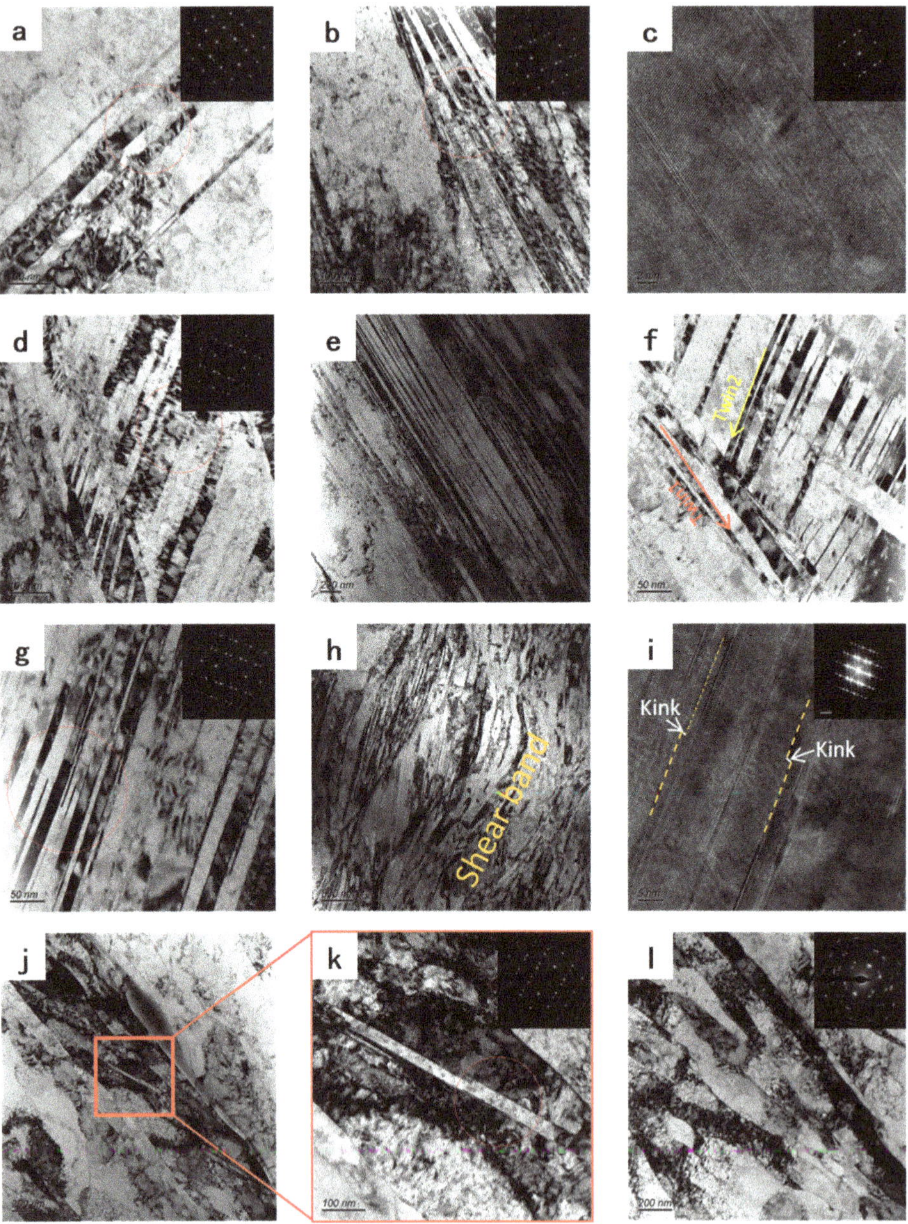

Figure 10. TEM images showing the microstructural evolution of CTR samples with different reductions: (**a–c**) CTR30, (**d–f**) CTR 60, (**g–i**) CTR 80, and (**j–l**) CTR 90.

4. Discussion

4.1. Effects of Deformation Temperature on the Texture of the Alloy

Rolling textures in face-centered cubic metals usually appear on α lines (<110> parallel to ND) and β lines (<110> tilted 60° in the RD direction) in the orientation space. The main textures on the α lines are Goss texture and Brass texture, while the main textures on the β line are Brass texture, S texture, and Copper texture. The Miller indices of the rolling texture components for FCC alloys are given in Table 4 [26]. The formation of a particular texture is closely related to the properties of the material, such as stacking fault energy (SFE). During the rolling process, the Copper texture is usually found in high SFE materials, while the Brass texture tends to form in low SFE materials [27]. In order to analyze the influence of rolling conditions on the evolution of the texture, the volume fraction of each texture under different rolling conditions was calculated using the Texture Call software. The volume fraction of these components with different rolling reductions is plotted in Figure 11.

Table 4. Important rolling texture components for FCC alloy [26].

Texture Component	Miller Indices	Euler Angles (°)		
		φ_1	Φ	φ_2
Cube	{001}<100>	0	0	0
Copper	{112}<111>	90	35	45
Brass	{011}<211>	55	90	45
Goss	{011}<100>	0	45	0
S	{123}<634>	59	37	63
R-Goss	{011}<011>	0	90	45
R-Copper	{112}<011>	0	35	45

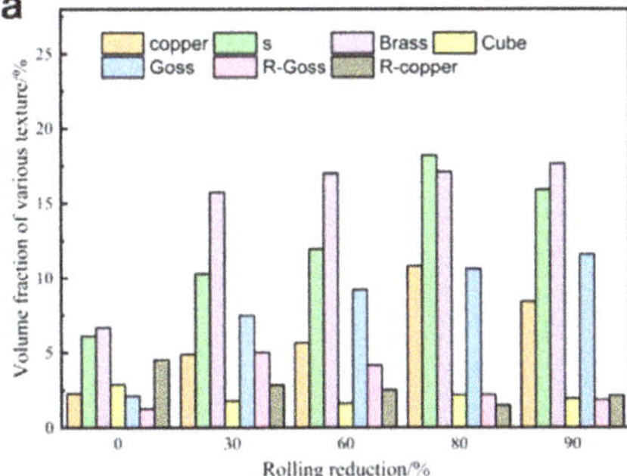
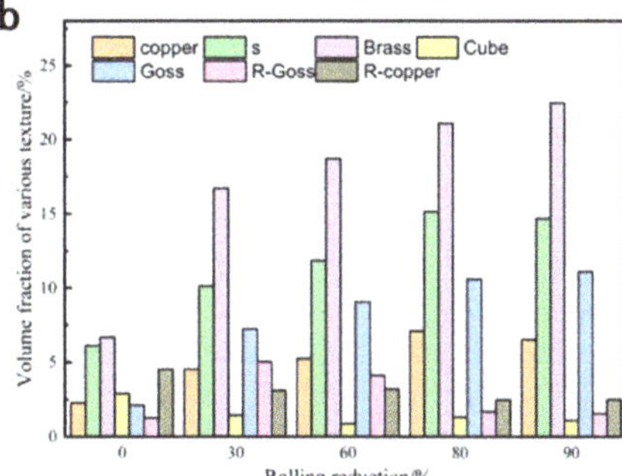

Figure 11. Variation in the volume fraction of various texture components with different rolling reductions at (**a**) room temperature and (**b**) cryogenic temperature.

As shown in Figure 11, the volume fractions of Brass {011}<211> texture and S {123}<634> texture in the solid-solution treated sample are relatively high, and the other components are relatively weak. During the rolling process, the Brass {011}<211> texture and Goss {011}<100> texture in both of the RTR and CTR samples increase with the increasing reduction. However, the volume fraction of the Brass texture in the CTR samples is consistently higher than that of the RTR samples at any given rolling reduction. The priority development of Brass texture during the CTR process could be attributed to extensive twins or inhibition of cross-slip [28]. As mentioned above, CTR samples have higher twin densities and smaller twin sizes than RTR samples. Besides, the cross-slip process is temperature sensitive and inhibited at cryogenic temperature, which could be the reason for the preferential development of Brass texture. At the beginning of deformation (30% reduction), the volume fraction of R-Goss texture increases, and then its volume fraction decreases with the increase of the rolling reduction. Meanwhile, the Cube {001}<100> texture remains at a low level after further reduction.

It is worth noting that the contents of the Copper texture and the S texture increase first and then decrease with the increase of rolling reduction, and both of them reach the peak values when the rolling reduction is 80%. In the plastic deformation of copper alloys, the Copper {112}<111> texture in the <111> direction has poor plastic deformation ability compared to that in the <100> and <110> directions due to the fewer slip systems and lower Schmid factor [29]. This could be attributed to the relatively low elongation of rolled samples at the rolling reduction of 80% (as shown in Table 2).

4.2. Effect of Rolling Temperature on Microstructure and Properties of the Alloy

There is a significant difference in microstructure and mechanical properties of the Cu-Ti-Cr-Mg alloys rolled at room temperature and cryogenic temperature. The yield strength, tensile strength, and work hardening rate of the composites increase with the decrease in temperature. Dislocation slips and deformation twins are two common deformation mechanisms in the plastic deformation of copper alloys. These two mechanisms are in competition during deformation, and the dominance mainly depends on the SFE of the material. Furthermore, the deformation mechanism also depends on the processing parameters, especially the strain rate and the deformation temperature [30]. Normally, the twin boundaries act as strong obstacles to the dislocation motion, leading to high strain-hardening. At the cryogenic temperature, the critical resolved shear stress (CRSS) of dislocation slip increases significantly, leading to the inhibition of dislocation motion, and the CRSS of the deformation twins is virtually unaffected by temperature [31]. Hence, deformation at the cryogenic temperature can promote the occurrence of deformation twins.

It has been reported that metals with relatively low SFEs (about 18~45 mJ/m^2) are conducive to the formation of deformation twins in FCC alloys [32]. The Cu-Ti alloy has a low SFEs of about 20 mJ/m^2 [33], and the deformation twins become their typical characteristic. In this study, deformation twins appear in both RTR and CTR samples, as shown in Figures 4 and 6. It indicates that even at room temperature, the SFs in the alloy could transition to twins. Furthermore, the evolution of twins during the CTR process is significantly faster than that of the RTR process, and there are more twins accumulated in the CTR60 sample than that in the RTR60 sample (Figures 6 and 7), which may be related to the decrease in the SEF value of the alloy at a cryogenic temperature [34].

On the other hand, cryogenic temperature inhibits the slip and annihilation of dislocations, preserving a high density of dislocations in the samples and enhancing the mechanical properties of the matrix. Figure 12 shows the X-ray diffraction results of the RTR and CTR samples. The improved Williamson-Hall method [35] was used to calculate the micro-strain of the material according to the width of the XRD diffraction peak. The formula is as follows:

$$\delta_{hkl}cos\theta_{hkl} = \frac{k\lambda}{d} + 2\varepsilon sin\theta_{hkl}$$

where δ_{hkl} is the half-height width of the diffraction peak, θ_{hkl} is the diffraction angle of the (hkl) crystal plane, k is a constant (about 0.9) [36], λ is the wavelength of the incident wave (0.154 nm of Cu-K$_\alpha$), d is the average grain size, and ε is the slope of the fitted line. The dislocation density can be calculated by [37]:

$$\rho = \frac{16.1\varepsilon^2}{b^2}$$

where ρ is the dislocation density of the alloy, and b is the Burgers vector of the matrix (0.255 nm) [38].

Figure 13 shows the variation of micro-strain and dislocation density of samples after different rolling reductions. With the increase of rolling reduction, the dislocation density shows an upward trend, however, the rate of increase of dislocation density gradually slows down. Micro-strain has the same trend as the dislocation density with the rolling reduction. It is worth noting that when the rolling reduction is 30%, the deformation energy storage of the RTR sample is not enough to cause dynamic recovery and dynamic recrystallization. Therefore, the dislocations are accumulated, and the hindrance effect of cryogenic temperature on dislocation motion is obvious. When the rolling reduction is higher than 30%, the dislocation density of the CTR samples is higher than that of the RTR samples, and the difference increases with the increase of the rolling reduction. After rolling by 90% reduction, the RTR and CTR samples have dislocation densities of 23.03×10^{14} m^{-2} and 29.98×10^{14} m^{-2}, respectively. With the increase of rolling reduction, the dynamic recovery in the RTR samples continuously increases, and the decrease in dislocation density originates from the recovery increases. However, the inhibition of dynamic recovery by cryogenic temperature allows more dislocations to be retained in CTR samples, resulting in a higher dislocation density.

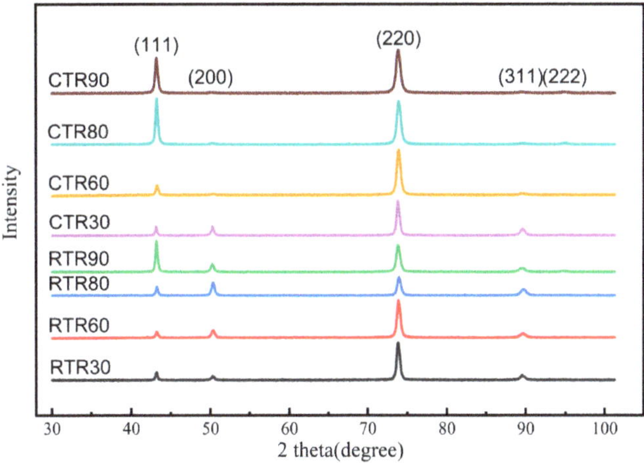

Figure 12. X-ray diffraction results of the RTR samples and CTR samples.

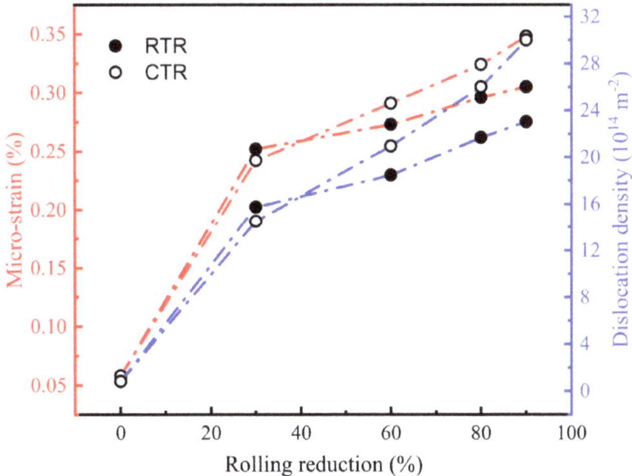

Figure 13. Variation of micro-strain and dislocation density after different rolling reductions.

According to the above results and discussion, it can be safely suspected that the cryogenic temperature significantly accelerates the evolution of microstructure during rolling, resulting in much more nano-twins and dislocations and better mechanical properties of the CTR alloy (as shown in Table 2).

5. Conclusions

1. The Cu-Ti-Cr-Mg alloy has a higher work hardening rate at cryogenic temperature. The yield strength and tensile strength of the CTR90 sample are 723 MPa and 796 MPa, increasing by about 25% and 23% compared with those of the RTR90 sample, respectively.
2. The solid solution-treated Cu-Ti-Cr-Mg alloy has weak Brass texture and S texture, and the textures change to Brass and Goss textures during the deformation. The cross-slip is inhibited at cryogenic temperature, leading to the priority development of Brass texture in cryogenic rolling. The microstructure evolution during the rolling deformation is related to dislocation slip and nano-twins deformation, and the cryogenic temperature promotes the deformation progress.
3. With the increase of rolling reduction, the dislocation density of Cu-Ti-Cr-Mg alloy increases significantly during CTR and RTR processes, finally reaching 29.98×10^{14} m^{-2} and 23.03×10^{14} m^{-2}, respectively. The difference in the dislocation densities between the CTR and RTR samples increases with the increasing rolling reduction. The enhanced mechanical properties of the alloy after cryogenic rolling are due to cryogenic temperature inhibiting the dynamic recovery and dynamic recrystallization during rolling.

Author Contributions: Study conception and design: R.L., Z.X., Z.L., X.W.; data collection: R.L., X.W.; analysis and interpretation of results: Z.X., X.M., Z.L.; draft manuscript preparation: R.L., Z.X., X.M. All authors have read and agreed to the published version of the manuscript.

Funding: The research is funded by the National Key Research and Development Program of China (2021YFB3803101), National Key Research and Development Program of China (No. U2202255), and grants from the Project of State Key Laboratory of Powder Metallurgy, Central South University, Changsha, China. The sponsor-ships are gratefully acknowledged.

Institutional Review Board Statement: Not applicable.

Informed Consent Statement: Not applicable.

Data Availability Statement: The data used to support the findings of this study are included within the article.

Conflicts of Interest: The authors declare no conflict of interest.

References

1. Huang, X.; Xie, G.; Liu, X.; Fu, H.; Shao, L.; Hao, Z. The influence of precipitation transformation on Young's modulus and strengthening mechanism of a Cu-Be binary alloy. *Mater. Sci. Eng. A* **2020**, *772*, 138592. [CrossRef]
2. Zhang, H.; Jiang, Y.; Xie, J.; Li, Y.; Yue, L. Precipitation behavior, microstructure and properties of aged Cu-1.7wt.%Be alloy. *J. Alloys Compd.* **2019**, *773*, 1121–1130. [CrossRef]
3. Zhang, H.T.; Fu, H.D.; Shen, Y.H.; Xie, J.X. Rapid design of secondary deformation-aging parameters for ultra-low Co content Cu-Ni-Co-Si-X alloy via Bayesian optimization machine learning. *Int. J. Miner. Metall. Mater.* **2022**, *29*, 1197–1205. [CrossRef]
4. Semboshi, S.; Ishikuro, M.; Sato, S.; Wagatsuma, K.; Takasugi, T. Extraction of precipitates from age-hardenable Cu-Ti alloys. *Mater. Charact.* **2013**, *82*, 23–31. [CrossRef]
5. Wei, H.; Wei, Y.; Hou, L.; Dang, N. Correlation of ageing precipitates with the corrosion behaviour of Cu-4wt.%Ti alloys in 3.5wt.% NaCl solution. *Corros. Sci.* **2016**, *111*, 382–390. [CrossRef]
6. Yang, H.; Bu, Y.; Wu, J.; Fang, Y.; Liu, J.; Huang, L.; Wang, H.T. High strength, high conductivity and good softening resistance Cu-Fe-Ti alloy. *J. Alloys Compd.* **2022**, *925*, 166595. [CrossRef]
7. Markandeya, R.; Nagarjuna, S.; Sarma, D. Effect of prior cold work on age hardening of Cu-3Ti-1Cr alloy. *Mater. Charact.* **2006**, *57*, 348–357. [CrossRef]
8. Li, S.; Li, Z.; Xiao, Z.; Li, S.; Shen, L.; Dong, Q. Microstructure and property of Cu-2.7Ti-0.15Mg-0.1Ce-0.1Zr alloy treated with a combined aging process. *Mater. Sci. Eng. A* **2016**, *650*, 345–353.
9. Rouxel, B.; Cayron, C.; Bornand, J.; Sanders, P.; Logé, R. Micro-addition of Fe in highly alloyed Cu-Ti alloys to improve both formability and strength. *Mater. Des.* **2022**, *213*, 110340. [CrossRef]
10. He, X.Q.; Fu, H.D.; Xie, J.X. Microstructure and properties evolution of in-situ fiber-reinforced Ag-Cu-Ni-Ce alloy during deformation and heat treatment. *Int. J. Miner. Metall. Mater.* **2022**, *29*, 2000–2011. [CrossRef]

11. Zhang, T.T.; Wang, W.X.; Zhang, J.; Yan, Z.F. Interfacial bonding characteristics and mechanical properties of H68/AZ31B clad plate. *Int. J. Miner. Metall. Mater.* **2022**, *29*, 1237–1248. [CrossRef]
12. Valiev, R.; Islamgaliev, R.; Alexandrov, I. Bulk Nanostructured Materials from Severe Plastic Deformation. *Progress Mater. Sci.* **2000**, *45*, 103–189. [CrossRef]
13. Valiev, R.; Langdon, T. Principles of equal-channel angular pressing as a processing tool for grain refinement. *Progress Mater. Sci.* **2006**, *51*, 881–981. [CrossRef]
14. Yang, H.B.; Chai, Y.F.; Jiang, B.; He, C.; Yang, Q.S.; Yaun, M. Enhanced mechanical properties of Mg-3Al-1Zn alloy sheets through slope extrysion. *Int. J. Miner. Metall. Mater.* **2022**, *29*, 1343–1350. [CrossRef]
15. Tsuji, N.; Saito, Y.; Lee, S.; Minamino, Y. ARB (Accumulative Roll-bonding) and other new techniques to produce bulk ultrafine grained materials. *Adv. Eng. Mater.* **2003**, *5*, 338–344. [CrossRef]
16. Zhilyaev, A.; Langdon, T. Using high-pressure torsion for metal processing: Fundamentals and applications. *Progress Mater. Sci.* **2008**, *53*, 893–979. [CrossRef]
17. Wang, P.F.; Liang, M.; Xu, X.Y.; Feng, J.Q.; Li, C.S.; Zhang, P.X.; Li, J.S. Effect of groove rolling on the microstructure and properties of Cu-Nb microcomposite wires. *Int. J. Miner. Metall. Mater.* **2021**, *28*, 279–284. [CrossRef]
18. Roy, B.; Kumar, N.; Nambissan, P.; Das, J. Evolution and interaction of twins, dislocations and stacking faults in rolled α-brass during nanostructuring at sub-zero temperature. *AIP Adv.* **2014**, *4*, 103–109. [CrossRef]
19. San, X.; Liang, X.; Cheng, L.; Li, C.; Zhu, X. Temperature effect on mechanical properties of Cu and Cu alloys. *Mater. Des.* **2012**, *35*, 480–483. [CrossRef]
20. Magalhães, D.; Hupalo, M.; Cintho, O. Natural aging behavior of AA7050 Al alloy after cryogenic rolling. *Mater. Sci. Eng. A* **2014**, *593*, 1–7. [CrossRef]
21. Li, G.; Zhang, J.; Wu, R.; Liu, S.; Song, B.; Jiao, Y.; Yang, Q.; Hou, L. Improving age hardening response and mechanical properties of a new Mg-RE alloy via simple pre-cold rolling. *J. Alloys Compd.* **2019**, *777*, 1375–1385. [CrossRef]
22. Liu, R.; Zhang, Z.; Li, L.; Zhang, Z. Microscopic mechanisms contribution to the synchronous improvement of strength and plasticity for TWIP copper alloys. *Sci. Rep.* **2015**, *5*, 9550–9556. [CrossRef]
23. Shanmugasundaram, T.; Murty, B.; Sarma, V. Development of ultrafine grained high strength Al- Cu alloy by cryorolling. *Scr. Mater.* **2006**, *54*, 2013–2017. [CrossRef]
24. Yu, H.; Du, Q.; Godbole, A.; Lu, C.; Kong, C. Improvement in strength and ductility of asymmetric-cryorolled copper sheets under low-temperature annealing. *Metall. Mater. Trans. A* **2018**, *49*, 4398–4403. [CrossRef]
25. Wang, P.; Jie, J.; Liu, C.; Guo, L.; Li, T. An effective method to obtain Cu-35Zn alloy with a good combination of strength and ductility through cryogenic rolling. *Mater. Sci. Eng. A* **2018**, *715*, 236–242. [CrossRef]
26. Hirsch, J.; Lücke, K. Overview no. 76: Mechanism of deformation and development of rolling textures in polycrystalline f.c.c. metals—II. Simulation and interpretation of experiments on the basis of Taylor-type theories. *Acta Metall.* **1988**, *36*, 2883–2904. [CrossRef]
27. Konkova, T.; Mironov, S.; Korznikov, A.; Semiatin, S. Microstructural response of pure copper to cryogenic rolling. *Acta Mater.* **2010**, *58*, 5262–5273. [CrossRef]
28. Dillamore, I.; Roberts, W. Rolling textures in f.c.c. and b.c.c. metals. *Acta Metall.* **1964**, *12*, 281–293. [CrossRef]
29. Fu, H.; Xu, S.; Li, W.; Xie, J.; Zhao, H.; Pan, Z. Effect of rolling and aging processes on microstructure and properties of Cu-Cr-Zr alloy. *Mater. Sci. Eng. A* **2017**, *700*, 107–115. [CrossRef]
30. Jiang, K.; Gan, B.; Li, J.; Dou, Q.; Suo, T. Towards strength-ductility synergy in a CrCoNi solid solution alloy via nanotwins. *Mater. Sci. Eng. A* **2021**, *816*, 141298. [CrossRef]
31. Zhang, K.; Zheng, J.H.; Huang, Y.; Pruncu, C.; Jiang, J. Evolution of twinning and shear bands in magnesium alloys during rolling at room and cryogenic temperature. *Mater. Des.* **2020**, *193*, 108793. [CrossRef]
32. Curtze, S.; Kuokkala, V. Dependence of tensile deformation behavior of TWIP steels on stacking fault energy, temperature and strain rate. *Acta Mater.* **2010**, *58*, 5129–5141. [CrossRef]
33. Hou, L.; Hou, L.; Wei, H.; Wei, Y. Thermodynamic calculation of stacking fault energy of copper alloy. *Chin. J. Nonferrous Met.* **2016**, *26*, 2363–2368.
34. Huang, S.; Li, W.; Lu, S.; Tian, F.; Shen, J.; Holmström, E.; Vitos, L. Temperature dependent stacking fault energy of FeCrCoNiMn high entropy alloy. *Scr. Mater.* **2015**, *108*, 44–47. [CrossRef]
35. Williamson, G.; Hall, W. X-ray line broadening from filed aluminium and wolfram. *Acta Metall.* **1953**, *1*, 22–31. [CrossRef]
36. Wang, X.; Xiao, Z.; Meng, X.; Yi, Y.; Chen, L. Microstructure and properties evolution of Cu-Ti-Cr-Mg alloy during equal channel angular pressing at room temperature and cryogenic temperature. *J. Alloys Compd.* **2022**, *927*, 166940. [CrossRef]
37. Xiao, Z.; Huang, Y.; Chen, C.; Li, Z.; Gong, S.; Huang, Y.; Zhang, C.; Zhang, X.X. Effects of thermal treatments on the residual stress and micro-yield strength of Al2O3 dispersion strengthened copper alloy. *J. Alloys Compd.* **2019**, *781*, 490–495. [CrossRef]
38. Chen, J.; Wang, J.; Xiao, X.; Wang, H.; Chen, H.; Yang, B. Contribution of Zr to strength and grain refinement in Cu-Cr-Zr alloy. *Mater. Sci. Eng. A* **2019**, *756*, 464–473. [CrossRef]

materials

Article

The Influence of Cryogenic Treatment on the Microstructure and Mechanical Characteristics of Aluminum Silicon Carbide Matrix Composites

Mingli Zhang [1,2], Ran Pan [3], Baosheng Liu [3], Kaixuan Gu [1,*], Zeju Weng [1], Chen Cui [1,2] and Junjie Wang [1,2,*]

1   CAS Key Laboratory of Cryogenics, Technical Institute of Physics and Chemistry, Beijing 100190, China
2   University of Chinese Academy of Sciences, Beijing 100049, China
3   AVIC Manufacturing Technology Institute, Beijing 100024, China
*   Correspondence: gukaixuan@mail.ipc.ac.cn (K.G.); wangjunjie@mail.ipc.ac.cn (J.W.);
    Tel.: +00-86-010-8254-3488 (K.G.); +00-86-010-8254-3758 (J.W.)

Abstract: Aluminum matrix composites have been widely used in aerospace and automotive fields due to their excellent physical properties. Cryogenic treatment was successfully adopted to improve the performance of aluminum alloy components, while its effect and mechanism on the aluminum matrix composite remained unclear. In this work, the effects of cryogenic treatment on the microstructure evolution and mechanical properties of 15%SiCp/2009 aluminum matrix composites were systematically investigated by means of Thermoelectric Power (TEP), Scanning Electron Microscopy (SEM) and Transmission Electron Microscopy (TEM). The results showed that TEP measurement can be an effective method for evaluating the precipitation characteristics of 15%SiCp/2009 aluminum matrix composites during aging. The addition of cryogenic treatment after solution and before aging treatment promoted the precipitation from the beginning stage of aging. Furthermore, the aging time for the maximum precipitation of the θ'' phase was about 4 h advanced, as the conduction of cryogenic treatment accelerates the aging kinetics. This was attributed to the great difference in the linear expansion coefficient between the aluminum alloy matrix and SiC-reinforced particles, which could induce high internal stress in their boundaries for precipitation. Moreover, the lattice contraction of the aluminum alloy matrix during cryogenic treatment led to the increase in dislocation density and micro defects near the boundaries, thus providing more nucleation sites for precipitation during the aging treatment. After undergoing artificial aging treatment for 20 h, the increase in dispersive, distributed precipitates after cryogenic treatment improved the hardness and yield strength by 4% and 16 MPa, respectively.

Keywords: 15%SiCp/2009 aluminum matrix composite; cryogenic treatment; thermoelectric power; aging behavior; tensile property

Citation: Zhang, M.; Pan, R.; Liu, B.; Gu, K.; Weng, Z.; Cui, C.; Wang, J. The Influence of Cryogenic Treatment on the Microstructure and Mechanical Characteristics of Aluminum Silicon Carbide Matrix Composites. *Materials* **2023**, *16*, 396. https://doi.org/10.3390/ma16010396

Academic Editor: Hajo Dieringa

Received: 12 December 2022
Revised: 23 December 2022
Accepted: 28 December 2022
Published: 1 January 2023

1. Introduction

In the past half century, the use of SiC particles as a reinforcing phase for the improvement of the machining and mechanical properties of aluminum alloys has been widely investigated [1–3]. When compared with the corresponding aluminum alloy, a particle-reinforced aluminum alloy matrix composite has a higher tensile strength [4,5], hardness [6] and fatigue strength [6]; qualities which have been widely used in the aerospace and automotive fields [7]. Similar to aluminum alloys, aluminum matrix composites can be subjected to solution and aging treatments for the enhancement of the mechanical properties of the material via solution and precipitation strengthening effects [8]. The aging behavior of aluminum matrix composites is quite different from that of the corresponding aluminum alloys due to the presence of the reinforced-particle phase. It was reported that the addition of SiC particles accelerated the formation of the S' phase (Al_2CuMg) in a SiCp-reinforced 2009 aluminum alloy [9]. However, research from Pal et al. [10] showed that it took a longer time for SiCp/Al-Cu-Mg composites to reach the peak of aging than it did for the corresponding alloy. Although SiC could affect the precipitation behavior and improve the mechanical properties of the composite's material, it would not change the aging sequence in an aluminum matrix when compared with the corresponding aluminum alloy [11]. Therefore, solution and aging treatment can also be the effective methods to modify the comprehensive properties of an aluminum alloy and its composite [12].

Cryogenic treatment, which is the process of subjecting materials to a low-temperature environment, has been combined with solution and aging treatment to improve the comprehensive performances of aluminum alloys. Steier et al. [13] found that deep cryogenic treatment promoted the formation of additional GP-zones in 6101 aluminum alloy as a result of improving the wear resistance. A deep cryogenic treatment with a long soaking time (45 h) had positive effects on the mechanical properties and machinability of a 5083 alloy [14]. Gao et al. [15] also found that deep cryogenic treatment improved the strength and plasticity of a 7A99 alloy through refining the precipitates and increasing the uniformity of the precipitate distribution. Park [16] revealed that cryogenic treatment followed by rapid reheating (uphill quenching) released the residual stress and obviously improved the tensile properties. It has been reported by Araghchi [17] and Wang et al. [9] that cryogenic treatment can promote the formation of small-size precipitates to improve the mechanical properties. It was first reported by Jovičević-Klug [18] et al. that deep cryogenic treatment facilitated the precipitation of the β'' phase while suppressing the formation of the β' phase with a large size during artificial aging treatment. Zhou et al. [19] attributed the improvement of aluminum alloys to the reduction of residual stress and a balanced state of residual stress in the material. Araghchi et al. [17] claimed a promotion of nucleation sites for the S' phase which resulted from the increase in dislocation density following cryogenic treatment in a 2024 aluminum alloy.

Owing to the positive effects of cryogenic treatment on aluminum alloys, investigation on cryogenically treated aluminum matrix composites has aroused concerns for both academia and industry in recent years. Li et al. [20] investigated the effects of cryogenic treatment on Al_2O_3/Al-Zn-Mg-Cu matrix composites and found that precipitation of the S phase was facilitated at $-108\,^{\circ}\text{C}$ when samples were recovered from cryogenic temperature to room temperature. Hong [21] et al. claimed that cryogenic treatment conducted in combination with aging for aluminum matrix composites promoted the formation of the S' phase around SiC particles and further improved their mechanical properties. The effects of cryogenic treatment on the aging behavior of aluminum matrix composites were further investigated by Wang [9] et al., who showed that deep cryogenic treatment induced the generation of thinner θ' precipitates in the grains with the increase in artificial aging time. It can be seen that the effects of cryogenic treatment have a close relationship with the subsequent aging parameters in aluminum matrix composites. However, the in-depth mechanisms surrounding the effects of cryogenic treatment on the microstructure evolution during subsequent aging in aluminum matrix composites remain unrevealed.

It can be concluded that the changes in precipitation characterizations are the main mechanism for the cryogenic treatment of aluminum alloys and composites and have great relationships with the type of alloys, the heat treatment parameters, and so on. However, the precipitation in aluminum alloys and composites induced by cryogenic treatment is not as obvious as the precipitation induced by aging treatment alone. The method of conventional microstructure characterization can be easily influenced by the selected areas and have a deficiency in the overall analysis of the microstructure evolution. Thermoelectric Power (TEP) is an intrinsic property of material which is related to electron scattering caused by the existence of crystalline defects such as the displacement of ions by thermal oscillations, the presence of accidental impurities or solute atoms, and imperfections including dislocations, grain boundaries, stacking faults, vacancies, and so on [22]. Therefore, a TEP measurement can be used for the precise and global evaluation of a microstructure evolution due to its microstructure sensitivity. Bakare et al. [23] revealed the formation behavior of GP-zones and the changing trend of dislocations with the increase in aging time in aluminum alloys by TEP measurement. Abdala et al. [24] found that TEP measurements were very sensitive to precipitation phenomena in a 6013 alloy, which allowed for the proposal of a precipitation sequence for this alloy. Massardier et al. [25] demonstrated that the TEP kinetics allowed an assessment of the residual concentration of solute in the final equilibrium state. However, investigation into the microstructure evolution of aluminum matrix composites induced by cryogenic treatment via TEP measurements and analysis has not been brought to the forefront.

Therefore, the present work aims to investigate the effects of cryogenic treatment on the microstructure changes and mechanical properties of 15%SiCp/2009Al aluminum matrix composites in different aging stages. Use of TEP measurements in conjunction with conventional microstructure characterization approaches are adopted for a global and systematical evaluation of the microstructure evolution. The influence of the combination sequence among solution, cryogenic, and aging treatments on the mechanical properties of aluminum matrix composites is also studied.

2. Materials and Methods

The aluminum matrix composite used in this study was the 15 vol% silicon-carbide-particle-reinforced 2009Al (SiCp/2009Al) composite. The material was provided in an as-forged state by the Institute of Metal Research of the Chinese Academy of Sciences (IMR). Its chemical composition is shown in Table 1.

Table 1. Chemical compositions of 15%SiCp/2009 aluminum matrix composite (wt%).

Cu	Mg	Si	Fe	Zn	O	Impurity	Al
3.44	1.38	0.29	0.06	0.05	0.15	0.15	Balance

A solution treatment (S) was carried out using a SX-5-12 box resistance furnace. The samples were heated to 510 °C and held for 2 h with subsequent water-quenching at room temperature. Cryogenic treatment is usually divided into shallow cryogenic treatment ($-40{\sim}-80$ °C) and deep cryogenic treatment ($-80{\sim}-196$ °C) according to the minimum temperature [14,26]. The previous research has demonstrated that the improving effects of cryogenic treatment can be promoted with a decrease in treating temperature [27]. In this work, cryogenic treatment (C) was conducted by cooling down quenched samples to -196 °C at the rate of 1 °C/min. The samples were then held for 12 h. The program-controlled SLX-80 cryogenic system was used for carrying out cryogenic treatment, as is shown in Figure 1. After 12 h of storage, samples were warmed up to room temperature in the atmosphere for 15 min. The artificial aging treatment (A) was carried out using a DZF-6210 vacuum-drying oven. During the aging process, the composite materials were heated to 170 °C and held for r2, 6, 10, and 20 h, respectively, before being cooled to room temperature. The time of cooling to room temperature was approximately 10 min. In this study, cryogenic treatment was conducted in combination with solution and aging treatments in various sequences. All combinations are presented with necessary illustrations in Table 2. The alphabetic order of each combination refers to the specific conducting sequence of the three basic treatments. For example, the process of SCA6 indicated that the samples were subjected to solution treatment, cryogenic treatment, and aging treatment for 6 h successively.

Figure 1. Program-controlled SLX-80 cryogenic system.

Table 2. Heat and cryogenic treatment schemes of the 15%SiCp/2009 aluminum matrix composites. The alphabetic order of each sample index indicates the execution sequence of treatments. For example, SCA2 indicates that the sample was treated by solution treatment, cryogenic treatment, and aging treatment for 2 h successively.

Treatment	Sample Index
Solution–Aging treatment	SA2, SA6, SA10, SA20
Solution–Cryogenic–Aging treatment	SCA2, SCA6, SCA10, SCA20
Solution–Aging–Cryogenic treatment	SA20C

The thermoelectric power (TEP) method was adopted for evaluating the microstructure evolution during aging. The principle of TEP measurement was to establish a temperature gradient at both ends of the samples and to measure the voltage difference between the two ends of the samples caused by the Seebeck effect. Based on the measurement principle of TEP, the apparatus was built according to the diagram shown in Figure 2. The sample size was $70 \times 5 \times 3$ mm^3, and its ends were fixed on the copper cap by bolts. The lower copper cap was regarded as the cold end (T), whose temperature was controlled by heating in ice water to maintain 15 °C. The upper copper cap was regarded as the hot end (T + ΔT), whose temperature was controlled by heating to maintain 25 °C. An Agilent 34975A data collector was used to record the voltage (ΔV) difference at both ends of the sample, and the acquisition step was 5 s. The relative TEP was defined as follows:

$$\Delta S = \frac{\Delta V}{\Delta T} \tag{1}$$

where ΔV is the low voltage arising from the Seeback effect between the two ends of the samples.

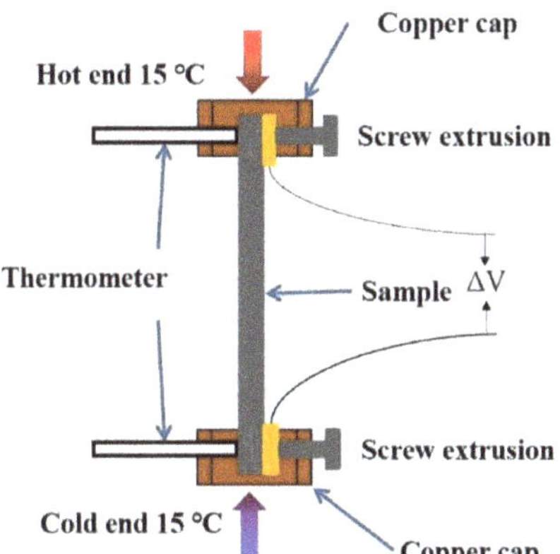

Figure 2. Schematic diagram of TEP measurement.

The Rockwell hardness (HRB) was tested according to the Chinese standard of GB/T 230.1-2018 with a SHBRV-187.5 digital Blackwell & Vickers hardness tester with a margin of error of $\pm2\%$. The loading force of the test was 100 Kg, and the duration was 5 s. Five points were tested for each sample, and the mean values were obtained as the final result. For further exploration of the effect of the sequence of cryogenic treatment and aging treatment on the mechanical properties of the composites, the electronic universal testing machine was used for the tensile test. Tensile specimens were prepared in accordance with the GB/T 228-2002 standard; the sample dimension is shown in Figure 3. The strain rate during the tensile process was 0.001/s. Three samples of each treatment (SA20, SCA20, SA20C) were tested and averaged as for a final result.

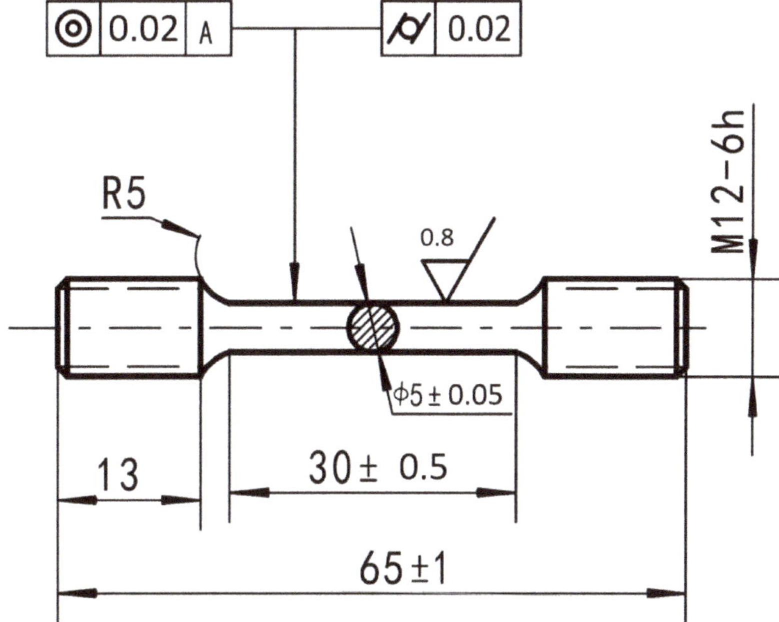

Figure 3. Size and geometry of the tensile test specimens.

For the purpose of investigating the change in microstructure of the composite after different treatments, the samples were ground on grit-abrasive papers with a mesh from 400 to 2000. The samples were then mechanically polished, and the polished surfaces were corroded for 10 s with Keller's reagent. The microstructure and tensile fracture surface were detected using a SU1510 scanning electron microscope (SEM) of, which was made by Hitachi in Japan. In order to further analyze the microstructure of the composites, the specimens were mechanically ground into thin slices with a thickness of 50 μm and then subjected to ion milling for examination by G20 transmission electron microscopy (TEM), which was made by FEI in the USA. The TEM and high-resolution images were Fourier-transformed by Digital Micrograph software, and the diffraction spots were calibrated using CaRIne software. The type of precipitates in the alloy could be determined using this account.

3. Results

3.1. Hardness and Tensile Properties

The hardness variations of the 15%SiCp/2009Al composites with different aging times are shown in Figure 4. It can be seen that the hardness increases with an extension of aging time. However, the hardness of both SA2 and SCA2 exhibited no obvious change when compared to that of the S and SC, respectively. The hardness of the sample treated with SC was higher than that of the sample treated by S. After aging for 6 h and 10 h, the hardness was obviously increased. With the increase of aging time to 20 h, the improvement in hardness induced by cryogenic treatment became more obvious and was approximately 2–4% higher than that of the samples without cryogenic treatment. In the 15%SiCp/2009Al composites, the Rockwell hardness was affected by the SiC particles and the aluminum matrix. As the SiC-reinforced phase cannot be altered by aging treatment and cryogenic treatment, the change in hardness can be mainly attributed to the microstructure evolution of the aluminum matrix.

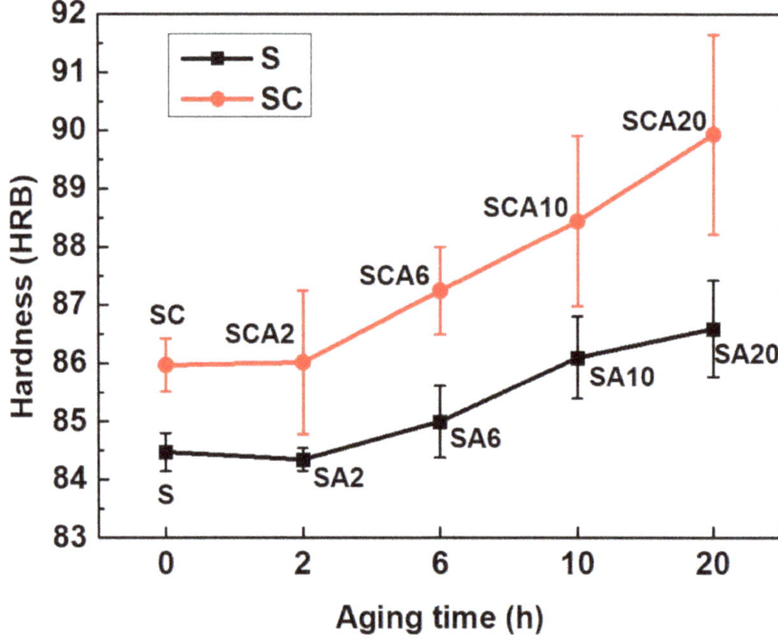

Figure 4. The hardness variations of 15%SiCp/2009Al composites with different aging times.

In order to investigate the effect of cryogenic treatment and its combination sequence with traditional heat treatment on 15%SiCp/2009Al composites, samples treated with SA20, SA20C, and SCA20 underwent tensile tests. The results are compared and shown in Figures 5 and 6. It can be seen from Figure 5 that the trend of stress–strain curves is consistent in the tensile process for all the specimens. The tensile strength, yield strength, and elongation can be obtained from these curves, as is shown in Figure 6. It can be observed that SA20C shows a slight improvement in tensile strength, while SCA20 demonstrates the opposite effect. However, this change in tensile strength is too small and can be ignored. Both SA20C and SCA20 can improve the yield strength of the 15%SiCp/2009Al composites. The improvement caused by SCA20 is more effective, being 16 MPa higher than that of the SA20 sample. The tendency of reduction in elongation can be observed. Due to the relatively poor plasticity of aluminum matrix composites and the discreteness of elongation, this reduction can also be ignored. Therefore, the most obvious effect caused by cryogenic treatment is the improvement in yield strength when cryogenic treatment is performed after solution and prior to aging treatment.

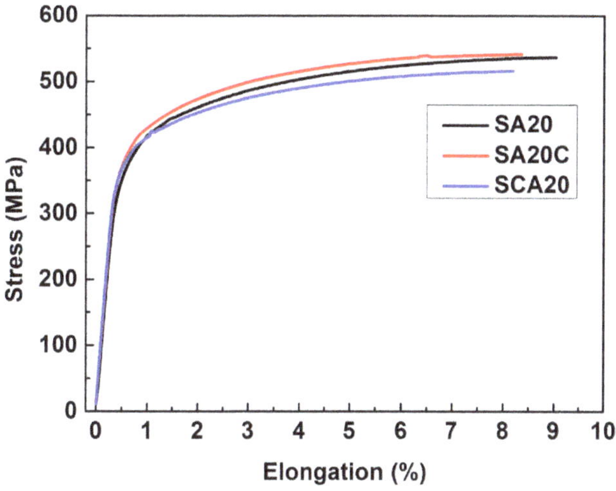

Figure 5. Stress–strain curves of specimens under different treatment conditions.

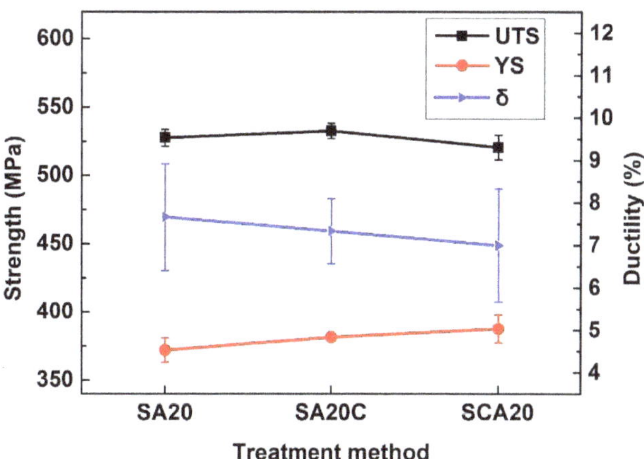

Figure 6. Tensile properties of specimens treated by different processes.

In order to further reveal the change in yield strength induced by cryogenic treatment, the macro and micro tensile fracture surfaces of SA20 and SCA20 were detected and compared, as is shown in Figures 7 and 8. It can be seen from Figure 7 that the tensile fracture surfaces of both SA20 and SCA20 are flat and have no obvious plastic deformation, which shows that the fracture of the 15%SiCp/2009Al composite is a typical brittle fracture. The micro-fracture surfaces show that the tensile fracture is comprises a majority of brittle dents and a small number of ductile dimples with a small size, as is shown in Figure 8a,d. The inner surface of the dent is relatively smooth, and the shape of the tear edge is consistent with the profile of the SiC particles, marked by the dotted line cycles in Figure 8b,c,f. A large number of studies have shown that there are three main fracture forms of aluminum matrix composites [28,29]: the fracture of reinforcement particles, the decohesion of SiC particles from the matrix, and the fracture of the aluminum matrix. Therefore, the fracture of the samples in both states is dominated by the decohesion of SiC particles from the matrix. There are also a small number of cracks propagating through the particles, which means that the fracture of reinforcement particles exists. Although the macroscopic appearance is that of a brittle fracture, there are a few small-sized dimples in a small range of aluminum alloy matrices, shown in Figure 8e,f, indicating that ductile tearing is the crack propagation mode in aluminum alloy matrices. There are relatively more ductile dimples in the matrix of the sample after cryogenic treatment when compared with that of the sample without cryogenic treatment (see a comparison of Figure 8b with Figure 8e). The tensile strength of the 15%SiCp/2009Al composite is mainly decided by the bond strength between the SiC particles and the aluminum matrix. The plastic deformation behavior of the matrix has a great influence on the yield strength. It can be seen that the variations of interface caused by cryogenic treatment are not enough to cause the change in tensile strength during tensile process. However, the changes in the aluminum matrix induced by cryogenic treatment will be the main reason for the improvement in yield strength.

Figure 7. Macro-fracture morphology of the SiCp/2009 aluminum composite, (**a**) SA20 and (**b**) SCA20.

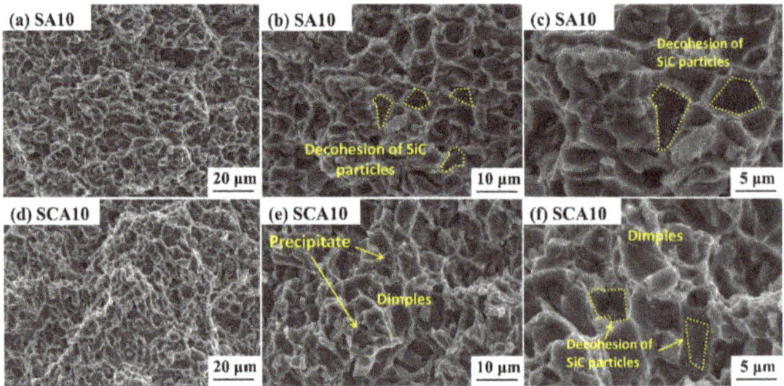

Figure 8. SEM micrographs of tensile fracture surface of samples treated by different processes: (**a–c**) without cryogenic treatment; (**d–f**) with cryogenic treatment.

3.2. Microstructure Evolution by SEM

In order to investigate the effect of cryogenic treatment on the precipitation behavior of the 15%SiCp/2009Al composites, the microstructures of the S and SC samples after aging for different times were detected by SEM, shown in Figure 9. It can be seen that the microstructure of the solution-treated (S) 15%SiCp/2009Al composites consisted of SiC particles with a size in the range of 4–10 μm and the recrystallized grain of the aluminum alloy matrix. There was a small amount of precipitates dispersed in the aluminum matrix of the solution-treated sample, appearing as the white particles shown in Figure 9a,d. P. Jin et al. [11] revealed that the remaining coarse particles in the solutionized SiCp/2009Al composites should be undissolved secondary phases, which were identified to be Al$_2$Cu (θphase) and Al$_7$Cu$_2$Fe phases. The volume fraction of the secondary phases is mainly determined by the solution temperature. The increase in solution temperature can reduce the content of remaining particles. There was no obvious difference in the microstructure between the S and SC samples.

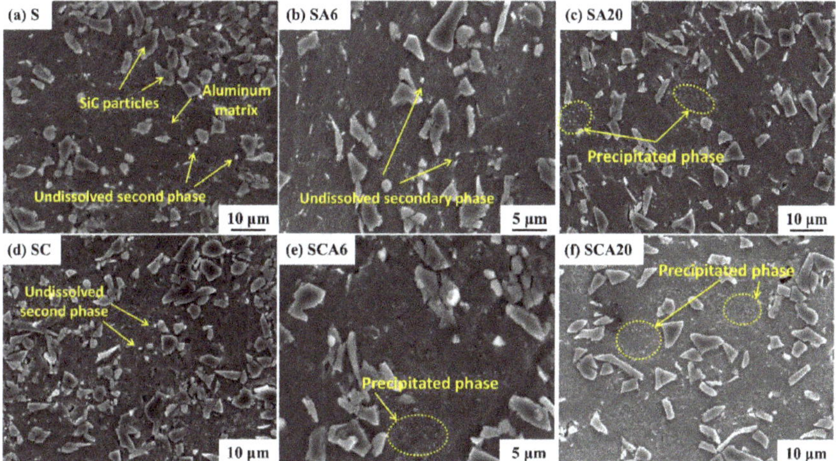

Figure 9. SEM micrographs of SiCp/2009Al composites treated by different processes: (**a**) S; (**b**) SA6; (**c**) SA20; (**d**) SC; (**e**) SCA6; and (**f**) SCA20.

After aging for 6 h (SA6), the white, small-size particles in the aluminum matrix were increased to a small extent when compared with that of the S sample. The aging treatment can promote the precipitation of more secondary particles, which depend on the aging temperature and time [10]. For the same aging time, the addition of cryogenic treatment (SCA6) had the trend of increasing the precipitation phases when compared to that of the SA6 samples, especially for the smaller-size particles, as is shown in Figure 9b,e. After aging for 20 h, the small-size precipitates increased obviously as the white particles shown in Figure 9c,f. With the addition of cryogenic treatment (SCA20), the volume fraction of precipitates appears to be higher than that of the SA20. In the 15%SiCp/2009Al composites, the equilibrium phases generated after aging were usually recognized as the θ (Al$_2$Cu) phase and the S (Al$_2$CuMg) phase, which depended on the ratio of Cu/Mg [30]. It can be inferred that cryogenic treatment after solution treatment can promote the precipitation of secondary phases. However, the type and precipitation stage of secondary phases induced by cryogenic treatment need further investigation.

3.3. ThermoElectric Power Measurement

In order to further confirm the change of precipitation from the global microstructure of the materials, the TEP measurements of S and SC samples after aging treatment with different times were tested and compared. Lattice defects that alter the electrical conductivity or elastic properties of the Al matrix, such as solute atoms or precipitates, have an effect on ΔS [25]. Therefore, the change of solute atoms and precipitates can be characterized sensitively by TEP. The relative TEP value of the as-quenched samples ($\Delta S0$) were taken as a reference, and the change value of ΔS was used to characterize the difference of precipitation and phase in the material; $\Delta(\Delta S) = \Delta St - \Delta S0$, where ΔSt is the TEP value of the sample after aging at 170 °C for different times. The TEP values of the 15%SiCp/2009 aluminum matrix composites after aging for different times are shown in Figure 10. It can be seen that the TEP curves of the 15%SiCp/2009Al composites can be divided into a descending stage and an ascending stage. With the increase in aging time, the TEP values were first decreased and then increased, and finally tended to be stable for both the S and SC samples. Cryogenic treatment followed with solution (SC) decreased the TEP value apparently when compared with that of the S sample. For the SC samples, the minimum value of the TEP was obtained after aging treatment for 6 h. The TEP of the S sample was also decreased with the increase in the aging time, and the minimum value of the SC sample was obtained after undergoing aging treatment for 10 h.

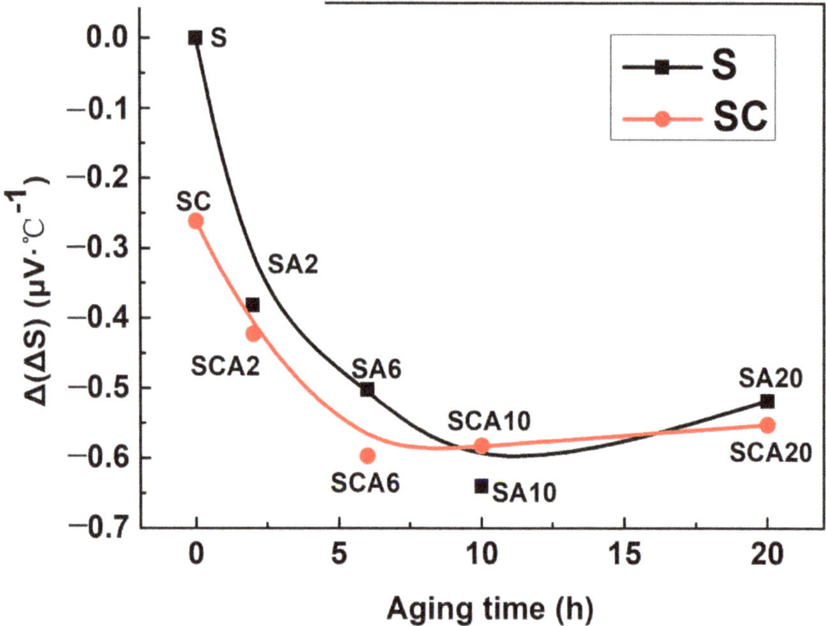

Figure 10. The TEP values of the S and SC samples change with the aging time.

For an aluminum alloy and its composites, there are two main factors that affect the TEP value during aging treatment [21,25]: (1) with the progress of precipitation, the decrease of solute atoms in the solution can change the value of TEP. The contributions of different solution elements to the TEP of aluminum alloy are varied, as Mg and Cu elements have a positive effect on the TEP while Si, Fe, and Mn elements have a negative effect on the TEP [25]. (2) Second, coherent or semi-coherent precipitation formed during aging has different effects on the TEP, depending on the type, size, and volume fraction of precipitation. In terms of coarse, incoherent precipitation, it is generally believed that it has no effect on the TEP of the alloy. Thus, the change in the TEP during the formation of such precipitation is basically only related to the decrease of solute content in the solution.

During the initial stage of aging, the formation of GP regions and transition phases occurs. The segregation of solute atoms increases lattice defects. Additionally, the precipitation of coherent Al_2Cu and Al_2CuMg phases reduces the content of Mg and Cu in the solid solution, which results in the reduction of the TEP. It can be seen that the TEP value of the SC sample was significantly lower than that of the S sample. During the descending stage, the TEP value of the SC sample was also lower than that of the S sample with the extension of aging time. After aging reaches a certain time, the TEP curve began to rise, indicating that the coherent phases had begun to transform into incoherent phases. It can be seen that the minimum values of S and SC are obtained by different aging times. Therefore, the change in aging behavior after cryogenic treatment can be reflected by the change in the TEP value.

3.4. Microstructure Detection by TEM

The samples of S and SC aged for 6 h and 20 h were adopted for TEM detection. It can be seen that the dislocation density in the aluminum matrix of the SCA6 sample was higher than that of SA6 sample, shown in Figure 11a,d. Many dislocation lines and some precipitates can be observed in the aluminum alloy and within the grain boundary in Figure 11d. It is well-acknowledged that the addition of SiC particles to the aluminum alloy can increase the dislocation density after quenching [11]. Cryogenic treatment can further increase the dislocation density. During the process of cryogenic treatment, excessive internal stress is generated at the interface due to the large difference of the expansion coefficient between the SiC and the aluminum matrix. The movement of dislocations is easily hindered by SiC particles, resulting in the stacking of dislocations at the interfaces and grain boundaries. The increase of dislocation density can promote the nucleating of precipitates. Some precipitates can be also seen in the grain boundary of SA6, as is shown in Figure 11b. The needle-like precipitates, with a size of about 20 nm, distribute homogeneously in the aluminum matrix in the SCA6 sample, as is shown in Figure 11e. Liu et al. [31] reported that these needle-like morphologies showed that the early-stage microstructure was dominated by GPI; zones and a small fraction of GPII zones. It can be inferred that more uniform nucleating sites can be produced by cryogenic treatment. After aging for 20 h, more precipitates can be generated in the aluminum matrix, interface, and grain boundaries in both SA20 (Figure 11c) and SCA20 (Figure 11f). It seems that the content of precipitates in SCA20 sample is higher than that of the SA20 sample. In order to determine the type of precipitates, a high-resolution transmission electron microscope (HRTEM) and a selected-area electron diffraction (SAED) were used to analyse the crystal structure of the precipitates. It can be seen that the aluminum alloy matrix has a face-centered cubic (fcc) lattice structure, shown in Figure 12c,e. The dark, blocky precipitate within the interface of the SiC and aluminum matrix is the θ' phase according to the HRTEM and the Fourier transform (FFT) patterns, as shown in Figure 12a,b,d. It has the tetragonal structure with lattice parameters of a = 0.404 nm and c = 0.580 nm [32]. This phase is recognized as the main strengthening phase in Al-Cu alloys [32]. It can also grow into large-sized precipitates with a size of close to 1 μm [31]. Therefore, the white particles in the SEM micrographs are mainly θ' phase. Precipitates in the grain boundaries are also identified to be θ' phase according to the results of SAED, as is shown in Figure 13a,b. Therefore, it can be concluded that cryogenic treatment after solution can promote the precipitates of the θ' phase in the 15%SiCp/2009Al composites.

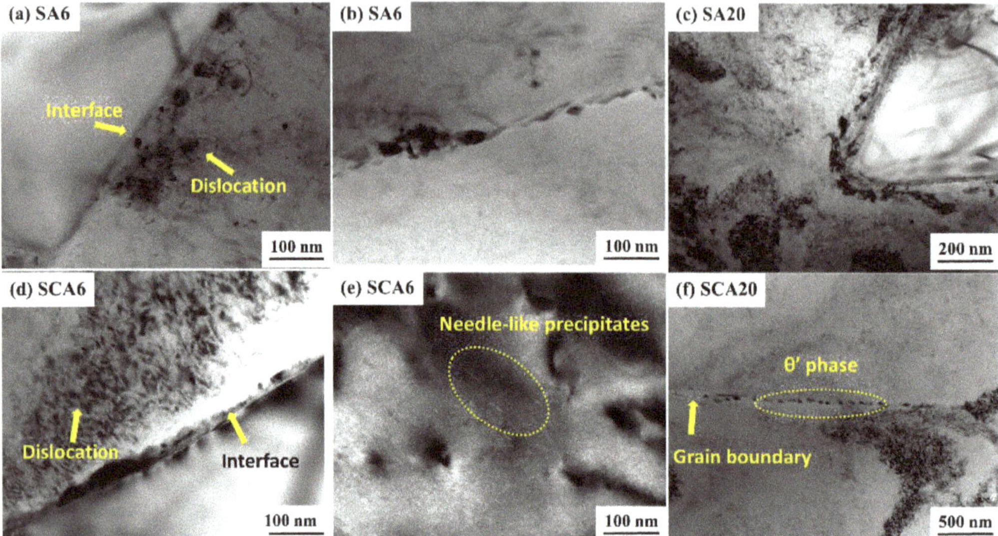

Figure 11. TEM micrographs of the 15%SiCp/2009Al composites treated by different processes: (a,b) SA6; (c) SA20; (d,e) SCA6; and (f) SCA20.

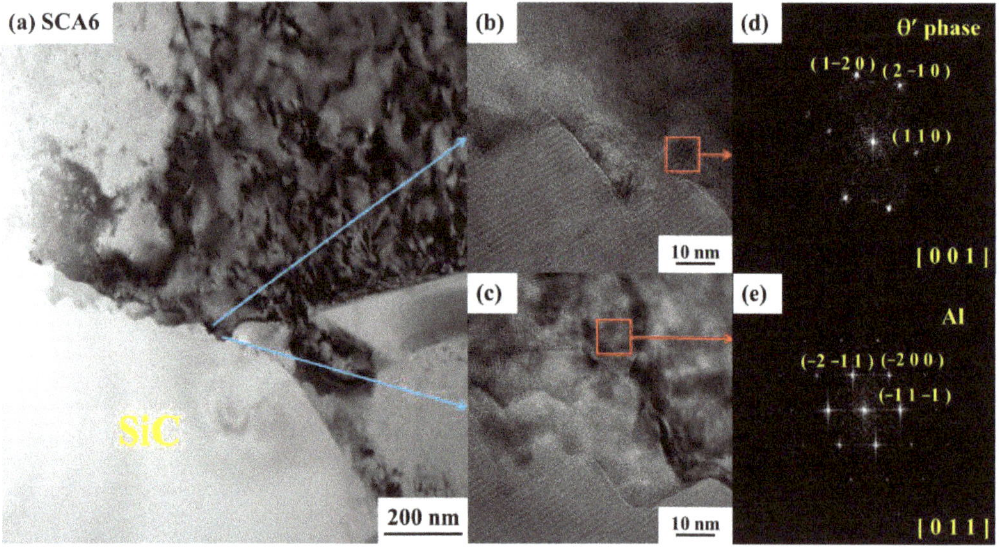

Figure 12. Micrographs of the 15%SiCp/2009Al composites treated by SCA6, (a) micrographs of SCA6 sample, (b,c) HRTEM micrographs of the selected areas in (a), (d,e) the FFT patterns of selected area in (b) and (c), separately.

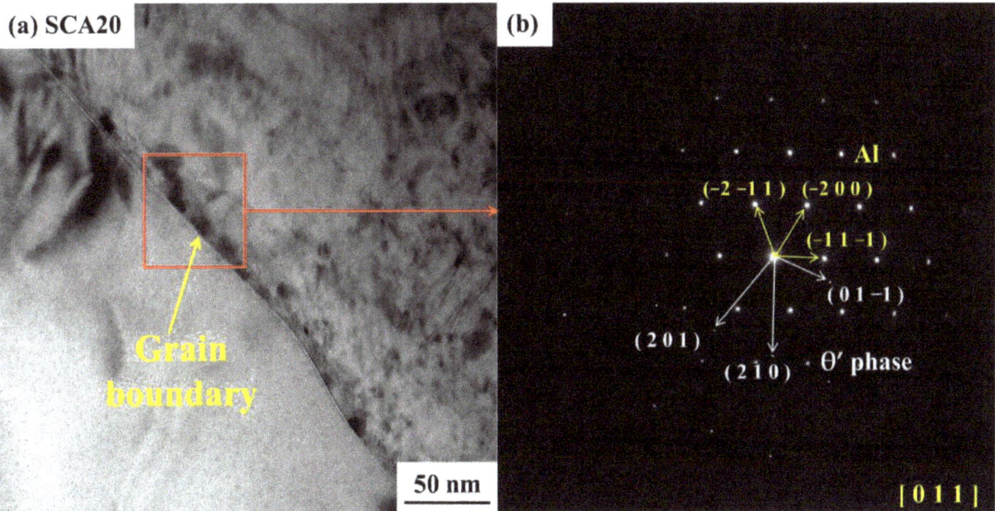

Figure 13. TEM micrographs (**a**) and selected-area electron diffraction (SAED) (**b**) of the 15%SiCp/2009Al composites treated by SCA20.

4. Discussion

It can be inferred from the above-mentioned results that cryogenic treatment after solution can obviously facilitate the aging process of 15%SiCp/2009 composites. The resultant higher hardness as well as the lower TEP of the SC sample when compared to that of the S sample can be attributed to the increase in dislocation density and the potential energy difference caused by compressive stress in the aluminum alloy matrix, induced by cryogenic treatment. These two factors are demonstrated by Klug et al. [33] as the main mechanisms for modification to the precipitates following cryogenic treatment. During the process of cryogenic treatment, the lattice shrinkage of the aluminum matrix is hindered by SiC particles, so it is easy to generate dislocation accumulation and improve the dislocation density near the interface. Furthermore, the lattice contraction of an aluminum matrix at cryogenic temperature occurs seriously, while the shrinkage of SiC particles is small. High, internal, residual stress in the interface and even within grains can be generated.

After aging for 2 h, the hardness exhibits no obvious change while the TEP reduction is obvious, meaning that the main change in the microstructure is the formation of GP-zones. There are no obvious equilibrium precipitates formed at this stage. However, the formation of a GP-zone and transitional phase in aluminum alloys can reduce the TEP by both increasing lattice distortion and precipitating Mg and Cu elements. This stage can be considered as the incubation period of aging process. Factors induced by cryogenic treatment followed with solution treatment, such as high dislocation density and internal residual stress, can act as a driving force for the precipitation of an aluminum alloy matrix. Therefore, the hardness of SCA2 is also higher than that of SA2.

After aging for 6 h, the hardness is increased which indicates that obvious precipitates are formed. Meanwhile, the TEP value decreases obviously, which shows that the precipitates in this stage are dominated by the coherent θ″ phase. The lower TEP and higher hardness of SCA6 show that a higher content of the θ″ phase can be induced by the addition of cryogenic treatment. With the extension of aging time, θ″ phases transform into the θ′ phase and θ phase, whose coherent relationships with the matrix are decreased and have little influence over the TEP value. Therefore, the TEP shows a little increase with the reduction of the θ″ phase after aging for longer than 6 h, with regard to the sample with cryogenic treatment. However, the TEP of a solution-treated sample will still keep descending after aging for 6 h, and it begins to rise after aging for 10 h. This indicates that cryogenic treatment increases the aging-precipitation driving force of 15%SiCp/2009 composites. As a result, the transformation of metastable phases in the aging process is advanced. With the extension of aging time, the transformation of θ'' phases into θ' and θ phases dominates the aging process; therefore, the hardness increases obviously. The higher hardness of the SCA sample compared to that of the SA sample can be attributed to the increase in precipitation content induced by cryogenic treatment. The increase of precipitates enhances its pinning effect on dislocations, as a result of enhancing the yield strength of the 15%SiCp/2009 composites.

5. Conclusions

The effects of deep cryogenic treatment on the microstructure and mechanical properties of 15%SiCp/2009 aluminum matrix composites were investigated in this work. The main conclusions are shown below.

(1) TEP measurement can be an effective method for evaluating the precipitation characteristics of 15%SiCp/2009 aluminum matrix composites during aging. Cryogenic treatment after solution promotes the precipitation from the start-stage of aging, which can advance the precipitation time of a stable, non-coherent phase. Cryogenic treatment increases dislocation density, which provides a greater driving force for precipitation during the aging process. It can also promote the precipitation of θ' phases after aging for 6 h. As a result, the content of the equilibrium precipitates formed after aging at the same time is higher than that of samples without cryogenic treatment.

(2) After cryogenic treatment, the hardness of 15%SiCp/2009 aluminum matrix composites is improved. The change in hardness induced by cryogenic treatment becomes more significant with the extension of aging time, which is improved by 4% after artificial aging treatment for 20 h.

(3) The cryogenic treatment combined with traditional heat treatment has little effect on the tensile strength and elongation of 15%SiCp/2009 aluminum matrix composites. Cryogenic treatment after solution and aging treatment (SAC) also has little effect on yield strength, while the conduction of cryogenic treatment after solution and prior to aging treatment (SCA) can increase the yield strength by 16 MPa.

Considering the applications of cryogenic treatment in the aerospace and automotive industries, it is more importance to investigate the effect of cryogenic treatment on fatigue properties, wear resistance, corrosion resistance, and the dimensional stability of aluminum matrix composites. Furthermore, as residual stress in aluminum matrix composites is a serious problem, the comprehensive modification of residual stress and mechanical properties has more important value for practical application.

Author Contributions: Conceptualization, M.Z. and K.G.; methodology, K.G. and M.Z.; validation, R.P. and B.L.; formal analysis, K.G. and M.Z.; investigation, Z.W.; resources, R.P. and J.W.; data curation, Z.W. and C.C.; writing—original draft preparation, M.Z.; writing—review and editing, K.G. and Z.W.; visualization, M.Z.; supervision, J.W.; project administration, B.L. and J.W; funding acquisition, R.P. All authors have read and agreed to the published version of the manuscript.

Funding: This research was funded by the Natural Science Foundation of Beijing, grant number 3214053, and the Projects for National Science and Technology Institutes of China, grant number 20190101.

Institutional Review Board Statement: Not applicable.

Informed Consent Statement: Not applicable.

Data Availability Statement: Not applicable.

Conflicts of Interest: The authors declare no conflict of interest.

References

1. Rathod, N.; Menghani, J. A Consequence of Reinforcements In Aluminum-based Metal Matrix Composites: A Literature Review. *Metall. Mater. Eng.* **2019**, *25*, 195–208. [CrossRef] [PubMed]
2. Lakshmikanthan, A.; Angadi, S.; Malik, V.; Saxena, K.; Prakash, C.; Dixit, S.; Mohammed, K.A. Mechanical and Tribological Properties of Aluminum-Based Metal-Matrix Composites. *Materials* **2022**, *15*, 6111. [CrossRef] [PubMed]
3. Wu, R.R.; Yuan, Z.; Li, Q.S. Microstructure and mechanical properties of 7075 Al alloy based composites with Al_2O_3 nanoparticles. *Int. J. Cast Met. Res.* **2017**, *30*, 337–340.
4. Wu, W.G.; Zeng, T.C.; Hao, W.F.; Jiang, S.P. Microstructure and Mechanical Properties of Aluminum Matrix Composites Reinforced With In-Situ TiB_2 Particles. *Front. Mater.* **2022**, *9*, 1–9. [CrossRef]
5. Galyshev, S.; Bulat, A. The Dependence of the Strength of a Carbon Fiber/Aluminum Matrix Composite on the Interface Shear Strength between the Matrix and Fiber. *Materials* **2022**, *12*, 1753–1771. [CrossRef]
6. Bhat, R.; Hegde, A.; Sharma, S.; Shankar, G.; Murthy, G. Hybrid Heat Treatment for Conventionally Treatable Steel Powder Reinforced Age Hardenable Aluminium Alloy Matrix Composites and Mechanical Property Evaluation. *Mater. Res.* **2021**, *25*, e20210429. [CrossRef]
7. Sagar, S.R.; Srikanth, K.M.; Jayasimha, R. Effect of cryogenic treatment and heat treatment on mechanical and tribological properties of A356 reinforced with SiC. *Mater. Today Proc.* **2021**, *45*, 184–190. [CrossRef]
8. Xiao, B.L.; Zhang, W.Y.; Zuo, T.; Fan, J.Z.; Shi, L.K. Effect of Heat Treatment on Mechanical Properties and Fracture Behavior of 15% SiCp/2009Al Composite. *Aerosp. Mater. Technol.* **2005**, *3*, 34–37.
9. Wang, Z.X.; Chen, J.; Liu, B.S.; Pan, R.; Guo, Y.; Li, Y. Effect of Deep Cryogenic Treatment on the Artificial Ageing Behavior of SiCp-AA2009 Composite. *Materials* **2022**, *12*, 1767–1781. [CrossRef]
10. Pal, S.; Mitra, R.; Bhanuprasad, V.V. Aging behaviour of Al-Cu-Mg alloy-SiC composites. *Mater. Sci. Eng. A* **2008**, *480*, 496–505. [CrossRef]
11. Jin, P.; Xiao, Q.Z.; Wang, Q.Z.; Ma, Z.Y.; Liu, Y.; Li, S. Effect of solution temperature on aging behavior and properties of SiCp/Al-Cu-Mg composites. *Mater. Sci. Eng. A* **2011**, *528*, 1504–1511. [CrossRef]
12. Araghchi, M.; Mansouri, H.; Vafaei, R.; Guo, Y. Optimization of the Mechanical Properties and Residual Stresses in 2024 Aluminum Alloy Through Heat Treatment. *J. Mater. Eng. Perform.* **2018**, *27*, 3234–3238. [CrossRef]
13. Steier, V.F.; Ashiuchi, E.S.; Reißig, L.; Araújo, J.A. Effect of a Deep Cryogenic Treatment on Wear and Microstructure of a 6101 Aluminum Alloy. *Adv. Mater. Sci. Eng.* **2016**, *2016*, 1582490.
14. Koklu, U. The drilling machinability of 5083 aluminum under shallow and deep cryogenic treatment. *Emerg. Mater. Res.* **2020**, *9*, 323–330. [CrossRef]
15. Gao, W.L.; Wang, X.J.; Chen, J.Z.; Ban, C.Y.; Cui, J.Z.; Liu, Z. Influence of Deep Cryogenic Treatment on Microstructure and Properties of 7A99 Ultra-High Strength Aluminum Alloy. *Metals* **2019**, *9*, 631. [CrossRef]
16. Park, K.; Ko, D.H.; Kim, B.M.; Lim, H.J.; Lee, J.M.; Cho, Y.R. Effects of Cryogenic Treatment on Residual Stress and Tensile Properties for 6061 Al Alloy. *Korean J. Met. Mater.* **2011**, *49*, 9–16. [CrossRef]
17. Araghchi, M.; Mansouri, H.; Vafaei, R.; Guo, Y. A novel cryogenic treatment for reduction of residual stresses in 2024 aluminum alloy. *Mater. Sci. Eng. A* **2017**, *689*, 48–52. [CrossRef]
18. Jovičević-Klug, M.; Rezar, R.; Jovičević-Klug, P.; Podgornik, B. Influence of deep cryogenic treatment on natural and artificial aging of Al-Mg-Si alloy EN AW 6026. *J. Alloys Compd.* **2022**, *899*, 163323. [CrossRef]
19. Zhou, C.; Sun, Q.; Qian, D.; Liu, J.; Sun, J.; Sun, Z. Effect of deep cryogenic treatment on mechanical properties and residual stress of AlSi10Mg alloy fabricated by laser powder bed fusion. *J. Mater. Process. Technol.* **2022**, *303*, 117543. [CrossRef]
20. Li, G.R.; Cheng, J.F.; Wang, H.M.; Li, C.Q. The influence of cryogenic-aging circular treatment on the microstructure and properties of aluminum matrix composites. *J. Alloys Compd.* **2017**, *695*, 1930–1945. [CrossRef]
21. Hong, T.R.; Shen, Y.W.; Geng, J.W.; Chen, D.; Li, X.F.; Zhou, C. Effect of cryogenic pre-treatment on aging behavior of in-situ TiB_2/Al–Cu–Mg composites. *Mater. Charact.* **2016**, *119*, 40–46. [CrossRef]
22. Pelletier, J.M.; Borrelly, R. Temperature and concentration dependences of thermoelectric power at high temperatures in some aluminium alloys. *Mater. Sci. Eng.* **1982**, *55*, 191–202. [CrossRef]
23. Bakare, F.; Alsubhi, Y.; Ragkousis, A.; Ebomwony, O.; Damisa, J.; Okunzuwa, S. The effect of pre-straining and pre-ageing on a novel thermomechanical treatment for improving the mechanical properties of AA2139 aerospace aluminium alloys. *Mater. Res. Express* **2017**, *4*, 076506. [CrossRef]
24. Abdala, M.R.W.S.; Garcia de Blas, J.C.; Barbosa, C.; Acselrad, O. Thermoelectrical power analysis of precipitation in 6013 aluminum alloy. *Mater. Charact.* **2008**, *9*, 271–277. [CrossRef]
25. Massardier, V.; Epicier, T.; Merle, P. Correlation between the microstructural evolution of a 6061 aluminium alloy and the evolution of its thermoelectric power. *Acta Mater.* **2000**, *48*, 2911–2924. [CrossRef]
26. LSenthilkumar, D. *Encyclopedia of Iron, Steel, and Their Alloys*; CRC Press-Taylor and Francis Group: New York, NY, USA, 2016; pp. 995–1007.
27. Jovičević-Klug, P.; Tóth, L.; Podgornik, B. Comparison of K340 Steel Microstructure and Mechanical Properties Using Shallow and Deep Cryogenic Treatment. *Coatings* **2022**, *12*, 1296. [CrossRef]
28. Chen, D.; Dong, J.F.; Ma, G.Z.; Ma, G.Z.; Wu, Z.; Zhang, X.L. The Research about Cryogenic Treatment Process of Nonferrous Alloys—A Review. *Mater. Rep.* **2010**, *24*, 1–4.

29. Cai, H.K.; Weng, Z.J.; Gu, K.X.; Wang, K.K.; Zheng, J.P.; Wang, J.J. Advances in Deep Cryogenic Treatment of Cemented Carbide. *Mater. Rep.* **2019**, *33*, 175–182.
30. Mochalova, M.V.; Avtokratova, E.V.; Kazakulov, I.Y.; Krymsky, S.V.; Mochalova, M.Y.; Murashkin, M.Y.; Sitdikov, O.S. Microstructure and Properties of an Aluminum D16 Alloy Subjected to Cryogenic Rolling. *Russ. Metall. (Met.)* **2011**, *4*, 364–369.
31. Liu, C.H.; Ma, Z.Y.; Ma, P.P.; Zhan, L.H.; Huang, M.H. Multiple precipitation reactions and formation of θ'-phase in a pre-deformed Al-Cu alloy. *Mater. Sci. Eng. A* **2018**, *733*, 28–38. [CrossRef]
32. Rodrigo, P.; Poza, P.; Utrilla, V.; Urena, A. Effect of reinforcement geometry on precipitation kinetics of powder metallurgy AA2009/SiC composites. *J. Alloys Compd.* **2009**, *479*, 451–456. [CrossRef]
33. Jovicevic-Klug, M.; Tegg, L.; Jovicevic-Klug, P.; Drazic, G.; Almasy, L.; Lim, B.; Cairney, J.M.; Podgornik, B. Multiscale modification of aluminum alloys with deep cryogenic treatment for advanced properties. *J. Mater. Res. Technol.* **2022**, *21*, 3062–3073. [CrossRef]

Article

Microstructure and Mechanical Properties of W-Al$_2$O$_3$ Alloy Plates Prepared by a Wet Chemical Method and Rolling Process

Changji Wang [1,2], Xiaonan Dong [1,3], Yao Liu [4], Shizhong Wei [2], Kunming Pan [2,*], Cheng Zhang [2], Mei Xiong [2], Feng Mao [2], Tao Jiang [2], Hua Yu [1], Xiaodong Wang [2] and Chong Chen [2,*]

[1] School of Materials Science and Engineering, Henan University of Science and Technology, Luoyang 471000, China
[2] Henan Key Laboratory of High-Temperature Structural and Functional Materials, Henan University of Science and Technology, Luoyang 471003, China
[3] Longmen Laboratory, Luoyang 471000, China
[4] Scientific Research Platform Service Center of Henan Province, Zhengzhou 450003, China
* Correspondence: pankunming2008@haust.edu.cn (K.P.); chongchen@haust.edu.cn (C.C.)

Abstract: The uneven distribution and large size of the second phase weakens the effect of dispersion strengthening in ODS-W alloys. In this article, the W-Al$_2$O$_3$ composite powders were fabricated using a wet chemical method, resulting in a finer powder and uniformly dispersed Al$_2$O$_3$ particles in the tungsten-based alloy. The particle size of the pure tungsten powder is 1.05 μm and the particle size of W-0.2 wt.%Al$_2$O$_3$ is 727 nm. Subsequently, the W-Al$_2$O$_3$ alloy plates were successfully obtained by induction sintering and rolling processes. Al$_2$O$_3$ effectively refined grain size from powder-making to sintering. The micro-hardness of the tungsten alloy plates reached 512 HV$_{0.2}$, which is 43.7% higher than that of pure tungsten plates. The nano-hardness reached 14.2 GPa, which is 24.1% higher than that of the pure tungsten plate; the compressive strength reached 2224 MPa, which is 37.2% higher than that of the pure tungsten.

Keywords: ODS-W; wet chemical method; Al$_2$O$_3$-reinforced; tungsten alloy plates

Citation: Wang, C.; Dong, X.; Liu, Y.; Wei, S.; Pan, K.; Zhang, C.; Xiong, M.; Mao, F.; Jiang, T.; Yu, H.; et al. Microstructure and Mechanical Properties of W-Al$_2$O$_3$ Alloy Plates Prepared by a Wet Chemical Method and Rolling Process. *Materials* **2022**, *15*, 7910. https://doi.org/10.3390/ma15227910

Academic Editors: Hailiang Yu, Zhilin Liu and Xiaohui Cui

Received: 14 October 2022
Accepted: 6 November 2022
Published: 9 November 2022

Publisher's Note: MDPI stays neutral with regard to jurisdictional claims in published maps and institutional affiliations.

1. Introduction

Tungsten or tungsten-base alloys are used in the aerospace industry to make nozzles for rocket propulsion and armor-piercing warheads in the defense military industry owing to a high melting point (3420 °C), high density, excellent thermal shock resistance, and high-temperature strength [1]. Because of the excellent sputtering resistance and low tritium retention, tungsten is also considered to be the most ideal material for use in the first wall to resist fusion plasma irradiation in the nuclear industry [2–6]. However, the room temperature brittleness, irradiation embrittlement, and recrystallization embrittlement of tungsten limit its wider applications [7].

The introduction of oxides and carbides into the tungsten matrix is considered to be an effective method to solve the above problems of tungsten [8–11]. Oxides such as La$_2$O$_3$ [12], Y$_2$O$_3$ [13], ZrO$_2$ [14], and Al$_2$O$_3$ [15] have been widely studied. Al$_2$O$_3$ has been added to metal matrix composites (such as Al, Mo, and Mg) as a strengthening phase because of the excellent high-temperature properties, good chemical stability, and low cost [16–19]. The introduction of a second phase using conventional mechanical alloying methods not only results in a non-uniform distribution of the second phase but also tends to introduce impurities during the alloying process; the use of wet chemical methods was able to circumvent these problems and prepare powders to the nanoscale [20–22]. In our previous work, the addition of finer Al$_2$O$_3$ to tungsten alloys using a wet chemical method was found to significantly improve hardness and toughness [15] and increase the recrystallization temperature of the alloys [23].

This study used a wet chemical method to achieve a homogeneous distribution of the second phase Al_2O_3 in the alloy. The alloy powder prepared using this method also allows for effective grain refinement, resulting in significant improvements in the properties and machinability of the alloy. The tungsten alloys were unidirectionally rolled into plates, and the mechanical properties of the tungsten alloy plates were investigated by conducting nano-hardness, micro-hardness, and room-temperature compression tests. The effects of the microstructure and weaving on the mechanical properties of the alloy plates were studied by performing scanning electron microscopy (SEM), electron backscatter diffraction (EBSD), and transmission electron microscopy (TEM).

2. Materials and Methods

The W-Al_2O_3 composite powders were prepared from aluminum nitrate ($Al(NO_3)_3$) and ammonium para tungstate ($(NH_4)10W_{12}O_{41}\cdot5H_2O$). Aluminum nitrate ($Al(NO_3)_3$) was dissolved in deionized water by stirring using a magnetic mixer. The obtained solution was added to a concentrated nitric acid solution to adjust the pH (pH = 0.5). The treated solution was poured into a hydrothermal reactor and placed in a drying oven for hydrothermal heating (170 °C, 15 h). Ammonium para tungstate was added to deionized water and dissolved completely. After the hydrothermal reaction, the precursor solution of Al_2O_3 was poured into the aqueous solution of ammonium para tungstate by applying magnetic stirring + ultrasonic shaking (20–24 h). The evenly stirred mixture was poured into a vessel and placed in a drying oven (80 °C). The dried composite powder was crushed using a pulverizer. The crushed powder was placed in a muffle furnace and calcined (500 °C, 2 h) to obtain the composite powder of WO_3 and Al_2O_3. The composite powder was placed in a hydrogen reduction furnace for two-stage reduction (620 °C, 6 h, 920 °C, 8 h). Finally, the W-Al_2O_3 composite powders were obtained. The W-Al_2O_3 composite powders were loaded into a rubber mold and rolled into a square billet by employing cold isostatic pressing (CIP). The billets were placed in an induction sintering furnace at a temperature of 2350 °C for a sintering time of 8 h. The hydrogen atmosphere was maintained throughout the sintering process to prevent oxidation of the alloy. The sintered body with a thickness of 24 mm was heated to 1500–1600 °C in the heating furnace and the tungsten alloy plates with a thickness of 3 mm were finally obtained after several unidirectional rolling. The total thickness reduction after rolling was 87.5%. Five compositions of tungsten alloy plates with Al_2O_3 contents of 0, 0.2 wt.%, 0.4 wt.%, 0.6 wt.%, and 0.8 wt.% were prepared to investigate the effect of different contents of Al_2O_3 on the properties and microstructure of the plates.

The physical phase of the prepared alloy powder was examined by performing X-ray diffraction (Bruker-D8) equipped with a CuKα target. The voltage and current used in XRD experiments were 40 kV and 40 mA, respectively. The microscopic morphology and the distribution of Al_2O_3 in the W-Al_2O_3 composite powder were analyzed by using JEOL JSM-IT 800 SHL scanning electron microscope, and the particle size of the W-Al_2O_3 composite powder was tested by using an LS-909 laser particle sizer (Zhuhai OMEC Instruments Co., Ltd., Zhuhai, China). After rolling the alloy, samples (6 mm × 6 mm) were cut from the rolled surface. After mounting and polishing, the micro-hardness of the alloy plates was tested using an HV-1000 Vickers micro-hardness tester (Laizhou Huayin Test Instrument Co. Ltd., Laizhou, China) with a load of 200 g and a holding time of 15 s. Each sample was tested 20 times, then the maximum and minimum values were removed, and the remaining values were averaged, which was conducted to minimize error. The nano-indentation hardness was characterized using a G200 Nano-Indenter instrument (Keysight Technologies) equipped with a Berkovich indenter. In order to ensure the accuracy of the test data, the tungsten samples were polished using the diamond suspensions and 36 points were tested for each sample under the constant strain rate of 0.05 s^{-1}. The pressing depth and dwell time were 1500 nm and 15 s, respectively. The maximum load in the nanoindentation tests was 870 mN. The hardness of tungsten alloys was measured by continuous stiffness method (CSM) and calculated by Oliver–Pharr method. Room-temperature compression samples (3 mm × 3 mm × 6 mm) were cut along the rolling direction of the plates, and room-temperature compression experiments were performed in a Instron 5582 Double Column Electronic Universal Testing Machine at a rate of 0.5 mm/min. Samples (3 mm × 8 mm) were cut in the tangential direction of the plates and polished in an argon-ion polishing machine (IB-19530CP, JEOL, Akishima-shi, Tokyo, Japan). After the sample was set, smoothed, and polished by using an argon-ion polisher, the weaving and microstructure of the deformed plates were observed using a JEOL JSM-IT 800 SHL scanning electron microscope (Akishima-shi, Tokyo, Japan) and an Oxford c-nano EBSD system with an EBSD scan step of 160 nm. Analysis of the interface and microstructure of the rolled sheet using FEI Talos F200X Transmission electron microscope. Figure 1 shows the schematic diagram of the examined sample.

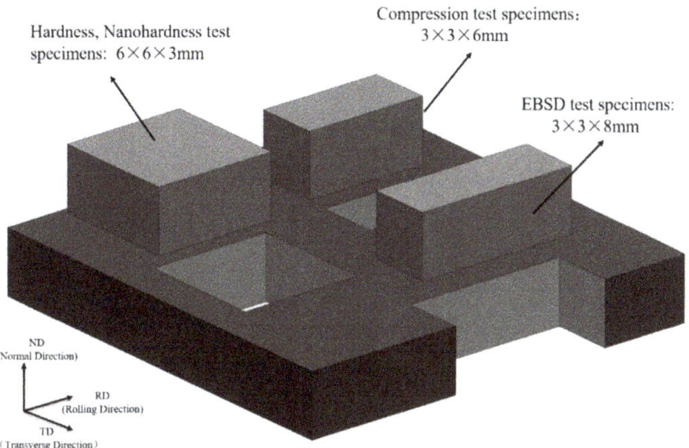

Figure 1. Schematic diagram of test specimens cut from the tungsten alloy plates.

3. Results and Discussion

3.1. Powder Characterization

Figure 2 shows the XRD patterns of the composite powders. The XRD peaks of pure tungsten and W-Al$_2$O$_3$ alloy powders with different compositions are the same, with no diffraction peaks of other phases except those of tungsten; this may be because the content of Al$_2$O$_3$ is extremely low to be detected in XRD. In addition, the intensity of the diffraction peaks is slightly different; this may be because the crystallinity of tungsten is affected by the addition of Al$_2$O$_3$.

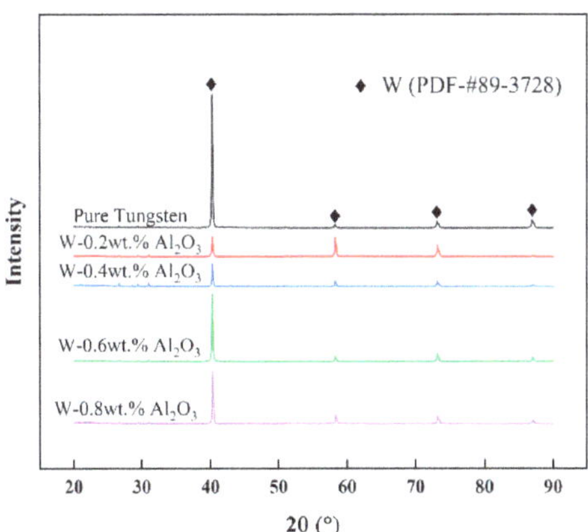

Figure 2. XRD patterns of the composite powders.

Figure 3 shows the SEM images of the composite powders with different compositions. The average particle size of the pure tungsten powder is 1.05 µm, as shown in Figure 3a. The particles appear to be bonded and the particles are significantly larger than those of the composite powders with other compositions. When the content of Al_2O_3 is 0.2 wt.% (Figure 3b), the particles are better than those of the powders with other compositions in terms of both morphology and size; the average particle size is only 727 nm, which is 300 nm smaller than that of pure tungsten powder. The particle sizes in the other compositions do not differ considerably are shown in Figure 3c–e. However, compared with those presented in Figure 3b, these compositions contain many large Al_2O_3 particles; this is one of the reasons for the deterioration of the performance with the increase in Al_2O_3 content. Figure 3f shows the average particle size of the powders with different compositions. The standard deviations of composite powders with Al_2O_3 content from 0 to 0.8 wt.% are 0.2, 0.09, 0.11, 0.12, and 0.12, respectively. The D50 of composite powders with Al_2O_3 content from 0 to 0.8 wt.% are 1.05, 0.73, 0.83, 0.94, and 0.93 µm, respectively. The particle size of the powder prepared using the wet chemical method reaches the nanometer level. A distinct difference between the particle size of pure tungsten and the other powders with Al_2O_3 can be observed, indicating that adding Al_2O_3 effectively reduces the particle size during the hydrothermal synthesis and the reduction process. This has a crucial effect on the subsequent sintering and rolling of the tungsten alloys.

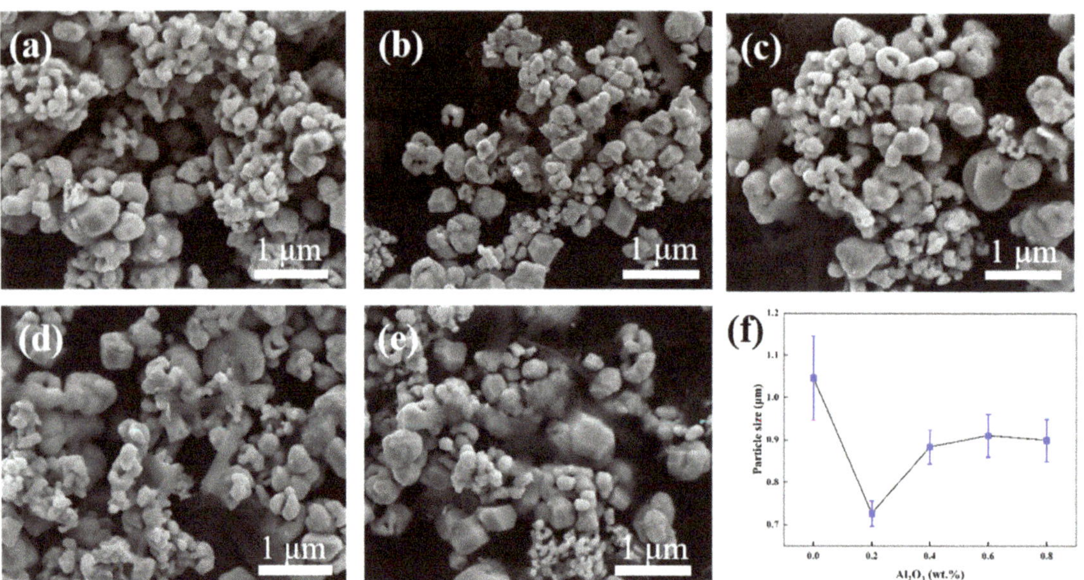

Figure 3. SEM images of the composite powders: (**a**) pure tungsten, (**b**) W-0.2 wt.%Al_2O_3, (**c**) W-0.4 wt.%Al_2O_3, (**d**) W-0.6 wt.%Al_2O_3, (**e**) W-0.8 wt.%Al_2O_3, (**f**) grain size of the composite powders.

3.2. Microstructure of Sintered and Rolled Alloys

After the blanks are pressed and sintered and the sample is cut for SEM observation, pores appear in varying degrees in each composition. The reason is that tungsten has a high melting point, and it is difficult to achieve full density even with high temperature and sintering over a long time period. Therefore, the rolling process was adopted to eliminate the pores. The SEM images of the tungsten alloys are shown in Figure 4 and the grain sizes were determined by the linear intercept method. At least 200 intercepts of grains were measured and at least 8 SEM micrographs from random and spread-out locations were analyzed for each condition. As shown in Figure 4a, the grain size of sintered pure tungsten is 24 μm. Figure 4b shows the grain size of sintered W-0.2 wt.%Al$_2$O$_3$ alloy is only 9 μm. When the content of Al$_2$O$_3$ is 0.4 wt.%, 0.6 wt.%, and 0.8 wt.% (Figure 4c–e), the grain sizes are 10 μm, 14 μm, and 12 μm, respectively, corresponding to the grain sizes of the alloy powder measured previously. Al$_2$O$_3$ distributed at the grain boundaries during the sintering process can hinder the diffusion of grain boundaries, thus inhibiting grain growth and achieving grain refinement. Excessive addition of Al$_2$O$_3$ grains leads to agglomeration, and the ability to inhibit W grain growth is weakened, but the grain size is substantially reduced compared with that of pure tungsten.

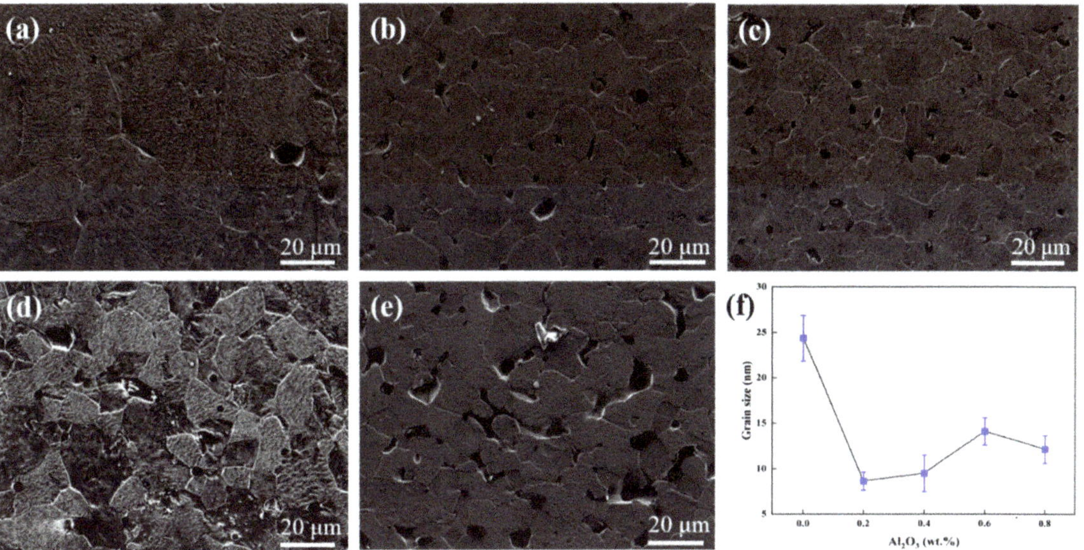

Figure 4. SEM images of the tungsten alloys: (**a**) pure tungsten, (**b**) W-0.2 wt.% Al$_2$O$_3$, (**c**) W-0.4 wt.% Al$_2$O$_3$, (**d**) W-0.6 wt.% Al$_2$O$_3$, (**e**) W-0.8 wt.%Al$_2$O$_3$, (**f**) grain size of the tungsten alloys.

As shown in Figure 5a–e, the grains of the sintered alloy are elongated after rolling. The degrees of elongation and the proportions of large-angle grain boundaries differ for plates with different compositions. Among these, the pure tungsten plate has the largest grains, maximum Feret diameter of 10.2 μm, and smallest percentage (only 39.6%) of large-angle grain boundaries. The W-0.2 wt.% Al$_2$O$_3$ alloy plate has a maximum Ferret diameter of 6.3 μm and a percentage of large-angle grain boundaries of 48.6%. The remaining plates of the other compositions have no significant advantage in these two aspects. Because of the large plate deformation under extreme stress, many dislocations in the form of dislocation walls and networks are widely distributed in the grain. With the movement of dislocations, a series of interactions between various dislocations cause the dislocation entanglement phenomenon. With the development of this entanglement phenomenon, the grains break into sub-grains; this is why the grains become smaller after rolling. The increase in the dislocation density and sub-grain boundaries gradually increases the deformation resistance of tungsten. The addition of Al$_2$O$_3$ enhances this phenomenon. Pure tungsten without Al$_2$O$_3$ impedes dislocations, with long strips of grains appearing after rolling. For several other compositions, aggregation, and uneven distribution due to excessive Al$_2$O$_3$ addition also cause such phenomena. Thus, the maximum Ferret diameter is the smallest for W-0.2 wt.% Al$_2$O$_3$. Figure 5f shows the phase distribution of the W-0.2 wt.% Al$_2$O$_3$ alloy plate. Al$_2$O$_3$ is distributed more uniformly in this plate compared with those in the plates with other compositions, and the Al$_2$O$_3$ particles are aggregated to the lowest degree. The uniform distribution of Al$_2$O$_3$ in the matrix can hinder the migration of grain boundaries during the sintering process, thus inhibiting grain growth. The finer the grains, the smaller the maximum Ferret diameter of the deformed grains. This is the reason for the highest percentage of large-angle grain boundaries, which can influence grain boundary strengthening, and one of the reasons for the higher hardness and compressive strength of the plates with this composition. The plates with other compositions have higher Al_2O_3 contents; thus, Al_2O_3 tends to aggregate and form large particles. This is why the mechanical properties of the W-0.2 wt.% Al_2O_3 alloy plates are better than the properties of the plates with other compositions.

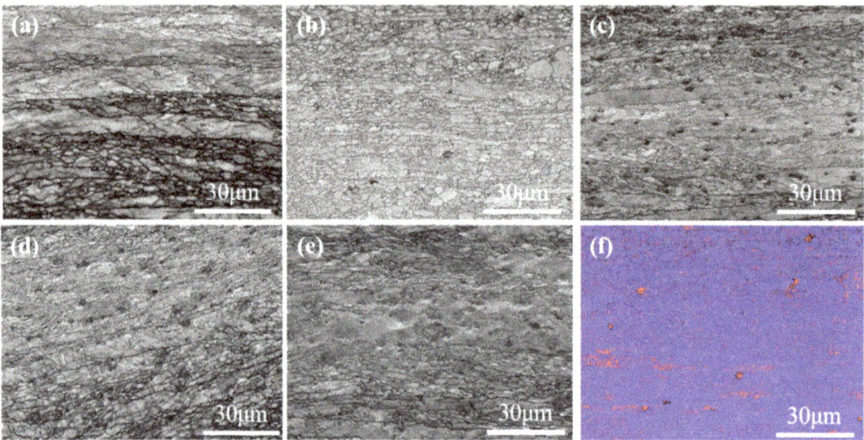

Figure 5. Grain boundaries of the plates: (**a**) pure tungsten, (**b**) W-0.2 wt.% Al_2O_3, (**c**) W-0.4 wt.% Al_2O_3, (**d**) W-0.6 wt.% Al_2O_3, (**e**) W-0.8 wt.% Al_2O_3, (**f**) phase distribution of W-0.2 wt.% Al_2O_3.

As shown in Figure 6a–f, the grain orientation distribution function (ODF) diagram of the alloy plates with different compositions after rolling shows a distinct texture. Tungsten is known to have a body-centered cubic (BCC) structure, the texture of which is usually more obvious at the Euler angle of $\varphi2 = 45°$. Figure 6a shows the standard ODF diagram of the BCC structure at the Euler angle of $\varphi2 = 45°$. The typical {0 0 1} <1 1 0> BCC plate texture appears in the ODF diagram of pure tungsten plates, and the addition of Al_2O_3 transforms the original texture into a {1 1 1} <1 1 0> texture (Figure 6c–f). The strength of the texture structure varies with the change in Al_2O_3 content. This knowledge helps us understand the evolution of texture during the rolling process. and allows us to adjust the process parameters and the second phase addition ratio to adapt the texture depending on the application needs. The texture structure mainly depends on the motion of slip and twin systems, and the Schmidt factor diagram theory can explain the evolution of the texture structure. Figure 7 shows the Schmidt factor diagrams for pure tungsten and alloy plates, combined with the orientation distribution function diagrams. The higher Schmidt factor means that the slip system is easier to slip. This is why the {1 1 1} <1 1 0> slip system is easier to open, changing the structure of the plate after adding Al_2O_3.

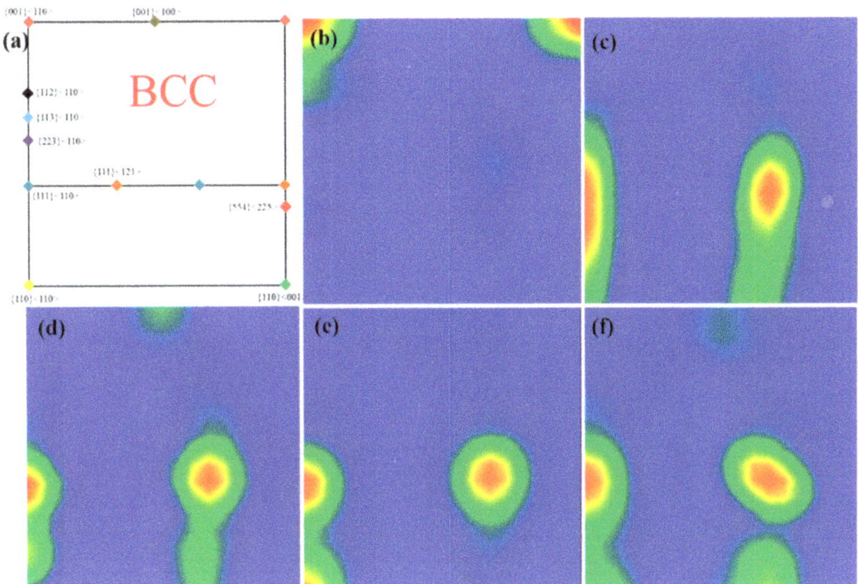

Figure 6. (**a**) standard ODF at $\varphi2 = 45°$. ODF diagram for different composition plates, (**b**) pure tungsten, (**c**) W-0.2 wt.% Al$_2$O$_3$, (**d**) W-0.4 wt.% Al$_2$O$_3$, (**e**) W-0.6 wt.% Al$_2$O$_3$, (**f**) W-0.8 wt.% Al$_2$O$_3$.

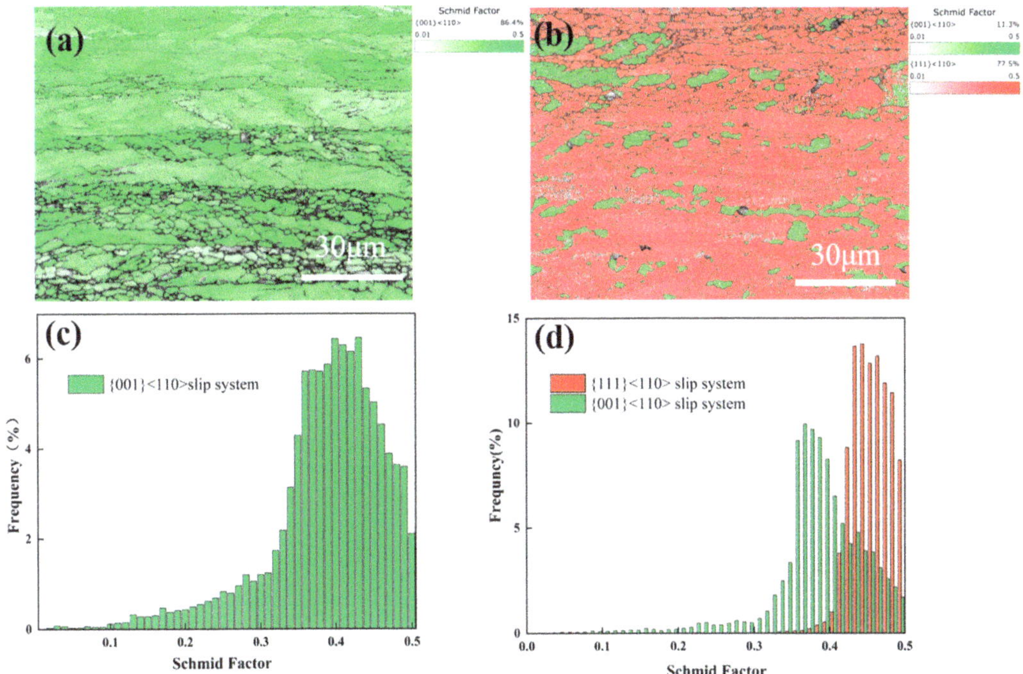

Figure 7. Schmidt coefficient plots and Schmidt coefficient frequency distributions for (**a**,**c**) pure tungsten, (**b**,**d**) W-0.2 wt.% Al$_2$O$_3$.

3.3. Microstructure and Interface Feature

The samples of the rolled plates were subjected to TEM imaging and most of the Al_2O_3 grains present in the plates and their fineness are visible in Figure 8a. Figure 8b shows the grain size distribution of the Al_2O_3 particles with an average grain size of approximately 73.7 nm. Point 1 in Figure 8c represents many sub-grain boundaries around Al_2O_3 particles, as indicated in energy-dispersive X-ray spectroscopy (Figure 8d). This is because Al_2O_3, as a high-hardness ceramic material, accumulates around Al_2O_3 during the rolling process. The dislocations cannot pass through Al_2O_3 during slip. Hence, many dislocations accumulate to form sub-grain boundaries and these sub-grain boundaries strongly impede the movement of dislocations. This is an important reason why Al_2O_3 can enhance the strength of the plate.

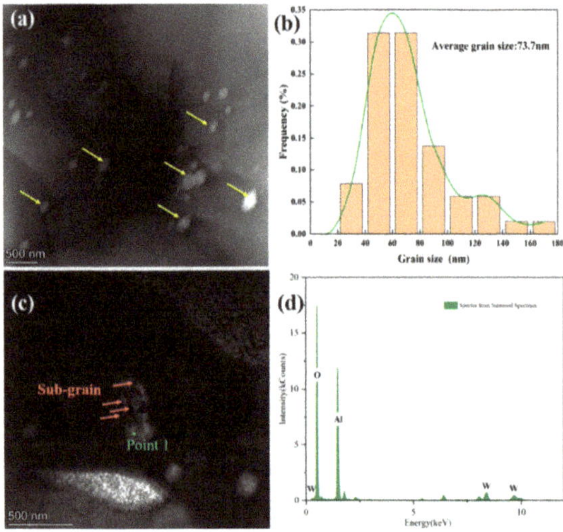

Figure 8. TEM images of the rolled W-0.2 wt.% Al_2O_3 plates: (**a**) Al_2O_3 present in the tungsten matrix, (**b**) Al_2O_3 average grain size, (**c**) sub-boundary around point 1, (**d**) EDS of point 1.

The selected area electron diffraction (SAED) pattern (Figure 9c) of the tungsten matrix part was analyzed. The crystal plane spacing was 0.159 nm, 0.222 nm, and 0.129 nm, corresponding to the crystal planes of ($\bar{2}$ 0 0), (0 $\bar{1}$ 1), and ($\bar{2}$ $\bar{1}$ 1), respectively. The angular relationship of each crystal plane was verified and determined to be consistent with the (0 1 1) crystal band axis parameter of W. The SAED pattern (Figure 9d) of the Al_2O_3 part was analyzed. The crystalline band axis was calculated based on the crystalline plane spacing and angle, and it was consistent with the ($\bar{2}$ 2 $\bar{1}$) crystalline band axis corresponding to α-Al_2O_3. Figure 9a shows a clear interface between Al_2O_3 and the tungsten matrix. The crystalline plane and the lattice stripe spacing are determined using the Fourier transform, as shown in Figure 9b. The (1 1 0) crystalline plane spacing in the tungsten matrix is 2.23 Å, and the (1 0 4) crystalline plane spacing in α-Al_2O_3 is 2.55 Å. According to the crystalline plane mismatch degree formula, the lattice mismatch degree of the two-phase interface is 33%, and the interface is a non-co-grained relationship. As Al_2O_3 is a high-hardness inorganic non-metallic material with plasticity very different from that of tungsten, a metallic material, the different deformation ability of the two phases leads to the non-coordinated deformation in the rolling process. A transition layer of approximately 3-5 nm appears at the interface of the two phases after rolling, but no similar transition layer is observed in the sintered state. The Al_2O_3 particles are strongly bonded to the W matrix and do not separate during the deformation process (compressive stress state). Therefore, the tungsten grains on the surface layer of the Al_2O_3 particles undergo a large slip along the surface of the Al_2O_3 particles, causing the tungsten atoms to deviate from their equilibrium position and show amorphous properties. This amorphous transition layer can effectively prevent dislocation movement because amorphous grains do not have many slip systems as in crystals or dense row surfaces in which dislocations can barely move, thus strengthening the tungsten matrix.

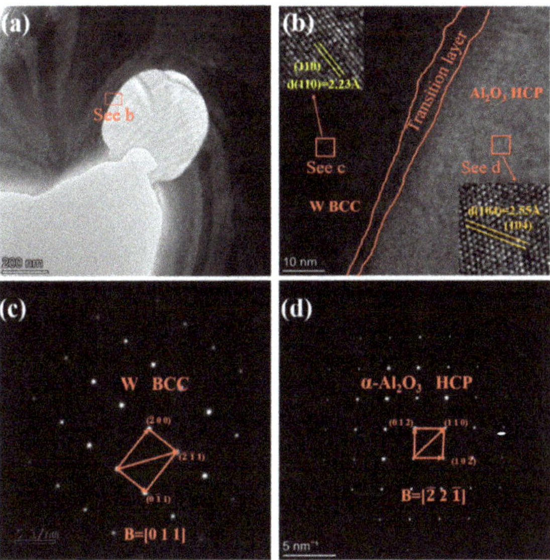

Figure 9. TEM images and selected area electron diffraction patterns analysis of W-0.2 wt.% Al_2O_3 plate: (**a**) bright field image, (**b**) HRTEM image of the interface, (**c**) SEAD of tungsten, (**d**) SEAD of Al_2O_3.

3.4. Mechanical Properties

The micro-hardness of tungsten alloy plates with different compositions is shown in Figure 10. The micro-hardness of pure tungsten plate is 356 $HV_{0.2}$, and the micro-hardness of the W-0.2 wt.% Al_2O_3 alloy plate is 512 $HV_{0.2}$, which is an increase of 43.7%. However, the micro-hardness of the W-0.4 wt.% Al_2O_3 alloy plate decreases to 495 $HV_{0.2}$, and with an increase in the Al_2O_3 content, the micro-hardness decreases further. The micro-hardness of the W-0.6 wt.% Al_2O_3 alloy plate is 491 $HV_{0.2}$. Among all the tested alloy plates, the microhardness of the W-0.8 wt.% Al_2O_3 alloy plate is the lowest (472 $HV_{0.2}$). The SEM images in Figure 4 of the composite powders show the low content of Al_2O_3 in the W-0.2 wt.% Al_2O_3 and Al_2O_3 dispersed around the tungsten grains. After sintering is performed, Al_2O_3 is uniformly distributed in the alloy. When subjected to external forces, dislocations are generated inside the alloy, and the fine Al_2O_3 particles dispersed inside the alloy play a role in pinning. Al_2O_3 particles dispersed in the alloy contribute to nailing dislocations and impede the movement of dislocations. However, when the content of Al_2O_3 increases, the fine Al_2O_3 particles agglomerate, thereby ineffectively inhibiting the grain growth during the sintering process and reducing the effect of nailing dislocations. The agglomeration of Al_2O_3 also results in the inhomogeneous microstructure of the tungsten alloys (see Figure 5), which eventually leads to the decrease in micro-hardness. Therefore, when the Al_2O_3 content exceeds a certain limit, the micro-hardness does not continue to increase but rather decreases to values even lower than that of pure tungsten.

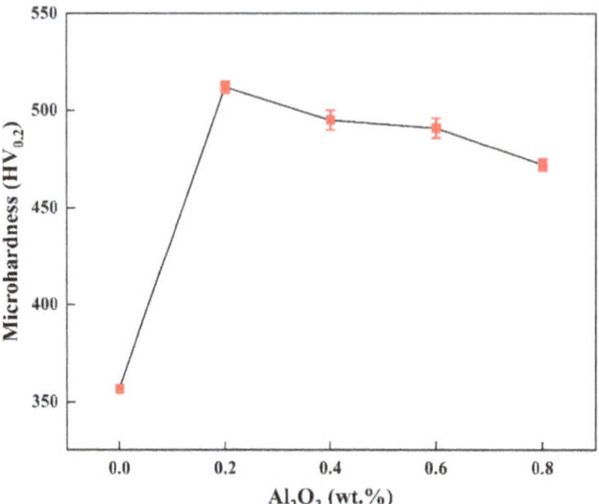

Figure 10. Microhardness of rolled samples with different Al_2O_3 contents.

The nano-indentation characterization technique allows the hardness of microscopic regions within individual grains to be measured and compared with micro-hardness to determine the variation of nano-hardness with indentation depth continuously. Hence, nano-hardness can eliminate the influence of grain boundaries on hardness and thus can directly reflect the effect of microstructure on hardness. Based on previously published work, the minimum indent spacing can be 10 times the indentation depth for a Berkovich tip [24]. In this work, the ratio of indent spacing to indent depth is 10, which can ensure that the data from the nano-indentation tests is valid. Figure 11a shows that the distance between each indentation is moderate and does not affect the test results. Figure 11b shows the nano-hardness of the plates with different compositions. The nano-hardness of pure tungsten is 11.44 GPa. After 0.2 wt.% Al_2O_3 is added, and the nano-hardness of the plates reaches 14.2 GPa, which is 24.1% higher than the nano-hardness of pure tungsten. After more Al_2O_3 is added, the nano-hardness gradually decreases, but even the lowest is 13.2% higher than that of pure tungsten. The increase in nano-hardness may be due to the interaction mechanism between the dispersed Al_2O_3 particles and dislocation motion, mainly the Orowan mechanism. During the downward pressure of the indenter, the dislocations caused by the load are blocked by the dispersed Al_2O_3 particles at the grain boundaries, causing an increase in nano-hardness [23]. However, when the Al_2O_3 content increases, the fine Al_2O_3 particles agglomerate to form large particles, and the uneven distribution reduces the hindering effect of Al_2O_3, resulting in a decrease in hardness.

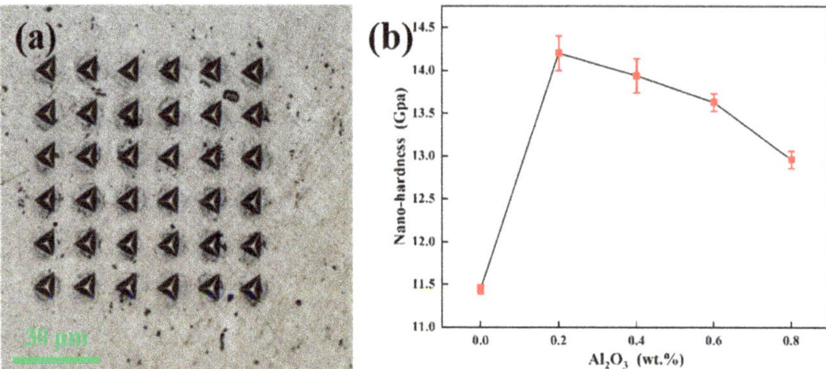

Figure 11. (a) Nanoindentation of W-0.2 wt.% Al_2O_3 alloy, (b) nano-hardness of rolled samples with different Al_2O_3 contents.

Figure 12 presents the room-temperature compressive stress–strain curves of tungsten alloy plates with different compositions and the corresponding maximum compressive strength and failure strain are shown in Figure 13. The curves reveal that the compressive strength of pure tungsten is 1620 MPa. The highest compressive strength of 2224 MPa is achieved when 0.2 wt.% Al_2O_3 is added, an increase of approximately 603 MPa or 37.2% compared with that of pure tungsten. When the Al_2O_3 content reaches 0.8 wt.%, the compressive strength is lower than that of pure tungsten, but the deformation of the W-0.8 wt.% Al_2O_3 alloy plates is greater than that of pure tungsten. Pure tungsten fractures at 13.9% deformation, but the fracture of the W-0.8 wt.% Al_2O_3 alloy plate occurs when the deformation reaches 20%. The deformation of the W-0.2 wt.% Al_2O_3 alloy plates is 29%. In general, the compressive strength is influenced by the Al_2O_3 particles. Under stress, the Al_2O_3 particles in the tungsten matrix can hinder the movement of dislocations and the expansion of cracks, and at the same time, because the grains are inhibited from growing during sintering and broken into many fine sub-crystals during rolling, the resistance to crack sprouting increases, causing the superior toughness and higher compressive strength of tungsten alloy plates. The compressive strength is about 604 MPa higher than that of pure tungsten plates, and the maximum strain at fracture increases from 13.9% to 29%, improving tungsten's ductility.

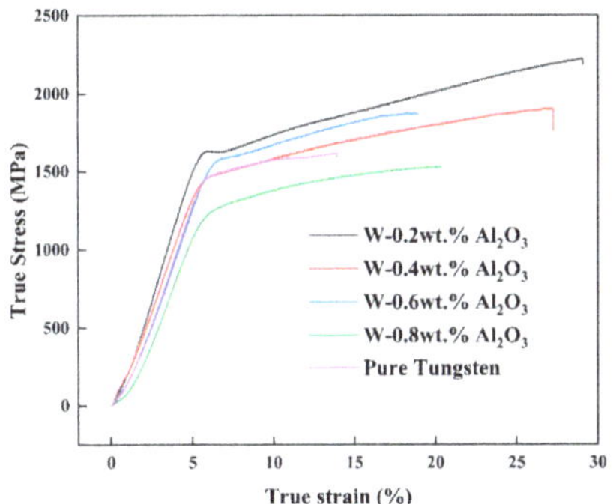

Figure 12. Compressive stress–strain curves of rolled samples with different Al_2O_3 contents.

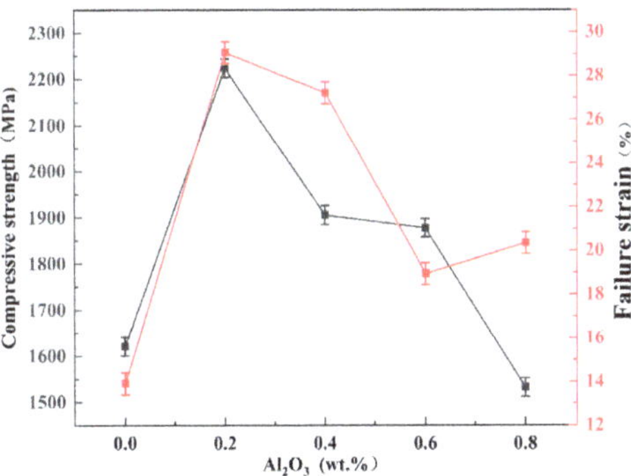

Figure 13. Compressive strength and failure strain of the samples with different Al_2O_3 contents.

The mechanical properties of some typical tungsten alloys from the literature are summarized in Table 1 and a visual comparison of these data graphically is shown in Figure 14, indicating that W-2 wt% HfC and W-10 wt.% Ta in the green shaded area has the same or even higher failure strains at room temperature than W-0.2 wt.% Al_2O_3 plates, but the strength is more than 240 MPa lower than W-0.2 wt.% Al_2O_3 plates. The compressive strength of WMoNbTaV-1ZrO$_2$ and $W_f/Zr_{41.2}Ti_{13.8}Cu_{12.5}Ni_{10}Be_{22.5}$ is slightly lower than that of the W-0.2 wt.% Al_2O_3 plate, but the failure strain is much lower than that of the W-0.2 wt.% Al_2O_3 plates. The rest of the tungsten alloys shaded in blue are far inferior to the W-0.2 wt.% Al_2O_3 plates in terms of failure strain and compressive strength. However, compared to ZrO_2, Al_2O_3 is cheaper and can be used to save costs in large batches for industrial production. Comparing the difference in compressive strength and failure strain between spin-forged and sintered W-1.5 ZrO_2, as well as the significant difference in compressive strength and failure strain between W-0.2 wt.% Al_2O_3 sheet and sintered W-0.25 wt.% Al_2O_3 alloy of similar composition, it can be seen that forging and extrusion are effective methods to increase the strength and ductility of the material.

Table 1. Compressive mechanical properties of typical tungsten alloys at room temperature.

Alloy	State	Compressive Strength	Elongation	References
95W-3.5Ni-1.5Fe	Swaged	1434 MPa	4.6%	[25]
WMoNbTaV	Sintered	1284.6 MPa	9.6%	[26]
WMoNbTaV-1ZrO$_2$	Sintered	2171.1 MPa	12.7%	[26]
WMoNbTaV	Sintered	1246 MPa	1.7%	[27]
WMoNbTa	Sintered	1058 MPa	2.1%	[27]
WMoNbTaTi$_{0.25}$	Sintered	1109 MPa	2.5%	[28]
WMoNbTaTi$_{0.5}$	Sintered	1211 MPa	5.9%	[28]
WMoNbTaTi$_{0.75}$	Sintered	1304 MPa	8.4%	[28]
WMoNbTaTi	Sintered	1343 MPa	14.1%	[29]
WMoNbTaVTi	Sintered	1515 MPa	10.6%	[29]
WNbTaV	Sintered	1530 MPa	12%	[30]
WNbTaVTi	Sintered	1420 MPa	20%	[30]
W-1.5 ZrO$_2$	Swaged	2235 MPa	38.2%	[31]
W-1.5 ZrO$_2$	Sintered	1628 MPa	21%	[32]
W-1.5% ZrO$_2$(Y)	Sintered	1680 MPa	23.9	[33]
W$_f$/Zr$_{41.2}$Ti$_{13.8}$Cu$_{12.5}$Ni$_{10}$Be$_{22.5}$	Sintered	2146 MPa	21.4	[34]
W-40 wt.% Ta	Sintered	1630 MPa	9.54%	[35]
W/2 wt.% HfC	Sintered	1980 MPa	34.7%	[36]
W-10 wt.% Ta	Sintered	1500 MPa	29%	[37]
W-0.25 wt.% Al$_2$O$_3$	Sintered	1318 MPa	23%	[15]
W-0.2 wt.% Al$_2$O$_3$	Rolled	2224 MPa	29%	This work

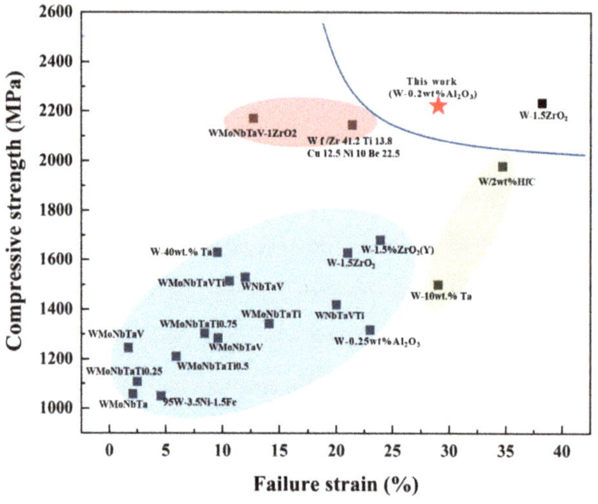

Figure 14. Compressive strength and failure strain diagrams of some tungsten alloys at room temperature.

4. Conclusions

1. Among the alloy powders of different compositions, the average grain size of the W-0.2 wt.% Al$_2$O$_3$ powder was only 727 nm, while the grain size of the rest of the powders was about 1 μm. The XRD pattern showed no diffraction peaks of Al$_2$O$_3$ due to the low Al$_2$O$_3$ content.
2. In the sintered specimens, the average grain size of pure tungsten was 24 μm, and that of W-0.2 wt.% Al$_2$O$_3$ was only 9 m. The addition of Al$_2$O$_3$ was effective in refining the grains.
3. The maximum Ferret diameter is 10 μm for pure tungsten and 6 μm for W-0.2 wt.% Al$_2$O$_3$ plates, with the same size trend as in the sintered state, due to dislocation cleavage of the grains during the rolling process. The texture of the plates changes from {001} <110> to {111} <110> after the addition of Al$_2$O$_3$.
4. As can be observed in the TEM image, the average grain size of the Al$_2$O$_3$ in the plate is approximately 100 nm, with a transition layer at the interface between the two phases and a non-co-grained interface relationship.
5. Comparing the mechanical properties of pure tungsten plates with those of tungsten alloy plates of different compositions, the microhardness has increased by up to 156 HV$_{0.2}$, an increase of 43%, while the nano-hardness increased by 2.75 GPa, an increase of 24.1%. The compressive strength of pure tungsten plates was 1620 MPa, but the addition of Al$_2$O$_3$ increased the compressive strength by up to 600 MPa, an increase of 37%. Pure tungsten plates fractured at a compression deflection of 13.9%, and at an Al$_2$O$_3$ content of 0.2 wt.%, after which the plates fractured only at a deflection of 29%. This proves that the doping of Al$_2$O$_3$ has enhanced the strength and toughness of the alloy plates.

Author Contributions: Conceptualization, C.W. and C.C.; Data curation, X.D.; Formal analysis, X.D. and M.X.; Investigation, F.M.; Methodology, C.W., Y.L. and X.W.; Project administration, Y.L., S.W., K.P. and C.C.; Resources, C.W.; Supervision, S.W., K.P., C.Z. and H.Y.; Validation, K.P. and C.Z.; Visualization, T.J.; Writing—original draft, C.W.; Writing—review & editing, K.P., C.Z., M.X., F.M., T.J., H.Y. and C.C. All authors have read and agreed to the published version of the manuscript.

Funding: This research was funded by the National Natural Science Foundation of China (Grant No. 52201118), Natural Science Foundation of Henan Province (Grant No. 222300420155), Program for Science & Technology Innovation Talents in the University of Henan Province (Grant No. 22HASTIT1006), Program for Central Plains Talents (Grant No. ZYYCYU202012172) and Ministry of Education, Singapore (Grant No. RG70/20 (2020-T1-001-023)).

Informed Consent Statement: Not applicable.

Acknowledgments: The authors greatly acknowledge the financial support from Provincial and Ministerial Co-construction of Collaborative Innovation Center for Non-ferrous Metal New Materials and Advanced Processing Technology.

Conflicts of Interest: The authors declare no conflict of interest.

References

1. Talignani, A.; Seede, R.; Whitt, A.; Zheng, S.; Ye, J.; Karaman, I.; Kirka, M.M.; Katoh, Y.; Wang, Y.M. A review on additive manufacturing of refractory tungsten and tungsten alloys. *Addit. Manuf.* **2022**, *58*, 103009. [CrossRef]
2. Zhou, H.-B.; Liu, Y.-L.; Jin, S.; Zhang, Y.; Luo, G.N.; Lu, G.-H. Towards suppressing H blistering by investigating the physical origin of the H–He interaction in W. *Nuclear Fusion* **2010**, *50*, 115010. [CrossRef]
3. Xu, Q.; Sato, K.; Cao, X.Z.; Zhang, P.; Wang, B.Y.; Yoshiie, T.; Watanabe, H.; Yoshida, N. Interaction of deuterium with vacancies induced by ion irradiation in W. *Nucl. Instrum. Methods Phys. Res. Sect. B Beam Interact. Mater. At.* **2013**, *315*, 146–148. [CrossRef]
4. Wang, H.; Gao, Y.; Fu, E.; Yang, T.; Xue, J.; Yan, S.; Chu, P.K.; Wang, Y. Irradiation effects on multilayered W/ZrO2 film under 4MeV Au ions. *J. Nucl. Mater.* **2014**, *455*, 86–90. [CrossRef]
5. Miao, S.; Xie, Z.M.; Zhang, T.; Wang, X.P.; Fang, Q.F.; Liu, C.S.; Luo, G.N.; Liu, X.; Lian, Y.Y. Mechanical properties and thermal stability of rolled W-0.5wt% TiC alloys. *Mater. Sci. Eng. A* **2016**, *671*, 87–95. [CrossRef]
6. Battabyal, M.; Schäublin, R.; Spätig, P.; Baluc, N. W–2wt.%Y2O3 composite: Microstructure and mechanical properties. *Mater. Sci. Eng. A* **2012**, *538*, 53–57. [CrossRef]

7.   Chakraborty, S.P.; Banerjee, S.; Sanyal, G.; Bhave, V.S.; Paul, B.; Sharma, I.G.; Suri, A.K. Studies on the synthesis of a Mo–30wt% W alloy by non-conventional approaches. *J. Alloys Compd.* **2010**, *501*, 211–217. [CrossRef]
8.   Veleva, L.; Oksiuta, Z.; Vogt, U.; Baluc, N. Sintering and characterization of W–Y and W–Y_2O_3 materials. *Fusion Eng. Des.* **2009**, *84*, 1920–1924. [CrossRef]
9.   Ryu, H.J.; Hong, S.H. Fabrication and properties of mechanically alloyed oxide-dispersed tungsten heavy alloys. *Mater. Sci. Eng. A* **2003**, *363*, 179–184. [CrossRef]
10. Kim, Y.; Lee, K.H.; Kim, E.-P.; Cheong, D.-I.; Hong, S.H. Fabrication of high temperature oxides dispersion strengthened tungsten composites by spark plasma sintering process. *Int. J. Refract. Met. Hard Mater.* **2009**, *27*, 842–846. [CrossRef]
11. Kim, Y.; Hong, M.-H.; Lee, S.H.; Kim, E.-P.; Lee, S.; Noh, J.-W. The effect of yttrium oxide on the sintering behavior and hardness of tungsten. *Met. Mater. Int.* **2006**, *12*, 245–248. [CrossRef]
12. Mabuchi, M.; Okamoto, K.; Saito, N.; Asahina, T.; Igarashi, T. Deformation behavior and strengthening mechanisms at intermediate temperatures in W-La_2O_3. *Mater. Sci. Eng. A* **1997**, *237*, 241–249. [CrossRef]
13. Zhao, M.; Zhou, Z.; Zhong, M.; Tan, J.; Lian, Y.; Liu, X. Thermal shock behavior of fine grained W–Y2O3 materials fabricated via two different manufacturing technologies. *J. Nucl. Mater.* **2016**, *470*, 236–243. [CrossRef]
14. Li, Z.; Xu, L.; Wei, S.; Chen, C.; Xiao, F. Fabrication and mechanical properties of tungsten alloys reinforced with c-ZrO2 particles. *J. Alloys Compd.* **2018**, *769*, 694–705. [CrossRef]
15. Wang, C.; Zhang, L.; Wei, S.; Pan, K.; Wu, X.; Li, Q. Preparation, microstructure, and constitutive equation of W-0.25 wt% Al2O3 alloy. *Mater. Sci. Eng. A* **2019**, *744*, 79–85. [CrossRef]
16. Zhou, Y.; Gao, Y.; Wei, S.; Pan, K.; Hu, Y. Preparation and characterization of Mo/Al_2O_3 composites. *Int. J. Refract. Met. Hard Mater.* **2016**, *54*, 186–195. [CrossRef]
17. Zabihi, M.; Toroghinejad, M.R.; Shafyei, A. Application of powder metallurgy and hot rolling processes for manufacturing aluminum/alumina composite strips. *Mater. Sci. Eng. A* **2013**, *560*, 567–574. [CrossRef]
18. Tian, B.; Liu, P.; Song, K.; Li, Y.; Liu, Y.; Ren, F.; Su, J. Microstructure and properties at elevated temperature of a nano-Al2O3 particles dispersion-strengthened copper base composite. *Mater. Sci. Eng. A* **2006**, *435–436*, 705–710. [CrossRef]
19. Rajkovic, V.; Bozic, D.; Jovanovic, M.T. Effects of copper and Al2O3 particles on characteristics of Cu–Al_2O_3 composites. *Mater. Des.* **2010**, *31*, 1962–1970. [CrossRef]
20. Wang, J.; Gan, X.; Li, Z.; Zhou, K. Microstructure and gas sensing property of porous spherical In_2O_3 particles prepared by hydrothermal method. *Powder Technol.* **2016**, *303*, 138–146. [CrossRef]
21. Visa, M. Synthesis and characterization of new zeolite materials obtained from fly ash for heavy metals removal in advanced wastewater treatment. *Powder Technol.* **2016**, *294*, 338–347. [CrossRef]
22. Selvarajan, S.; Suganthi, A.; Rajarajan, M.; Arunprasath, K. Highly efficient BiVO4/WO3 nanocomposite towards superior photocatalytic performance. *Powder Technol.* **2017**, *307*, 203–212. [CrossRef]
23. Wang, C.; Zhang, L.; Pan, K.; Wei, S.; Wu, X.; Li, Q. Effect of Al_2O_3 content and swaging on microstructure and mechanical properties of Al2O3/W alloys. *Int. J. Refract. Met. Hard Mater.* **2020**, *86*, 105082. [CrossRef]
24. Sudharshan Phani, P.; Oliver, W.C. A critical assessment of the effect of indentation spacing on the measurement of hardness and modulus using instrumented indentation testing. *Mater. Des.* **2019**, *164*, 107563. [CrossRef]
25. Debata, M.; Acharya, T.S.; Sengupta, P.; Acharya, P.P.; Bajpai, S.; Jayasankar, K. Effect of high energy ball milling on structure and properties of 95W-3.5Ni-1.5Fe heavy alloys. *Int. J. Refract. Met. Hard Mater.* **2017**, *69*, 170–179. [CrossRef]
26. Zong, L.; Xu, L.; Luo, C.; Li, Z.; Zhao, Y.; Xu, Z.; Zhu, C.; Wei, S. Fabrication of nano-ZrO2 strengthened WMoNbTaV refractory high-entropy alloy by spark plasma sintering. *Mater. Sci. Eng. A* **2022**, *843*, 143113. [CrossRef]
27. Senkov, O.N.; Wilks, G.B.; Scott, J.M.; Miracle, D.B. Mechanical properties of Nb25Mo25Ta25W25 and V20Nb20Mo20Ta20W20 refractory high entropy alloys. *Intermetallics* **2011**, *19*, 698–706. [CrossRef]
28. Han, Z.D.; Luan, H.W.; Liu, X.; Chen, N.; Li, X.Y.; Shao, Y.; Yao, K.F. Microstructures and mechanical properties of TixNbMoTaW refractory high-entropy alloys. *Mater. Sci. Eng. A* **2018**, *712*, 380–385. [CrossRef]
29. Han, Z.D.; Chen, N.; Zhao, S.F.; Fan, L.W.; Yang, G.N.; Shao, Y.; Yao, K.F. Effect of Ti additions on mechanical properties of NbMoTaW and VNbMoTaW refractory high entropy alloys. *Intermetallics* **2017**, *84*, 153–157. [CrossRef]
30. Yao, H.W.; Qiao, J.W.; Gao, M.C.; Hawk, J.A.; Ma, S.G.; Zhou, H.F.; Zhang, Y. NbTaV-(Ti,W) refractory high-entropy alloys: Experiments and modeling. *Mater. Sci. Eng. A* **2016**, *674*, 203–211. [CrossRef]
31. Li, Z.; Chen, Y.; Wei, S.; Pan, K.; Shen, H.; Xu, L. Effect of rotary swaging and subsequent annealing on microstructure and mechanical properties of W-1.5ZrO2 alloys. *J. Alloy. Compd.* **2021**, *875*, 160041. [CrossRef]
32. Wang, C.; Zhang, L.; Wei, S.; Pan, K.; Wu, X.; Li, Q. Effect of ZrO2 content on microstructure and mechanical properties of W alloys fabricated by spark plasma sintering. *Int. J. Refract. Met. Hard Mater.* **2019**, *79*, 79–89. [CrossRef]
33. Xiao, F.; Xu, L.; Zhou, Y.; Pan, K.; Li, J.; Liu, W.; Wei, S. A hybrid microstructure design strategy achieving W-ZrO2(Y) alloy with high compressive strength and critical failure strain. *J. Alloys Compd.* **2017**, *708*, 202–212. [CrossRef]
34. Zong, H.T.; Ma, M.Z.; Liu, L.; Zhang, X.Y.; Bai, B.W.; Yu, P.F.; Qi, L.; Jing, Q.; Li, G.; Liu, R.P. Wf/Zr41.2Ti13.8Cu12.5Ni10Be22.5 bulk metallic glass composites prepared by a new melt infiltrating method. *J. Alloys Compd.* **2010**, *504*, S106–S109. [CrossRef]
35. Duan, X.; Huang, Y.; Liu, W.; Cai, Q.; Liu, W.; Ma, Y. Effect of Ta on the microstructure and mechanical properties of WTa alloys prepared by arc melting. *Mater. Charact.* **2022**, *188*, 111823. [CrossRef]

36. Zhang, J.; Ma, S.; Zhu, J.; Kang, K.; Luo, G.; Wu, C.; Shen, Q.; Zhang, L. Microstructure and Compression Strength of W/HfC Composites Synthesized by Plasma Activated Sintering. *Met. Mater. Int.* **2018**, *25*, 416–424. [CrossRef]
37. Liu, W.; Huang, Y.; Wang, Y.; Zhang, Y.; Duan, X.; Liu, W.; Ma, Y. Microstructure and mechanical properties of W-10 wt-%Ta alloys prepared by spark plasma sintering. *Mater. Sci. Technol.* **2022**, *38*, 159–168. [CrossRef]

Article

Synergistic Effects of La and Y on the Microstructure and Mechanical Properties of Cast Al-Si-Cu Alloys

Luming Shuai [1], Xiuliang Zou [1], Yuqiang Rao [1], Xiaobin Lu [2] and Hong Yan [1,*]

[1] Department of Material Processing Engineering, College of Advanced Manufacturing, Nanchang University, Nanchang 330031, China
[2] Luxe Machinery (Gao'an) Co., Ltd., Gaoan 330800, China
* Correspondence: hyan@ncu.edu.cn; Tel.: +86-0791-8396-9622

Abstract: The effects of La and Y on the microstructure and mechanical properties of cast Al-Si-Cu alloys were investigated by X-ray diffractometer (XRD), optical microscope (OM), and scanning electron microscope (SEM). The results indicated that the addition of La and Y had a great effect on the refinement of α-Al grains, the modification of eutectic Si phase, and the reduction of β-Al$_5$FeSi length in Al-Si-Cu alloys. The A380 + 0.6 wt.% La/Y alloy exhibited the best microstructure and mechanical properties. The UTS and EI of the A380 + 0.6 wt.% La/Y alloy were 215.3 MPa and 5.1%, which were 22.9% and 37.8% higher than those of the matrix alloy, respectively. In addition, neither Al$_{11}$La$_3$ nor Al$_3$Y generated by the addition of La and Y could not serve as the nucleation core of α-Al grains, so the grain refinement of α-Al originated from the growth limitation and constitutional supercooling. Since La and Y promote twinning generation and constitutional supercooling, the eutectic Si phase also changed from stripe-like to short fibrous or even granular and was significantly refined. Furthermore, thermodynamic calculations indicated that the Al$_{11}$La$_3$ phase was formed first and the Al$_3$Y phase was generated on the Al$_{11}$La$_3$ phase.

Keywords: mixed rare earth; Al-Si-Cu alloy; microstructure; mechanical properties; thermodynamic calculation

Citation: Shuai, L.; Zou, X.; Rao, Y.; Lu, X.; Yan, H. Synergistic Effects of La and Y on the Microstructure and Mechanical Properties of Cast Al-Si-Cu Alloys. *Materials* **2022**, *15*, 7283. https://doi.org/10.3390/ma15207283

Academic Editors: Hailiang Yu, Zhilin Liu and Xiaohui Cui

Received: 22 September 2022
Accepted: 17 October 2022
Published: 18 October 2022

Publisher's Note: MDPI stays neutral with regard to jurisdictional claims in published maps and institutional affiliations.

1. Introduction

Al-Si alloys are widely used in automotive and aerospace industries due to excellent casting properties and high specific strength [1]. However, the traditional cast Al-Si alloys exhibited poor mechanical properties because of coarse α-Al dendrites, flake eutectic Si, and long needle-like β-Al$_5$FeSi [2,3]. In order to improve the mechanical properties of cast Al-Si alloys, grain refinement and modification techniques have been widely studied and applied in recent years [4,5].

The chemical modification method has been shown to be very effective in extensive research. Rare earths can be used as both grain refiners and chemical modifiers, such as La [6], Y [7], Ce [8], Eu [9]. Especially, as the most economical rare earth metal, La has received much attention in the research. Related literature [10–13] investigated the modification effect of La on eutectic Si and β-Al$_5$FeSi phase of Al-Si-Cu alloy. Mehdi et al. [10] reported that 0.2 wt.% La could reduce the amount of β-Al$_5$FeSi phase and refine the size of α-Al grains in the alloy. Tsai et al. [11] indicated that La was not able to refine the Si phase well until the additions reached 1.0 wt.%. The highest fracture elongation of the modified alloy was achieved at 0.6 wt.% La addition. Moreover, according to the previous research [12,13], the degree of morphological alteration of microstructure by La was unsatisfactory because of the generation of needle-like La-rich compounds. Therefore, some researchers have focused on the modification effect of the combination addition of La with other elements on Al-Si alloys [14,15]. Qiu et al. [14] investigated the synergistic effect of La and Sr on the Al-Si alloy. It indicated that La not only limited the growth of eutectic Si, but also changed the morphology of the eutectic Si by entering the grooves of the eutectic

Si. Cao et al. [15] reported the effects of La+Zr on Al-Si alloys. The results demonstrated that the heterogeneous core of α-Al was provided by Zr, which refined the α-Al grains and improved the tensile strength of the alloy. Meanwhile, La modified the eutectic Si phase and improved the fracture elongation of the alloy. In addition, Y also exhibited a powerful modification effect on eutectic Si. Liu et al. [16] reported that 0.2 wt.% Y could transform the morphology of Si phase from stripe to short fibrous or even granular. It also refined the α-Al grains and β-Al$_5$FeSi phase at the same time. Li et al. [17] investigated the different effects of Fe, Mn, and Y on Al-Si alloys. It was shown that Y transformed the dendritic α-Al grains into equiaxed grains, and the grain size also became smaller by improving the nucleation supercooling. The enhancement of mechanical properties of the alloys with the addition of La or Y is mainly due to grain refining and the second phase strengthening. On the other hand, both La and Y could modify eutectic Si and refine α-Al grains, but there were some differences in the content used and the modification effect. Therefore, it would be interesting to investigate the effect of the combined addition of La and Y on the microstructure of Al-Si alloys. Previous literature [13,18] has reported the effect of La or Y alone in Al-Si alloys, respectively. However, there were few studies on the effect of combined La and Y additions on Al-Si alloys. Understanding the modification mechanism of La and Y on the microstructure of Al-Si alloys is contributing to the research of modifiers and broadening the applications of Al-Si alloys. However, the mechanism of the synergistic effect of La and Y in the Al-Si alloy is not clear.

This paper investigated synergistic effects of La and Y on the microstructure and mechanical properties of cast Al-Si-Cu alloys. In addition, the Gibbs free energy changes of Al$_{11}$La$_3$ and Al$_3$Y produced by the reaction of La+Y and Al were also calculated. Meanwhile, the generation mechanisms of Al$_{11}$La$_3$ and Al$_3$Y were analyzed based on the results of calculations and experiments. Furthermore, the mechanism of the microstructure evolution was discussed.

2. Experimental

2.1. Original Materials

The commercial A380 alloy was chosen as the matrix. The composition of the used A380 alloy is shown in Table 1.

Table 1. Composition of A380 alloy.

Elements	Si	Cu	Mg	Zn	Fe	Mn	Al
(wt.%)	10.22	2.40	0.86	0.23	1.21	0.27	Bal.

2.2. Preparation of Master Alloys

To avoid the high burn rate of direct rare earth addition, the Al-Si-Cu alloy was modified with a homemade Al-10% La and Al-10% Y master alloys. Industrial pure Al (purity ≥ 99.87%), La, and Y were selected as raw materials to prepare rare earth master alloys by high-energy ultrasonic vibration method.

After the graphite crucible was preheated to 400 °C with the resistance furnace, the weighed and dried aluminum block was placed in the graphite crucible. At the casting temperature of 750 °C, La and Y were added with a tin foil wrapper. In order to obtain better dispersion [19], the preheated ultrasonic amplitude rod was probed into the molten liquid surface at 10–20 mm for intermittent ultrasonic vibration. The ultrasonic frequency was set to 20 kHz and the ultrasonic time was set to 15 min. The Al-10La and Al-10Y master alloys were obtained by pouring after ultrasonic completion.

2.3. Preparation of Modified Aluminum Alloys

The graphite crucible was placed in a resistance furnace and preheated to about 400 °C. The weighed lump of A380 alloy was put into the graphite crucible and heated up to 750 °C to melt it completely. Then, the previously prepared Al-10%La and Al-10%Y master alloys were added to the melt alternately to obtain rare earth aluminum alloy samples with different contents of La/Y (0 wt.%, 0.3 wt.%, 0.6 wt.%, 0.9 wt.%). Meanwhile, the melt was also treated with ultrasound for 15 min. After cooling to 720 °C, the melt was skimmed and poured into a preheated mold (300 °C), as shown in Figure 1a. During the entire experimental melting process, Argon gas was used to protect the melt from oxidation. The samples were taken from the bottom part of the casting after it cooled. The corresponding chemical compositions of the prepared alloys with different modifiers are shown in Table 2.

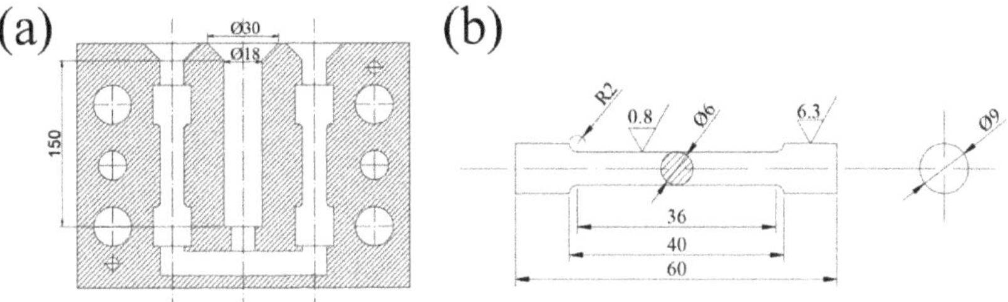

Figure 1. (**a**) mold drawing (unit: mm) and (**b**) tensile specimen bar (unit: mm).

Table 2. Chemical compositions of experimental alloys (wt.%).

Alloys	Si	Cu	Mg	Fe	La	Y	Al
A380 + 0 La/Y	10.22	2.40	0.86	1.21	0	0	Bal.
A380 + 0.3 La/Y	10.14	2.39	0.82	1.22	0.16	0.14	Bal.
A380 + 0.6 La/Y	10.09	2.39	0.83	1.18	0.33	0.31	Bal.
A380 + 0.9 La/Y	10.01	2.38	0.84	1.17	0.42	0.46	Bal.

2.4. Microstructure Characterization

After all samples were etched by 0.5% HF solution, the microstructure and the fracture surface was observed by optical microscopy (OM, OLYMPUS BX51 microscope, Olympus Metrology, Inc., Tokyo, Japan) and scanning electron microscopy (SEM, Quanta 200, FEI Metrology, Inc., Hillsboro, OR, USA) equipped with energy dispersive spectroscopy (EDS, FEI Metrology, Inc., Hillsboro, OR, USA). The quantitative data analysis was also performed using the software IPP6.0. To determine the specific phases in the alloy specimens, X-ray diffractometer (XRD, D8 advance, Bruker, Inc., Karlsruhe, Germany) was used for the tests and the data obtained were analyzed using jade5 software.

According to the Chinese standard GB/T228-2002, these metal rods with different compositions were processed into tensile specimen bars with a diameter of 6 mm and a standard distance of 36 mm, as shown in Figure 1b. Tensile testing was performed on an electronic universal tensile test machine, model SUNS UTM5105. The starting pitch of the samples was measured at 50 mm and the tensile rate was set at 1 mm/min. To avoid errors in the experiments, five sample bars were selected from each tensile sample bar.

3. Results

3.1. Effect of Adding La+Y on the Microstructure of the Alloy

Figure 2 shows the OM images of matrix alloys with different contents of La+Y. As indicated by the arrow in Figure 2a, the matrix alloy is mainly composed of coarse α-Al grains, striped eutectic Si phase, and long needle-like β-Al$_5$FeSi phase. In contrast, the addition of La and Y had a significant effect on modification of the cast microstructure. As exhibited in Figure 2b–d, the morphology of the eutectic Si phase in the alloy changes from strip-like to short fibrous or even granular. Meanwhile, both α-Al grains and β-Al$_5$FeSi phase are well refined. The secondary dendrite arm spacing (SDAS) of α-Al grains, the area of Si phase, and the length of β-Al$_5$FeSi phase were measured by IPP software, as shown in

Figure 3. The results show that the A380 + 0.6 wt.% La/Y alloy exhibits the best refinement of microstructure. The SDAS and the length of β-Al$_5$FeSi is decreased from 233 μm and 77.56 μm to 58 μm and 24.95 μm, respectively. The area of eutectic Si is also decreased from 289 μm^2 to 6 μm^2. Moreover, Figure 2c,d exhibit similar microstructure rather than improvement, which indicate that the increase of La+Y is not in favor of the microstructure optimization of the alloy anymore.

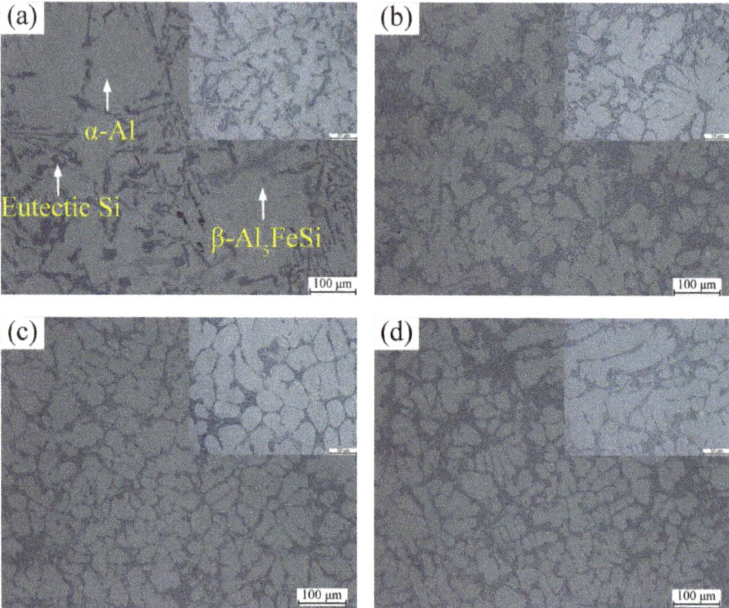

Figure 2. High magnification OM images of cast A380 alloys with different (La+Y) contents: (**a**) unmodified alloy; (**b**) 0.3 wt.%; (**c**) 0.6 wt.%; (**d**) 0.9 wt.%.

Figure 4 shows the SEM images of matrix alloys with different (La+Y) contents. The gray long needle-like phase and bright spot-like phase are observed in Figure 4a, which are inferred to be β-Al$_5$FeSi phase and Al$_2$Cu phase by combining with EDS and XRD analysis (Figure 5). As shown in Figure 4b, some bright short rod-like phases are also observed, which are inferred to be Al$_{11}$La$_3$ and Al$_3$Y phases in combination with EDS analysis and corresponding binary phase diagram [20,21]. Comparing the EDS of Fe with La and Y, as shown in Figure 4a–d, it can be seen that the rare earth phase of the alloy does not contain Fe elements. So, the red circle in Figure 4b shows some rare earth phases attached to the needle-like β-Al$_5$FeSi phase. According to the previous research [22], the rare earth elements can hinder its growth by attached to the β-Al$_5$FeSi phase. Moreover, some letter-like α-Al$_8$Fe$_2$Si phases can be observed in Figure 4c. The mechanical properties of the alloy are much less impaired by the α-Al$_8$Fe$_2$Si phase compared to the needle-like β-Al$_5$FeSi phase. In addition, as shown in Figure 4d, the bulk gray La-rich phases are observed in the A380 + 0.9 La/Y alloy, which can affect the mechanical properties of the alloy. On the other hand, the equilibrium partition coefficient of copper atoms is reduced because the enriched La and Y in the solid–liquid interface front hinders the diffusion of Cu atoms. Therefore, the solubility gradient of Cu in the liquid phase at the interface front increases, which increases the compositional supercooling of the alloy. So, the intensity of the peaks from the Al$_2$Cu phase noticeably decreases with the addition of alloying elements, which is consistent with the XRD (Figure 5).

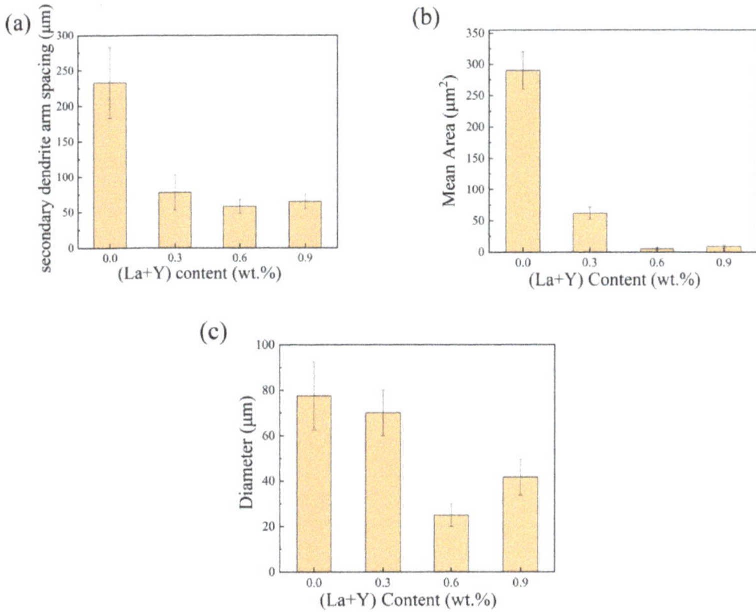

Figure 3. (**a**) The SDAS of α-Al grain, (**b**) the area of eutectic Si, and (**c**) the length of β-Al$_5$FeSi phase of A380 alloys with different (La+Y) contents.

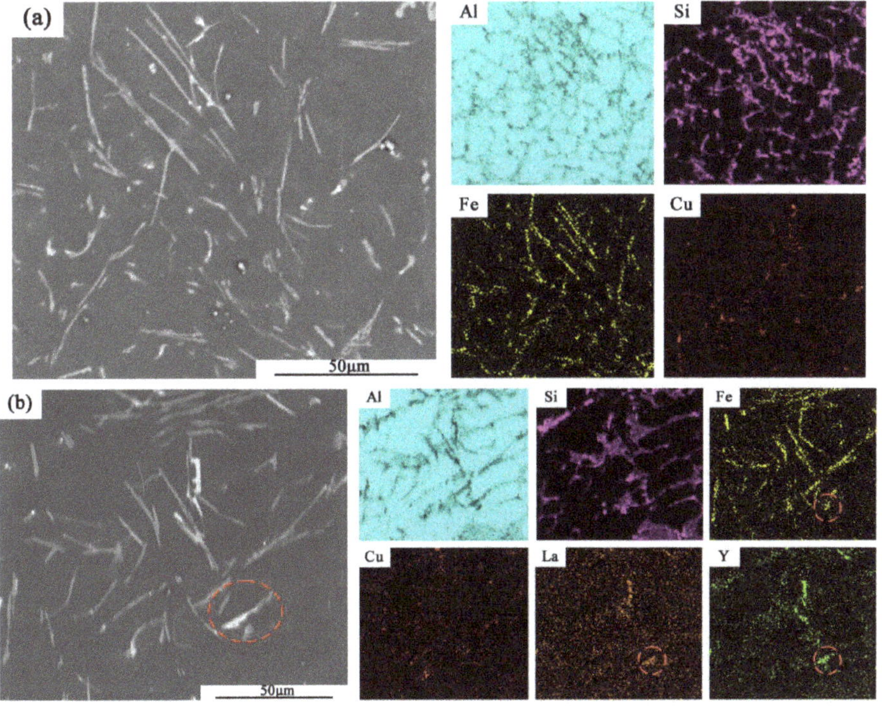

Figure 4. *Cont.*

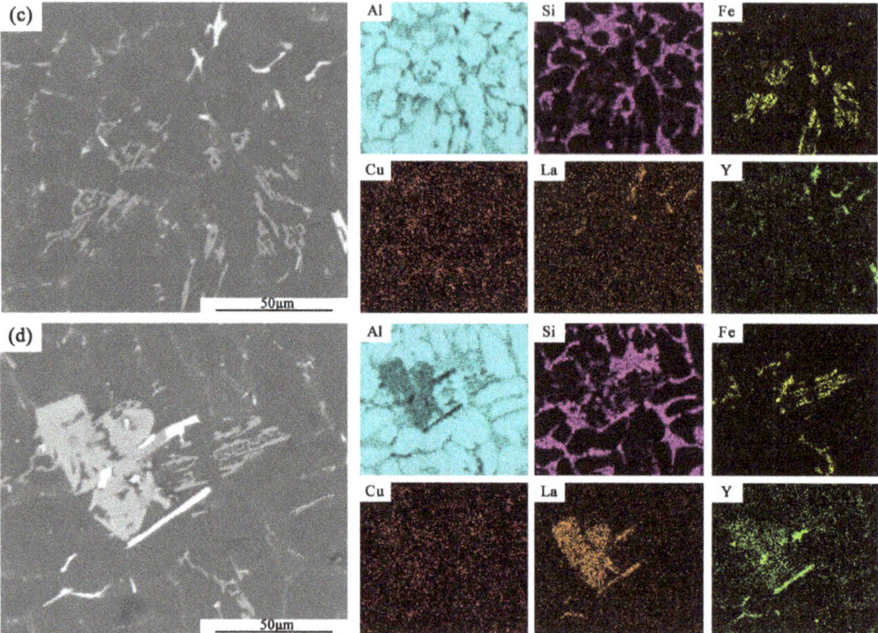

Figure 4. Scanning electron micrographs of cast A380 alloys with different (La+Y) additions: (**a**) matrix alloy; (**b**) 0.3 wt.%; (**c**) 0.6 wt.%; (**d**) 0.9 wt.%.

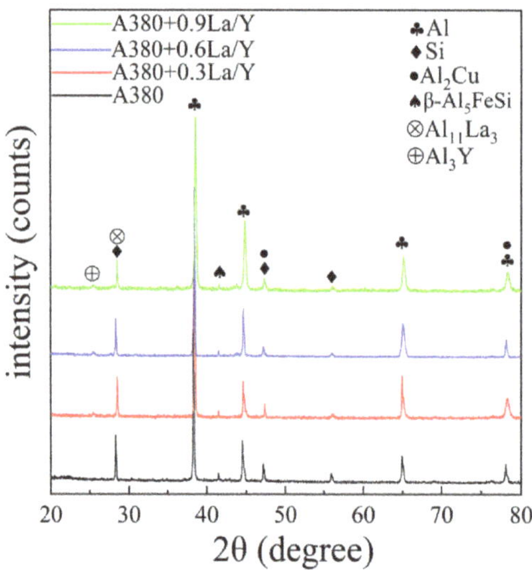

Figure 5. XRD patterns of A380 alloys with different (La+Y) contents.

3.2. Mechanical Properties

The mechanical properties of matrix alloys with different contents of La+Y are tested, as shown in Figure 6. The ultimate tensile strength (UTS) and elongation index (EI) of the matrix alloy are 175.2 MPa and 3.7%, respectively. The A380 + 0.6 wt.% La/Y alloy has the best UTS and EI, which reach 215.3 MPa and 5.1%, respectively. These represent an increase of 22.9% and 37.8%, respectively, compared to the base alloy. However, the UTS and EI start to decrease when the addition of La+Y reached 0.9 wt.%.

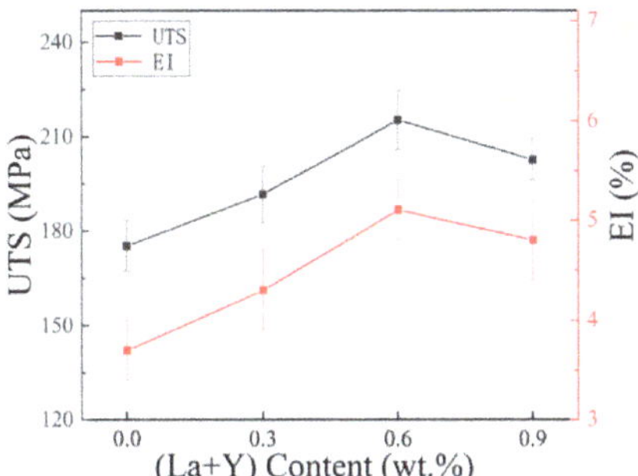

Figure 6. Mechanical properties of A380 alloys with different (La+Y) contents.

The mechanical behavior of the alloy is controlled by the microstructure of the alloy. The addition of 0.6 wt.% La+Y results in a significant refinement of the α-Al grains and Si phase, which affect the UTS and EI of the alloy. When the contents of La+Y is excessive, the rare earth phase becomes coarse. The coarse phase causes concentration of stress, which adversely affects the tensile strength and elongation of the alloy.

3.3. Fracture Analysis

Figure 7 shows the SEM images of the fractures of the matrix alloys with different La+Y additions. As indicated by the arrow in Figure 7a, the fracture surface of the unmodified alloy is distributed with large cleavage platform and a few dimples, so brittle fracture is the main mode of fracture of the unmodified alloy. However, La+Y leads to a trend of fracture mode alteration from brittle fracture to ductile fracture. When the addition of La+Y reaches 0.6 wt.%, as shown in Figure 7c, more dimples and few cleavage platforms can be observed on the fracture surface, so ductile fracture becomes the main fracture mode of the alloy at this time, which is beneficial to the improvement of the mechanical properties of the alloy. However, when the rare earth addition reaches 0.9 wt.%, large cleavage platforms appear at many places, as shown in Figure 7d, which are typically brittle fractures. According to the SEM image (Figure 4d), rare earth phases form agglomerates, which may become a source of fracture in the alloy and damages the mechanical properties of the alloy, which is consistent with the results in Figure 6.

Figure 7e shows an enlarged image of the fracture in the A380 alloy with 0.6 wt.% La+Y additions. The elemental distributions of point 1 and point 2 are obtained by point scanning. The addition of 0.6 wt.% La+Y results in a small-sized massive Fe-rich phase (point 2), which limits the impairment of the relative mechanical properties by the long needle-like β-Al$_5$FeSi in the alloy. Point 1 is a rare earth phase, which is small enough to benefit the mechanical properties of the alloy as well.

The Si phase is an important factor affecting the fracture mechanism of Al-Si alloys [23]. The coarse needle-like Si phase promotes stress concentration and distributes all over the alloy. The coarse Si phase is brittle, which can crack and break away from the matrix. When the alloy is subjected to external forces, the needle-like eutectic Si phase will rotate, resulting in internal stress concentration around the eutectic Si phase. When the concentrated internal stress exceeds the stress required for fracture, the Si phase causes the alloy to fracture [24]. After the addition of La+Y, the Si phase is obviously refined and the mechanical properties are improved.

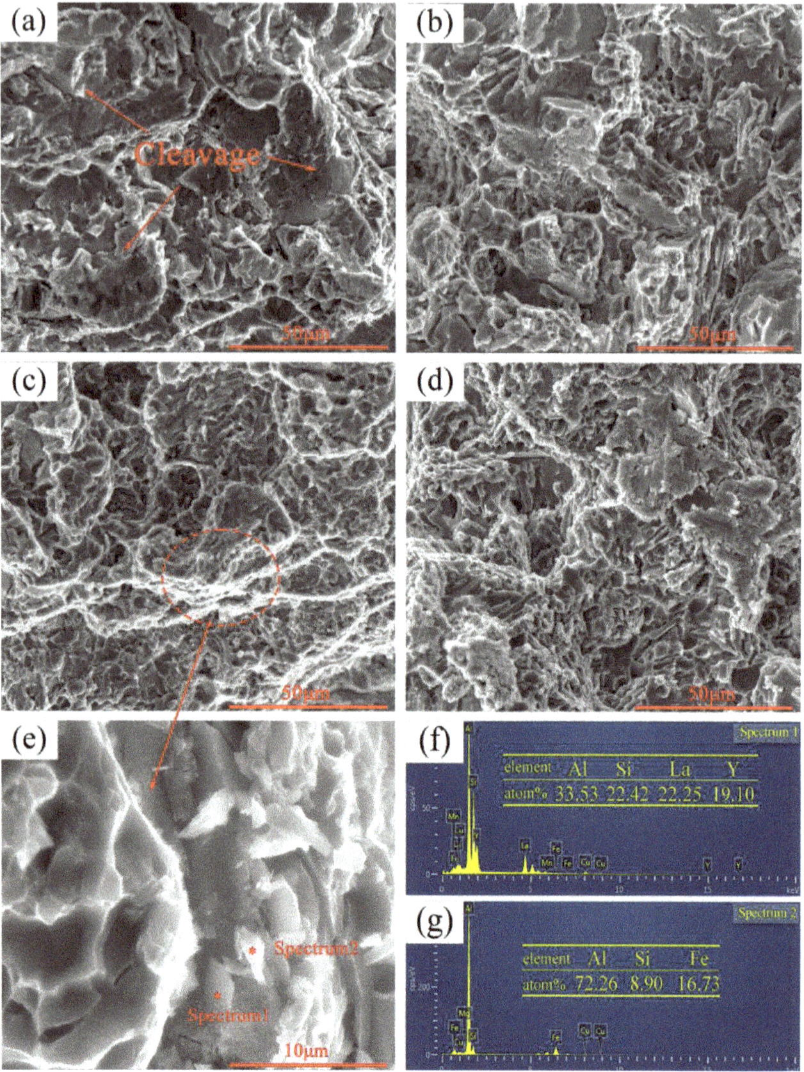

Figure 7. SEM of fractures of cast A380 alloys with different (La+Y) contents: (**a**) matrix alloy; (**b**) 0.3 wt.%; (**c**) 0.6 wt.%; (**d**) 0.9 wt.%; (**e**) enlarged view of the selected area in (**c**); (**f**) elemental point scan at point 1; (**g**) elemental point scan at point 2.

4. Discussion

4.1. Effect of La and Y on α-Al Grains

During solidification, La and Y are enriched at the solid–liquid interface front due to the low solid solution of La and Y in the Al-Si-Cu alloy. Moreover, the La and Y are surface active elements, which make the surface tension and critical nucleation radius of the solution smaller [14]. Meanwhile, La and Y form a surface-active film between the grains and the solution, which inhibits the grain growth and refines the grains, as shown in Figure 8.

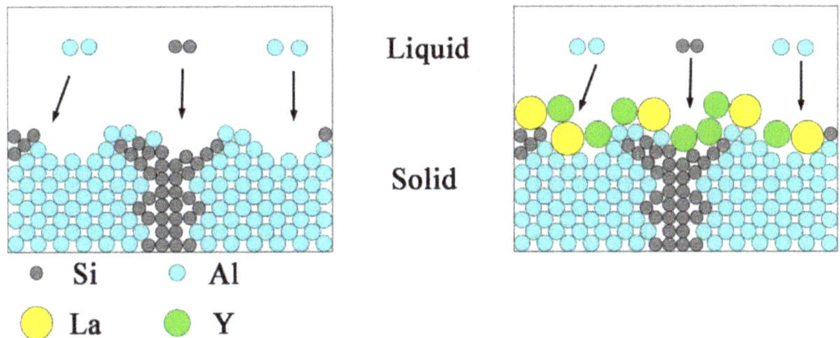

Figure 8. Limitation of the growth of α-Al and eutectic Si phases by La and Y.

On the other hand, compositional supercooling is generated through the aggregation of La and Y elements at the solid–liquid interface fronts during the solidification of α-Al primordial crystals. The solid–liquid interface changes from flat interface to regular interface and irregular interface due to the compositional supercooling during solidification. Without considering convection, the presence of compositional supercooling in solution can be determined by the following equation [12]:

$$\frac{G_L}{v} \leq \frac{m_L C_0 (1 - k_0)}{D_L k_0} \tag{1}$$

$$k_0 = c_S^* / c_L^* \tag{2}$$

where G_L is the temperature gradient at the solid-liquid interface front ($^\circ$C·mm^{-1}), m_L is the slope of the liquid phase line ($^\circ$C·(%·mm)$^{-1}$), C_0 is the original composition of solute in the alloy (%), k_0 is the solute partition coefficient, D_L is the solute diffusion coefficient, v is the solidification rate, c_S^* is the solute solubility in the solid phase, c_L^* is the solute solubility in the liquid phase.

From Equations (1) and (2), it can be seen that the k_0 is small because of the low solubility of La and Y in aluminum, so that the compositional supercooling is easily formed during the solidification of the alloy. The nucleation rate of the alloy is directly proportional to the supercooling degree, with the larger the supercooling degree, the higher the nucleation rate because of the lower nucleation work required for nucleation. In the case of supercooling, the increase in C_0 causes an increase in c_L^* and thus k_0 decreases, which leads to the increase in compositional supercooling resulting in a change in the solidification interface morphology. Especially, the compositional supercooling zone at the interface front will gradually widen as C_0 increases with the decrease of the ratio of temperature gradient to solidification rate G_L/v and the enrichment of rare earths during the solidification process. If the compositional supercooling is higher than the supercooling degree required for nucleation of the effective substrate for heterogeneous nucleation, a large number of free equiaxed crystal nuclei may be generated directly in the compositional supercooling zone. Therefore, compositional supercooling can increase the nucleation rate and refine the grains. In addition, the compositional supercooling also changes the dendrite morphology to smaller-sized equiaxed crystals. This is because the deeper supercooling causes the melting of the dendrite roots during the growth of dendrites, resulting in the presence of a large number of free nuclei, which changes the growth of dendrites, and the transformation of α-Al grains from developed dendrites to smaller-sized equiaxed crystals.

Figure 9 show the DSC curves of the A380 alloys at a 10 $^\circ$C/min heating rate. Peak 1 corresponds to the melting of eutectic Si, and peak 2 corresponds to the melting of primary α-Al. Table 3 shows the results of the analysis of the DSC curves. Where ΔT_1 is the nucleation supercooling of eutectic Si and ΔT_2 is the nucleation supercooling of primary α-Al. The results show that La and Y can increase ΔT_1 and ΔT_2, which improve the nucleation rate. Finally, the grains in the alloy are refined.

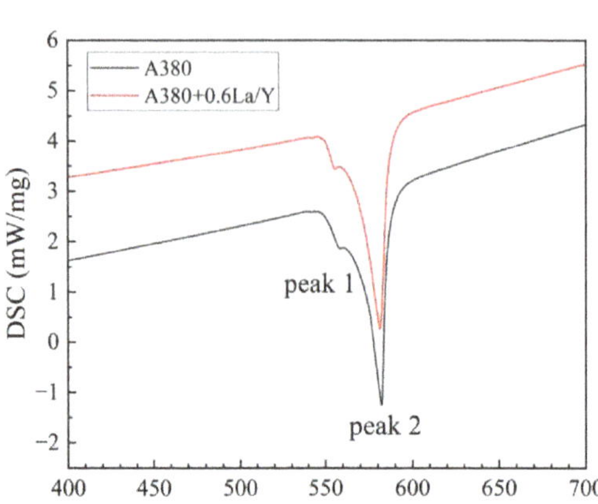

Figure 9. DSC curves for the La/Y-free and La/Y-contained A380 alloys.

Table 3. The Results of DSC Analysis.

No.	Eutectic Si		Primary α-Al	
	Peak 1 (°C)	ΔT_1 (°C)	Peak 2 (°C)	ΔT_2 (°C)
A380	556.50	47.30	582.25	21.55
A380 + 0.6 La/Y	555.43	50.97	581.14	25.26

4.2. Modification of Eutectic Si Phase by La and Y

In the study of modifiers for Al-Si alloys, the modification of eutectic Si phase is the focus of modifier initiation [25]. The modifier atoms generally take effect by adsorption on the eutectic Si growth platform. Depending on the adsorption location, two types of modification effects are possible [26]: (i) by adsorption along the re-entrant edge, thereby preventing further attachment of Si atoms to the growing crystal and growth in that direction, and (ii) by adsorption on the Si{111} growth platform and promoting a shift in the Si phase stacking order. The first possibility is a generalization of the poisoning case of the twin plane re-entrant mechanism, while the latter possibility is a modification effect of the impurity-induced twinning theory (IIT) mechanism. The adsorption of modified atoms on the {111} step surface of Si and the changes in the stacking order promote the frequent formation of twins.

According to the IIT [27], if the ratio of the atomic radius of La/Y and Si is 1.646, then La/Y can induce twin crystal formation on the Si phase. The ratio of La, Y atoms to Si atoms is 1.59 and 1.55, respectively, which is very close to 1.646. Therefore, the elements La and Y can promote the twinning phenomenon. A large number of twin crystals are generated on the Si phase growth platform, resulting in a shift from anisotropic to isotropic growth of the Si phase. The morphology of eutectic Si is transformed from slat-like to fibrous. In addition, the modifier makes the Si crystal decelerate in the original growth direction, so the Si phase is no longer growing faster than the α-Al phase [28]. As shown in Figure 2c, the eutectic Si is transformed from long strips to short fibers and partially spheroidized. The size of the eutectic Si is significantly reduced and no longer seriously damages the mechanical properties of the alloy.

In addition, according to Hume–Rothery criterion [29], the solid solution will be very small when the difference of atomic radii is more than 15%. Therefore, rare earth elements will be gathered at the solid–liquid interface front during solidification. The radii and relative atomic masses of La and Y atoms are very different from those of Al and Si atoms. It is difficult for La and Y atoms to enter the lattice of α-Al and eutectic Si. Therefore, the atoms of La and Y distribute on the Si surface and dendrite surface at the front of the solid–liquid interface can act as a hindrance to the growth of Si phase and α-Al dendrites, as shown in Figure 8. As a result, solute diffusion and exchange will be inhibited, leading to a more moderate accumulation of solutes on the Si surface. Finally, the growth rate of eutectic Si is slowed down and the coordinated growth of α-Al and Si is achieved.

4.3. Effect of La and Y on Fe-Rich Phase

In Al-Si-Cu alloys, the modifiers that are effective for the Si phase usually work well for β-Al$_5$FeSi as well. La and Y will also be adsorbed on the surface of β-Al$_5$FeSi after entering the melt. This phenomenon will result in the formation of atomic films that inhibit the diffusion and precipitation of Fe atoms, which finally limited the growth of the β-Al$_5$FeSi phase [30]. In addition, according to Chen et al. [22], the addition of La generates the Fe$_9$LaSi$_4$ phase in the Al-Si-Cu alloy, which consumes Fe elements. On the other hand, a part of Y also reacts with Si to form the Al-Si-Cu-Y phase [31], which attaches to β-Al$_5$FeSi, consuming the atoms of nearby Si and hindering its growth, as in Figure 4b. Moreover, the length of the β-Al$_5$FeSi phase decreases due to the restrained and consumed of Fe atoms in the melt, which could be observed from the Fe distribution of the EDS surface scan in Figure 4a–d. The reduction of the length of β-Al$_5$FeSi phase and the appearance of bulk α-Al$_8$Fe$_2$Si phase in the alloy will reduce the negative impact of the original long needle-like β-Al$_5$FeSi on the mechanical properties.

4.4. Thermodynamic Analysis of Rare Earth Phase Generation

According to the binary phase diagrams of Al-Y and Al-La, Al$_3$Y [20] and Al$_{11}$La$_3$ [21] are produced in the alloy, with melting points of 637 °C and 628 °C, respectively. Combined with the analysis of EDS and XRD, the fine rod-like bright phases in the modified alloy can be recognized as Al$_{11}$La$_3$ and Al$_3$Y formed by the reaction of La and Y with Al. In order to analyze the generation capacity of Al$_{11}$La$_3$ and Al$_3$Y, the Gibbs free energy of the two phases can be obtained using classical thermodynamic analysis [32].

According to the above discussion, Al$_{11}$La$_3$ and Al$_3$Y are generated in the alloy with the following reaction equations:

$$3Al + Y = Al_3Y, \tag{3}$$

$$11Al + 3La = Al_{11}La_3. \tag{4}$$

According to Kirchhoff's law, the relationship between the standard enthalpy of production and the isobaric heat capacity in a chemical reaction is as follows [33]:

$$\left[\frac{\partial \left(\Delta H_T^0 \right)}{\partial T} \right]_P = \Delta c_p \tag{5}$$

Under isobaric conditions, the equation can be transformed into:

$$\Delta H_T^0 = \Delta H_{T_0}^0 + \int_{T_0}^{T} \Delta c_p dT \tag{6}$$

The standard entropy of the reaction can be calculated using the following equation:

$$\Delta S_T^0 = \Delta S_{T_0}^0 + \int_{T_0}^{T} \frac{\Delta c_p}{T} dT \tag{7}$$

Thus, the standard free energy of production for a pure matter reaction can be expressed as:

$$\Delta G_T^0 = \Delta H_{T_0}^0 - T\Delta S_{T_0}^0 + \int_{T_0}^{T} \Delta c_p dT - T \int_{T_0}^{T} \frac{\Delta c_p}{T} dT \tag{8}$$

where T_0 is the reaction starting temperature and T is the reaction ending temperature. The Gibbs free energy of Reaction (3) and Reaction (4) with respect to temperature T can be obtained from Equation (8), as shown in Table 4.

Table 4. Gibbs free energy of Al_3Y and $Al_{11}La_3$.

Reaction	3Al + Y = Al_3Y	11Al + 3La = $Al_{11}La_3$
ΔG_T^0 (J·mol^{-1})	−185,600 + 83.89 T	−574,001 + 64.313 T

From the results, the Gibbs free energy change is negative for both $Al_{11}La_3$ and Al_3Y in the experimental temperature range of 500 to 780 °C so that the reaction can proceed spontaneously due to the decrease in the total energy of the system during the reaction. The stability and formation ability of $Al_{11}La_3$ are much higher than that of Al_3Y due to the large difference between the Gibbs free energy change of $Al_{11}La_3$ and Al_3Y. Therefore, at the solid–liquid interface front during solidification, the $Al_{11}La_3$ phase should be formed by the reaction of La with Al first. Then, the Al_3Y phase should be formed by the reaction of Y with Al. $Al_{11}La_3$ and Al_3Y are aggregated at the grain boundaries due to the low solubility of La and Y in Al and Si. So, Al_3Y is attached to the $Al_{11}La_3$ phase to generate because of the sequence of generation, which is consistent with that shown in Figure 4.

4.5. Effects of La and Y on the Mechanical Properties of A380 Alloys

La, Y can improve the mechanical properties of A380 alloy, as shown in Figure 6. The strengthening mechanism is dominated by grain refinement and the second phase strengthening. Grain refinement leads to more grain boundaries, which increase the resistance to dislocation movement and improve the resistance to deformation [34]. The contribution of grain refinement on mechanical properties can be measured by the Hall–Petch formula [35]:

$$\Delta\sigma_{Hall-Petch} = Kd^{-1/2} \tag{9}$$

where $\Delta\sigma_{Hall\text{-}Petch}$ is the yield strength increment, K is a parameter on influence extent of grain boundaries to yield strength, and d is the average grain size. Therefore, grain refinement, which means a decrease in d-value, will lead to an improvement in mechanical properties.

In addition, Al_3Y and $Al_{11}La_3$ in the alloy will also provide second-phase strengthening by hindering dislocation movement in tensile experiments. The second phase strengthening in an alloy can be measured by the Orowan mechanism with the following equation [36]:

$$\Delta\sigma_P = \frac{2MGb}{1.18 \times 4\pi \times |\lambda - d|} \ln\frac{d}{2b} \tag{10}$$

where M is the Taylor factor, G is the shear modulus of α-Mg, b is the Burgers vector, λ is the particle interspacing, d is the particle size. The interaction of Al_3Y and $Al_{11}La_3$ with dislocations generates dislocation loops through the Orowan bypassing mechanism, which hinders the dislocation movements in the alloy and reinforces the mechanical properties.

5. Conclusions

The effects of La and Y on the microstructural evolution and mechanical properties of cast Al-Si-Cu alloys were investigated in this work. The main conclusions drawn from this work are as follows:

La+Y can be used as an effective modifier for Al-Si-Cu alloys. With the optimum content of 0.6 wt.%, the alloy has the best microstructure. The coarse dendritic primary

α-Al grains are fully refined and transformed to equiaxed grains. Meanwhile, the eutectic Si is modified from strips to short fibrous or even granular. In addition, the length of long needle-like β-Al$_5$FeSi is reduced and bulk α-Al$_8$Fe$_2$Si is observed.

The refinement effect of La and Y on the microstructure is mainly derived from the phenomenon of compositional supercooling and the limitation of solute diffusion. Meanwhile, La and Y can promote the generation of twins on the Si phase and modify the morphology through the IIT mechanism.

The Gibbs free energy changes of the generated Al$_{11}$La$_3$ phase and Al$_3$Y phase are both negative in the alloys. Al$_{11}$La$_3$ has more negative Gibbs free energy change than the Al$_3$Y phase, thus Al$_{11}$La$_3$ is more stable. At low La+Y content, Al$_3$Y is generated on Al$_{11}$La$_3$ and covers it. However, the uncovered Al$_{11}$La$_3$ is observed with an La+Y content of 0.9 wt.% in the alloy.

The addition of 0.6 wt.% La+Y result in UTS and EI of 215.3 MPa and 5.1%, respectively. These values increase by 22.9% and 37.8%, respectively, compared to the unmodified alloy. The fracture test reveals that the main fracture mode is changed from brittle fracture to ductile fracture.

Author Contributions: Writing—original draft preparation, L.S.; writing—review and editing, X.Z.; investigation, Y.R. and X.L.; supervision and resources, H.Y. All authors have read and agreed to the published version of the manuscript.

Funding: This research was funded by the National Natural Science foundation of China (No.51965040).

Institutional Review Board Statement: Not applicable.

Informed Consent Statement: Not applicable.

Data Availability Statement: The data presented in this study are available on request from the corresponding author.

Acknowledgments: The authors thank the National Natural Science foundation of China (No.51965040).

Conflicts of Interest: The authors declare no conflict of interest.

References

1. Okayasu, M.; Ota, K.; Takeuchi, S.; Ohfuji, H.; Shiraishi, T. Influence of microstructural characteristics on mechanical properties of ADC12 aluminum alloy. *Mater. Sci. Eng. A* **2014**, *592*, 189–200. [CrossRef]
2. Shankar, S.; Apelian, D. Die soldering: Mechanism of the interface reaction between molten aluminum alloy and tool steel. *Metall. Mater. Trans. B* **2002**, *33*, 465–476. [CrossRef]
3. Tang, P.; Li, W.; Wang, K.; Du, J.; Chen, X.; Zhao, Y.; Li, W. Effect of Al-Ti-C master alloy addition on microstructures and mechanical properties of cast eutectic Al-Si-Fe-Cu alloy. *Mater. Des.* **2017**, *115*, 147–157. [CrossRef]
4. Pan, S.; Chen, X.; Zhou, X.; Wang, Z.; Chen, K.; Cao, Y.; Lu, F.; Li, S. Micro-alloying effect of Er and Zr on microstructural evolution and yield strength of Al-3Cu (wt.%) binary alloys. *Mater. Sci. Eng. A* **2020**, *790*, 139–391. [CrossRef]
5. Liao, H.; Xu, H.; Hu, Y. Effect of RE addition on solidification process and high-temperature strength of Al-12%Si-4%Cu-1.6%Mn heat-resistant alloy. *Trans. Nonferr. Metal. Soc.* **2019**, *29*, 1117–1126. [CrossRef]
6. Li, Z.; Yan, H.; Hu, Z.; Song, X. Fluidity of ADC12 + xLa aluminum alloys. *Rare Met.* **2021**, *40*, 1191–1197. [CrossRef]
7. Mao, G.; Yan, H.; Zhu, C.; Wu, Z.; Gao, W. The varied mechanisms of yttrium (Y) modifying a hypoeutectic Al–Si alloy under conditions of different cooling rates. *J. Alloys Compd.* **2019**, *806*, 909–916. [CrossRef]
8. Kang, J.; Su, R.; Wu, D.; Liu, C.; Li, T.; Wang, L.; Narayanaswamy, B. Synergistic effects of Ce and Mg on the microstructure and tensile properties of Al-7Si-0.3Mg-0.2Fe alloy. *J. Alloys Compd.* **2019**, *796*, 267–278. [CrossRef]
9. Li, J.; Wang, X.; Ludwig, T.; Tsunekawa, Y.; Arnberg, L.; Jiang, J.; Schumacher, P. Modification of eutectic Si in Al-Si-Cu alloys with Eu addition. *Acta Mater.* **2015**, *84*, 153–163. [CrossRef]
10. Hosseinifar, M.; Malakhov, D.V. Effect of Ce and La on microstructure and properties of a 6xxx series type aluminum alloy. *J. Mater. Sci.* **2008**, *43*, 7157–7164. [CrossRef]
11. Tsai, Y.; Chou, C.; Lee, S.; Lin, C.; Lin, J.; Lim, S. Effect of trace La addition on the microstructures and mechanical properties of A356 (Al–7Si–0.35Mg) aluminum alloys. *J. Alloys Compd.* **2009**, *487*, 157–162. [CrossRef]
12. Song, X.; Yan, H.; Chen, F. Impact of rare earth element La on microstructure and hot crack resistance of ADC12 alloy. *J. Wuhan Univ. Technol-Mat. Sci. Ed.* **2018**, *33*, 193–197. [CrossRef]
13. Huang, X.; Yan, H. Effect of trace La addition on the microstructure and mechanical property of as-cast ADC12 Al-Alloy. *J. Wuhan Univ. Technol-Mat. Sci. Ed.* **2013**, *28*, 202–205. [CrossRef]

14. Qiu, C.; Miao, S.; Li, X.; Xia, X.; Ding, J.; Wang, Y.; Zhao, W. Synergistic effect of Sr and La on the microstructure and mechanical properties of A356.2 alloy. *Mater. Des.* **2017**, *114*, 563–571. [CrossRef]

15. Cao, Y.; Chen, X.; Wang, Z.; Chen, K.; Pan, S.; Zhu, Y.; Wang, Y. Synergistic influence of La and Zr on microstructure and mechanical performance of an Al-Si-Mg alloy at casting state. *J. Alloys Compd.* **2022**, *902*, 163829. [CrossRef]

16. Liu, J.; Wu, Q.; Yan, H.; Zhong, S.; Huang, Z. Effect of Trace Yttrium Addition and Heat Treatment on the Microstructure and Mechanical Properties of As-Cast ADC12 Aluminum Alloy. *Appl. Sci.* **2018**, *9*, 53. [CrossRef]

17. Li, Q.; Zhu, Y.; Zhao, S.; Lan, Y.; Zhou, H. Influences of Fe, Mn and Y additions on microstructure and mechanical properties of hypoeutectic Al–7%Si alloy. *Intermetallics* **2020**, *120*, 106768. [CrossRef]

18. Wei, Z.; Lei, Y.; Yan, H.; Xu, X.; He, J. Microstructure and mechanical properties of A356 alloy with yttrium addition processed by hot extrusion. *J. Rare Earths* **2019**, *37*, 659–667. [CrossRef]

19. Lakshmanan, M.; SelwinRajadurai, J.; Chakkravarthy, V.; Rajakarunakaran, S. Wear and EBSD studies on (SiC/NiTi) reinforced Al7075 composite. *Mater. Lett.* **2020**, *272*, 127879. [CrossRef]

20. Liu, S.; Du, Y.; Chen, H. A thermodynamic reassessment of the Al–Y system. *Calphad* **2006**, *30*, 334–340. [CrossRef]

21. Zhou, S.; Napolitano, R. Phase equilibria and thermodynamic limits for partitionless crystallization in the Al–La binary system. *Acta Mater.* **2006**, *54*, 831–840. [CrossRef]

22. Chen, X.; Chen, Y.; Tang, Y.; Xiao, D. Effects of the excess iron on phase and magnetocaloric property of LaFe11.6–xSi1.4 alloys. *J. Rare Earths* **2015**, *33*, 1293–1297. [CrossRef]

23. Li, Q.; Xia, T.; Lan, Y.; Li, P.; Fan, L. Effects of rare earth Er addition on microstructure and mechanical properties of hypereutectic Al–20% Si alloy. *Mater. Sci. Eng. A* **2013**, *588*, 97–102. [CrossRef]

24. Wu, D.; Kang, J.; Feng, Z.; Su, R.; Liu, C.; Li, T.; Wang, L. Utilizing a novel modifier to realize multi-refinement and optimized heat treatment of A356 alloy. *J. Alloys Compd.* **2019**, *791*, 628–640. [CrossRef]

25. Li, L.; Li, D.; Mao, F.; Feng, J.; Zhang, Y.; Kang, Y. Effect of cooling rate on eutectic Si in Al-7.0Si-0.3Mg alloys modified by La additions. *J. Alloys Compd.* **2020**, *826*, 154–206. [CrossRef]

26. Timpel, M.; Wanderka, N.; Schlesiger, R.; Yamamoto, T.; Banhart, J. The role of strontium in modifying aluminium–silicon alloys. *Acta Mater.* **2012**, *60*, 3920–3928. [CrossRef]

27. Lu, S.; Hellawell, A. The mechanism of silicon modifification in aluminum–silicon alloys: Impurity-induced twinning. *Metall. Mater. Trans. A* **1987**, *18*, 1721–1733. [CrossRef]

28. Moniri, S.; Xiao, X.; Shahani, A. The mechanism of eutectic modification by trace impurities. *Sci. Rep.* **2019**, *9*, 3381. [CrossRef]

29. Saccone, A.; Cacciamani, G.; Maccio, D.; Borzone, G.; Ferro, R. Contribution to the study of the alloys and intermetallic compounds of aluminium with the rare-earth metals. *Intermetallics* **1998**, *6*, 201–215. [CrossRef]

30. Rao, Y.; Yan, H.; Hu, Z. Modification of eutectic silicon and β-Al_5FeSi phases in as-cast ADC12 alloys by using samarium addition. *J. Rare Earths* **2013**, *31*, 916–922. [CrossRef]

31. Wan, B.; Chen, W.; Liu, L.; Cao, X.; Zhou, L.; Fu, Z. Effect of trace yttrium addition on the microstructure and tensile properties of recycled Al–7Si–0.3Mg–1.0Fe casting alloys. *Mater. Sci. Eng. A* **2016**, *666*, 165–175. [CrossRef]

32. Binnewies, M.; Mike, E. *Thermochemical Data of Elements and Compounds*; Wiley-VCH Verlag Gmbh: Weinheim, Germany, 2002; p. 783.

33. Robert Mortimer, G. *Physical Chemistry*; Elsevier Academic Press: Memphis, TN, USA; Amsterdam, The Netherlands, 2008; p. 173.

34. Chakkravarthy, V.; Lakshmanan, M.; Manojkumar, P.; Prabhakaran, R. Crystallographic orientation and wear characteristics of TiN, SiC, Nb embedded Al7075 composite. *Mater. Lett.* **2022**, *306*, 130936. [CrossRef]

35. Wyrzykowski, J.W.; Grabski, M.W. The Hall–Petch relation in aluminium and its dependence on the grain boundary structure. *Philos. Mag. A* **1986**, *53*, 505–520. [CrossRef]

36. Morris, D.G.; Muñoz-Morris, M.A. High creep strength, dispersion-strengthened iron aluminide prepared by multidirectional high-strain forging. *Acta Mater.* **2010**, *58*, 6080–6089. [CrossRef]

Article

A Quasi In-Situ Study on the Microstructural Evolution of 2195 Al-Cu-Li Alloy during Homogenization

Hao Huang [1], Wei Xiong [1,*], Zhen Jiang [1] and Jin Zhang [1,2]

[1]   Light Alloy Research Institute, Central South University, Changsha 410083, China
[2]   State Key Laboratory of High Performance and Complex Manufacturing, Central South University, Changsha 410083, China
*   Correspondence: wxiong@csu.edu.cn

Abstract: An optimized homogenization process for Al alloy ingots is key to subsequent material manufacturing, as it largely reduces metallurgical defects, such as segregation and secondary phases. However, studies on their exact microstructural evolution at different homogenization temperatures are scarce, especially for complex systems, such as the 2195 Al-Cu-Li alloy. The present work aims to elucidate the microstructural evolution of the 2195 Al-Cu-Li alloy during homogenization, including the dissolution and precipitation behavior of the T_B (Al_7Cu_4Li) phase and S (Al_2CuMg) phase at different homogenization temperatures. The results show that there are Cu segregation zones (Cu-SZ) at the dendrite boundaries with θ (Al_2Cu) and S eutectic phases. When the temperature rises from 300 °C to 400 °C, fine T_B phases precipitate at the Cu-SZ, and the Mg and Ag in the S phases gradually diffuse into the matrix. Upon further increasing the temperature to 450 °C, T_B and θ phases at the grain boundaries are coarsened, and an S-θ phase transition is observed. Finally, at 500 °C, all T_B and S phases are dissolved, leaving only θ phases at triangular grain boundaries. This work provides guidance for optimizing the homogenization procedure in 2195 alloys.

Keywords: Al-Cu-Li alloy; homogenization; microstructural evolution; quasi in situ

Citation: Huang, H.; Xiong, W.; Jiang, Z.; Zhang, J. A Quasi In-Situ Study on the Microstructural Evolution of 2195 Al-Cu-Li Alloy during Homogenization. *Materials* **2022**, *15*, 6573. https://doi.org/10.3390/ma15196573

Academic Editor: Hideki Hosoda

Received: 27 August 2022
Accepted: 20 September 2022
Published: 22 September 2022

Publisher's Note: MDPI stays neutral with regard to jurisdictional claims in published maps and institutional affiliations.

1. Introduction

Al-Cu-Li alloys possess higher specific strength and specific stiffness compared with traditional aluminum alloys, enabling considerable weight reductions while maintaining good mechanical properties in aerospace applications [1–6]. They are considered to be desirable structural parts for the aerospace industry now [7–12]. However, existing studies show that large structural portions of Al-Cu-Li alloys are inclined to cracking during severe deformation and, therefore, have poor formability [13,14]. To solve these problems, cast ingots of Al-Cu-Li alloys are usually homogenized prior to manufacturing. The microstructural evolution of these ingots during homogenization has a significant impact on their deformation, recovery and recrystallization behaviors in subsequent processes, such as rolling and forging [15–20]. For example, T_B and θ phases on the grain boundaries will dissolve in the appropriate homogenization process, thus inhibiting the initiation of cracks during subsequent processing [21,22] and improving their mechanical properties after aging [23–25]. Therefore, an in-depth investigation of the microstructural evolution during the homogenization of Al-Cu-Li alloys is significant for understanding their material properties and optimizing subsequent processing. The dissolution behavior of T_B and S phases especially needs to be elucidated as these eutectic phases commonly exist in Al-Cu-Li alloys, and their distribution at grain boundaries severely deteriorates the formability of the materials.

Liu et al. [26] researched the impacts of homogenization on the T_B, θ and S phases in Al-3.8Cu-1.28Li-0.4Mg alloy. It was found these eutectic phases are completely dissolved at 530 °C with extended homogenization time. Li et al. [27,28] reported that T_B and S phases are only partially dissolved after the first-step homogenization at 400 °C for 8 h, and their complete dissolution occurs during second-step homogenization. Liu and Li only studied the dissolution of primary eutectic T_B and S phases in the ingot during homogenization; they did not notice that the T_B phase will also precipitate during homogenization and that the S phase will transform into the θ phase. Yang et al. [29] found that there are plenty of θ phases and little of S phases on the grain boundary of the Al-3.52Cu-1.28Li-0.38Mg ingot, and the S phases are preferentially dissolved during homogenization as their melting point is lower. Although Yang thought that the S phase dissolves first compared to the θ phase, he did not find that the S phase is first transformed into the θ phase and then dissolved. Chen et al. [30] studied the Al-5Cu-1Li-0.6Mg alloy and observed that θ and T_1 phases are coarsened during homogenization, and T_B and θ phases are dissolved at elevated temperatures. Chen noticed that the T_B phase coarsened during homogenization, but he did not find that the T_B phase will precipitate during homogenization and did not distinguish the primary eutectic T_B phase from the T_B phase that precipitated later. The above studies on the microstructural evolution in the homogenization of Al-Cu-Li alloys are not comprehensive enough. To date, there has been little research on the detailed precipitation and dissolution behavior of T_B and S phases during homogenization, such as the S phase transforming into the θ phase and the newly precipitated T_B phase during homogenization.

Some researchers have studied the nucleation and growing of second phase particles during the cooling of homogenized aluminum alloy, the micro-segregation of alloying elements during the homogenization of aluminum alloy, and stress analysis by computational simulation methods [31–33]. In this study, the microstructural evolution of the 2195 Al-Cu-Li alloy during hominization is obtained via direct experimental observations, which is facile and commonly adopted by the research community, this could provide an experimental basis, such as boundaries conditions for subsequent computational simulation research.

To more intuitively investigate the precipitation and dissolution behavior of T_B and S phases, a combination of methods, including quasi in situ scanning electron microscopy (SEM), transmission electron microscopy (TEM), X-ray diffraction (XRD), Vickers hardness tests and energy-dispersive X-ray spectroscopy (EDS), were applied to study the phase composition, microstructure and elemental distribution of the 2195 alloy at different homogenization temperatures (25–500 °C). We demonstrate that, by raising the homogenization temperature from 300 °C to 450 °C, a fine T_B phase forms by precipitating from Cu segregation zones and then coarsens; these are continuously distributed on the grain boundaries, while the S phase first transforms into the θ phase and then dissolves. These results can be used to optimize the homogenization procedure for 2195 alloys.

2. Materials and Methods

The experimental material employed is an Al-Cu-Li alloy ingot made by the central south university (Ø180 mm × 500 mm). The elemental ingredients were verified by the inductively coupled plasma-atomic emission spectroscopy (ICP-AES) (SPECTRO BLUE SOP, Kleve, Germany) (see Table 1).

Table 1. Elemental ingredients of the 2195 ingot (wt. %).

Cu	Li	Mg	Ag	Zr	Zn	Mn	Fe	Si	Al
3.99	0.90	0.27	0.28	0.15	0.02	0.03	<0.02	<0.04	Bal.

A 10 × 10 × 5 mm sample was taken from the ingot. We cut a small-sized sample to reduce the temperature difference between the interior and exterior of the sample during homogenization. The sample was put into the muffle furnace (SG-XS1700, Shanghai Sager Furnace Co., Limited, Shanghai, China) and heated from 25 °C to 500 °C; the heating rate of the muffle furnace was 1 °C/min and the temperature error was controlled within ±1 °C. The sample was quenched with water (less than 5 s) after reaching 100 °C, 200 °C, 300 °C,

400 °C, 450 °C and 500 °C, followed by quasi in situ SEM observations. The observation area was located by Vickers hardness indentation on the sample. The temperature of the furnace was maintained until the sample was put back after finishing quasi in situ SEM observation. The homogenization procedure is presented in Figure 1, and the workflow diagram is displayed in Figure 2.

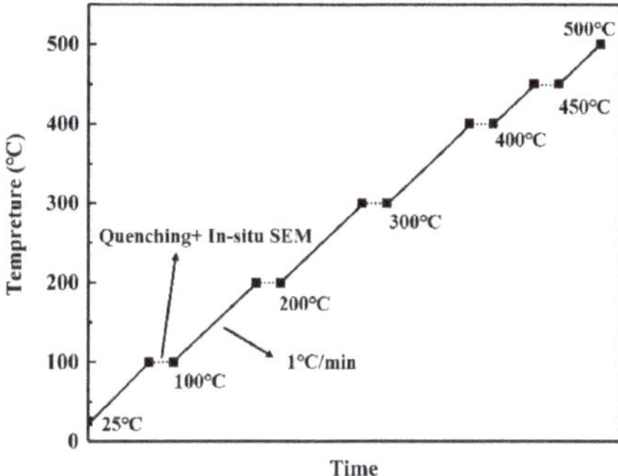

Figure 1. Schematic diagram of homogenization procedure.

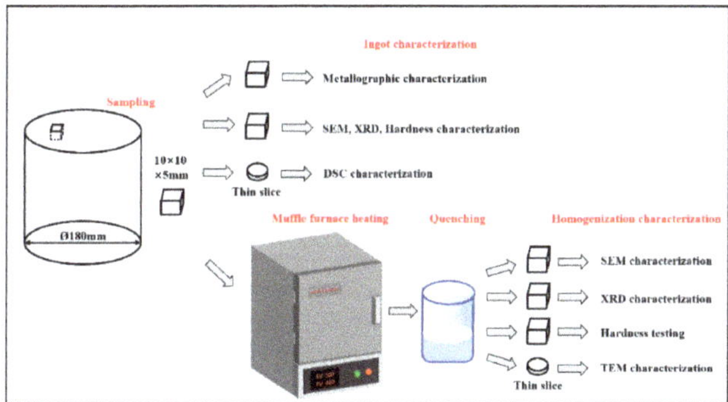

Figure 2. Schematic diagram of the workflow.

X-ray diffraction (XRD) (D/Max 2550VB, Rigaku, Tokyo, Japan) was used to identify the secondary phases, the scanning rate was 4°/min and 2θ has a range of 20° to 100°. The power of the X-ray generator is 18 kW and the stability error of the X-ray generator is within ±0.01%. The Differential Scanning Calorimetry (DSC) (DSC8000, PerkinElmer, Waltham, MA, USA) was applied to verify the non-equilibrium solidus temperature of the ingot by heating the sample from 0 to 600 °C at 10 k/min. The precipitation behavior was detected using transmission electron microscopy (TEM) (Tecnai G2F20, FEI, Lincoln, NE, USA). The acceleration voltage was 200 KV. A Vickers hardness tester (YZHV-1000P, Yizong Precision Instrument Co., Ltd., Shanghai, China) was applied to test the hardness of the sample.

The evolution of the microstructure during homogenization was investigated with a Scanning Electron Microscopy (SEM) (EVO MA10, ZEISS, Oberkochen, Germany) instrument possessing an Energy Dispersive Spectrometer (EDS) detector; the accelerating voltage is 20 kV, and the elemental distribution was studied using the EDS line-scanning (LS) and map-scanning (MS) modes. The SEM resolution was 3.0 nm.

3. Results and Discussion

3.1. Microstructural Characterization of 2195 Ingot

Figure 3a displays the metallographic image of the 2195 ingot, where obvious dendritic structures are observed, and the mean grain size is 150–250 μm. Figure 3b displays the backscattered SEM image of the ingot, and the boxed area in the Figure is the quasi in situ observation zone (QISOZ). As shown by the arrows, there is a high number of gray regions (these regions were proved to be Cu segregation zones in Figure 4); the non-equilibrium eutectic phases are distributed in the gray regions. The size and shape of these gray regions are similar to the corroded zones in the metallographic image. Figure 3c,d show the QISOZ backscattered SEM images at 500× and 1000× magnification, respectively. The EDS results of each point in Figure 3d are shown in Table 2. There are θ and S phases in the ingot, and a certain amount of Ag exists in the S phase. This is because the interactive energy between Ag and Mg atoms was strong (−0.0632 eV) [34]; hence, Ag is easily combined with Mg to form Mg-Ag clusters [35–39]. Figure 3e presents the XRD result of 2195 ingot. Besides α(Al), θ phases and S phases also exist in the ingot, which is identical to the EDS results in Figure 3d. Figure 3f displays the DSC result of the ingot. There is an apparent endothermic peak at 525 °C in the ingot, which led to the melting of some low-melting-point eutectic phases, indicating that the non-equilibrium solidus temperature of the ingot is 525 °C. Therefore, in this study, the maximum temperature of the homogenization was chosen to be 500 °C to avoid over-burning the sample.

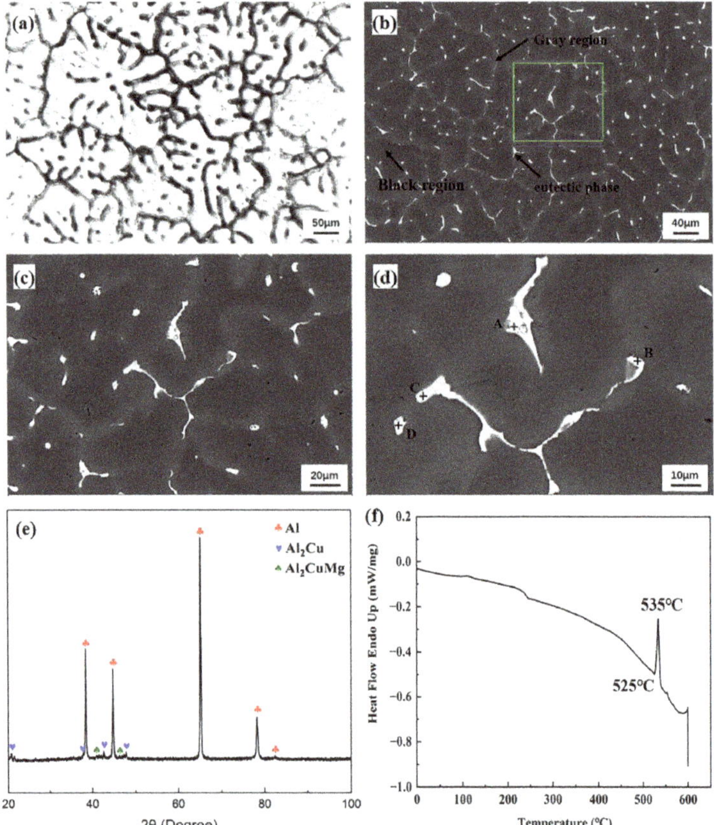

Figure 3. 2195 ingot microstructure: (**a**) ingot metallographic image; (**b**) in situ observation zone; (**c,d**) 500× and 1000× magnification of in situ observation zone; (**e**) ingot XRD; (**f**) ingot DSC.

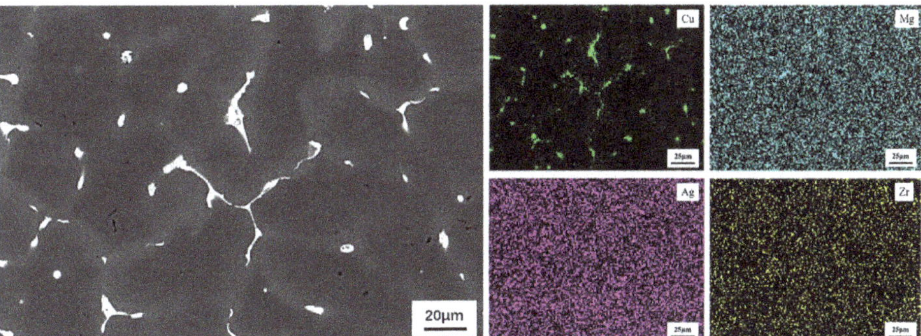

Figure 4. EDS elemental distribution of QISOZ.

Table 2. EDS results for each point in Figure 2d (at. %).

Point	Al	Cu	Mg	Ag	Closest Phase
A	68.8	30.6	0.38		Al_2Cu
B	66.8	31.1	1.5		Al_2Cu
C	71.6	13.2	10.6	4.5	Al_2CuMg
D	78.7	10.7	7.1	3.4	Al_2CuMg

The EDS elemental distribution of Cu, Mg, Ag and Zr on the QISOZ is displayed in Figure 4. Cu is segregated in the gray regions, so the gray regions are named Cu segregation zones (Cu-SZ). Mg, Ag and Zr were homogeneously distributed, while Cu shows severe segregation.

3.2. Microstructural Characterization of the Sample at 100 °C, 200 °C and 300 °C

Figure 5 presents the backscattered SEM pictures of QISOZ, magnified by 500 times and 1000 times at 100 °C, 200 °C and 300 °C, where L1 and L2 in Figure 5b represent the line scan of the S phase mentioned in Figure 2d and Table 2. Comparing Figure 5a,c,e and Figure 5b,d,f, respectively, the size and morphology of the eutectic phases at 100 °C, 200 °C and 300 °C show no obvious changes, and even the very small eutectic phases circled in Figure 5b,d,f remained identical, and the gray zones of Cu-SZ showed no obvious changes in size and shape. Figure 6 shows the EDS elemental distribution of Cu with a QISOZ magnification of 500 times when the sample is heated to 100 °C, 200 °C and 300 °C. It can be seen that there is no obvious difference in the elemental distribution of Cu at these three temperatures. Cu is still slightly segregated in the gray region and heavily segregated in the eutectic phase. This shows that the diffusion of Cu is not obvious before 300 °C.

Figure 7 shows the elemental distribution along L1 and L2 at 100 °C, 200 °C and 300 °C. The concentrations of Cu, Mg and Ag do not significantly change from 100 °C to 300 °C, the Cu/Mg is still close to 1:1, and slight segregation of Ag still exists in the S phase, demonstrating that the diffusion of Cu, Mg and Ag is still limited before reaching 300 °C.

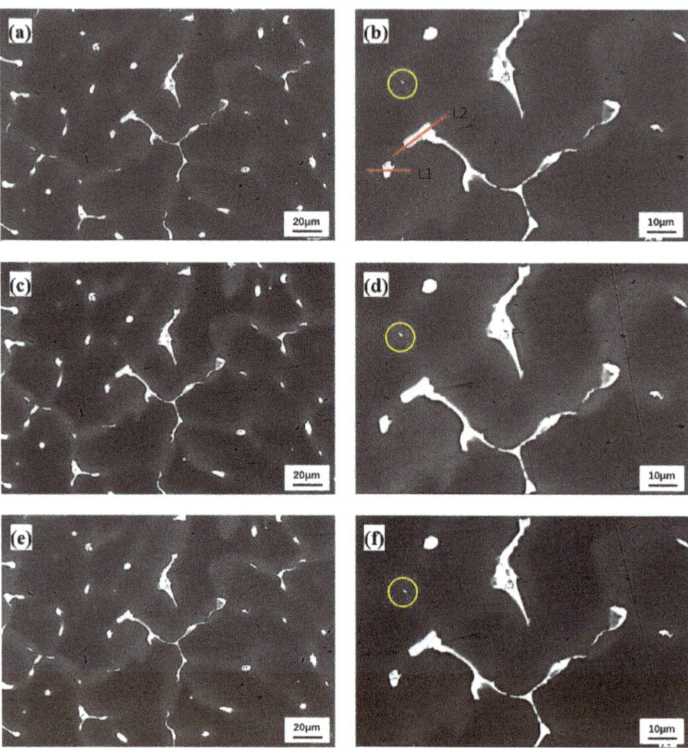

Figure 5. QISOZ backscattered SEM images at 500 times (**a**,**c**,**e**) and 1000 times (**b**,**d**,**f**) magnification at 100 °C (**a**,**b**), 200 °C (**c**,**d**) and 300 °C (**e**,**f**).

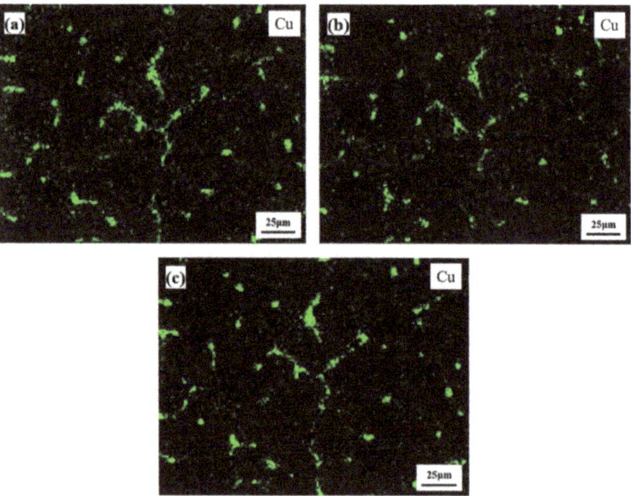

Figure 6. EDS elemental distribution of Cu with QISOZ magnification of 500 times at 100 °C (**a**), 200 °C (**b**) and 300 °C (**c**).

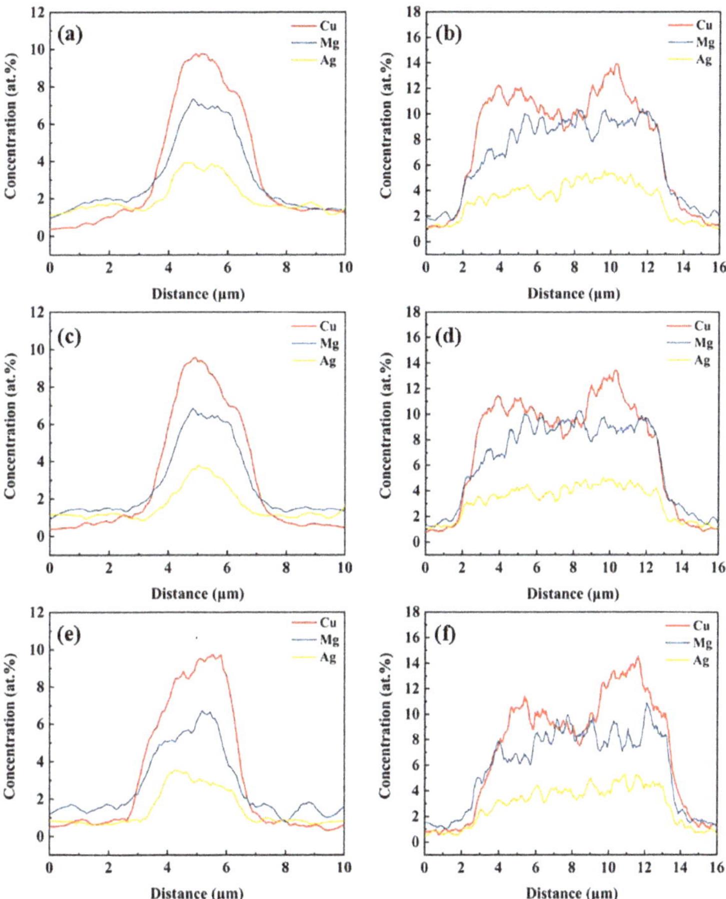

Figure 7. Elemental distribution along L1 (**a,c,e**) and L2 (**b,d,f**) at 100 °C (**a,b**), 200 °C (**c,d**) and 300 °C (**e,f**) of the sample.

The diffusion coefficient D follows the Arrhenius equation [40]:

$$D = D_0 \exp\left(-\frac{Q}{RT}\right) \tag{1}$$

where D_0 is the diffusion constant(m^2/s), Q is the activation energy per mole of atoms(J/mol), T is the thermodynamic temperature(K), and R is the gas constant, which is 8.314 J/(mol· K).

According to Xie et al. [41], the diffusion constant of solute Cu and Mg in aluminum is 4.8×10^{-5} m^2/s and 1.2×10^{-4} m^2/s, respectively, and their diffusional activation energies are 133.6×10^3 J/mol ($Q_{(Cu)}$) and 131×10^3 J/mol ($Q_{(Mg)}$), respectively. Their diffusion coefficients at 300 °C can be calculated using Equation (1) ($D_{(Cu)} = 1.92 \times 10^{-17}$ m^2/s, $D_{(Mg)} = 1.37 \times 10^{-16}$ m^2/s); such slow diffusion kinetics below 300 °C are consistent with the experimental observations.

3.3. Microstructural Characterization of the Sample at 400 °C, 450 °C and 500 °C

Figure 8 displays the backscattered SEM pictures of the sample at 400 °C, 450 °C and 500 °C, with QISOZ magnifications of 500 and 1000 times. Although the morphology of the eutectic phase does not significantly change at 400 °C, and Cu-SZ is still observed, small precipitates are noticed in the Cu-SZ for the first time. These small precipitates are probably Cu-containing phases (confirmed as T_B phases in Figure 9), as they are formed within the Cu-SZ. After further increasing the temperature to 450 °C, the Cu-SZ disappears, and the precipitates are roughened and uninterruptedly distributed on the grain boundary. The small precipitates that were previously observed in the grain interior are partially dissolved (circled in the picture), and the θ phases on the grain boundary are coarsened. This phenomenon can be elucidated by the Ostwald ripening theory [42–45]: because the fine T_B phases dissolve and the large T_B phases grow, the specific interface energy per unit quality decreases, and the total free energy of the system decreases. When the temperature reaches 500 °C, all the precipitates located on the grain boundaries are dissolved, and most of the small-sized eutectic phases dispersed within the grains are dissolved, leaving only large eutectic phases at the triangular grain boundary owing to their relatively slow dissolution speed. Additionally, some over-burnt pits can be found in the region with θ phases. This is because these samples were directly transferred to the furnace at 450 °C, and underwent subsequent heating from 450 °C to 500 °C over 50 min. It was reported [46] that when the temperature of the alloy rises rapidly, and beyond the melting point of the θ phase, θ phases will rapidly dissolve and leave obvious over-burnt pits.

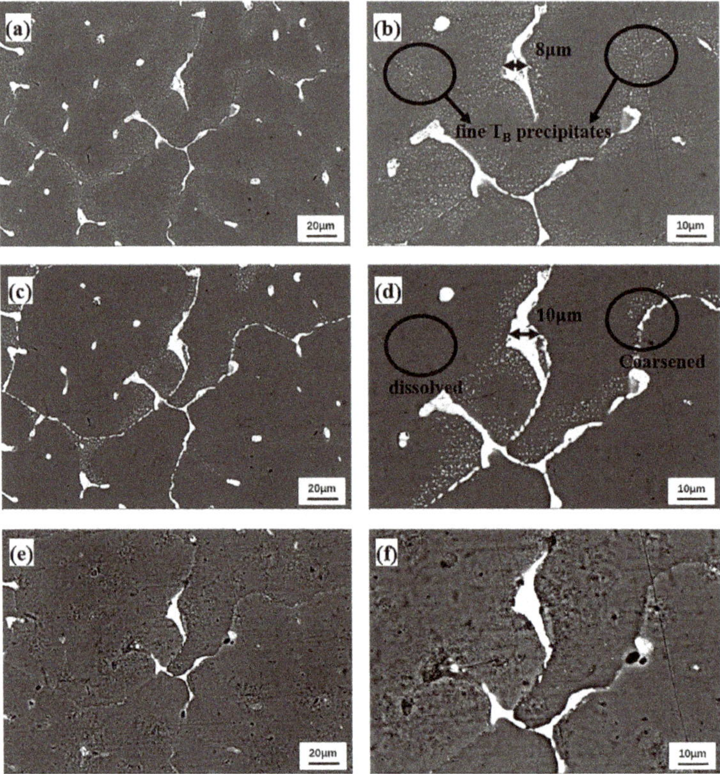

Figure 8. QISOZ backscattered SEM images at 500× (**a,c,e**) and 1000× (**b,d,f**) magnifications at 400 °C (**a,b**), 450 °C (**c,d**) and 500 °C (**e,f**).

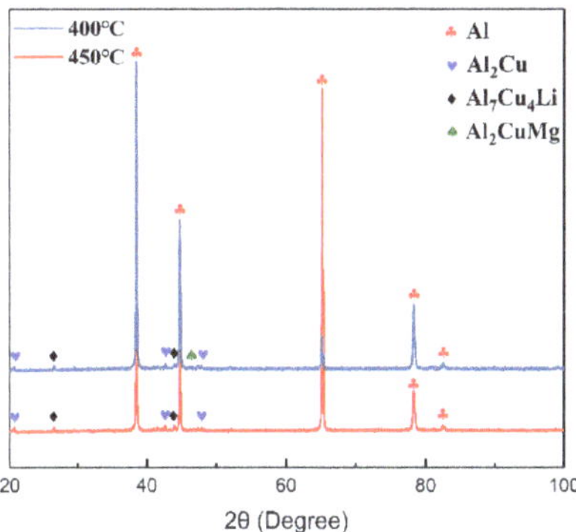

Figure 9. XRD pattern of the sample at 400 °C and 450 °C.

Figure 9 shows the XRD result of the sample at 400 °C and 450 °C. When contrasted with the XRD result of the ingot at 400 °C, the diffraction peak of the T_B phase appears, and the peak strength of the S phase is weakened. At 450 °C, there is no diffraction peak of the S phase. Therefore, the fine precipitates observed at 400 °C are T_B phases, and the T_B phases coarsened at 450 °C.

Figure 10 shows the TEM images along the <110>$_{Al}$ zone axis of the sample heated to 400 °C, 450 °C and 500 °C. At 400 °C, fine T_B phases precipitate at and close to the grain boundary, and the T_B phases are coarsened at 450 °C. When the temperature increases to 500 °C, the T_B phases dissolve in the matrix. The TEM results are in keeping with the SEM results above.

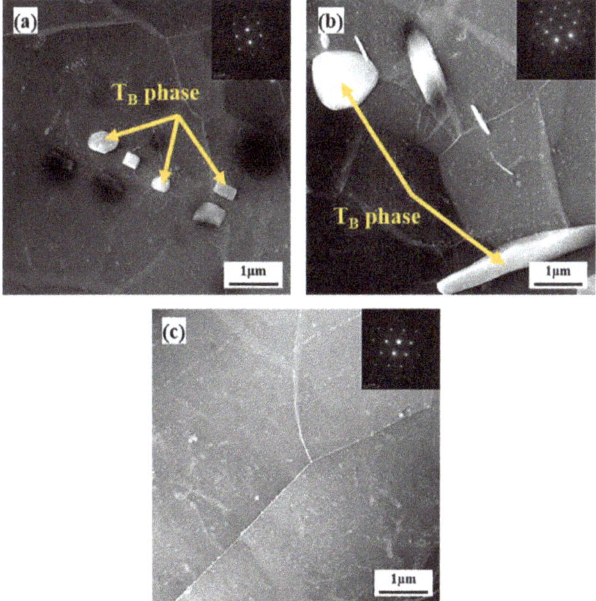

Figure 10. TEM images along <110>$_{Al}$ zone axis of the samples at dissimilar temperatures: (a) 400 °C; (b) 450 °C; (c) 500 °C.

Figure 11 shows the EDS elemental distribution of Cu when the sample is heated to 400 °C, 450 °C and 500 °C. Cu becomes increasingly dispersed with elevated temperatures, and the θ phases on the grain boundary are first coarsened at 450 °C and then dissolved at 500 °C. This indicates that while the temperature was increased from 400 °C to 450 °C, the dissolution speed of the θ phase located on the grain boundary is less than the diffusion speed of Cu to the grain boundary in the Cu-SZ, and part of the Cu in the Cu-SZ directly diffuses into the grain, so the dispersion degree of Cu increases. The distribution of Cu at 500 °C is more dispersive than that at 450 °C because T_B phases and most of the θ phases are dissolved and increase the supersaturation of Cu in the alloy matrix.

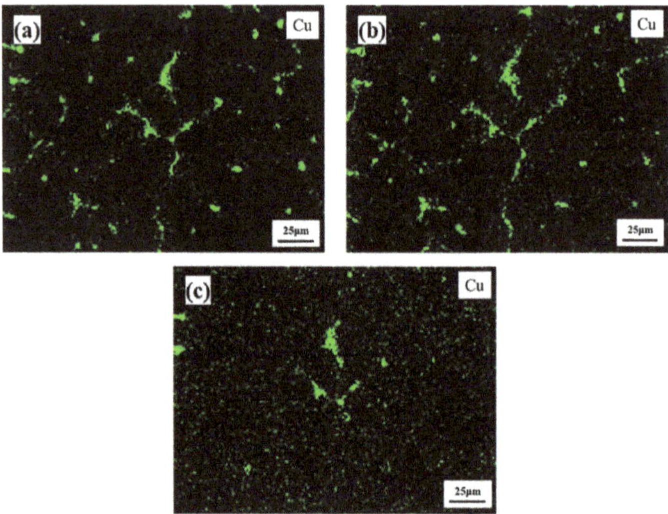

Figure 11. EDS elemental distribution of Cu with QISOZ magnification of 500 times at 400 °C (**a**), 450 °C (**b**) and 500 °C (**c**).

Figure 12 shows the elemental distributions along L1 and L2 at 400 °C, 450 °C and 500 °C. Compared with low-temperature data (see Figure 7e–f), Mg and Ag concentrations along L1 and L2 are decreased at 400 °C and the segregation of Ag disappears. When the temperature rises to 450 °C, the segregation of Mg along L1 and L2 is also removed, which is in keeping with the diminished XRD peaks in the S phases at 450 °C (see Figure 9). This suggests that diffusion of Ag and Mg from the S phase into the matrix is probably only activated above 300 °C and completed at 450 °C, transforming the S phase into the θ phase. Figure 12e,f show that the θ phases are dissolved when homogenizing at 500 °C.

At 400 °C, the diffusion coefficient of Mg in the Al matrix is 1.23×10^{-14} m^2/s, which is nearly eight times faster than that of Cu (1.59×10^{-15} m^2/s). Such a large discrepancy suggests that the diffusion of Mg from the S phase into the matrix occurs much more readily than Cu. Table 3 displays the dissolution temperature of S phases in Al-Cu-Li alloys that have dissimilar Cu/Mg ratios. When the Cu/Mg ratio increases, the dissolution temperature of the S phase decreases.

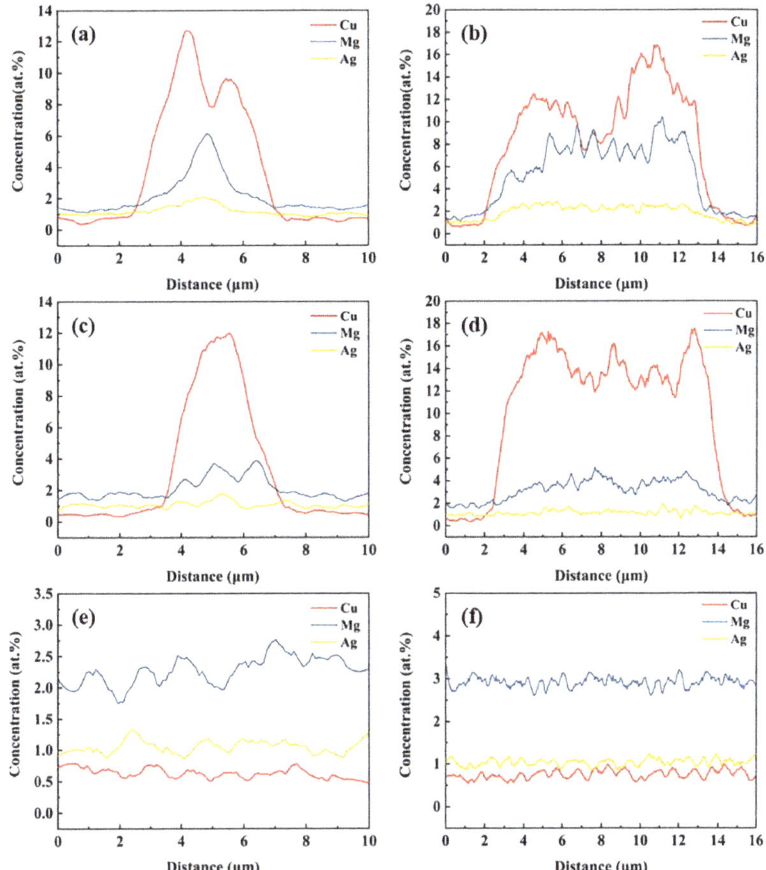

Figure 12. Elemental distributions along L1 (**a,c,e**) and L2 (**b,d,f**) at 400 °C (**a,b**), 450 °C (**c,d**) and 500 °C (**e,f**).

Table 3. Effect of Cu/Mg ratio on the dissolution temperature of S phase.

	Cu/(wt. %)	Mg/(wt. %)	Cu/Mg	T$_{(S\ dissolve)}$/°C
Li [27]	3.5	0.5	7	500
Yang [29]	3.52	0.38	9.26	495
Liu [26]	3.8	0.4	9.5	470
Present work	3.99	0.27	14.8	450

Figure 13 summarizes the evolution of the elemental concentration in the S phase during homogenization. The concentrations of Cu, Mg and Ag show almost no changes before 300 °C; from 300 °C to 400 °C, Mg and Ag decrease while Cu slightly increases, suggesting that the S phase gradually transforms into the θ phase. Upon further increasing the temperature to 450 °C, the quantity of Cu starts to decrease as the S phase is continuously transformed into the θ phase and the θ phase is gradually dissolved within this temperature range; finally, from 450 °C to 500 °C, the transformed θ phase dissolves into the matrix, leading to a sharp decrease in Cu concentration.

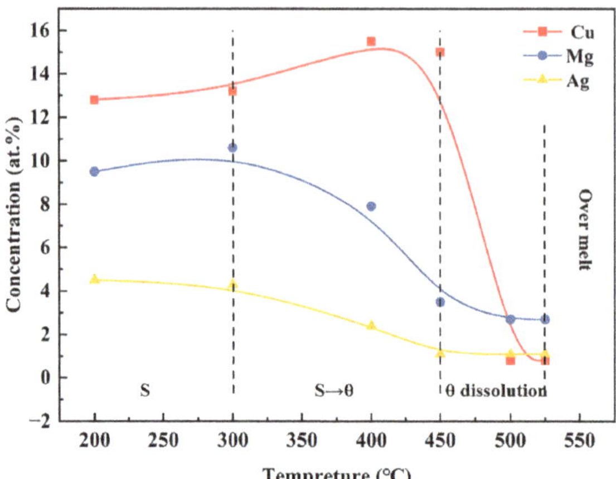

Figure 13. Evolution of elemental concentrations of Cu, Mg and Ag in S phase during homogenization.

Figure 14 shows the evolution of sample hardness during homogenization. Before 300 °C, the hardness barely changes at around 66 HV, which is consistent with the invariant phase configuration below 300 °C. When the temperature is improved from 300 °C to 450 °C, the hardness of the sample reduces from 66 HV to 54 HV, probably due to annealing-induced stress relief and recrystallization at this temperature [47]. The hardness of the sample rises from 54 HV to 71 HV when the temperature is increased to 500 °C, which is in keeping with our experimental observation (Figure 11) that most eutectic phases dissolve at this temperature and the supersaturation of Cu in the matrix has an apparent increase, leading to enhanced hardness.

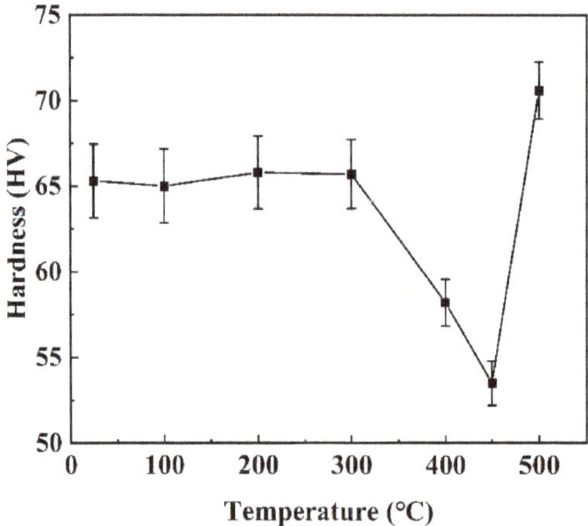

Figure 14. Evolution of sample hardness during homogenization.

Based on the above observations, Figure 15 illustrates the possible phase evolution process of the 2195 alloy during homogenization. As shown in Figure 15a, the microstructure of the sample does not alter when the temperature is elevated from 25 °C to 300 °C. When the temperature is improved to 400 °C (see Figure 15b), T_B phases precipitate from the Cu-SZ; this phenomenon was not found in previous studies. Ag and Mg in S phases diffuse into the matrix, and the segregation of Ag disappears. When further heated to 450 °C (see Figure 15c), the Cu-SZ disappears, the T_B phases and θ phases on the grain boundary are coarsened, Mg in the S phases continuously diffuses towards the matrix, transforming the S phase into the θ phase, and the newly formed θ phase starts to dissolve, which is consistent to the XRD results (see Figure 9) and aforementioned diffusivity calculation. Chen et al. [30] previously found that T_B and θ phases coarsened during the homogenization of the Al-5Cu-1Li-0.6Mg alloy, and Yang et al. [29] previously found that the S phase dissolved before the θ phase in the Al-3.52Cu-1.28Li-0.38Mg alloy. However, Chen did not distinguish the primary T_B eutectic phase from the later precipitated T_B phase, and neither of them found that the S phase transformed into the θ phase during homogenization. Some researchers have found the η-S phase transformation during the homogenization of the 7-series alloy (Al-Zn-Mg-Cu) by calculating the atomic diffusion velocity, SEM observation, TEM observation and other means [48,49], the S-θ phase transformation in this work could have similar behaviors. In further research, we will perform a more in-depth study on the S-θ phase transition, including the kinetics of such transformation and the effects of alloying elements and pretreatment parameters on this process, and apply computational simulations to study the microstructural evolution of Al-Cu-Li alloys. At 500 °C (see Figure 15d), the T_B phases located on the crystal boundary and the small-sized θ phases in the grain interior were all dissolved, and the large θ phases were partially dissolved, while the θ phases at the triangular crystal boundary dissolved more slowly.

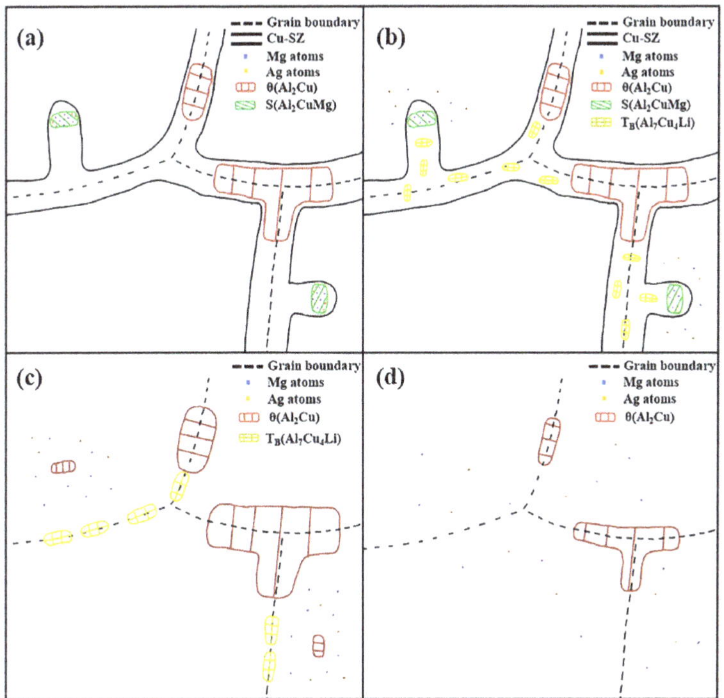

Figure 15. Schematic diagram of microstructural evolution during homogenization (**a**) 25–300 °C; (**b**) 400 °C; (**c**) 450 °C; (**d**) 500 °C.

4. Conclusions

In this paper, the evolution of the secondary phases (mainly the T_B phase and S phase), the main alloying elements and the mechanical properties of 2195 Al-Cu-Li alloys during homogenization were investigated. Cu-SZ was found at and near the grain boundary of 2195 ingot, and θ phases and S phases were distributed on Cu-SZ. When the sample was heated from 25 °C to 300 °C, the morphology and elemental distribution of the eutectic phase did not change, due to the insufficient diffusion kinetics at low temperatures. When the temperature was improved from 300 °C to 450 °C, the T_B phase first precipitated and then coarsened at the Cu-SZ due to Ostwald ripening, the Mg and Ag in the S phase diffused into the matrix, and the S phase was probably transformed into the θ phase. The hardness of the sample barely changed until heated above 300 °C. The sample was first softened when heating to 450 °C due to annealing, and then significantly increased from 450 °C to 500 °C due to the dissolution of most eutectic phases and enhanced solid-solution strengthening. The above results suggest that the homogenization temperature for the 2195 alloy can directly begin at 300 °C and should rapidly increase from 400 °C to 450 °C, avoiding the extension of the homogenization time owing to the coarsening of the θ and T_B phases on the grain boundaries. In the future, the phase evolution of 2195 Al-Cu-Li alloys will be investigated by means of computational simulations based on the experimental data.

Author Contributions: H.H.: investigation, methodology, writing–original draft. W.X.: resources, conceptualization, supervision, writing—review and editing. Z.J.: investigation, visualization, data curation. J.Z.: formal analysis, funding acquisition, data curation. All authors have read and agreed to the published version of the manuscript.

Funding: This work was financially supported by the National Key R&D Program of China (No. 2020YFA0711104), the National Natural Science Foundation of China (Grant No. U21B6004), Major Project of Scientific Innovation of Hunan Province (Grant No. 2021GK1040).

Informed Consent Statement: Not applicable.

Data Availability Statement: Not applicable.

Acknowledgments: The authors would like to take this opportunity to express their appreciation to the National Key R&D Program of China (No. 2020YFA0711104), the National Natural Science Foundation of China (Grant No. U21B6004), Major Project of Scientific Innovation of Hunan Province (Grant No. 2021GK1040).

Conflicts of Interest: The authors declare that they have no known competing financial interests or personal relationships that could have appeared to influence the work reported in this paper.

References

1. Rioja, R.J. Fabrication methods to manufacture isotropic Al-Li alloys and products for space and aerospace applications. *Mater. Sci. Eng. A* **1998**, *257*, 100–107. [CrossRef]
2. Rioja, R.J.; Liu, J. The Evolution of Al-Li Base Products for Aerospace and Space Applications. *Metall. Mater. Trans. A* **2012**, *43*, 3325–3337. [CrossRef]
3. Abd El-Aty, A.; Xu, Y.; Guo, X.; Zhang, S.H.; Ma, Y.; Chen, D. Strengthening mechanisms, deformation behavior, and anisotropic mechanical properties of Al-Li alloys: A review. *J. Adv. Res.* **2018**, *10*, 49–67. [CrossRef] [PubMed]
4. Gupta, R.K.; Nayan, N.; Nagasireesha, G.; Sharma, S.C. Development and characterization of Al–Li alloys. *Mater. Sci. Eng. A* **2006**, *420*, 228–234. [CrossRef]
5. Xu, J.; Deng, Y.; Chen, J. Enhancing the Corrosion Resistance of Al-Cu-Li Alloys through Regulating Precipitation. *Materials* **2020**, *13*, 2628. [CrossRef]
6. Yang, X.; Zhang, L.; Sobolev, S.; Du, Y. Kinetic Phase Diagrams of Ternary Al-Cu-Li System during Rapid Solidification: A Phase-Field Study. *Materials* **2018**, *11*, 260. [CrossRef] [PubMed]
7. Yu, N.; Shang, J.; Cao, Y.; Ma, D.; Liu, Q. Comparative Analysis of Al-Li Alloy and Aluminum Honeycomb Panel for Aerospace Application by Structural Optimization. *Math. Probl. Eng.* **2015**, *2015*, 815257. [CrossRef]
8. Dursun, T.; Soutis, C. Recent developments in advanced aircraft aluminium alloys. *Mater. Des.* **2014**, *56*, 862–871. [CrossRef]
9. Yu, X.; Zhao, Z.; Shi, D.; Dai, H.; Sun, J.; Dong, X. Enhanced High-Temperature Mechanical Properties of Al-Cu-Li Alloy through T1 Coarsening Inhibition and Ce-Containing Intermetallic Refinement. *Materials* **2019**, *12*, 1521. [CrossRef] [PubMed]
10. Wang, X.; Jiang, J.; Li, G.; Shao, W.; Zhen, L. Precipitation during Quenching in 2A97 Aluminum Alloy and the Influences from Grain Structure. *Materials* **2021**, *14*, 2802. [CrossRef] [PubMed]
11. Zhang, J.; Jiang, Z.; Xu, F.; Chen, M. Effects of Pre-Stretching on Creep Behavior, Mechanical Property and Microstructure in Creep Aging of Al-Cu-Li Alloy. *Materials* **2019**, *12*, 333. [CrossRef]
12. Zhang, W.; Mao, Y.; Yang, P.; Li, N.; Ke, L.; Chen, Y. Effect of Welding Speed on Microstructure Evolution and Mechanical Properties of Friction Stir Welded 2198 Al-Cu-Li Alloy Joints. *Materials* **2022**, *15*, 969. [CrossRef] [PubMed]
13. Mori, K.-i.; Ishiguro, M.; Isomura, Y. Hot shear spinning of cast aluminium alloy parts. *J. Mater. Process. Technol.* **2009**, *209*, 3621–3627. [CrossRef]

14. Fan, X.; He, Z.; Lin, P.; Yuan, S. Microstructure, texture and hardness of Al-Cu-Li alloy sheet during hot gas forming with integrated heat treatment. *Mater. Des.* **2016**, *94*, 449–456. [CrossRef]
15. Deng, Y.; Xu, J.; Chen, J.; Guo, X. Effect of double-step homogenization treatments on the microstructure and mechanical properties of Al-Cu-Li–Zr alloy. *Mater. Sci. Eng. A* **2020**, *795*, 139975. [CrossRef]
16. Liu, Q.; Fan, G.; Tan, Z.; Li, Z.; Zhang, D.; Wang, J.; Zhang, H. Precipitation of Al3Zr by two-step homogenization and its effect on the recrystallization and mechanical property in 2195 Al-Cu-Li alloys. *Mater. Sci. Eng. A* **2021**, *821*, 141637. [CrossRef]
17. Tsivoulas, D.; Robson, J.D. Heterogeneous Zr solute segregation and Al$_3$Zr dispersoid distributions in Al-Cu-Li alloys. *Acta Mater.* **2015**, *93*, 73–86. [CrossRef]
18. Malikov, A.; Orishich, A.; Vitoshkin, I.; Bulina, N.; Karpov, E.; Gutakovskii, A.; Batsanov, S.; Ancharov, A.; Tabakaev, R. Effect of the structure and the phase composition on the mechanical properties of Al-Cu-Li alloy laser welds. *Mater. Sci. Eng. A* **2021**, *809*, 140947. [CrossRef]
19. Zhang, S.; Li, M.; Wang, H.; Cheng, W.; Lei, W.; Liu, Y.; Liang, W. Microstructure and Tensile Properties of ECAPed Mg-9Al-1Si-1SiC Composites: The Influence of Initial Microstructures. *Materials* **2018**, *11*, 136. [CrossRef]
20. Wang, J.; Lu, Y.; Zhou, D.; Sun, L.; Li, R.; Xu, W. Influence of Homogenization on Microstructural Response and Mechanical Property of Al-Cu-Mn Alloy. *Materials* **2018**, *11*, 914. [CrossRef]
21. Schöbel, M.; Baumgartner, G.; Gerth, S.; Bernardi, J.; Hofmann, M. Microstresses and crack formation in AlSi7MgCu and AlSi17Cu4 alloys for engine components. *Acta Mater.* **2014**, *81*, 401–408. [CrossRef]
22. Han, B.; Chen, Y.; Tao, W.; Li, H.; Li, L. Microstructural evolution and interfacial crack corrosion behavior of double-sided laser beam welded 2060/2099 Al-Li alloys T-joints. *Mater. Des.* **2017**, *135*, 353–365. [CrossRef]
23. Xiang, K.; Lei, X.; Ding, L.; Jia, Z.; Yang, X.; Liu, Q. Optimizing mechanical property of spray formed Al-Zn-Mg-Cu alloy by combination of homogenization and warm-rolling. *Mater. Sci. Eng. A* **2022**, *846*, 143248. [CrossRef]
24. Aal, M.I.A.E. Influence of the pre-homogenization treatment on the microstructure evolution and the mechanical properties of Al–Cu alloys processed by ECAP. *Mater. Sci. Eng. A* **2011**, *528*, 6946–6957. [CrossRef]
25. Li, L.; Cao, H.; Qi, F.; Wang, Q.; Zhao, N.; Liu, Y.; Ye, X.; Ouyang, X. Effect of Heat Treatment on Microstructure and Mechanical Properties of Mg-5Zn-1Mn Alloy Tube. *Metals* **2020**, *10*, 301. [CrossRef]
26. Liu, Q.; Zhu, R.-h.; Li, J.-f.; Chen, Y.-l.; Zhang, X.-h.; Zhang, L.; Zheng, Z.-q. Microstructural evolution of Mg, Ag and Zn micro-alloyed Al-Cu-Li alloy during homogenization. *Trans. Nonferrous Met. Soc. China* **2016**, *26*, 607–619. [CrossRef]
27. Li, H.; Yu, W.; Wang, X.; Du, R.; You, W. Investigation on Microstructural Evolution and Properties of an Al-Cu-Li Alloy with Mg and Zn Microalloying during Homogenization. *Metals* **2018**, *8*, 1010. [CrossRef]
28. Li, H.-Y.; Su, X.-J.; Yin, H.; Huang, D.-S. Microstructural evolution during homogenization of Al-Cu-Li-Mn-Zr-Ti alloy. *Trans. Nonferrous Met. Soc. China* **2013**, *23*, 2543–2550. [CrossRef]
29. Shengli, Y.; Jian, S.; Xiaodong, Y.; Xiwu, L.; Fei, Z.; Baoqing, S. Homogenization Treatment Parameter Optimization and Microstructural Evolution of Al-Cu-Li Alloy. *Rare Met. Mater. Eng.* **2017**, *46*, 0028–0034. [CrossRef]
30. Chen, W.; Zhang, W.; Ding, D.; Xiao, D. Microstructure of Al-5Cu-1Li-0.6Mg-0.5Ag-0.5Mn Alloys. *Metals* **2021**, *11*, 37. [CrossRef]
31. Du, Q.; Jia, L.; Tang, K.; Holmedal, B. Modelling and experimental validation of microstructure evolution during the cooling stage of homogenization heat treatment of Al–Mg–Si alloys. *Materialia* **2018**, *4*, 70–80. [CrossRef]
32. Hu, Z.; Li, P.; Lu, Z.; Ma, B.; Ding, J.; Zhao, Y.; Tang, P.; Huang, Q. DICTRA (R) Simulation of Mg and Mn Micro-segregations in Al-Mg5. 0-Mn0. 5 Alloy During Homogenization Annealing. *Rare Met. Mater. Eng.* **2021**, *50*, 23–28.
33. Ammarullah, M.I.; Afif, I.Y.; Maula, M.I.; Winarni, T.I.; Tauviqirrahman, M.; Jamari, J. Tresca stress evaluation of Metal-on-UHMWPE total hip arthroplasty during peak loading from normal walking activity. *Mater. Today Proc.* **2022**, *63*, S143–S146. [CrossRef]
34. Weng, Y.; Jia, Z.; Ding, L.; Liu, M.; Wu, X.; Liu, Q. Combined effect of pre-aging and Ag/Cu addition on the natural aging and bake hardening in Al-Mg-Si alloys. *Prog. Nat. Sci. Mater. Int.* **2018**, *28*, 363–370. [CrossRef]
35. Mukhopadhyay, A.K.; Rao, V.V.R. Characterization of S (Al$_2$CuMg) phase particles present in as-cast and annealed Al–Cu–Mg(–Li)–Ag alloys. *Mater. Sci. Eng. A* **1999**, *268*, 8–14. [CrossRef]
36. Guo, C.; Zhang, H.; Li, J. Influence of Zn and/or Ag additions on microstructure and properties of Al-Mg based alloys. *J. Alloys Compd.* **2022**, *904*, 163998. [CrossRef]
37. Guo, C.; Zhang, H.; Wu, Z.; Shen, X.; Wang, P.; Li, B.; Cui, J.; Nagaumi, H. An atomic-resolution investigation of precipitation evolution in Al-Mg-Ag alloys. *Mater. Lett.* **2019**, *248*, 231–235. [CrossRef]
38. Reich, L.; Murayama, M.; Hono, K. Evolution of Ω phase in an Al–Cu–Mg–Ag alloy—A three-dimensional atom probe study. *Acta Mater.* **1998**, *46*, 6053–6062. [CrossRef]
39. Gumbmann, E.; Lefebvre, W.; De Geuser, F.; Sigli, C.; Deschamps, A. The effect of minor solute additions on the precipitation path of an Al Cu Li alloy. *Acta Mater.* **2016**, *115*, 104–114. [CrossRef]
40. Jensrud, O.; Ryum, N. The development of microstructures in Al-Li alloys. *Mater. Sci. Eng.* **1984**, *64*, 229–236. [CrossRef]
41. XIE, F.-Y.; KRAFT, T.; ZUO, Y.; MOON, C.-H.; CHANG, Y.A. Microstructure and microsegregation in Al-rich Al–Cu–Mg alloys. *Acta Mater.* **1999**, *47*, 489–500. [CrossRef]
42. Lifshitz, I.M.; Slyozov, V.V. The kinetics of precipitation from supersaturated solid solutions. *J. Phys. Chem. Solids* **1961**, *19*, 35–50. [CrossRef]

43. Gu, B.P.; Liedl, G.L.; Kulwicki, J.H.; Sanders, T.H. Coarsening of δ' (Al3Li) precipitates in an Al-2.8Li0.3Mn alloy. *Mater. Sci. Eng.* **1985**, *70*, 217–228. [CrossRef]
44. Wu, G.; Zhang, X.; Zhang, L.; Wang, Y.; Shi, C.; Li, P.; Ren, G.; Ding, W. An insight into the precipitate evolution and mechanical properties of a novel high-performance cast Al-Li-Cu-Mg-X alloy. *J. Alloys Compd.* **2021**, *875*, 159996. [CrossRef]
45. Ye, L.; Liu, X.; Tang, J.; Liu, S.; Zhang, Y. Ostwald ripening and stability of precipitates during two successive overaging in an Al-Mg-Li alloy. *Mater. Lett.* **2021**, *291*, 129616. [CrossRef]
46. Wang, Y.; Ma, X.; Zhao, G.; Xu, X.; Chen, X.; Zhang, C. Microstructure evolution of spray deposited and as-cast 2195 Al-Li alloys during homogenization. *J. Mater. Sci. Technol.* **2021**, *82*, 161–178. [CrossRef]
47. Shen, J.; Chen, B.; Wan, J.; Shen, J.; Li, J. Effect of annealing on microstructure and mechanical properties of an Al–Mg-Sc-Zr alloy. *Mater. Sci. Eng. A* **2022**, *838*, 142821. [CrossRef]
48. Xu, D.; Li, Z.; Wang, G.; Li, X.; Lv, X.; Zhang, Y.a.; Fan, Y.; Xiong, B. Phase transformation and microstructure evolution of an ultra-high strength Al-Zn-Mg-Cu alloy during homogenization. *Mater. Charact.* **2017**, *131*, 285–297. [CrossRef]
49. Liu, Y.; Jiang, D.; Xie, W.; Hu, J.; Ma, B. Solidification phases and their evolution during homogenization of a DC cast Al–8.35Zn–2.5Mg–2.25Cu alloy. *Mater. Charact.* **2014**, *93*, 173–183. [CrossRef]

Article

Interface Microstructure and Mechanical Properties of Al/Steel Bimetallic Composites Fabricated by Liquid-Solid Casting with Rare Earth Eu Additions

Feng Mao [1,2,*,†], Po Zhang [3,†], Shizhong Wei [1,*], Chong Chen [1,2], Guoshang Zhang [2,3], Mei Xiong [1], Tao Wang [3], Junliang Guo [3] and Changji Wang [1,2]

[1] National Joint Engineering Research Center for Abrasion Control and Molding of Metal Materials, Henan University of Science and Technology, Luoyang 471003, China
[2] Longmen Laboratory, Luoyang 471000, China
[3] School of Materials Science and Engineering, Henan University of Science and Technology, Luoyang 471003, China
* Correspondence: maofeng718@163.com (F.M.); hnwsz@126.com (S.W.); Tel.: +86-183-3671-2258 (F.M.); +86-138-0379-2970 (S.W.)
† These authors contributed equally to this work and should be considered co-first authors.

Abstract: To improve the Al/Steel bimetallic interface, Eu was firstly added to the Al/Steel bimetallic interface made by liquid-solid casting. The effects of Eu addition on the microstructure, mechanical capacities, and rupture behavior of the Al/Steel bimetallic interface was studied in detail. As the addition of 0.1 wt.% Eu, the morphology of eutectic Si changed from coarse plate-like to fine fibrous and granular in Al-Si alloys, and the average thickness of the intermetallic compounds layer decreased to a minimum value of 7.96 μm. In addition, there was a more sudden drop of Fe in steel side and the Si in Al side was observed to be more than the other conditions. The addition of Eu did not change the kinds of intermetallic compounds in the Al/steel reaction layer, which was composed of Al_5Fe_2, τ_1-(Al, Si)$_5$Fe$_3$, $Al_{13}Fe_4$, τ_5-Al$_7$Fe$_2$Si, and τ_6-Al$_9$Fe$_2$Si$_2$ phases. The addition of the element Eu did not change the preferential orientation of the Al_5Fe_2, τ_1-(Al, Si)$_5$Fe$_3$, $Al_{13}Fe_4$, τ_5-Al$_7$Fe$_2$Si, and τ_6-Al$_9$Fe$_2$Si$_2$ phases, but refined the grain size of each phase and decreased the polar density of Al_5Fe_2 phase. Eu was mainly enriched in the front of the ternary compound layer (τ_6-Al$_9$Fe$_2$Si$_2$) near the Al side and steel matrix. The Fe and Al element distribution area tended to narrow in the interface after the addition of 0.1 wt.% Eu, which is probably because that Eu inhibits the spread of Al atoms along the c-axis direction of the Al_5Fe_2 phase and the growth of $Al_{13}Fe_4$, τ_5-Al$_7$Fe$_2$Si, and τ_6-Al$_9$Fe$_2$Si$_2$ phases. When the Eu content was 0.1 wt.%, the shear strength of the Al/Steel bimetal achieved a maximum of 31.21 MPa, which was 47% higher than the bimetal without Eu.

Keywords: Al/steel bimetal; bonding; Eu; compound casting; EBSD; EPMA

Citation: Mao, F.; Zhang, P.; Wei, S.; Chen, C.; Zhang, G.; Xiong, M.; Wang, T.; Guo, J.; Wang, C. Interface Microstructure and Mechanical Properties of Al/Steel Bimetallic Composites Fabricated by Liquid-Solid Casting with Rare Earth Eu Additions. *Materials* **2022**, *15*, 6507. https://doi.org/10.3390/ma15196507

Academic Editors: Hailiang Yu, Zhilin Liu and Xiaohui Cui

Received: 3 September 2022
Accepted: 16 September 2022
Published: 20 September 2022

Publisher's Note: MDPI stays neutral with regard to jurisdictional claims in published maps and institutional affiliations.

1. Introduction

In recent years, Al/steel bimetallic has been widely studied [1–6] because it combines the excellent characteristics of aluminum alloy and steel [7–14], such as high intensity of steel and lower density of aluminum alloy. Therefore, it is widely used in the field of fuel economy and is lightweight. The liquid-solid composite method is a frequent method for the preparation of Al/Steel bimetallic composites because of its simple preparation process and low requirements in material selection. Due to the huge difference in chemical and physical properties of aluminum and steel, the poor wettability of liquid aluminum to steel and the easy oxidation of steel are major challenges for Al/Steel composite casting [15]. However, the formation of brittle intermetallic compounds (Al_5Fe_2) in the reaction region is another difficulty for Al/steel composites [16–18].

Current research has focused on improving the Al/steel bimetallic interface by adding different types of intermediate coatings between aluminum and steel to prevent the formation of Al-Fe intermetallic compounds. Jiang et al. [19–21] found that hot-dipped aluminum or zinc plating on the steel surface could improve the wettability of liquid aluminum on the steel surface, which can enhance the bonding strength of the Al/Steel bimetallic interface. Some scholars have tried to decrease the thickness of the brittle intermetallic phase of Al/steel bimetals by adding alloying elements to the hot-dipped aluminum melt. Cheng et al. [22] investigated the effect of hot dip coating of aluminum alloys with different Si contents on the intermetallic compound layer. The results indicated that the solid solution of Si in the Al_5Fe_2 phase increases with the increase of Si content. The Si atoms occupied their vacant position on the C-axis of the Al_5Fe_2 phase forming a ternary Al-Fe-Si intermetallic (Al_7Fe_2Si), which reduced the spread rate of Al atoms to Al_5Fe_2 and caused a gradual decrease in the thickness of Al_5Fe_2. Chen et al. [23] found that the addition of Ni element into the hot dip coating of aluminum alloy melt could reduce the growth rate of the Al_5Fe_2 layer, which effectively controlled the growth of Al_5Fe_2 hard and brittle phase.

Rare earth elements have long been used to refine and strengthen alloys, which can modify the eutectic Si and refine the α-Al grains in Al-Si alloys [24–27]. Li et al. [28] investigated the variation of eutectic Si in A356 alloys with various Eu contents. The eutectic Si morphology of A356 alloys was transformed from flakes to fibers with the increase of Eu content, which substantially enhanced the mechanical properties. Muhammad et al. [29] showed that at the Sc content of 0.4 wt.% Sc, the eutectic Si in the A357 alloy changed from acicular to fibers and the grain size in the alloy was reduced by 80%, which make the ultimate tensile strength increase by 28%. Shi et al. [30] showed that the addition of 0.3 wt.% Er had a significant influence on the refinement of α-Al grains and the morphological changes of eutectic Si. Li et al. [31] indicated the effect of adding different contents of Y on the microstructure of Al-7Si-0.5 Mg alloy. It was found that the eutectic Si morphology changed from sheet-like to fine-branched, and the tensile properties were significantly enhanced. From previous research work, the study of RE modification has mainly focused on a single alloy; there are very few studies on the influence of RE addition on the Al/steel bimetallic interface.

Eu was the only rare earth element capable of producing fibrous eutectic Si in Al-Si alloy, which maintains similarity with Sr and Na. However, the effect of rare earth Eu on the interface microstructure and mechanical properties of Al/steel bimetallic composites has not yet been widely investigated. In this paper, rare earth element Eu was first added to the liquid-solid composite of Al/steel bimetal, and the interface microstructures of Al/Steel bimetallic composites and the effect mechanism was studied by SEM, EPMA and EBSD. Then, the effect of Eu on the mechanical capacities and rupture behavior of the Al/steel bimetallic interface was also studied. Many results are firstly reported and are very novel.

2. Experimental

2.1. Materials

The Al-7Si alloys with different Eu contents alloy and 45 steel were respectively selected as casting alloys and solid matrix materials to produce Al/Steel bimetallic, and the Al-7Si alloys with different Eu contents were also used as hot-dip aluminum plating materials for 45 steel matrix. The chemical composition of the experimental materials were listed in Table 1.

2.2. Experimental Procedure

The steel substrates had a diameter of 36 mm, a height of 115 mm, and a wall thickness of 3 mm. Before the experiment, the steel substrates were ground by silicon carbide paper and then were immersed in 15 wt.% sodium hydroxide solution at 45 °C for 20 min to remove oil contamination from the steel substrate surface. Next, the steel substrates were immersed in 10 wt.% hydrochloric acid to remove rust from the surface of the steel substrates, and finally were soaked in 5 wt.%K_2ZrF_6 solution at 90 °C for 10 min. Hot dipping aluminum was firstly conducted by immersing steel pipe 45 into Al-7Si alloy with different Eu contents melted at 730 °C for 5 min. Subsequently, steel pipe after hot dipped aluminum was quickly put into a metal mold with a preheated temperature of 400 °C, and the Al-7Si alloys melt of the same composition as the hot-dip alloys with a temperature of 730 °C were poured into the metal mold. Al/Steel bimetallic composites were finally obtained after casting solidification. Figure 1 shows a schematic diagram of bimetallic casting.

Table 1. Chemical compositions of the experimental materials (wt.%).

Alloys	Compositions							
	Al	Si	Eu	Fe	Cu	C	Mn	Cr
45 steel	-	0.24	-	Bal.	0.21	0.45	0.57	0.17
Al-7Si	Bal.	6.97	0	0.07	<0.01	-	<0.01	-
Al-7Si-0.05Eu	Bal.	6.98	0.047	0.06	<0.01	-	<0.01	-
Al-7Si-0.1Eu	Bal.	7.03	0.09	0.06	<0.01	-	<0.01	-
Al-7Si-0.2Eu	Bal.	6.95	0.198	0.08	<0.01	-	<0.01	-

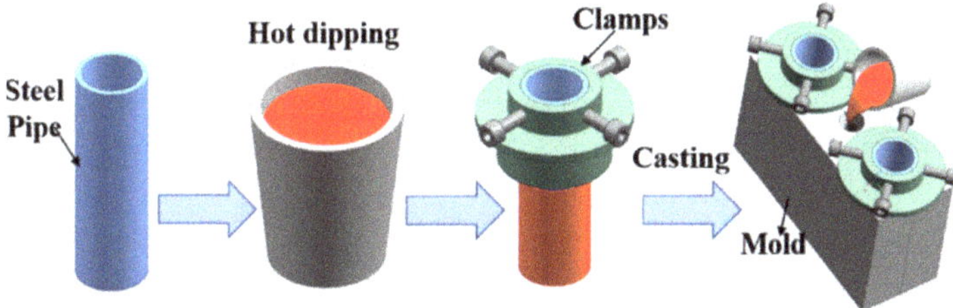

Figure 1. Schematic diagram of bimetallic casting.

2.3. Microstructural Characterizations

The interface microstructure of Al/Steel bimetallic castings was observed using JSM-IT800 (JEOL, Tokyo, Japan) scanning electron microscope (SEM) with energy-dispersive X-ray spectroscopy (EDS) and an optical microscopy (OM). Samples were ion-polished by an argon ion polishing instrument (JEOL IB-19530CP) (JEOL, Tokyo, Japan) at 6 kV for 30 min and then analyzed using EBSD analysis (Oxford C-nano) (Oxford, London, UK) for the verification of the phases formed at the Al/Steel bimetallic interface. Electron probe microanalysis (EPMA; JXA-8230) (JEOL, Tokyo, Japan) was performed to analyze the elemental distribution at the Al/steel bimetallic interface. The thickness of the intermetallic compounds at the Al/Steel bimetallic interface was calculated using the Image Pro software. To reduce the error, five SEM photographs of each Eu content and 10 places were measured in each photograph.

2.4. Mechanical Characterizations

The shear strength of the specimens was measured using a WDW-300 microcomputer-controlled electronic universal testing machine. As seen in Figure 2, the shear sample was fixed on the testing machine. The indenter was allowed to make contact with the test piece, which was then loaded from top to bottom at a loading speed of 1 mm/min so that the test piece broke along the interface of the Al/Steel composite. To ensure the precision of the test results, five samples were selected for the shear strength test in each group of processes, and the average values were considered as the shear strength of the samples.

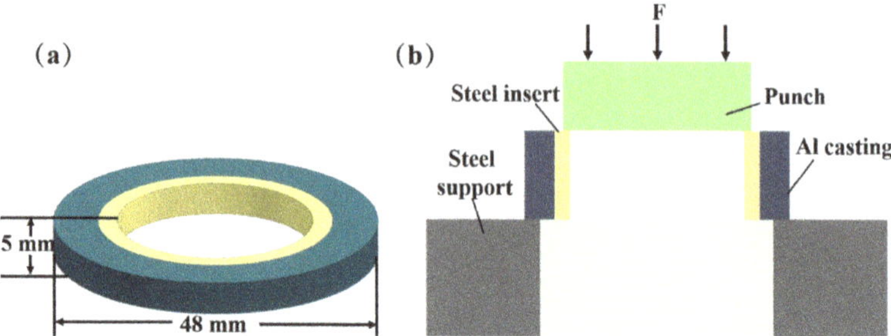

Figure 2. Size and shear diagram of shear specimen: (**a**) Shear specimen size (**b**) Shear diagram.

3. Results

3.1. Effect of Eu on Eutectic Si of Al-Si Alloys

Figure 3 reveals the morphology of eutectic Si and elements distributions in Al-7Si alloys with different Eu contents. Due to the small solid solution of Si in Al, the Si element was mainly present in the form of plate-like eutectic silicon in the Al matrix, as indicated in Figure 3a–c, which easily splits the alloy matrix and thus severely weakens its mechanical properties. As seen in Figure 3d–f, after the addition of 0.1 wt.% Eu, the morphology of the eutectic silicon in the alloy changed significantly from plate-like to fine fibers. In addition, the distribution of eutectic silicon became more homogeneous and continuous with each other.

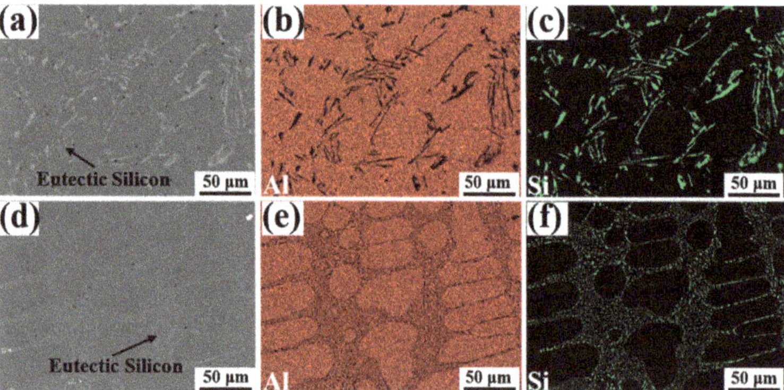

Figure 3. The morphology of eutectic Si and elements distributions of Al and Si elements in Al-7Si alloys with different Eu contents: (**a–c**) 0 wt.% Eu; (**d–f**) 0.1 wt.% Eu.

3.2. Effect of Eu Contents on Interface Microstructures of Al/Steel Bimetallic Composites

Figure 4 depicts the SEM micrographs of the Al/Steel interface with different Eu contents, showing a uniform and dense intermediate layer at the interface. As presented in Figure 4a–c, the thickness of the intermetallic compound layer decreased as the Eu content increased from 0 wt.% to 0.1 wt.%, and the intermetallic compound layer near the steel substrate side became gentle. With the Eu content increasing from 0.1 wt.% to 0.2 wt.%, the thickness of the intermetallic compound layer began to increase, and the undulating shape intermetallic compound layer again appeared on the steel substrate side. The measured results of the average thickness of the intermetallic compound layer at different Eu contents are shown in Figure 5. The average thickness of the intermetallic compound layer reached a minimum value of 7.96 μm at 0.1 wt.% Eu.

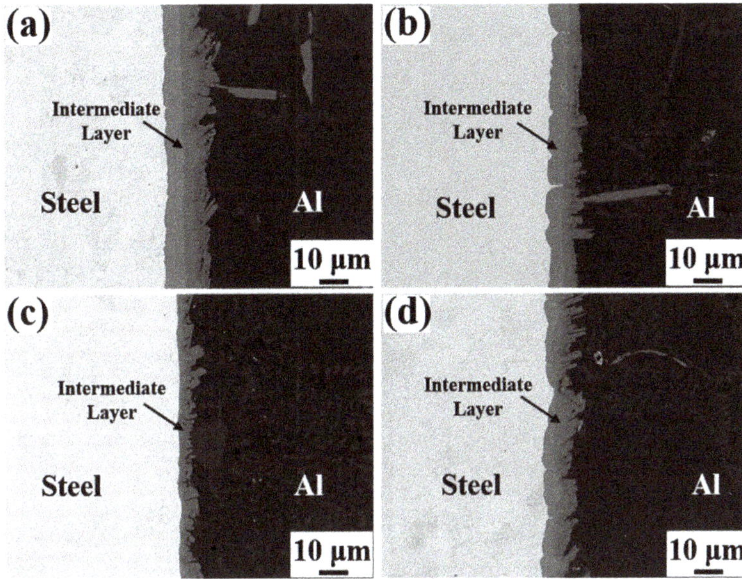

Figure 4. SEM micrographs of Al/Steel bimetallic interfaces with different Eu contents: (**a**) 0 wt.% Eu; (**b**) 0.05 wt.% Eu; (**c**) 0.1 wt.% Eu; (**d**) 0.2 wt.% Eu.

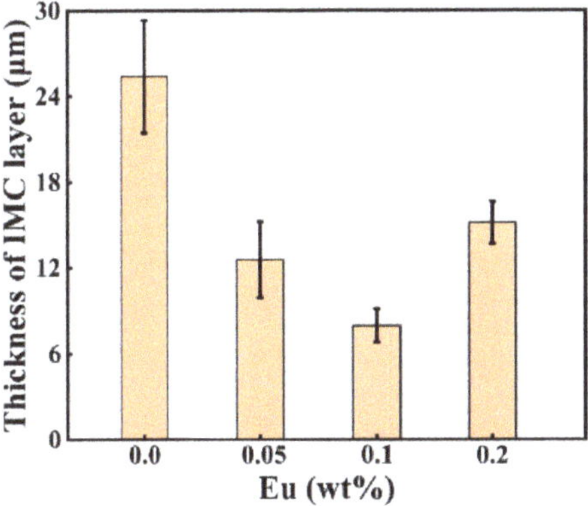

Figure 5. Intermediate layer thickness at the Al/Steel bimetallic interface with different Eu contents.

Figure 6 displays the magnified SEM micrographs and EDS line analysis of the Al/Steel bimetallic interface with different Eu contents. The obvious delamination was observed in the intermediate layers of bimetallic interfaces with different Eu contents, indicating that there are many kinds of Al-Fe-Si binary or ternary compounds in the interlayer, as displayed in Figure 6a–d. To identify the chemical composition of each point in the interface of Figure 6a–d, Tables 2–5 show the results of energy dispersive spectrometer (EDS) point analysis in the interface. The Al_5Fe_2, τ_1-$(Al, Si)_5Fe_3$, $Al_{13}Fe_4$, τ_5-Al_7Fe_2Si, and τ_6-$Al_9Fe_2Si_2$ phases were found in the intermetallic layers from the steel matrix to the Al matrix in the Al/steel bimetallic interface. In addition, the thickness of the Al_5Fe_2 layer at the interface decreased with the addition of Eu, as seen in Figure 6a–d. The EDS line analysis indicates that Al, Si, Fe, and Eu elements are diffused at the interface and the fluctuation of the element distribution line in the interface becomes gentle after adding Eu, as depicted in Figure 6e–h. At 0.1 wt.% Eu, there was a more sudden drop of Fe in steel side and the Si in

Al side was observed to be more than the other conditions, which is because that the Al_5Fe_2 and $Al_{13}Fe_4$ phases are relatively thin, and the intermetallic layer is mainly consisted of $\tau_5\text{-}Al_7Fe_2Si$ and $\tau_6\text{-}Al_9Fe_2Si_2$ phases.

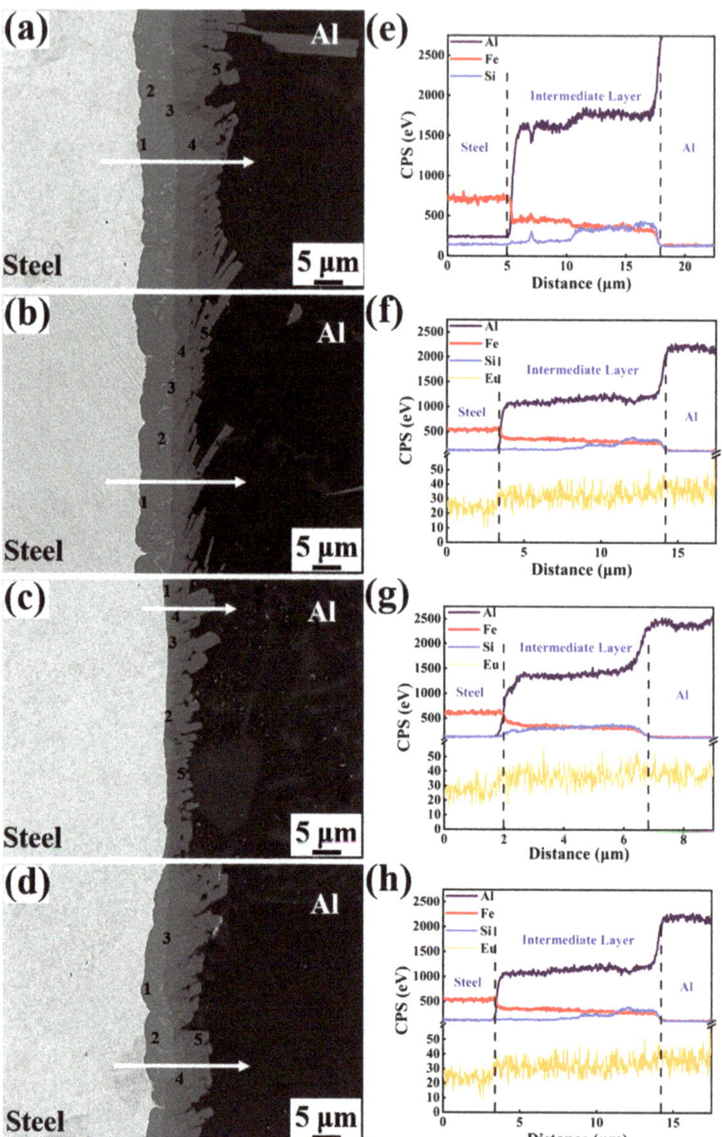

Figure 6. SEM micrographs and EDS line analysis of the Al/Steel bimetallic interface with different Eu contents: (**a**,**e**) 0 wt.% Eu; (**b**,**f**) 0.05 wt.% Eu; (**c**,**g**) 0.1 wt.% Eu; (**d**,**h**) 0.2 wt.% Eu.

Table 2. Energy spectrum analysis results of each point in Figure 6a.

Points	Compositions (at%)			Possible Phase
	Al	Fe	Si	
1	68.62	28.74	2.64	Al_5Fe_2
2	48.18	37.71	14.11	τ_1-(Al, Fe)$_5$Si$_3$
3	73.93	23.34	2.73	$Al_{13}Fe_4$
4	68.95	20.86	10.19	τ_5-Al$_7$Fe$_2$Si
5	70.69	16.03	13.28	τ_6-Al$_9$Fe$_2$Si$_2$

Table 3. Energy spectrum analysis results of each point in Figure 6b.

Points	Compositions (at%)				Possible Phase
	Al	Fe	Si	Eu	
1	69.16	29.44	1.39	0.01	Al_5Fe_2
2	46.37	39.03	14.58	0.02	τ_1-(Al, Fe)$_5$Si$_3$
3	71.01	24.34	3.99	0.06	$Al_{13}Fe_4$
4	67.09	21.71	11.12	0.08	τ_5-Al$_7$Fe$_2$Si
5	69.49	15.24	15.08	0.19	τ_6-Al$_9$Fe$_2$Si$_2$

Table 4. Energy spectrum analysis results of each point in Figure 6c.

Points	Compositions (at%)				Possible Phase
	Al	Fe	Si	Eu	
1	67.11	29.75	3.14	0.00	Al_5Fe_2
2	53.89	34.95	11.15	0.01	τ_1-(Al, Fe)$_5$Si$_3$
3	70.74	25.67	3.54	0.05	$Al_{13}Fe_4$
4	68.37	20.64	10.89	0.1	τ_5-Al$_7$Fe$_2$Si
5	67.48	15.56	16.32	0.64	τ_6-Al$_9$Fe$_2$Si$_2$

Table 5. Energy spectrum analysis results of each point in Figure 6d.

Points	Compositions (at%)				Possible Phase
	Al	Fe	Si	Eu	
1	68.18	29.02	2.79	0.01	Al_5Fe_2
2	45.51	15.25	39.18	0.06	τ_1-(Al, Fe)$_5$Si$_3$
3	73.95	23.57	2.41	0.07	$Al_{13}Fe_4$
4	67.69	20.94	11.19	0.18	τ_5-Al$_7$Fe$_2$Si
5	65.88	16.91	16.78	0.43	τ_6-Al$_9$Fe$_2$Si$_2$

The thermo-Cal software was used to analyze the phase composition of the Al/Steel bimetallic interface based on the thermodynamic database of the Al-Fe-Si system established by Du et al. [32]. Figure 7 indicates the calculated isothermal sections of the Al-Fe-Si system at 730 °C and 650 °C. It can be seen that there are many kinds of binary and ternary compounds in the Al-Fe-Si system, which the τ_6-Al$_9$Fe$_2$Si$_2$ phase is formed below 650 °C. According to the experimental results, the diffusion paths of Al-7Si alloy and 45 steel were plotted with red lines in the isothermal sections of the Al-Fe-Si system at 730 °C and 650 °C calculated by Thermo-Calc software, as illustrated in Figure 7. The τ_6-Al$_9$Fe$_2$Si$_2$, τ_5-Al$_7$Fe$_2$Si, Al$_{13}$Fe$_4$, Al$_5$Fe$_2$, and τ_1-(Al, Si)$_5$Fe$_3$ phases are formed from Al side to steel side due to the mutual diffusion of elements, which is consistent with the results of SEM and EDS point analysis in Figure 6 and Tables 2–5.

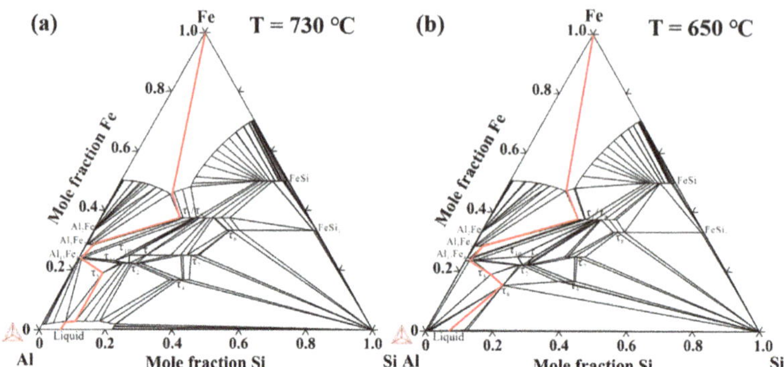

Figure 7. Isothermal sections of Al-Fe-Si system at (**a**) 730 °C and (**b**) 650 °C along with the diffusion path of Al-Si/Fe bimetal.

In order to accurately identify the intermetallic phase [33–37], the EBSD phase distribution diagram of the intermetallic layer with different Eu contents are shown in Figure 8. It can be determined that the phases from the steel side to the Al side were Al_5Fe_2 layer, dispersed τ_1-(Al, Si)$_5$Fe$_3$ in Al_5Fe_2, $Al_{13}Fe_4$ layer, τ_5-Al$_7$Fe$_2$Si layer, and τ_6-Al$_9$Fe$_2$Si$_2$ layer in the Al/Steel bimetallic interface layer without Eu addition and with 0.2 wt.% Eu content. The phase type of the intermetallic compound layer did not change with the addition of rare earth Eu, which is also consistent with the results of SEM and EDS point analysis in Figure 6 and Tables 2–5.

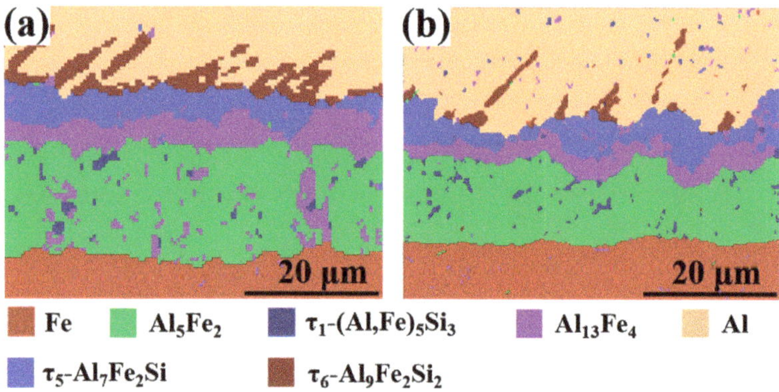

Figure 8. EBSD phase distribution diagram of the intermetallic layer with different Eu contents: (**a**) 0 wt.% Eu; (**b**) 0.2 wt.% Eu.

3.3. Effect Mechanism of Eu on Liquid-Solid Al/Steel Bimetallic Interface

Figure 9 shows the Al, Fe, Si, and Eu elements distribution of the Al/Steel bimetallic interface with the addition of 0 wt.% and 0.1 wt.% Eu contents. It can be from Figure 9b,c,f,g, that the Fe content decreased from the steel matrix to the aluminum matrix, which was the highest in the Al_5Fe_2 phase of the intermetallic compound layer. The Al content decreased gradually from the aluminum matrix to the steel matrix, which was the highest in the ternary phases (τ_6-Al$_9$Fe$_2$Si$_2$, τ_5-Al$_7$Fe$_2$Si,) of the intermetallic compound layer. However, after the addition of 0.1 wt.% Eu, the Fe, and Al element distribution area tended to narrow in the interface. The Si element was mainly distributed in ternary compounds (τ_6-Al$_9$Fe$_2$Si$_2$, τ_5-Al$_7$Fe$_2$Si, τ_1-(Al, Si)$_5$Fe$_3$) and the aluminum matrix, as seen in Figure 9d,h. It should be pointed out that the Eu element was found to be mainly distributed at the front of the ternary compound (τ_6-Al$_9$Fe$_2$Si$_2$) and steel matrix, as displayed in Figure 9i.

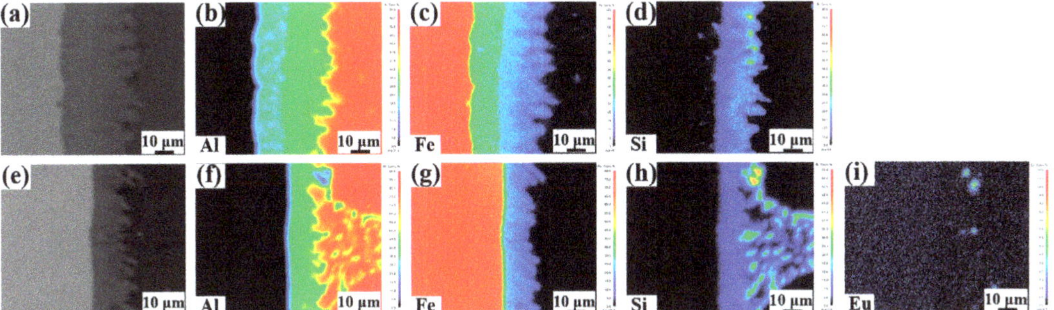

Figure 9. Elements distributions of the Al/Steel interface with the addition of 0 wt.% and 0.1 wt.% Eu contents: (**a,e**) The morphology of the interface; (**b,f**) Al; (**c,g**) Fe; (**d,h**) Si; (**i**) Eu.

Since the hard and brittle Al_5Fe_2 phase has the greatest influence on the interfacial properties, the Al_5Fe_2 phase in different Eu contents was analyzed by EBSD. The grain orientation diagrams of the different phases and the polar pole diagram of the Al_5Fe_2 phase in the Al/steel bimetallic interfacial layer at 0 wt.% and 0.2 wt.% Eu contents were plotted using Aztec Crystal software, as shown in Figure 10. It can be seen from Figure 10a,c that most of the Al_5Fe_2 grains at the interface grew in the form of columnar grains perpendicular to the bimetallic interface. Figure 10b shows that the Al_5Fe_2 grains exhibited obvious preferential orientation in the <001> direction, and the polar density was 23.11. With the addition of 0.2 wt.% Eu, the Al_5Fe_2 grains also had a preferred orientation in the <001> direction, with a polar density of 19.72, as seen in Figure 10d. The addition of the Eu element does not change the preferential orientation of the Al_5Fe_2 phase in the direction but reduces the extreme density value of the preferred orientation. Figure 11 illustrates the grain size of each phase in the intermetallic compound layer at 0 wt.% and 0.2 wt.% Eu contents according to EBSD statistics. The results show that the grain sizes of each phase were refined in the interface reaction layer at 0.2 wt.% Eu.

From the above results, it can be seen that the addition of Eu substantially reduced the thickness of the Al_5Fe_2 phase in the Al/steel bimetallic interface reaction layer. The results of existing studies [38,39] show that the growth of the Al_5Fe_2 phase in the reaction layer is typically controlled by atomic diffusion from the Al side to the steel side. The atomic stacking model of Al_5Fe_2 is a face-centered orthogonal structure, and there are many vacancies and gaps in the c-axis <001> direction of its crystal lattice. Al atoms preferentially diffuse through the c-axis direction of the Al_5Fe_2 phase, resulting in the rapid growth of the Al_5Fe_2 phase [40–42]. The most probable reason for the lower thickness and polar density of the Al_5Fe_2 phase is that Eu may inhibit the diffusion of Al atoms along the c-axis direction of the Al_5Fe_2 phase.

As the τ_6-$Al_9Fe_2Si_2$ phase is formed below 650 °C, the formation of $Al_{13}Fe_4$, τ_5-Al_7Fe_2Si, and τ_6-$Al_9Fe_2Si_2$ layers is affected by the Fe diffusion from the steel side to the Al side and the solidification of τ_6-$Al_9Fe_2Si_2$ phase. Due to the small solid solution of element Eu in the intermetallic compound during solidification, Eu will be pushed out at the front of the binary or ternary compound layers ($Al_{13}Fe_4$, τ_5-Al_7Fe_2Si, and τ_6-$Al_9Fe_2Si_2$), which can be proved by the enrichment of Eu element at the front of the ternary compound layer (τ_6-$Al_9Fe_2Si_2$) near the aluminum side in Figure 9i. The enrichment of the Eu element at the front of the binary or ternary compound layers may hinder the growth of $Al_{13}Fe_4$, τ_5-Al_7Fe_2Si, and τ_6-$Al_9Fe_2Si_2$ phases, resulting in the thinner Fe element distribution area on the Al side. The thickness of the intermetallic compound layer began to increase at 0.2 wt.% Eu. Eu element may have an larger diffusion rate from Al side to steel side in this condition, leading to the lower Eu content in Al side, which may weaken the ability to restrict the diffusion of Al in C-axis of Al_5Fe_2 phase and the growth of $Al_{13}Fe_4$, τ_5-Al_7Fe_2Si, and τ_6-$Al_9Fe_2Si_2$ phases.

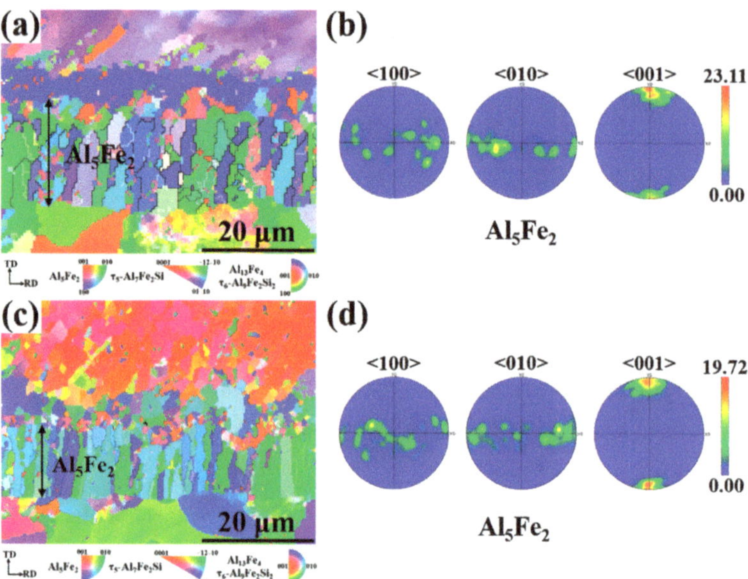

Figure 10. Grain orientation distribution map of bimetallic interface and the grain boundary and polar pole diagram of Al5Fe2 phase: (**a,b**) 0 wt.% Eu; (**c,d**) 0.2 wt.% Eu.

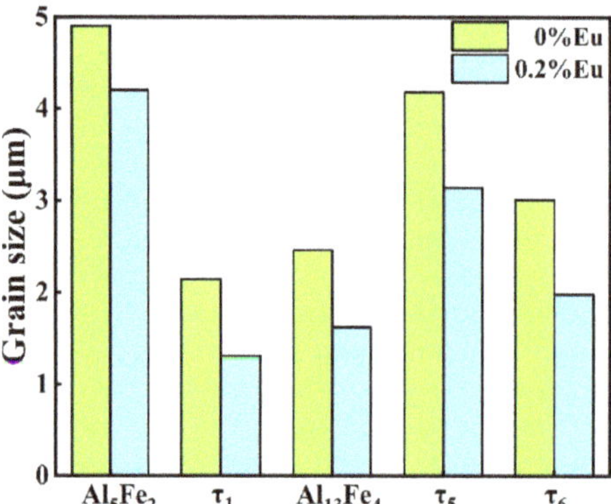

Figure 11. The grain size of each phase in the intermetallic compound layer.

3.4. Mechanical Properties of the Al/Steel Bimetal

Figure 12a illustrates the shear strength of Al/Steel bimetal specimens added with different Eu contents. The results indicate that when the Eu content is 0.1 wt.%, the shear strength of the bimetallic specimen is the highest, which is 31.21MPa and 47% higher than that of the bimetallic specimen without Eu addition. However, when the Eu content was increased to 0.2 wt.%, the shear strength of the samples decreased to 26.39MPa. Figure 12b shows the comparison of shear strength results of the Al/Steel bimetal interface under different surface treatments [14,43,44], and the newly developed Al/Steel bimetal of this study has the highest shear strength values.

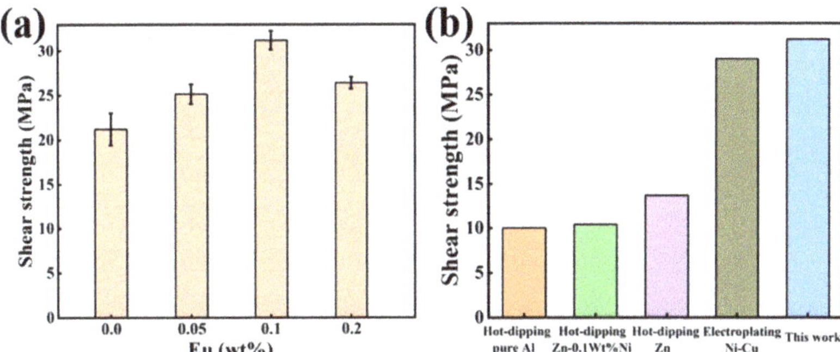

Figure 12. (**a**) Shear strength of Al/Steel bimetal with different Eu contents; (**b**) The comparison of shear strength results at the Al/Steel bimetal interface for different sur-face treatment.

Figure 13 is the SEM images of fracture morphologies of Al/steel bimetal samples on the steel substrate side with different Eu contents. It can be seen from the figure that there are cleavage steps and tear edges in the fracture morphology of the samples, and no obvious plastic deformation occurs, indicating that the samples belong to brittle fractures. As illustrated in Table 6, EDS results show that the fracture structure is mainly the Al_5Fe_2 phase. As the Eu content increased from 0 wt.% to 0.1 wt.%, the number of tear edges at the fracture increased, as seen in Figure 13a–c. However, as seen in Figure 13d, when the Eu content increased to 0.2 wt.%, the number of tear edges on the fracture surface begins to decrease. The possible reasons for improving the mechanical properties are as follows: (1) The Al_5Fe_2 phase layer near the steel side is gentle and has a refiner grain with the addition of 0.1 wt.% Eu. (2) The reaction layer's thickness is thinner with the addition of 0.1 wt.% Eu. (3) The modification of eutectic Si in the Al-Si alloy matrix after Eu addition. The above reasons can decrease the stress concentration in the reaction layer and improve the shear strength of the Al/steel bimetal.

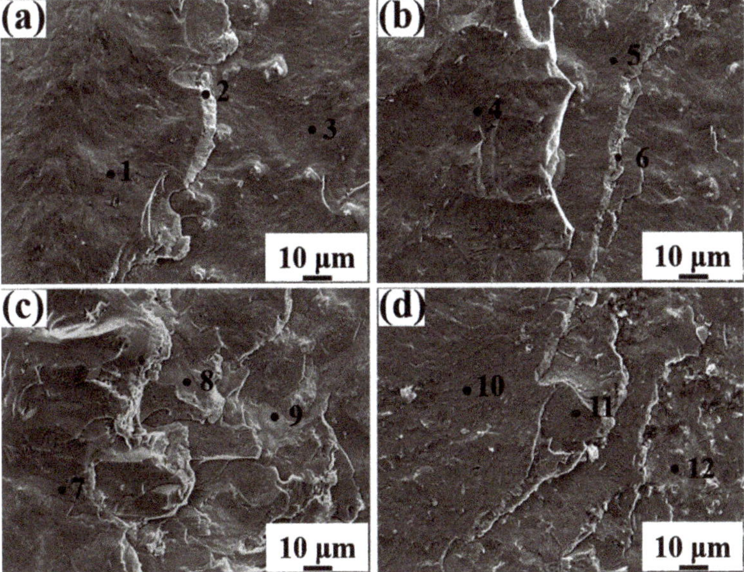

Figure 13. Shear fracture morphology at steel matrix side with different Eu contents: (**a**) 0 wt.% Eu; (**b**) 0.05 wt.% Eu; (**c**) 0.1 wt.% Eu; (**d**) 0.2 wt.% Eu.

Table 6. Energy spectrum analysis results of each point in Figure 13.

Points	Compositions (at%)				Phase
	Al	Fe	Si	Eu	
1	70.59	27.51	1.90	0.00	Al_5Fe_2
2	69.09	28.29	2.62	0.00	Al_5Fe_2
3	69.81	27.85	2.34	0.00	Al_5Fe_2
4	69.58	28.07	2.18	0.17	Al_5Fe_2
5	69.20	28.10	2.63	0.07	Al_5Fe_2
6	69.64	28.90	1.31	0.15	Al_5Fe_2
7	67.36	29.93	2.50	0.21	Al_5Fe_2
8	68.34	27.34	3.86	0.46	Al_5Fe_2
9	70.61	27.87	1.39	0.13	Al_5Fe_2
10	69.76	28.44	1.75	0.06	Al_5Fe_2
11	70.51	28.09	1.29	0.11	Al_5Fe_2
12	67.92	29.14	2.86	0.08	Al_5Fe_2

4. Conclusions

(1) With the addition of 0.1 wt.% Eu, the morphology of eutectic silicon changed from plate-like to fine fibers with a more uniform distribution in the Al-Si alloy.

(2) The Al/steel bimetallic interfacial reaction layer was composed of Al_5Fe_2, τ_1-(Al, Si)$_5$Fe$_3$, $Al_{13}Fe_4$, τ_5-Al_7Fe_2Si, and τ_6-$Al_9Fe_2Si_2$ phases. The intermetallic compound species in the reaction layer were unaffected by the addition of Eu.

(3) When the Eu content is 0.1 wt.%, the average thickness of the reaction layer and the Al_5Fe_2 layer decreased to the minimum value. In addition, there was a more sudden drop of Fe in steel side and the Si in Al side was observed to be more than the other conditions.

(4) The thickness of Al and Fe elements distribution in the reaction layer decreased as the content of Eu reached 0.1 wt.%. Eu was mainly enriched in the front of the ternary compound layer (τ_6-$Al_9Fe_2Si_2$) near the Al side and steel matrix.

(5) The addition of the element Eu did not change the preferred orientation of the Al_5Fe_2, τ_1-(Al, Si)$_5$Fe$_3$, $Al_{13}Fe_4$, τ_5-Al_7Fe_2Si, and τ_6-$Al_9Fe_2Si_2$ phases, but refined the grain size of each phase in the interfacial reaction layer.

(6) The highest shear strength of bimetallic specimens was obtained when the Eu content was 0.1 wt.%, which was 47% higher than that of bimetallic specimens without Eu addition.

Author Contributions: Data curation, P.Z.; Formal analysis, G.Z.; Funding acquisition, F.M. and S.W.; Investigation, P.Z. and C.W.; Methodology, F.M., P.Z., S.W., C.C., G.Z., M.X., T.W. and J.G.; Software, C.C., M.X., T.W., J.G. and C.W.; Writing—original draft, F.M. and P.Z.; Writing—review & editing, F.M. and S.W. All authors have read and agreed to the published version of the manuscript.

Funding: This research was funded by National Key R&D Program of China [2020YFB2008400], Opening Project of Shanxi Key Laboratory of Controlled Metal Solidification and Precision Manufacturing, North University of China [MSPM201904].

Institutional Review Board Statement: Not applicable.

Informed Consent Statement: Not applicable.

Data Availability Statement: The data presented in this study are available on request from the corresponding author.

Acknowledgments: The authors thank the National Key R&D Program of China (2020YFB2008400) and the Opening Project of Shanxi Key Laboratory of Controlled Metal Solidification and Precision Manufacturing, North University of China (MSPM201904). The authors also wish to take this opportunity to thank the support of the Provincial and Ministerial Co-construction of Collaborative Innovation Center for Non-ferrous Metal New Materials and Advanced Processing Technology.

Conflicts of Interest: The authors declare no conflict of interest.

References

1. Ramadan, M.; Alghamdi, A.S.; Hafez, K.M.; Subhani, T.; Abdel Halim, K.S. Development and Optimization of Tin/Flux Mixture for Direct Tinning and Interfacial Bonding in Aluminum/Steel Bimetallic Compound Casting. *Materials* **2020**, *13*, 5642. [CrossRef] [PubMed]
2. Chen, G.; Chang, X.S.; Liu, G.B.; Chen, Q.; Han, F.; Zhang, S.; Zhao, Z.D. Formation of metallurgical bonding interface in aluminum-steel bimetal parts by thixotropic-core compound forging. *J. Mater. Process. Technol.* **2020**, *283*, 116710. [CrossRef]
3. Ramadan, M.; Alghamdi, A.S.; Subhani, T.; Halim, K.S.A. Fabrication and Characterization of Sn-Based Babbitt Alloy Nanocomposite Reinforced with Al2O3 Nanoparticles/Carbon Steel Bimetallic Material. *Materials* **2020**, *13*, 2759. [CrossRef] [PubMed]
4. Fan, Z.S.; Huang, S.T.; Deng, J.H. Cladding of aluminum alloy 6061-T6 to mild steel by an electromagnetic tube bulging process: Finite element modeling. *Adv. Manuf.* **2019**, *7*, 73–83. [CrossRef]
5. Guo, Z.L.; Liu, M.; Bian, X.F.; Liu, M.J.; Li, J.G. An Al–7Si alloy/cast iron bimetallic composite with super-high shear strength. *J. Mater. Res. Technol.* **2019**, *8*, 3126–3136. [CrossRef]
6. Zhang, Y.; Bandyopadhyay, A. Influence of Compositionally Graded Interface on Microstructure and Compressive Deformation of 316L Stainless Steel to Al12Si Aluminum Alloy Bimetallic Structures. *ACS Appl. Mater. Interfaces* **2021**, *13*, 9174–9185. [CrossRef]
7. Yu, F.; Wang, X.; Huang, T.; Chao, D. Effect of Grain Refiner on Fracture Toughness of 7050 Ingot and Plate. *Materials* **2021**, *14*, 6705. [CrossRef]
8. Sui, Y.D.; Wang, Q.D.; Ye, B.; Zhang, L.; Jiang, H.Y.; Ding, W.J. Effect of solidification sequence on the microstructure and mechanical properties of die-cast Al–11Si–2Cu–Fe alloy. *J. Alloys Compd.* **2015**, *649*, 679–686. [CrossRef]
9. Shaha, S.K.; Czerwinski, F.; Kasprzak, W.; Friedman, J.; Chen, D.L. Effect of Zr, V and Ti on hot compression behavior of the Al–Si cast alloy for powertrain applications. *J. Alloys Compd.* **2014**, *615*, 1019–1031. [CrossRef]
10. Jiang, W.M.; Fan, Z.T.; Dai, Y.C.; Li, C. Effects of rare earth elements addition on microstructures, tensile properties and fractography of A357 alloy. *Mater. Sci. Eng. A* **2014**, *597*, 237–244. [CrossRef]
11. Mao, F.; Yan, G.Y.; Xuan, Z.J.; Cao, Z.Q.; Wang, T.M. Effect of Eu addition on the microstructures and mechanical properties of A356 aluminum alloys. *J. Alloys Compd.* **2015**, *650*, 896–906. [CrossRef]
12. Yao, Z.S.; Xu, G.; Jiang, Z.Y.; Tian, J.Y.; Yuan, Q.; Ma, H.W. Effects of Ni and Cr on Cryogenic Impact Toughness of Bainite/Martensite Multiphase Steels. *Met. Mater. Int.* **2019**, *25*, 1151–1160. [CrossRef]
13. Yoon, J.I.; Jung, J.; Lee, H.H.; Kim, J.Y.; Kim, H.S. Relationships Between Stretch-Flangeability and Microstructure-Mechanical Properties in Ultra-High-Strength Dual-Phase Steels. *Met. Mater. Int.* **2019**, *25*, 1161–1169. [CrossRef]
14. Jiang, W.M.; Li, G.Y.; Wu, Y.; Liu, X.W.; Fan, Z.T. Effect of heat treatment on bonding strength of aluminum/steel bimetal produced by a compound casting. *J. Mater. Process. Technol.* **2018**, *258*, 239–250. [CrossRef]
15. Khoonsari, E.M.; Jalilian, F.; Paray, F.; Emadi, D.; Drew, R.A.L. Interaction of 308 stainless steel insert with A319 aluminium casting alloy. *Mater. Sci. Technol.* **2013**, *26*, 833–841. [CrossRef]
16. Shao, L.; Shi, Y.; Huang, J.K.; Wu, S.J. Effect of joining parameters on microstructure of dissimilar metal joints between aluminum and galvanized steel. *Mater. Des.* **2015**, *66*, 453–458. [CrossRef]
17. Pontevichi, S.; Bosselet, F.; Barbeau, F.; Peronnet, M.; Viala, J.C. Solid-Liquid Phase Equilibria in the Al-Fe-Si System at 727 °C. *J. Phase Equilib. Diffus.* **2004**, *25*, 528–537. [CrossRef]
18. Springer, H.; Kostka, A.; Payton, E.J.; Raabe, D.; Kaysser-Pyzalla, A.; Eggeler, G. On the formation and growth of intermetallic phases during interdiffusion between low-carbon steel and aluminum alloys. *Acta Mater.* **2011**, *59*, 1586–1600. [CrossRef]
19. Jiang, W.M.; Fan, Z.T.; Li, C. Improved steel/aluminum bonding in bimetallic castings by a compound casting process. *J. Mater. Process. Technol.* **2015**, *226*, 25–31. [CrossRef]
20. Jiang, W.M.; Fan, Z.T.; Li, G.Y.; Liu, X.W.; Liu, F.C. Effects of hot-dip galvanizing and aluminizing on interfacial microstructures and mechanical properties of aluminum/iron bimetallic composites. *J. Alloys Compd.* **2016**, *688*, 742–751. [CrossRef]
21. Jiang, W.M.; Fan, Z.T.; Li, G.Y.; Li, C. Effects of zinc coating on interfacial microstructures and mechanical properties of aluminum/steel bimetallic composites. *J. Alloys Compd.* **2016**, *678*, 249–257. [CrossRef]
22. Cheng, W.J.; Wang, C.J. Effect of silicon on the formation of intermetallic phases in aluminide coating on mild steel. *Intermetallics* **2011**, *19*, 1455–1460. [CrossRef]
23. Chen, X.C.; Peng, H.P.; Su, X.P.; Liu, Y.; Wu, C.J.; Chen, H.R. Effect of Nickel on the Microstructures of Coating in Hot-Dipped Aluminide Steel. *Mater. Manuf. Process.* **2015**, *31*, 1261–1268. [CrossRef]
24. Mingo, B.; Arrabal, R.; Mohedano, M.; Mendis, C.L.; del Olmo, R.; Matykina, E.; Hort, N.; Merino, M.C.; Pardo, A. Corrosion of Mg-9Al alloy with minor alloying elements (Mn, Nd, Ca, Y and Sn). *Mater. Des.* **2017**, *130*, 48–58. [CrossRef]
25. Liu, H.H.; Fu, P.X.; Liu, H.W.; Cao, Y.F.; Sun, C.; Du, N.Y.; Li, D.Z. Effects of Rare Earth elements on microstructure evolution and mechanical properties of 718H pre-hardened mold steel. *J. Mater. Sci. Technol.* **2020**, *50*, 245–256. [CrossRef]
26. Chen, L.; Hu, M.Y.; Guo, J.; Chong, X.Y.; Feng, J. Mechanical and thermal properties of RETaO4 (RE = Yb, Lu, Sc) ceramics with monoclinic-prime phase. *J. Mater. Sci. Technol.* **2020**, *52*, 20–28. [CrossRef]
27. Cai, Q.; Zhai, C.; Luo, Q.; Zhang, T.Y.; Li, Q. Effects of magnetic field on the microstructure and mechanical property of Mg-Al-Gd alloys. *Mater. Charact.* **2019**, *154*, 233–240. [CrossRef]

28. Li, J.H.; Wang, X.D.; Ludwig, T.H.; Tsunekawa, Y.; Arnberg, L.; Jiang, J.Z.; Schumacher, P. Modification of eutectic Si in Al–Si alloys with Eu addition. *Acta Mater.* **2015**, *84*, 153–163. [CrossRef]
29. Muhammad, A.; Xu, C.; Wang, X.J.; Hanada, S.J.; Yamagata, H.; Hao, L.R.; Ma, C.L. High strength aluminum cast alloy: A Sc modification of a standard Al–Si–Mg cast alloy. *Mater. Sci. Eng. A* **2014**, *604*, 122–126. [CrossRef]
30. Shi, Z.M.; Wang, Q.; Zhao, G.; Zhang, R.Y. Effects of erbium modification on the microstructure and mechanical properties of A356 aluminum alloys. *Mater. Sci. Eng. A* **2015**, *626*, 102–107. [CrossRef]
31. Li, B.; Wang, H.W.; Jie, J.C.; Wei, Z.J. Effects of yttrium and heat treatment on the microstructure and tensile properties of Al–7.5Si–0.5Mg alloy. *Mater. Des.* **2011**, *32*, 1617–1622. [CrossRef]
32. Du, Y.; Schuster, J.C.; Liu, Z.K.; Hu, R.X.; Nash, P.; Sun, W.H.; Zhang, W.W.; Wang, J.; Zhang, L.J.; Tang, C.Y.; et al. A thermodynamic description of the Al–Fe–Si system over the whole composition and temperature ranges via a hybrid approach of CALPHAD and key experiments. *Intermetallics* **2008**, *16*, 554–570. [CrossRef]
33. Burkhardt, U.; Grin, Y.; Ellner, M.; Peters, K. Structure Refinement of the Iron-Aluminium Phase with the Approximate Composition Fe_2Al_5. *Acta Crystall. Sect. B Struct. Sci.* **1994**, *50*, 313–316. [CrossRef]
34. Grin, J.; Burkhardt, U.; Ellner, M.; Peters, K. Refinement of the Fe_4Al_{13} structure and its relationship to the quasihomological homeotypical structures. *Z. Kristallogr. Cryst. Mater.* **1994**, *209*, 479–487. [CrossRef]
35. Yanson, T.I.; Manyako, M.B.; Bodak, O.I.; German, N.V.; Zarechnyuk, O.S.; Cerný, R.; Pacheco, J.V.; Yvon, K. Triclinic $Fe_3Al_2Si_3$ and orthorhombic $Fe_3Al_2Si_4$ with new structure types. *Acta Crystallogr. Sect. C Cryst. Struct. Commun.* **1996**, *52*, 2964–2967. [CrossRef]
36. Roger, J.; Bosselet, F.; Viala, J.C. X-rays structural analysis and thermal stability studies of the ternary compound α-AlFeSi. *J. Solid State Chem.* **2011**, *184*, 1120–1128. [CrossRef]
37. Rømming, C.; Hansen, V.; Gjønnes, J. Crystal structure of β-Al$_{4.5}$FeSi. *Acta Crystall. Sect. B Struct. Sci.* **1994**, *50*, 307–312. [CrossRef]
38. Li, Y.L.; Liu, Y.R.; Yang, J. First principle calculations and mechanical properties of the intermetallic compounds in a laser welded steel/aluminum joint. *Opt. Laser Technol.* **2020**, *122*, 105875. [CrossRef]
39. Takata, N.; Nishimoto, M.; Kobayashi, S.; Takeyama, M. Crystallography of Fe2Al5 phase at the interface between solid Fe and liquid Al. *Intermetallics* **2015**, *67*, 1–11. [CrossRef]
40. Takata, N.; Nishimoto, M.; Kobayashi, S.; Takeyama, M. Morphology and formation of Fe–Al intermetallic layers on iron hot-dipped in Al–Mg–Si alloy melt. *Intermetallics* **2014**, *54*, 136–142. [CrossRef]
41. Richards, R.W.; Jones, R.D.; Clements, P.D.; Clarke, H. Metallurgy of continuous hot dip aluminizing. *Int. Mater. Rev.* **1994**, *39*, 191–212. [CrossRef]
42. Bahadur, A.; Mohanty, O. Structural Studies of Hot Dip Aluminized Coatings on Mild Steel. *Mater. Trans. JIM* **1991**, *32*, 1053–1061. [CrossRef]
43. Dezellus, O.; Digonnet, B.; Sacerdote-Peronnet, M.; Bosselet, F.; Rouby, D.; Viala, J.C. Mechanical testing of steel/aluminium–silicon interfaces by pushout. *Int. J. Adhes. Adhes.* **2007**, *27*, 417–421. [CrossRef]
44. Aguado, E.; Baquedano, A.; Uribe, U.; Fernandez-Calvo, A.I.; Niklas, A. Comparative Study of Different Interfaces of Steel Inserts in Aluminium Castings. *Mater. Sci. Forum* **2013**, *765*, 711–715. [CrossRef]

Article

Effect of Annealing Temperature on the Interfacial Microstructure and Bonding Strength of Cu/Al Clad Sheets with a Stainless Steel Interlayer

Haitao Gao [1], Hao Gu [1], Sai Wang [2,*], Yanni Xuan [3] and Hailiang Yu [1]

[1] State Key Laboratory of High Performance Complex Manufacturing, Light Alloys Research Institute, College of Mechanical and Electrical Engineering, Central South University, Changsha 410083, China; gaohaitao@csu.edu.cn (H.G.); guhao0927@csu.edu.cn (H.G.); yuhailiang@csu.edu.cn (H.Y.)
[2] State Key Laboratory of Rolling and Automation, Northeastern University, Shenyang 110819, China
[3] Department of Energy and Power Engineering, School of Energy and Power Engineering, Changsha University of Science and Technology, Changsha 410114, China; xuanyanni@csut.edu.cn
* Correspondence: 1510213@stu.neu.edu.cn; Tel.: +86-130-5750-2965

Abstract: To explore the influence of annealing temperatures on the interfacial structure and peeling strength of Cu/Al clad sheets with a 304 stainless steel foil interlayer, an intermediate annealing treatment was performed at temperatures of 450 °C, 550 °C, and 600 °C, separately. The experimental results indicate that the interfacial atomic diffusion is significantly enhanced by increasing the intermediate annealing temperature. The average peeling strength of the clad sheets annealed at 550 °C can reach 34.3 N/mm and the crack propagation is along the steel/Cu interface, Cu-Al intermetallic compounds layer, and Al matrix. However, after high-temperature annealing treatment (600 °C), the liquid phase is formed at the bonding interface and the clear Cu/steel/Al interface is replaced by the chaotic composite interfaces. The clad sheet broke completely in the unduly thick intermetallic compounds layer, resulting in a sharp decrease in the interfacial bonding strength.

Keywords: Cu/Al clad sheet; interlayer; annealing temperature; interfacial reaction; bonding strength

Citation: Gao, H.; Gu, H.; Wang, S.; Xuan, Y.; Yu, H. Effect of Annealing Temperature on the Interfacial Microstructure and Bonding Strength of Cu/Al Clad Sheets with a Stainless Steel Interlayer. *Materials* **2022**, *15*, 2119. https://doi.org/10.3390/ma15062119

Academic Editor: Amir Mostafaei

Received: 10 February 2022
Accepted: 10 March 2022
Published: 13 March 2022

Publisher's Note: MDPI stays neutral with regard to jurisdictional claims in published maps and institutional affiliations.

1. Introduction

Cu/Al clad sheets have been widely applied in many fields, such as power electronics, aerospace, and electronic communication [1]. The common methods to produce Cu/Al clad sheets are rolling bonding [2], twin-roll casting bonding [3], explosive bonding [4], and diffusion bonding/TLP bonding/diffusion brazing [5]. Different from these traditional techniques, the powder-in-tube method exhibits dominant advantages to fabricate metallic clad sheets with high interfacial bonding strength, which can easily achieve the closure of local defects around the bonding interface and the regulation in thickness and structure of intermetallic compounds (IMCs) [6]. It is a promising and environmentally friendly method to fabricate metallic clad sheets.

The annealing treatment has a significant influence on the interfacial structure and mechanical performance of metallic clad sheets [7,8]. Cu-Al IMCs are easily formed at the bonding interface due to the strong chemical affinity between Cu and Al [9]. As such, a considerable amount of research has been carried out to regulate the structure of Cu-Al IMCs. In the annealing process, Gao et al. [10] found that the $CuAl_2$ phase was in possession of the priority formation rather than other Cu-Al IMCs, whose formation sequence was in the order of $CuAl_2$, Cu_9Al_4, $CuAl$, and Cu_4Al_3 (or Cu_3Al_2) [2]. The mathematical relationship between the IMCs layer thickness, annealing time, and annealing temperature was built by Pelzer et al. [11]. Similarly, Lee et al. [12] indicated that the growth of Cu-Al IMCs could be controlled by a diffusion mechanism. Li et al. [13] showed that a thicker IMCs layer induced by the high annealing temperature usually led to a decrease in interfacial bonding strength. Mao et al. [14] indicated that the thickness of the IMCs layer in Cu/Al clad sheet could be controlled within 550 nm after annealing at 250 °C and the peeling strength could reach 39 N/mm. Therefore, effectively controlling the thickness and structure of IMCs is a critical factor to obtaining a high-performance Cu/Al clad sheet.

Our previous research found that the introduction of a 304 stainless steel (SUS304) interlayer could significantly improve the interfacial strength of the Cu/Al clad sheet by optimizing the type and structure of IMCs [15]. In this paper, Cu/Al clad sheets with a SUS304 interlayer were prepared by the powder-in-tube method. The influence of annealing temperature on the interfacial structure and bonding strength was investigated. Furthermore, the interfacial strengthening mechanism was systematically discussed.

2. Materials and Methods

The initial materials are the commercial pure copper tube (99.9%, Guangfeng Metal Materials Co., Ltd., Dongguan, China) with an outer diameter of 10 mm and a wall thickness of 1 mm and atomized aluminum powder with an average diameter of 10 μm. The cold-rolled SUS304 foils (Guangfeng Metal Materials Co., Ltd., Dongguan, China) with a thickness of 30 μm are chosen as the interlayer materials. The mechanical parameters of the SUS304 foils have been illustrated in our previous research [15]. Before roll cladding, the oxide layer on the inner surface of the Cu tube is removed by the wire brush. The detailed preparation process of the Cu/Al clad sheets with the SUS304 interlayer is illustrated in Figure 1. Firstly, the Cu cube, SUS304 foil, and Al powder (Hunan Jinhao New Material Technology Co., Ltd., Miluo, China) are manually assembled together to form the composite tube billet. Secondly, the Cu/Al clad sheets are rolled to 1.5 mm after multi-pass cold rolling and then annealed at different temperatures of 450 °C, 550 °C, and 600 °C, which are designated as IFR-450, IFR-550, and IFR-600, respectively. Finally, these annealed samples are further rolled to 0.5 mm.

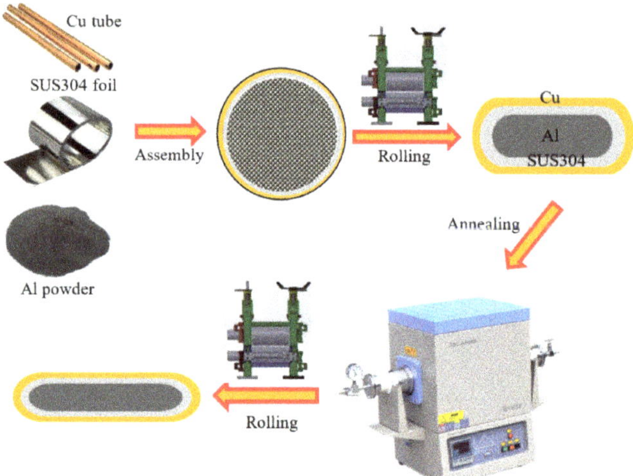

Figure 1. Schematic illustration of the preparation process of Cu/Al clad sheets with SUS304 interlayer.

To observe the microstructure of the bonding interface and peeling surface, scanning electron microscopy (SEM) was performed using an FEI Quanta 250F (FEI Company, Hillsboro, OR, USA) device with an acceleration voltage of 20 kV and equipped with energy dispersive spectrometry (EDS, Oxford Instruments Group, London, UK). The Al-Cu phase diagram is plotted by the software 'Binary Alloy Phase Diagrams' (ASM International, Almere, the Netherlands). The crystalline phases on the peeling surface are detected by D8 ADVANCE X-ray diffraction (XRD, ceramic X-ray tube, Brooke company, karlsruhe,

Germany), using Cu K$_\alpha$ radiation ($\lambda \approx 1.54$ Å), equipped with a Linx one-dimensional array detector (Liwei wisdom international Co., Ltd., HongKong, China). The angle step, time step, and scanning range are 0.01°, 0.1 s, and 10–100°, respectively. T-peel tests on a TH5000 universal testing machine (Xintianhui Electronic Technology Co., Ltd., Yangzhou, China) are adopted to test the interfacial bonding strength of Cu/Al clad sheets under a crosshead speed of 1 mm/min.

3. Results and Discussion

3.1. Interfacial Microstructure

Prior research indicated that the peeling strength of Cu/Al clad sheets was mainly determined by the bonding strength between the SUS304 interlayer and the Cu/Al matrix [15]. Figure 2 exhibits the interfacial microstructure of IFR-450 and the corresponding EDS mapping results. A flat bonding interface without visible cracks between the SUS304 interlayer and Cu/Al matrix is formed (Figure 2a), and the IMCs are seldom formed due to the lower intermediate annealing temperature [16], which is proved by the EDS mapping results of the Al/SUS304/Cu bonding interface (Figure 2b–d). The position of the SUS304 interlayer is marked by the distribution of the Fe and Cr elements (Figure 2e,f). In the rolling process, the SUS304 fragments are squeezed into the Cu/Al matrix. Under this situation, the interfacial bonding strength is mainly contributed by the mechanical joggles [17].

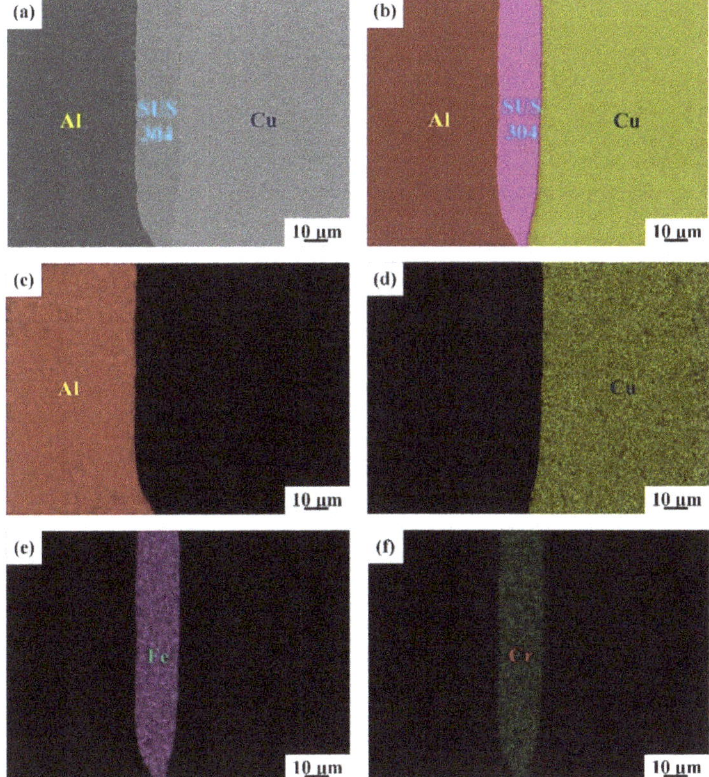

Figure 2. SEM images of the bonding interface for IFR-450 (**a**) and the corresponding EDS mapping: (**b**) EDS layered image, (**c**) Al element, (**d**) Cu element, (**e**) Fe element, (**f**) Cr element.

When the intermediate annealing temperature increases to 550 °C, obvious IMCs with an average thickness of 10 µm are formed at the bonding interface of the SUS304 interlayer and Al matrix [18], as shown in Figure 3a, which is also strongly proved by the corresponding EDS mapping results (Figure 3b). A small increase in the annealing temperature may result in a significant increment in the diffusion coefficient [19] due to their exponential relationship (Figure 3c). In contrast, the chemical compound type formed between the SUS304 interlayer and Cu matrix [20] is the solid solution (Figure 3d). The position of the SUS304 interlayer is marked by the distribution of Fe and Cr elements (Figure 3e,f). For Cu/Al clad sheets without an interlayer, the Cu-Al IMCs are broken into fragments in the rolling process [6]. Nevertheless, the IMCs formed between the SUS304 interlayer and Al matrix can retain their continuity. The existence of an SUS304 interlayer with a weak deformation capacity can significantly inhibit the crush of Al-SUS304 IMCs and enhance the interfacial bonding strength.

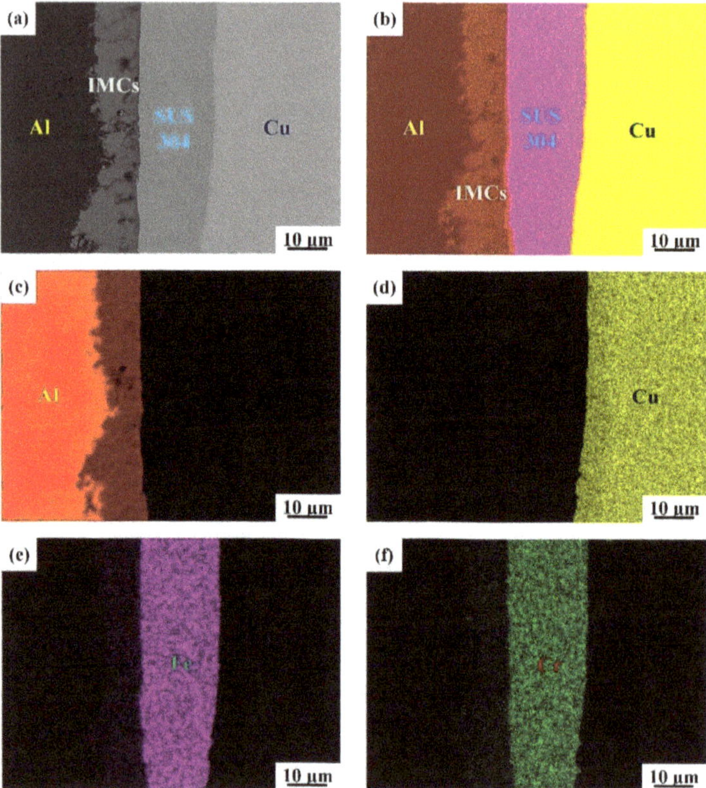

Figure 3. SEM images of the bonding interface for IFR-550 (**a**) and the corresponding EDS mapping: (**b**) EDS layered image, (**c**) Al element, (**d**) Cu element, (**e**) Fe element, (**f**) Cr element.

The interfacial microstructure of IFR-600 is exhibited in Figure 4. As the intermediate annealing temperature reaches 600 °C, the clear Cu/SUS304/Al interface disappears and is replaced by the chaotic composite interfaces (Figure 4a). According to the Al-Cu phase diagram (Figure 5), the liquid phase will be formed at the bonding interface with the annealing temperature of 600 °C (red line in Figure 5). In this case, the Cu/SUS304/Al interface is destroyed by the disturbing force induced from the formation process of the liquid phase, resulting in the exfoliation of SUS304 fragments from the Cu matrix and being involved in the liquid phases. Due to the relatively high annealing temperature, the diffusional degree of IFR-600 is much larger than those in IFR-450 and IFR-550 (Figures 2 and 3), which is also proven by our prior research [21]. Based on the EDS mapping results (Figure 4b–d), it can be observed that the SUS304 fragments are surrounded by thick Cu-Al IMCs. The position of the SUS304 interlayer is marked by the distribution of Fe and Cr elements (Figure 4e,f). Moreover, the hardness of Cu-Al IMCs is higher than the Cu/Al matrix, leading to the increase in the thickness reduction of the SUS304 interlayer in the further cold rolling process. For IFR-600, the residual thickness of the SUS304 interlayer is just 5.9 µm, around a third of those in IFR-450 and IFR-550.

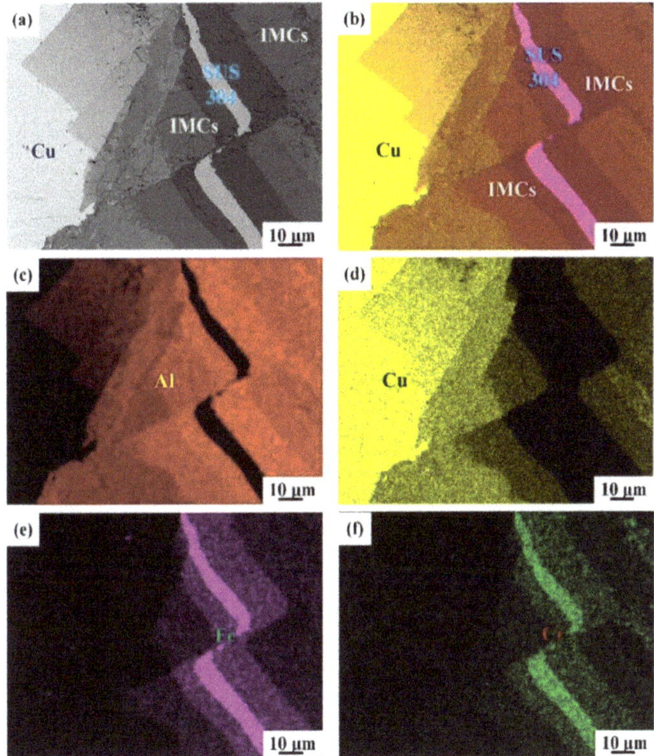

Figure 4. SEM images of bonding interface for IFR-600 (**a**) and the corresponding EDS mapping: (**b**) EDS layered image, (**c**) Al element, (**d**) Cu element, (**e**) Fe element, (**f**) Cr element.

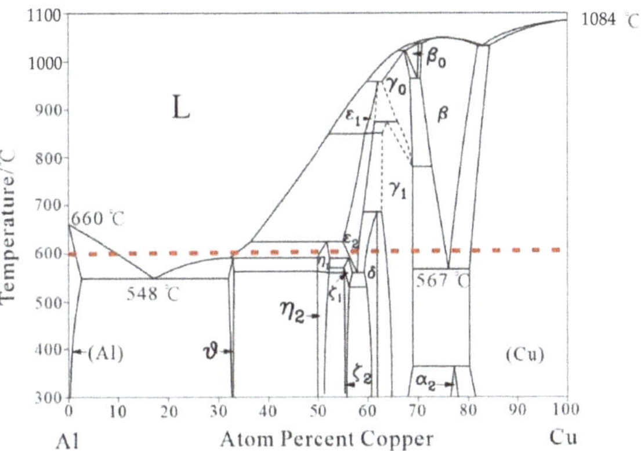

Figure 5. Binary alloy phase diagram of Al-Cu.

3.2. Elemental Diffusion across the Bonding Interface

The EDS line results across the Cu/SUS304/Al interface are provided in Figure 6. After roll cladding, the thickness reduction in the SUS304 interlayers in IFR-450, IFR-550, and IFR-600 are 11.6 µm, 10.2 µm, and 24.1 µm, respectively. In general, the thickness reduction in the SUS304 interlayers should decrease with the increment of intermediate annealing temperature due to the aggravate softening of the Cu/Al matrix. Nevertheless, the formation of liquid phases at the bonding interface leads to the SUS304 interlayer being surrounded by the Cu-Al IMCs and a large thickness reduction in the SUS304 interlayer (Figure 4). On the other hand, the interfacial atomic diffusion is also enhanced by the increase in annealing temperature, which may also increase the corrosion resistance of the clad sheets [22]. Compared with IFR-450, the width of the diffusional layer for the Al/SUS304 interface and Cu/SUS304 interface in IFR-550 is increased from 2.2 µm to 20.1 µm and from 1.1 µm to 3.5 µm (Figure 6a,b). As for IFR-600, the SUS304 fragments are totally surrounded by the thick Cu-Al IMCs, which is strongly proven by the EDS line results (Figure 6c).

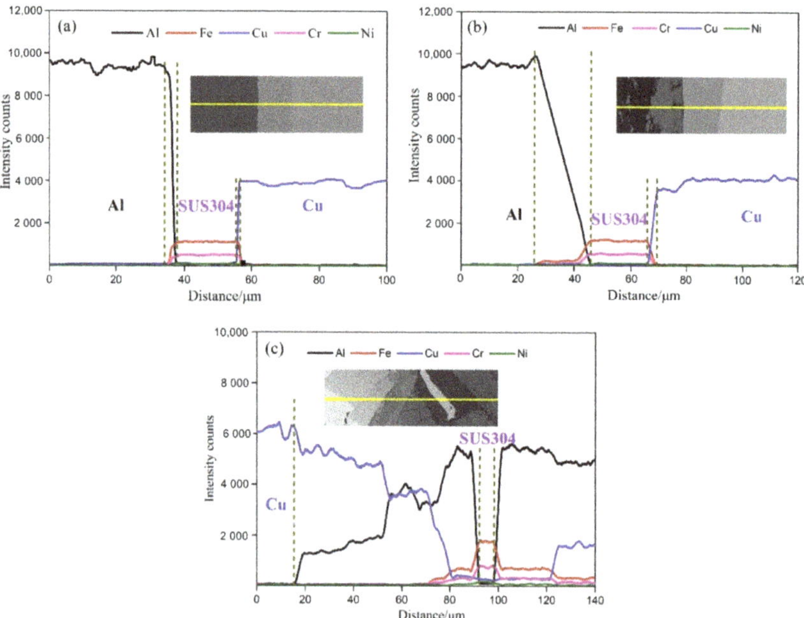

Figure 6. EDS line scan analysis across the Cu/SUS304/Al interface: (**a**) IFR-450, (**b**) IFR-550, (**c**) IFR-600.

3.3. Peeling Surface of the Clad Sheets

The XRD patterns of the peeling surface of IFR samples are provided in Figure 7. It can be seen that the intermediate annealing temperature has a significant impact on the type and content of crystalline phases on the peeling surface. The main crystalline phases in the peeling surface of the Cu side for IFR-450 and IFR-550 are Cu, Al, Cu_9Al_4, and $CuAl_2$. The interfacial elemental diffusion is enhanced by the increase in the intermediate annealing temperature, resulting in the increment of Cu-Al IMCs, especially the Cu_9Al_4 phase (Figure 7a). In addition to the Al, Fe, $CuAl_2$ phases, a new Al_7Cu_2Fe phase is observed at the Al side of IFR-550 (Figure 7b). As the intermediate annealing temperature reaches 600 °C, the main crystalline phases in the peeling surface of the Cu side for IFR-600 are transformed into $CuAl_2$, Al, Cu_9Al_4, and FeAl. As to the Al side of IFR-600, the main crystalline phases are $CuAl_2$, Fe_3Al, and Al.

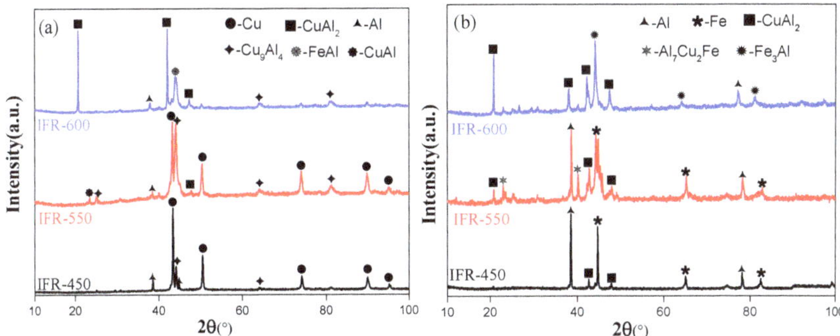

Figure 7. XRD patterns of peeling surface of IFR samples: (**a**) Cu side, (**b**) Al side.

The morphologies of the peeling surface for IFR-450 and the chemical compositions of corresponding crystalline phases are shown in Figure 8 and Table 1. There are two entirely different peeling morphologies on the Cu side and Al side. The ridge-shaped mixture of the Al matrix and Cu-Al IMCs and invaginated Cu matrix can be observed on the Cu side (Figure 8a), which is identified by the EDS results (Table 1). Moreover, many distinct wrinkles are formed at the invaginated Cu matrix, which resulted from the severe shear deformation induced by the plastic difference of the Cu matrix and SUS304 interlayer (Figure 8c). As to the Al side, many SUS304 fragments with uneven sizes are tightly embedded into the Al matrix, which are formed through the random fracture of SUS304 foil in the roll cladding process (Figure 8b). Moreover, these SUS304 fragments' gaps are filled with a mixture of reticulated Al and Cu-Al IMCs (Figure 8d).

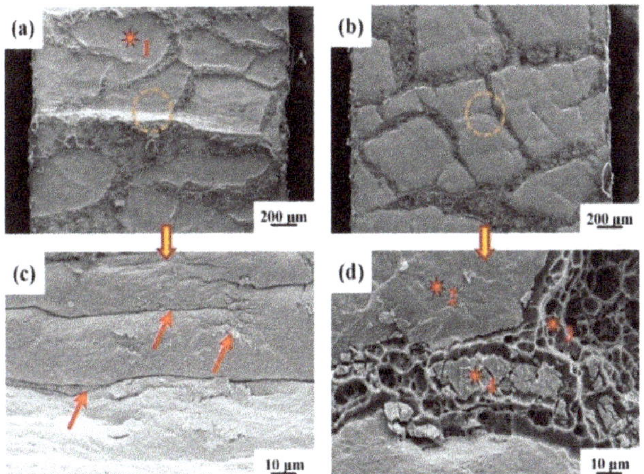

Figure 8. SEM images of Cu and Al surfaces for IFR-450 after peeling test: (**a**) Cu side, (**b**) Al side, (**c,d**) enlarged images of the circles in the (**a,b**).

Table 1. Chemical composition of corresponding points at the peeling surface of IFR-450 (at. %).

Points	Cu	Al	Fe	Cr	Ni	Phase
1	97.3	2.7	-	-	-	Cu
2	-	-	78.6	16.1	5.3	SUS304
3	1.6	98.4	-	-	-	Al
4	67.2	32.8	-	-	-	Cu$_9$Al$_4$

The morphologies of the peeling surface for IFR-550 and the chemical compositions of corresponding crystalline phases are shown in Figure 9 and Table 2. The crystalline phases in the peeling surface of the Cu side for IFR-550 are folded Cu matrix, reticulate Al matrix, and the Cu_9Al_4 phase with obvious cracks (Figure 9a), which is consistent with the XRD results. Compared with IFR-450, the content of the Cu_9Al_4 phase exhibits a significant increase. The increase in the intermediate annealing temperature leads to the formation of a thick Cu-Al IMCs layer, which is brittle and will be broken in the further rolling process (Figure 9c). SUS304 fragments can still be clearly observed on the peeling surface of the Al side for IFR-550 and their height is smaller than the Cu-Al IMCs (Figure 9b). Furthermore, obvious wrinkles appear on the surface of the SUS304 fragments and a new Al_7Cu_2Fe phase is detected on the edge of the SUS304 fragments (Figure 9d). Those above phenomena indicate that the interfacial elemental diffusion has been enhanced by increasing the intermediate annealing temperature.

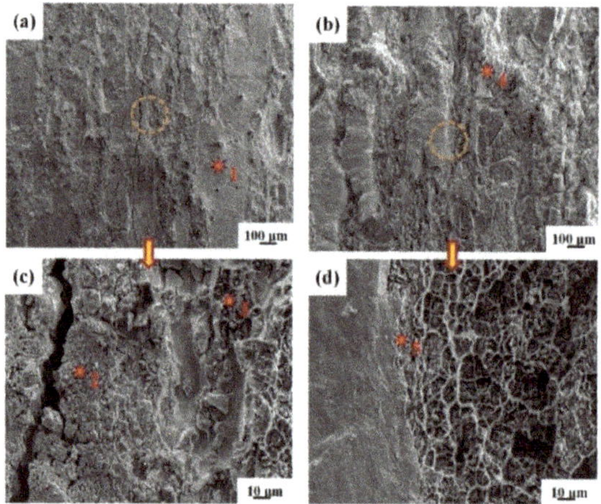

Figure 9. SEM images of Cu and Al surfaces for IFR-550 after peeling test: (**a**) Cu side, (**b**) Al side, (**c,d**) enlarged images of the circles in the (**a,b**).

Table 2. Chemical composition of corresponding points at the peeling surface of IFR-550 (at. %).

Points	Cu	Al	Fe	Cr	Ni	Phase
1	92.6	7.4	-	-	-	Cu
2	63.7	36.3	-	-	-	Cu_9Al_4
3	3.2	96.8	-	-	-	Al
4	72.5	27.5	-	-	-	Cu_9Al_4
5	68.2	19.3	9.1	2.5	0.9	Al_7Cu_2Fe

The morphologies of the peeling surface for IFR-600 and the chemical compositions of corresponding crystalline phases are shown in Figure 10 and Table 3. The main crystalline phases on the peeling surface of the Cu side for IFR-600 are the $CuAl_2$ phase with a coarse surface and the vitreous FeAl phase (Figure 10a,c). Compared with IFR-450 and IFR-550, the Cu matrix is undetectable for IFR-600, which implies the transformation of the crack propagation path from the initial Cu/SUS304 interface to the Cu-Al IMCs and Fe-Al IMCs layers. High intermediate annealing temperature results in the formation of an unduly thick IMCs layer and Kirkendall voids, which promotes crack propagation and reduces the interfacial bonding interface [23,24]. Similarly, the SUS304 fragments cannot be observed on the peeling surface of the Al side, only some castle peaks of the $CuAl_2$ phase and rock-like Fe_3Al phase (Figure 10b,d). Furthermore, the reticular Al matrix also disappears. The above results also prove that the interfacial cracks are propagated along the Cu-Al IMCs and Fe-Al IMCs layer and the bonding strength of the Al matrix is higher than these IMCs.

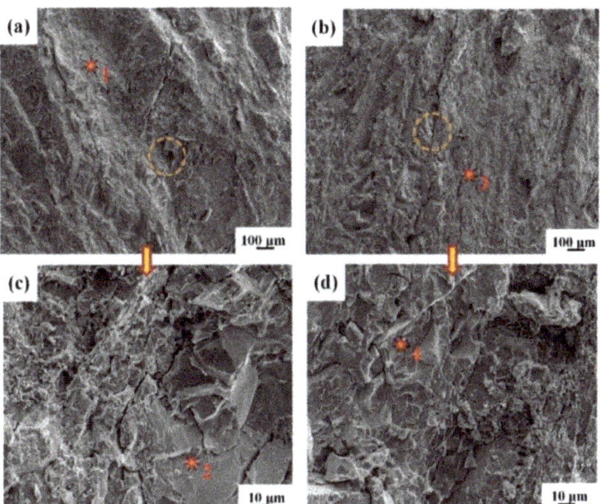

Figure 10. SEM images of Cu and Al surfaces for IFR-600 after peeling test: (**a**) Cu side, (**b**) Al side, (**c,d**) enlarged images of the circles in the (**a,b**).

Table 3. Chemical composition of corresponding points at the peeling surface of IFR-600 (at. %).

Points	Cu	Al	Fe	Cr	Ni	Phase
1	26.8	73.2	-	-	-	$CuAl_2$
2	5.2	45.6	43.2	4.3	1.7	FeAl
3	39.5	60.5	-	-	-	$CuAl_2$
4	4.3	18.7	59.6	13.8	3.6	Fe_3Al

3.4. Peeling Strength of the Clad Sheets

Figure 11 shows the peeling strength curves and average peeling strength of the Cu/Al clad sheets with the SUS304 interlayer annealed at different temperatures. To highlight the strengthening effect of the SUS304 interlayer, the peeling curve of the clad sheets without the SUS304 interlayer (SR-0) is also introduced. With the increase in the intermediate annealing temperature, the peeling strength of the clad sheets firstly increase and then decrease (Figure 10a). Compared with IFR-450, the peeling strength of IFR-550 is increased by 11%, from 30.9 N/mm to 34.3 N/mm. Moreover, the fluctuation of the peeling strength curve for IFR-550 is larger than that of IFR-450, indicating the difference in the bonding strength among the SUS304/Cu interface, Cu-Al IMCs layer, and Al matrix being enlarged.

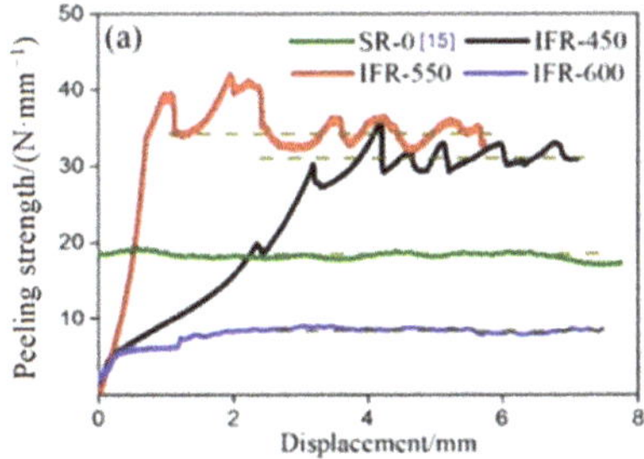
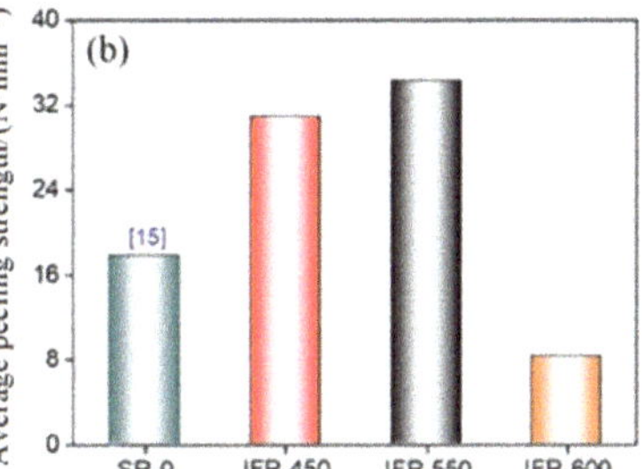

Figure 11. Peeling strength curves (**a**) and average peeling strength (**b**) of Cu/Al clad sheets with SUS304 interlayer annealed at different temperatures.

4. Discussions

According to the above results in Figures 7–10, it can be deduced that the interfacial cracks propagate along the Cu/SUS304 interface, Al matrix, and Cu-Al IMCs layer for IFR-450 and IFR-550. While for IFR-600, the propagation path of interfacial cracks is changed, which is along the Al/SUS304 interface, Al matrix, and Cu-Al IMCs layer. The high intermediate annealing temperature leads to an excessive formation of Al-Fe IMCs and Cu-Al IMCs, resulting in a decrease in the bonding strength for the Al/SUS304 interface and Cu/Al interface. Moreover, Chen et al. [25] indicated that the bonding strength of the Cu/steel interface would be enhanced by increasing the annealing temperature in the range of 400 °C–1000 °C. Thus, the interfacial cracks are more likely formed and prolongated along the Al/SUS304 interface instead of the Cu/SUS304 interface for IFR-600.

The diffusion process across the bonding interface of Cu/Al clad sheets has been verified to conform to the vacancy diffusion [26]. The energy input for interfacial atomic diffusion from two aspects: one is the thermal energy from the intermediate annealing treatment; the other is the shear strain energy provided by the ductility difference between the SUS304 interlayer and Cu/Al matrix, as shown in Figure 12. Our previous research has indicated that increasing the thickness of the SUS304 interlayer can improve the interfacial bonding strength by enlarging the shear strain energy at the bonding interface [15]. In this study, the interfacial atomic diffusion, as well as the interfacial bonding strength, is enhanced by raising the intermediate annealing temperature. While for IFR-600, the peeling strength sharply decreases to 8.3 N/mm, which is even lower than the clad sheets without the SUS304 interlayer, as shown in Figure 11. A high intermediate annealing temperature leads to the formation of unduly thick IMCs layers, which promote the crack propagation and decrease the interfacial bonding strength.

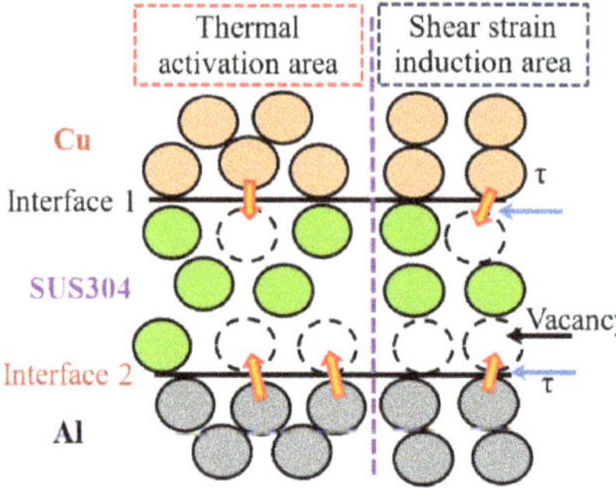

Figure 12. Schematic diagram of vacancy diffusion around the bonding interface.

5. Conclusions

In this work, the influence of an intermediate annealing temperature on the interfacial microstructure, elemental diffusion, and peeling strength of Cu/Al clad sheet containing a SUS304 interlayer is investigated. The main conclusions are as follows:

(1) The interfacial atomic diffusion is significantly enhanced by increasing the intermediate annealing temperature. Nevertheless, after a high-temperature annealing treatment (IFR-600), a liquid phase is formed at the bonding interface and the clear Cu/SUS304/Al interface in IFR-450 and IFR-550 is replaced by the chaotic composite interfaces.

(2) For IFR-450 and IFR-550, the interfacial crack is propagated along the Cu/SUS304 interface, Cu-Al IMCs layer, and Al matrix. Compared with IFR-450, the interfacial bonding strength for IFR-550 is improved by 11%, from 30.9 N/mm to 34.3 N/mm, which is proven by the obvious wrinkles on the surface of SUS304 fragments and the formation of a new Al_7Cu_2Fe phase.

(3)   As the intermediate annealing temperature is increased to 600 °C (IFR-600), the propagation path of interfacial crack is changed into the Cu-Al IMCs layer and Fe-Al IMCs layer. The clad sheets broke completely in the unduly thick IMCs layer, resulting in the sharp decrease of interfacial bonding strength, only 8.3 N/mm.

Author Contributions: Conceptualization, S.W. and H.G. (Haitao Gao); methodology, S.W.; software, Y.X.; validation, S.W. and H.Y.; formal analysis, H.G. (Hao Gu); investigation, H.G. (Hao Gu); data curation, Y.X.; writing—original draft preparation, H.G. (Haitao Gao) and Y.X.; writing—review and editing, H.G. (Haitao Gao) and S.W.; supervision, H.G. (Haitao Gao) and H.Y.; project administration, S.W.; funding acquisition, H.G. (Hao Gu) and Y.X. All authors have read and agreed to the published version of the manuscript.

Funding: The authors thank the Fundamental Research Funds for their financial support and the Central Universities of Central South University (Hao Gu, Grant No.: 2021zzts0150) and the Foundation of State Key Laboratory of High-efficiency Utilization of Coal and Green Chemical Engineering (Yanni Xuan, Grant No. 2022-K54).

Institutional Review Board Statement: Not applicable.

Informed Consent Statement: Not applicable.

Data Availability Statement: The data presented in this study are available on request from the corresponding author.

Acknowledgments: We also thank Yu Zhoufeng from Shiyanjia Lab (www.shiyanjia.com) for the EDS analysis.

Conflicts of Interest: The authors declare no conflict of interest.

References

1.   Kim, M.J.; Lee, K.S.; Han, S.H.; Hong, S.I. Interface strengthening of a roll-bonded two-ply Al/Cu sheet by short annealing. *Mater. Charact.* **2021**, *174*, 111021. [CrossRef]
2.   Sheng, L.Y.; Yang, F.; Xi, T.F.; Lai, C.; Ye, H.Q. Influence of heat treatment on interface of Cu/Al bimetal composite fabricated by cold rolling. *Compos. Part B* **2011**, *42*, 1468–1473. [CrossRef]
3.   Huang, H.G.; Dong, Y.K.; Yan, M.; Du, F.S. Evolution of bonding interface in solid–liquid cast-rolling bonding of Cu/Al clad strip. *Trans. Nonferrous Met. Soc.* **2017**, *27*, 1019–1025. [CrossRef]
4.   Elango, E.; Saravanan, S.; Raghukandan, K. Experimental and numerical studies on aluminum-stainless steel explosive cladding. *J. Cent. South Univ.* **2020**, *27*, 1742–1753. [CrossRef]
5.   Alikhani, A.; Beygi, R.; Zarezadeh-Mehrizi, M.; Nematzadeh, F.; Galvao, I. Effect of Mg and Si on intermetallic formation and fracture behavior of pure aluminum-galvanized carbon-steel joints made by weld-brazing. *J. Cent. South Univ.* **2021**, *28*, 3626–3638. [CrossRef]
6.   Gao, H.T.; Liu, X.H.; Qi, J.L.; Ai, Z.R.; Liu, L.Z. Microstructure and mechanical properties of Cu/Al/Cu clad strip processed by the powder-in-tube method. *J. Mater. Process. Technol.* **2018**, *251*, 1–11. [CrossRef]
7.   Simsir, M.; Kumruoğlu, L.C.; Özer, A. An investigation into stainless-steel/structural-alloy-steel bimetal produced by shell mould casting. *Mater. Des.* **2009**, *30*, 264–270. [CrossRef]
8.   Lee, K.S.; Yoon, D.H.; Kim, H.K.; Kwon, Y.N.; Lee, Y.S. Effect of annealing on the interface microstructure and mechanical properties of a STS–Al–Mg 3-ply clad sheet. *Mater. Sci. Eng. A* **2012**, *556*, 319–330. [CrossRef]
9.   Lee, S.; Son, I.S.; Lee, J.K.; Lee, J.S.; Kim, Y.B.; Lee, G.A.; Lee, S.P.; Cho, Y.R.; Bae, D.S. Effect of aging treatment on bonding interface properties of hot-pressed Cu/Al clad material. *Int. J. Precis. Eng. Manuf.* **2015**, *16*, 525–530. [CrossRef]
10.   Gao, K.; Song, S.J.; Li, S.M.; Fu, H.Z. Characterization of microstructures and growth orientation deviating of Al$_2$Cu phase dendrite at different directional solidification rates. *J. Alloys Compd.* **2016**, *660*, 73–79. [CrossRef]
11.   Pelzer, R.; Nelhiebel, M.; Zink, R.; Woehlert, S.; Lassnig, A.; Khatibi, G. High temperature storage reliability investigation of the Al–Cu wire bond interface. *Microelectron. Reliab.* **2012**, *52*, 1966–1970. [CrossRef]
12.   Lee, W.B.; Bang, K.S.; Jung, S.B. Effects of intermetallic compound on the electrical and mechanical properties of friction welded Cu/Al bimetallic joints during annealing. *J. Alloys Compd.* **2005**, *390*, 212–219. [CrossRef]
13.   Li, H.Y.; Chen, W.G.; Dong, L.L.; Shi, Y.G.; Liu, J.; Fu, Y.Q. Interfacial bonding mechanism and annealing effect on Cu-Al joint produced by solid-liquid compound casting. *J. Mater. Process. Technol.* **2018**, *252*, 795–803. [CrossRef]
14.   Mao, Z.P.; Xie, J.P.; Wang, A.Q.; Wang, W.Y.; Ma, D.Q.; Liu, P. Effects of annealing temperature on the interfacial microstructure and bonding strength of Cu/Al clad sheets produced by twin-roll casting and rolling. *J. Mater. Process. Technol.* **2020**, *285*, 116804. [CrossRef]

15. Gao, H.T.; Wang, L.; Liu, S.L.; Li, J.; Kong, C.; Yu, H.L. Effects of a stainless steel interlayer on the interfacial microstructure and bonding strength of Cu/Al clad sheets prepared via the powder-in-tube method. *J. Mater. Res. Technol.* **2021**, *15*, 3514–3524. [CrossRef]
16. Chen, G.; Xu, G.M. Interfacial reaction in twin-roll cast AA1100/409L clad sheet during different sequence of cold rolling and annealing. *Met. Mater. Int.* **2021**, *27*, 3013–3025. [CrossRef]
17. Liu, J.; Wu, Y.Z.; Wang, L.; Kong, C.; Pesin, A.; Zhilyaev, A.P.; Yu, H.L. Fabrication and characterization of high bonding strength Al/Ti/Al-laminated composites via cryorolling. *Acta Metall. Sin.* **2020**, *33*, 871–880. [CrossRef]
18. Hasanniah, A.; Movahedi, M. Welding of Al-Mg aluminum alloy to aluminum clad steel sheet using pulsed gas tungsten arc process. *J. Manuf. Process.* **2018**, *31*, 494–501. [CrossRef]
19. Jannot, Y.; Bal, H.M.; Degiovanni, A.; Moyne, C. Influence of heat transfer on the estimation of water vapor diffusion coefficient in transient regime. *Int. J. Heat Mass Transf.* **2021**, *177*, 121558. [CrossRef]
20. Ai-Ghamdi, K.A.; Hussain, G. Parameter-formability relationship in ISF of tri-layered Cu-Steel-Cu composite sheet metal: Response surface and microscopic analyses. *Int. J. Precis. Eng. Manuf.* **2016**, *17*, 1633–1642. [CrossRef]
21. Gao, H.T.; Li, J.; Lei, G.; Song, L.L.; Kong, C.; Yu, H.L. High strength and thermal stability of multilayered Cu/Al composites fabricated through accumulative roll bonding and cryorolling. *Metall. Mater. Trans. A* **2022**. [CrossRef]
22. Sahar, R.; Ali, A.; Stanislav, J.; Alireza, G.K.; Fredrick, M.; Carlos, L.; Dinara, S.; Sławomir, K.; Reza, S.; Miroslaw, B.; et al. Effect of annealing on the micromorphology and corrosion properties of Ti/SS thin films. *Superlattice Microstruct.* **2020**, *146*, 106681.
23. Peng, X.K.; Wuhrer, R.; Heness, G.; Yeung, W.Y. Rolling strain effects on the interlaminar properties of roll bonded copper/aluminium metal laminates. *J. Mater. Sci.* **2000**, *35*, 4357–4363. [CrossRef]
24. Chen, C.Y.; Chen, H.L.; Hwang, W.S. Influence of interfacial structure development on the fracture mechanism and bond strength of aluminum/copper bimetal plate. *Mater. Trans.* **2006**, *47*, 1232–1239. [CrossRef]
25. Chen, J.Q.; Liu, X.H.; Yan, S.; Yu, Q.B. Experimental study on bending capacity of copper/steel/copper cold rolled composite strip. *J. Northeast Univ. (Nat. Sci.)* **2019**, *40*, 647–652.
26. Yu, Q.B.; Liu, X.H.; Sun, Y.; Qi, J.L. Deformation-induced reaction diffusion of Cu/Al multilayered composite by rolling at room temperature. *Sci. Sin. Technol.* **2016**, *46*, 1166–1174.

Article

Analysis of Mechanical Parameters of Asymmetrical Rolling Dealing with Three Region Percentages in Deformation Zones

Qilin Zhao [1], Xianghua Liu [1,2,*] and Xiangkun Sun [2]

[1] School of Materials Science and Engineering, Northeastern University, Shenyang 110819, China; zhaoql73@163.com
[2] State Key Laboratory of Rolling and Automation, Northeastern University, Shenyang 110819, China; sxk20081647@163.com
* Correspondence: liuxh@mail.neu.edu.cn; Tel.: +86-024-83682273

Abstract: A series of mathematical models were proposed to calculate the roll force, torque and power for cold strip asymmetrical rolling by means of the slab method, taking the percentages of the forward-slip, backward-slip and cross-shear zones into account. The friction power, plastic work and total energy consumption can be obtained by the models. The effects of variable rolling parameters— such as the speed ratio, entry thickness, friction coefficient and front and back tension—on the process of asymmetrical rolling are analyzed. In all cases, an increase in speed ratio leads to an increase in friction work and its proportions. The increase in entry thickness and deformation resistance causes both friction work and plastic deformation work to increase. The proportion of friction work decreases with increasing deformation resistance, entry thickness, front tension and back tension. In the circumstances of a thin strip being rolled with a large speed ratio, the proportion of friction work could exceed that of plastic deformation work. The concept of a threshold point of friction work was proposed to explain this phenomenon. As an example, threshold points T1, T2, T3 with the effect of the entry thickness and S1, S2, S3 with the effect of the friction coefficient have been obtained by computation. Finally, the experiment of the strip asymmetrical rolling was conducted, and a maximum error of 9.7% and an RMS error of 5.9% were found in the comparison of roll forces between experimental measurement values and calculated ones.

Keywords: strip asymmetrical rolling; mathematic models; percentage of cross-shear region; friction work; energy consumption; slab method

Citation: Zhao, Q.; Liu, X.; Sun, X. Analysis of Mechanical Parameters of Asymmetrical Rolling Dealing with Three Region Percentages in Deformation Zones. *Materials* **2022**, *15*, 1219. https://doi.org/10.3390/ma15031219

Academic Editor: Adam Grajcar

Received: 31 December 2021
Accepted: 2 February 2022
Published: 6 February 2022

Publisher's Note: MDPI stays neutral with regard to jurisdictional claims in published maps and institutional affiliations.

1. Introduction

In asymmetrical rolling, the neutral point of the upper roll and lower roll is not aligned vertically due to there being different velocities on the upper and lower surface of the workpiece; this could be caused by the different diameters of the rolls, their different peripheral velocities or the different friction conditions between the workpiece and the rolls. Thus, within the deformation zone and beside the backward-slip zone and the forward-slip zone in a common rolling, there is a unique zone in the asymmetrical rolling process called the cross-shear zone. One obvious feature of the cross-shear zone is that the direction of friction force on the upper and lower surfaces of the strip is reversed.

Since the asymmetrical rolling theory was first introduced [1], extensive experimental investigations and theoretical analyses have been conducted on asymmetrical rolling. The curvature of a workpiece caused by asymmetrical rolling aroused great interest. The curvature of rolled material was experimentally investigated [2], and the effects of different work roll diameters and reduction ratios on the bending of the workpiece were analyzed by the finite element method [3]. Analytical models based on the slab method [4] and the finite element method [5] were built to predict strip curvature.

Since one of the advantages of asymmetrical rolling is that less roll force and less roll torque are required, various models are thus built to investigate the deformation in asymmetrical rolling and to analyze the influences of different rolling parameters. Analytical models were built by the slab method to investigate the deformation mechanism of the sheet at the roll gap during asymmetrical cold strip rolling [6]; an experimental study was also conducted [7], and the stream function method [8] and finite element method [9] were employed in further analysis. The influence of the friction coefficient ratio on shear deformation, rolling pressure and torque was investigated using slab analysis [10]. The plane strain asymmetrical rolling was analyzed by a model based on the slab method that considered the contact arc as the parabola [11]. Analytical models considering the shear stress along the vertical sides of each slab were built to calculate the roll force and torque [12]. An analytical model was built by the slab method to analyze the effects of the work roll radius, roll speed and the friction coefficient on rolling pressure, roll force and roll torque [13], and to study the relationship between the asymmetrical rolling deformation zone configuration and rolling parameters [14]. Analytical models were also built for the numerical study of multi-layer sheet rolling [15–17].

Asymmetrical rolling is beneficial to enlarging deformation, resulting in its outstanding capability in thickness reduction. The minimum thickness limit of symmetrical rolling was broken in the asymmetrical rolling experiment [18]. Analytical models were built to investigate the minimum thickness in asymmetrical cold rolling [19,20] by the slab method. A semi-empirical formula was built to calculate the minimum thickness on the basis of experimental and theoretical studies [21]. Analysis of the relationship between the deformation zone configuration and rolling parameters shows that the minimum thickness can be reached when the forward-slip disappears, and the rolling parameters keep the deformation zone configuration as a cross-shear zone and backward-slip zone (C + B), or an all-cross-shear zone (AC) [22]. The softening phenomenon [23] and size effect [24] in asymmetrical rolling of pure copper foil were studied. An analytical model based on the slab method, which considers the percentage of three regions in the plastic deformation zone, was used to study the effect of rolling parameters on the deformation zone configuration, and provided an accurate calculation of roll force and roll torque [25,26]. The main parameters of rolling a particular asymmetric regime were obtained from an asymmetric rolling process simulation for establishing a link between the peripheral speeds of the rolls, pressure and contact length [27].

Asymmetrical rolling technology is beneficial to grain refinement and texture control, so it is used as a method to improve the mechanical properties of materials. Ultrafine grain with an average size of 0.5μm was obtained by asymmetrical cold rolling of an AA1016 aluminum strip [28], and refined surface grain with an average size of ~3μm was obtained by asymmetrical hot rolling of non-magnetic austenitic steel [29]. After a two-time asymmetrical rolling and heat treatment of the AA1050 Al alloy sheet, the R-value of the sheet increased from 0.61 to 1.3 [30]. A single-pass asymmetric rolling was carried out on extra-low-carbon steel to investigate the influence of thickness reduction per pass on texture evolutions [31]. Models (FEM coupled with microstructure evolution models and cellular automata models) were also built to study the microstructure evolution of plates during asymmetrical rolling [32]. The effects of processing parameters of asymmetrical rolling on the mechanical properties of aluminum alloy AA6061 were investigated experimentally [33].

Though various models were built for asymmetrical rolling, most of them were suitable for plate and sheet rolling, and only several models were suitable for thin strip and foil rolling. The deformation behavior of a workpiece in thin strip and foil rolling are different from that in plate and sheet rolling; the friction and the tension play important roles in thin strip and foil rolling. Energy consumption, especially in work used to overcome the friction resistance, is important for the industrial application of asymmetrical rolling technology to thin strip and foil rolling. In this paper, analytical models based on the slab method are proposed for asymmetrical strip rolling, considering the three region (backward-slip zone, forward-slip zone and cross-shear zone) percentages in the deformation zone. The roll pressure, roll force, roll torque and roll power are calculated to obtain the friction work and plastic deformation work. The effects of rolling parameters, such as the speed ratio, entry thickness, front and back tension, the friction coefficient and the deformation resistance of the strip on friction work in asymmetrical thin strip rolling, are analyzed.

2. Mathematical Models

2.1. Basic Assumptions

To simplify the derivation of analytical models, the following assumptions are employed:

- The rolls are rigid bodies; the strips being rolled are rigid-plastic material.
- The friction coefficients between the strip and the roll are constant, but may be different on the upper and lower surface of the strip.
- The von Mises criterion of yield is adopted.
- The plastic deformation is a plane strain.
- Plane sections perpendicular to the direction of the rolling remain plane; stresses are uniformly distributed within each slab element.
- The contact arc is simplified as a string.

Figure 1 illustrates the schematic of a typical deformation zone in asymmetrical strip rolling. v_1 and v_2 are the peripheral speeds of the upper roll and lower roll, respectively, and $v_1 > v_2$. The deformation zone is divided into three regions according to the direction of frictional stresses between the rolls and the strip. The region between the exit of the deformation zone and the neutral point is a forward-slip zone (F), the region between the upper neutral point and the lower neutral point is a cross-shear zone (C) and the region between the lower neutral point and the entrance of the deformation zone is a backward-slip zone (B).

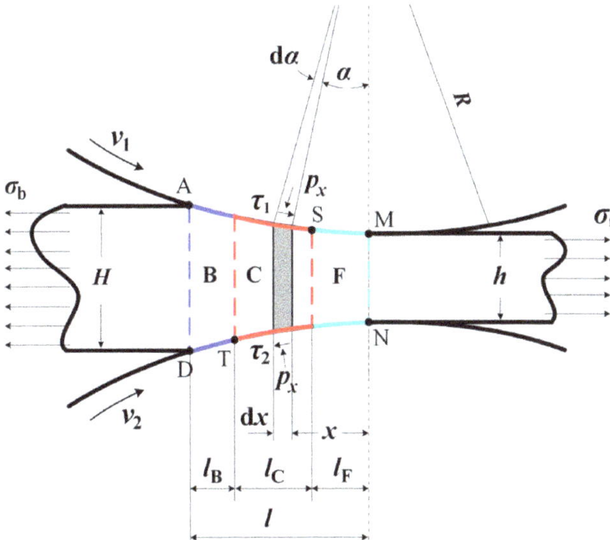

Figure 1. Schematic of deformation zone in asymmetrical strip rolling.

The lengths of the forward-slip zone, cross-shear zone and backward-slip zone are denoted by l_F, l_C, l_B and the length of contact $l = l_F + l_C + l_B$. The percentages of the three regions are expressed as $Q_B = l_B/l$, $Q_C = l_C/l$ and $Q_F = l_F/l$.

2.2. Rolling Pressure

Stresses on a slab in three regions are illustrated in Figure 2. From the horizontal force equilibrium on a slab in the cross-shear zone, we obtain

$$(\sigma_x + d\sigma_x)(h_x + dh_x) - \sigma_x h_x - 2p_x R d\alpha \sin\alpha + (\tau_1 - \tau_2)R d\alpha \cos\alpha = 0 \tag{1}$$

where $\tau_1 = p_x f_1$, $\tau_2 = p_x f_2$ are friction stresses on the upper and lower surfaces of the strip, respectively.

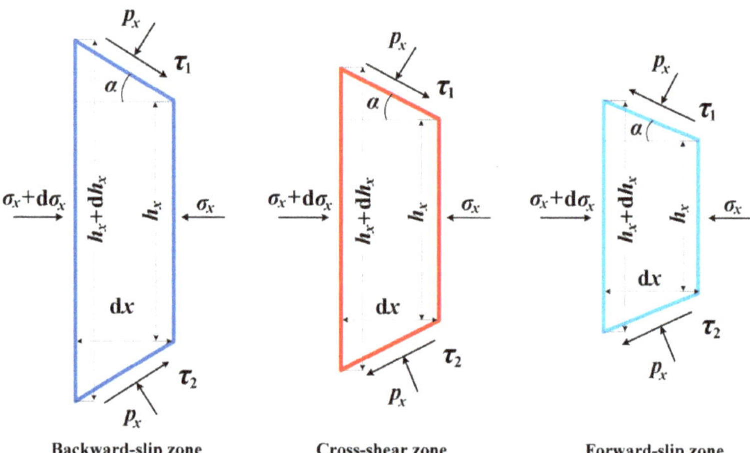

Figure 2. Stresses analysis on the slab in three deformation zones.

In cold strip rolling, contact angle α is very small. From geometries shown in Figures 1 and 2, it can be known that $R d\alpha = dx / \cos\alpha$, $dh_x = 2dx \tan\alpha$.

The yield criterion for the plane strain can be expressed as

$$p_x - \sigma_x = K \tag{2}$$

Referring to geometry in Figure 1, the thickness of a strip at a distance x from the center of the rolls can be expressed as

$$h_x = h + \frac{\Delta h}{l}x \tag{3}$$

Substituting Equations (2) and (3) into Equation (1) and rearranging it, we have

$$dp_x - \frac{dh_x}{h_x}(K - \delta_2 p_x) = 0 \tag{4}$$

Integrating Equation (4) with respect to x, we obtain

$$\frac{1}{\delta_2}\ln(\delta_2 p_x - K) - \ln\frac{1}{h_x} = C_C^* \tag{5}$$

Using the same method, we obtain similar equations for slabs in the forward-slip zone and backward-slip zone,

$$\frac{1}{\delta_1}\ln(\delta_1 p_x + K) - \ln h_x = C_F^* \tag{6}$$

$$\frac{1}{\delta_1}\ln(\delta_1 p_x - K) + \ln h_x = C_B^* \tag{7}$$

where $\delta_1 = \frac{(f_1+f_2)l}{\Delta h}$, $\delta_2 = \frac{(f_1-f_2)l}{\Delta h}$. C_F^*, C_C^* and C_B^* are integral constants for the forward-slip zone, cross-shear zone and backward-slip zone, respectively.

The boundary conditions on the entrance and exit of the deformation zone are: $h_x = H$, $p_x = K - \sigma_b$ and $h_x = h$, $p_x = K - \sigma_f$. Substituting these boundary conditions into Equations (6) and (7), integral constants C_B^* and C_F^* are obtained. The rolling pressure of the forward-slip zone and backward-slip zone can be expressed as

$$p_F = \frac{1}{\delta_1}\left[\left(\frac{h_x}{h}\right)^{\delta_1}(\delta_1(K-\sigma_f)+K)-K\right] \tag{8}$$

$$p_B = \frac{1}{\delta_1}\left[\left(\frac{H}{h_x}\right)^{\delta_1}(\delta_1(K-\sigma_b)-K)+K\right] \tag{9}$$

The boundary conditions on the interface between the cross-shear zone and backward-slip are: $h_x = H - Q_B\Delta h$, $p_x = p_B$. Substituting these boundary conditions into Equation (5), integral constant C_C^* can be obtained. Thus, rolling pressure of the cross shear zone is

$$p_C = \frac{K}{\delta_2}\left(\frac{H-Q_B\Delta h}{h_x}\right)^{\delta_2}\left[\frac{\delta_2}{\delta_1}(\delta_1\varepsilon_1-1){\mu_b}^{\delta_1}+\frac{\delta_2}{\delta_1}-1\right]+\frac{K}{\delta_2} \tag{10}$$

where $\mu_b = \frac{H}{H-Q_B\Delta h}$, $\mu_f = \frac{h+Q_F\Delta h}{h}$, $\varepsilon_1 = \frac{K-\sigma_b}{K}$.

2.3. Three Region Percentages

On the interface between the cross-shear zone and forward-slip zone, $p_C = p_F$, the thickness of the strip on the interface is $h_x = h + Q_F\Delta h$. Combing Equation (8), Equation (10) and the boundary conditions, we obtain

$${\mu_f}^{\delta_1}(\delta_1\varepsilon_2+1)-1 = i^{\delta_2}\left[(\delta_1\varepsilon_1-1)\left(\frac{\mu}{i}\right)^{\delta_1}\frac{1}{{\mu_f}^{\delta_1}}+1-\frac{\delta_1}{\delta_2}\right]+\frac{\delta_1}{\delta_2} \tag{11}$$

By solving Equation (11), the percentages of the forward-slip zone, cross-shear zone and backward-slip zone can be obtained.

$$Q_F = \frac{hX^{\frac{1}{\delta_1}}-h}{\Delta h} \tag{12}$$

$$Q_C = \frac{(i-1)hX^{\frac{1}{\delta_1}}}{\Delta h} \tag{13}$$

$$Q_B = \frac{H-ihX^{\frac{1}{\delta_1}}}{\Delta h} \tag{14}$$

$$X = \frac{\left(i^{\delta_2}-\frac{\delta_1}{\delta_2}i^{\delta_2}+\frac{\delta_1}{\delta_2}+1\right)+\sqrt{\left(i^{\delta_2}-\frac{\delta_1}{\delta_2}i^{\delta_2}+\frac{\delta_1}{\delta_2}+1\right)^2+4(\delta_1\varepsilon_2+1)(\delta_1\varepsilon_1-1)i^{\delta_2}\left(\frac{\mu}{i}\right)^{\delta_1}}}{2(\delta_1\varepsilon_2+1)} \tag{15}$$

where $\varepsilon_2 = \frac{K-\sigma_f}{K}$.

2.4. Roll Force

By integrating the normal rolling pressure along the arc of contact, the roll force can be obtained. Thus, the rolling force per unit width can be expressed as

$$P = \int_{l-Q_Bl}^{l} p_B dx + \int_{Q_Fl}^{l-Q_Bl} p_C dx + \int_0^{Q_Fl} p_F dx \tag{16}$$

Substituting Equations (8)–(10) into Equation (16), we have

$$P = Kl\left[\frac{H(\delta_1\varepsilon_1-1)\left(\mu_b^{\delta_1-1}-1\right)}{\delta_1\Delta h(\delta_1-1)} + \frac{Q_B}{\delta_1}\right] + Kl\left\{\frac{H-Q_B\Delta h}{\delta_2\Delta h(\delta_2-1)}\left(\mu_c^{\delta_2-1}-1\right)\left[\frac{\delta_2}{\delta_1}(\delta_1\varepsilon_1-1)\mu_b^{\delta_1}+\frac{\delta_2}{\delta_1}-1\right]+\frac{Q_C}{\delta_2}\right\}$$
$$+Kl\left[\frac{h(\delta_1\varepsilon_2+1)\left(\mu_f^{\delta_1+1}-1\right)}{\delta_1\Delta h(\delta_1+1)}-\frac{Q_F}{\delta_1}\right] \tag{17}$$

The roll force is $P_T = P * B_0$.

2.5. Roll Torque

The torque acting upon one roll can be obtained by integrating the torque acting upon the roll by the friction force of the unit area along the arc of contact. Frictional stress is $p_F f_1$, $p_C f_1$, $p_B f_1$ on the upper surfaces of the three regions illustrated in Figure 1, and $p_F f_2$, $p_C f_2$, $p_B f_2$ on the lower surfaces. The roll torque acting upon the upper work roll and lower work roll is

$$T_1 = R^2 \int_{\alpha_2}^{\alpha_l} p_B f_1 d\alpha + R^2 \int_{\alpha_1}^{\alpha_2} p_C f_1 d\alpha - R^2 \int_0^{\alpha_1} p_F f_1 d\alpha \tag{18}$$

$$T_2 = R^2 \int_{\alpha_2}^{\alpha_l} p_B f_2 d\alpha + R^2 \int_{\alpha_1}^{\alpha_2} p_C f_2 d\alpha - R^2 \int_0^{\alpha_1} p_F f_2 d\alpha \tag{19}$$

Substituting Equations (8)–(10) into Equations (18) and (19), and integrating Equations (18) and (19), we have

$$T_1 = \frac{KRlf_1}{\delta_1\Delta h}\left[\frac{H(\delta_1\varepsilon_1-1)}{1-\delta_1}\left(1-\mu_b^{\delta_1-1}\right)+\Delta hQ_B\right] + \frac{KRlf_1(H-Q_B\Delta h)\left(\mu_c^{\delta_2-1}-1\right)}{\delta_2(\delta_2-1)\Delta h}\left[\frac{\delta_2}{\delta_1}(\delta_1\varepsilon_1-1)\mu_b^{\delta_1}+\frac{\delta_2}{\delta_1}-1\right]$$
$$+\frac{Q_C KRlf_1}{\delta_2} - \frac{KRlf_1}{\delta_1\Delta h}\left[\frac{h(\delta_1\varepsilon_2+1)}{\delta_1+1}\left(\mu_f^{\delta_1+1}-1\right)-\Delta hQ_F\right] \tag{20}$$

$$T_2 = \frac{KRlf_2}{\delta_1\Delta h}\left[\frac{H(\delta_1\varepsilon_1-1)}{1-\delta_1}\left(1-\mu_b^{\delta_1-1}\right)+\Delta hQ_B\right] + \frac{KRlf_2(H-Q_B\Delta h)\left(\mu_c^{\delta_2-1}-1\right)}{\delta_2(\delta_2-1)\Delta h}\left[\frac{\delta_2}{\delta_1}(\delta_1\varepsilon_1-1)\mu_b^{\delta_1}+\frac{\delta_2}{\delta_1}-1\right]$$
$$+\frac{Q_C KRlf_2}{\delta_2} - \frac{KRlf_2}{\delta_1\Delta h}\left[\frac{h(\delta_1\varepsilon_2+1)}{\delta_1+1}\left(\mu_f^{\delta_1+1}-1\right)-\Delta hQ_F\right] \tag{21}$$

where $\mu_c = \frac{H-Q_B\Delta h}{h+Q_F\Delta h}$.

The total roll torque per unit width is $T = T_1 + T_2$.

2.6. Roll Power

In rotational systems, power is the product of the torque and angular velocity. The roll power of one roll can be obtained by the product of torque acting upon the roll and roll angular velocity. Roll power of the upper roll and lower roll can be expressed as

$$A_1 = T_1\frac{v_1}{R} = \frac{Klf_1v_1}{\delta_1\Delta h}\left[\frac{H(\delta_1\varepsilon_1-1)}{1-\delta_1}\left(1-\mu_b^{\delta_1-1}\right)+\Delta hQ_B\right]$$
$$+\frac{Klf_1v_1(H-Q_B\Delta h)\left(\mu_c^{\delta_2-1}-1\right)}{\delta_2(\delta_2-1)\Delta h}\left[\frac{\delta_2}{\delta_1}(\delta_1\varepsilon_1-1)\mu_b^{\delta_1}+\frac{\delta_2}{\delta_1}-1\right]+\frac{Q_C Klf_1v_1}{\delta_2}$$
$$-\frac{Klf_1v_1}{\delta_1\Delta h}\left[\frac{h(\delta_1\varepsilon_2+1)}{\delta_1+1}\left(\mu_f^{\delta_1+1}-1\right)-\Delta hQ_F\right] \tag{22}$$

$$A_2 = T_2\frac{v_2}{R} = \frac{Klf_2v_2}{\delta_1\Delta h}\left[\frac{H(\delta_1\varepsilon_1-1)}{1-\delta_1}\left(1-\mu_b^{\delta_1-1}\right)+\Delta hQ_B\right]$$
$$+\frac{Klf_2v_2(H-Q_B\Delta h)\left(\mu_c^{\delta_2-1}-1\right)}{\delta_2(\delta_2-1)\Delta h}\left[\frac{\delta_2}{\delta_1}(\delta_1\varepsilon_1-1)\mu_b^{\delta_1}+\frac{\delta_2}{\delta_1}-1\right]+\frac{Q_C Klf_2v_2}{\delta_2}$$
$$-\frac{Klf_2v_2}{\delta_1\Delta h}\left[\frac{h(\delta_1\varepsilon_2+1)}{\delta_1+1}\left(\mu_f^{\delta_1+1}-1\right)-\Delta hQ_F\right] \tag{23}$$

In asymmetrical strip rolling, tensions exerted by the coiler motors also contributed to the plastic deformation of the strip. The tensile power per unit area can be obtained by the product of tensile stress and the velocity of the strip. The power of the coiler and uncoiler due to the tensions are

$$A_{\mathrm{rf}} = \sigma_{\mathrm{f}} h v_{\mathrm{h}} \tag{24}$$

$$A_{\mathrm{rb}} = -\sigma_{\mathrm{b}} H v_{\mathrm{H}} \tag{25}$$

The total roll power per unit width required is $A_{\mathrm{T}} = A_1 + A_2 + A_{\mathrm{rf}} + A_{\mathrm{rb}}$.

2.7. Friction Power

Friction power on the unit area is equivalent to the product of friction stress and relative slipping velocity, and total friction power can be obtained by integrating it on the whole friction surface. Friction power on the upper surface of the strip in the deformation zone is

$$A_{\mathrm{f1}} = \int_0^{l_{\mathrm{F}}} p_{\mathrm{F}} f_1 \left(\frac{v_x}{\cos \alpha} - v_1 \right) dx + \int_{l_{\mathrm{F}}}^{l_{\mathrm{F}}+l_{\mathrm{C}}} p_{\mathrm{C}} f_1 \left(v_1 - \frac{v_x}{\cos \alpha} \right) dx + \int_{l_{\mathrm{F}}+l_{\mathrm{C}}}^{l} p_{\mathrm{B}} f_1 \left(v_1 - \frac{v_x}{\cos \alpha} \right) dx \tag{26}$$

In cold rolling of a thin strip, the contact angle α is very small; thus, $\cos \alpha \approx 1$. The mass flow relationship in the deformation zone can be expressed as $h_x v_x = h_1 v_1 = h_2 v_2$. Substituting Equations (8)–(10) into Equation (26), we obtain

$$
\begin{aligned}
A_{\mathrm{f1}} = {} & \frac{Kl f_1}{\delta_1 \Delta h} \left[\frac{h_2 v_2 (\delta_1 \varepsilon_2 + 1)\left(\mu_{\mathrm{f}}^{\delta_1} - 1 \right)}{\delta_1} - h_2 v_2 \ln \mu_{\mathrm{f}} - \frac{h v_1 (\delta_1 \varepsilon_2 + 1)\left(\mu_{\mathrm{f}}^{\delta_1+1} - 1 \right)}{\delta_1+1} + v_1 (h_1 - h) \right] \\[4pt]
& + \frac{Kl f_1}{\delta_2 \Delta h} \left\{ \frac{h_2 v_1 \left(\mu_{\mathrm{c}}^{\delta_2-1} - 1 \right)}{\delta_2-1} \left[\frac{\delta_2}{\delta_1}(\delta_1 \varepsilon_1 - 1)\mu_{\mathrm{b}}^{\delta_1} + \frac{\delta_2}{\delta_1} - 1 \right] + v_1 (h_2 - h_1) \right. \\[4pt]
& \left. \quad - \frac{h_2 v_2 \left(\mu_{\mathrm{c}}^{\delta_2} - 1 \right)}{\delta_2} \left[\frac{\delta_2}{\delta_1}(\delta_1 \varepsilon_1 - 1)\mu_{\mathrm{b}}^{\delta_1} + \frac{\delta_2}{\delta_1} - 1 \right] - h_2 v_2 \ln \mu_{\mathrm{c}} \right\} \\[4pt]
& + \frac{Kl f_1}{\delta_1 \Delta h} \left[\frac{H v_1 (\delta_1 \varepsilon_1 - 1)\left(\mu_{\mathrm{b}}^{\delta_1-1} - 1 \right)}{\delta_1-1} + v_1 (H - h_2) - \frac{h_2 v_2 (\delta_1 \varepsilon_1 - 1)\left(\mu_{\mathrm{b}}^{\delta_1} - 1 \right)}{\delta_1} \right. \\[4pt]
& \left. \quad - h_2 v_2 \ln \mu_{\mathrm{b}} \right]
\end{aligned}
\tag{27}
$$

Friction power on the lower surface of the strip in the deformation zone can be obtained using the same method.

$$
\begin{aligned}
A_{\mathrm{f2}} = {} & \frac{Kl f_2}{\delta_1 \Delta h} \left[\frac{h_2 v_2 (\delta_1 \varepsilon_2 + 1)\left(\mu_{\mathrm{f}}^{\delta_1} - 1 \right)}{\delta_1} - h_2 v_2 \ln \mu_{\mathrm{f}} - \frac{h v_2 (\delta_1 \varepsilon_2 + 1)\left(\mu_{\mathrm{f}}^{\delta_1+1} - 1 \right)}{\delta_1+1} + v_2 (h_1 - h) \right] \\[4pt]
& + \frac{Kl f_2}{\delta_2 \Delta h} \left\{ -\frac{h_2 v_1 \left(\mu_{\mathrm{c}}^{\delta_2-1} - 1 \right)}{\delta_2-1} \left[\frac{\delta_2}{\delta_1}(\delta_1 \varepsilon_1 - 1)\mu_{\mathrm{b}}^{\delta_1} + \frac{\delta_2}{\delta_1} - 1 \right] + v_2 (h_2 - h_1) \right. \\[4pt]
& \left. \quad + \frac{h_2 v_2 \left(\mu_{\mathrm{c}}^{\delta_2} - 1 \right)}{\delta_2} \left[\frac{\delta_2}{\delta_1}(\delta_1 \varepsilon_1 - 1)\mu_{\mathrm{b}}^{\delta_1} + \frac{\delta_2}{\delta_1} - 1 \right] - h_2 v_2 \ln \mu_{\mathrm{c}} \right\} \\[4pt]
& + \frac{Kl f_2}{\delta_1 \Delta h} \left[\frac{H v_2 (\delta_1 \varepsilon_1 - 1)\left(\mu_{\mathrm{b}}^{\delta_1-1} - 1 \right)}{\delta_1-1} + v_2 (H - h_2) - \frac{h_2 v_2 (\delta_1 \varepsilon_1 - 1)\left(\mu_{\mathrm{b}}^{\delta_1} - 1 \right)}{\delta_1} \right. \\[4pt]
& \left. \quad - h_2 v_2 \ln \mu_{\mathrm{b}} \right]
\end{aligned}
\tag{28}
$$

The friction power per unit width can be expressed as $A_{\mathrm{f}} = A_{\mathrm{f1}} + A_{\mathrm{f2}}$.

2.8. Energy Consumption in Asymmetrical Rolling

The energy consumption in asymmetrical strip rolling refers to the work consumed during the process of rolling a strip of a certain size to target thickness. In cold strip manufacturing, the main energy-consuming process is rolling. Therefore, only energy consumption in the rolling process of asymmetrical strip rolling is investigated in this work. The energy consumption in the rolling process mainly consists of energy consumption in rolling, energy consumption of auxiliary equipment and energy consumption due to friction in the drive system. To simplify the analysis, only the energy consumption in rolling is taken into account in this work.

The energy consumption in rolling is equivalent to the work completed by the motors of the rolling mill and the motors of the coiler and uncoiler. The energy consumption in one pass can be expressed as

$$W_\mathrm{T} = W_1 + W_2 + W_\mathrm{rf} + W_\mathrm{rb} \tag{29}$$

where $W_1 = \frac{A_1 B_0 t}{60000}$, $W_2 = \frac{A_2 B_0 t}{60000}$, $W_\mathrm{rf} = \frac{A_\mathrm{rf} B_0 t}{60000}$, $W_\mathrm{rb} = \frac{A_\mathrm{rb} B_0 t}{60000}$.

The work performed in the deformation during plastic working is equal to the work of external forces. In a rolling pass, rolling work (W_T) is equivalent to the sum of the plastic deformation work of the strip (W_d) and the friction work on the contact surface between the strip and rolls (W_f).

$$W_\mathrm{T} = W_\mathrm{d} + W_\mathrm{f} \tag{30}$$

Friction work W_f consists of W_f1 (on the upper surface) and W_f2 (on the lower surface)

$$W_\mathrm{f} = W_\mathrm{f1} + W_\mathrm{f2} \tag{31}$$

where $W_\mathrm{f1} = \frac{A_\mathrm{f1} B_0 t}{60000}$, $W_\mathrm{f2} = \frac{A_\mathrm{f2} B_0 t}{60000}$.

The proportion of friction work (Q_f) and proportion of plastic deformation work (Q_d) can be expressed as $Q_\mathrm{f} = \frac{Q_\mathrm{f}}{W_\mathrm{T}} \times 100\%$ and $Q_\mathrm{d} = \frac{Q_\mathrm{d}}{W_\mathrm{T}} \times 100\%$, respectively.

3. Results and Discussion

Several asymmetrical strip rolling experiments have been conducted on the asymmetrical mill, with a work roll diameter of 90 mm, to verify the validity of the proposed models. The lower roll served as the slow roll and was set to a peripheral velocity of 4.0 m/min, while the peripheral velocity of the upper roll varied with the speed ratio. Emulsion was used as a lubricant. The 430 stainless steel strips, with 80.0 mm in width, were rolled from 0.150 mm to 0.105 mm under different speed ratios. By fitting the tensile test data of 430 stainless steel strips with different reduction ratios, the plane deformation resistance can be expressed as

$$K = 407.9 + 588.6\bar{\varepsilon}^{0.718} \tag{32}$$

where $\bar{\varepsilon} = 0.4\left(1 - \frac{H}{H_0}\right) + 0.6\left(1 - \frac{h}{H_0}\right)$.

The model-calculated roll force is compared with the experiment-measured roll force. A comparison of roll force between experimental measurement values and those calculated by the models is illustrated in Figure 3a. It was found that both the experimental and the calculated roll force decreased with an increasing speed ratio. The maximum error and RMS error between the experimentally measured roll forces and calculated ones are 9.7% and 5.9%, respectively. The model-calculated forces are also compared with the results of models from the literature in Figure 3b. Models from the literature cannot finish calculations in the third point. All model-calculated roll forces are less than experimental values, but the present model has better accuracy than models from the literature. Data used in the calculation are listed in Table 1.

The proposed models are used to calculate rolling pressure, roll force, roll torque and roll power, as well as friction work and plastic deformation work in asymmetrical strip rolling. A work roll diameter of 88 mm, and 430 stainless steel strips of initial length L_0 = 1000 m and initial width B_0 = 100 mm, are used in the analysis. From the calculation results, the effects of rolling parameters on friction work and plastic deformation work are analyzed.

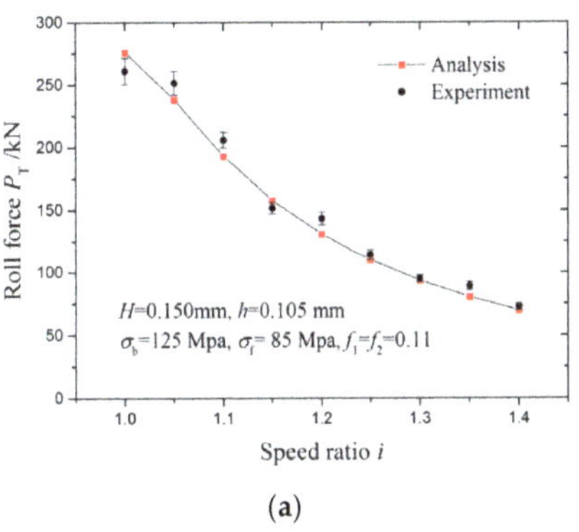

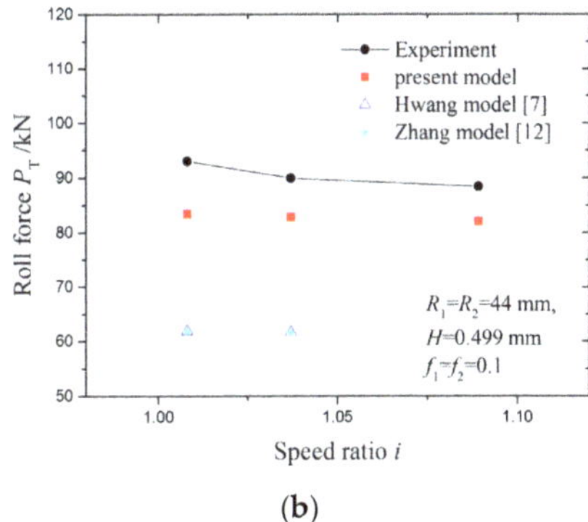

(a)   (b)

Figure 3. Comparisons of model-calculated roll force with: (**a**) experimental results; (**b**) experimental results and results of literature models.

Table 1. Data for roll force calculation.

Entry Thickness, mm	Reduction, %	Back Tension, MPa	Front Tension, MPa	Speed Ratio	Yield Shear Strength, MPa	Experimental Force, kN
	14.33	111.35	85.92	1.008	257.0	93.07
0.499	14.79	111.78	99.72	1.037	258.2	89.95
	15.87	116.73	98.52	1.089	261.2	88.40

Figure 4 illustrates the effects of the speed ratio on friction work, plastic deformation work and rolling work in a single rolling pass with the same reduction ratio. It can be seen that the upper friction work increases significantly with an increasing speed ratio, but the lower friction work only increases slightly. This is because the percentage of cross-shear zone increases with an increasing speed ratio. As a result, the upper and lower neutral points move toward the exit and the entrance of the deformation zone, respectively, thus increasing the slipping distance between the upper roll and the strip. Therefore, the friction work increases. The plastic deformation work remains unchanged under a different speed ratio because the deformation degree is the same.

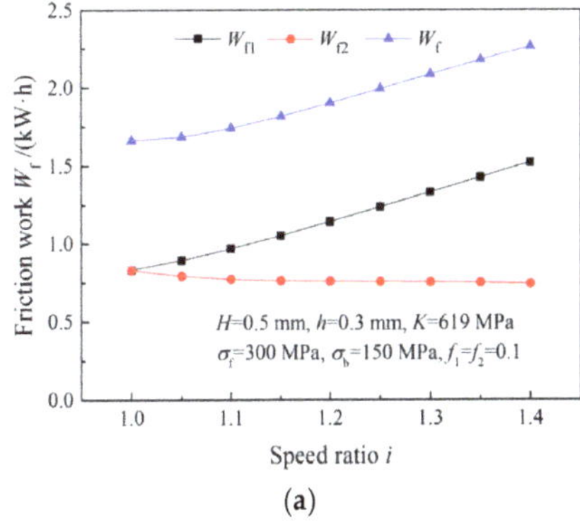

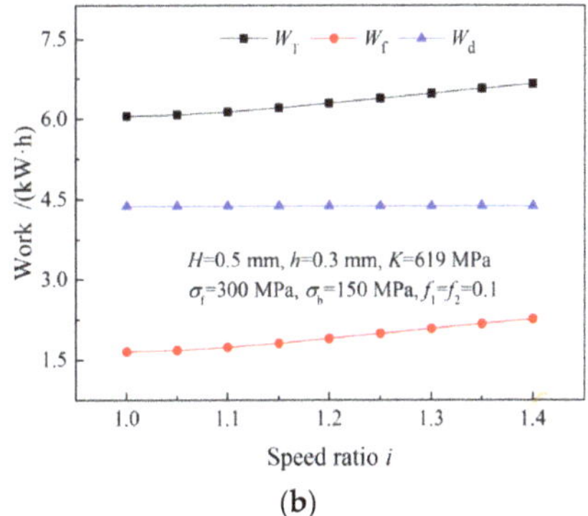

(a)   (b)

Figure 4. Effects of speed ratio on: (**a**) friction work, (**b**) plastic deformation work and rolling work, in a single pass.

Figure 5 illustrates the effects of the speed ratio on total friction work, total plastic deformation work, total rolling work, the proportion of total friction work and the proportion of total plastic deformation work in multi-pass rolling. The curves in Figure 5a,b were obtained from five-pass rolling with the same reduction ratio, while the curves in Figure 5c,d were obtained from multi-pass rolling with the same roll force. It is seen that both the total friction work and the total rolling work increase as the speed ratio increases, but the total plastic deformation work remains constant, as in Figure 5a,c. Owing to the same level of accumulated reduction, the total plastic deformation work is not affected by the speed ratio. The sum of accumulated relative velocity between rolls and strips and the sum of the slipping distance increase with an increasing speed ratio. Consequently, the total friction work increases, leading to an increase in the total rolling work. Correspondingly, the proportion of total friction work gradually increases, and the proportion of total plastic deformation work gradually decreases with an increasing speed ratio, as shown in Figure 5b,d. The proportion of total friction work exceeds the proportion of total plastic deformation work at a certain speed ratio.

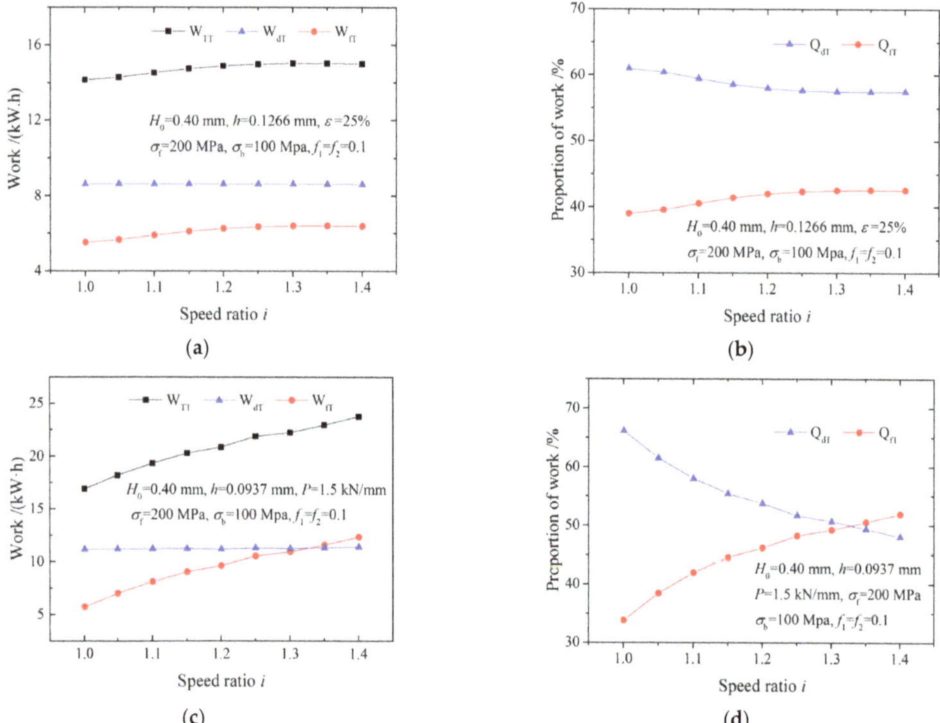

Figure 5. Sum of works and proportions of friction work and plastic deformation work in multi-pass rolling: Work (**a**), proportion of friction work and plastic deformation work (**b**) in a five-pass rolling with same reduction ratio; work (**c**), proportion of friction work and plastic deformation work (**d**) in multi-pass rolling with same roll force.

Figure 6 illustrates the effects of the speed ratio and entry thickness on friction work, plastic deformation work and its proportions. According to Figure 6, we know that the plastic deformation work increases with both an increasing speed ratio and an increasing entry thickness. This is because both an increase in entry thickness and an increase in speed ratio increase the magnitude of deformation, thus resulting in an increase in plastic deformation work. Friction work increases with an increasing speed ratio for the same entry thickness, and also increases with increasing thickness for the same speed ratio. As the speed ratio increases, the percentage of cross-shear zone increases, and accumulated relative velocity between rolls and strips consequently increases. As a result, friction work increases.

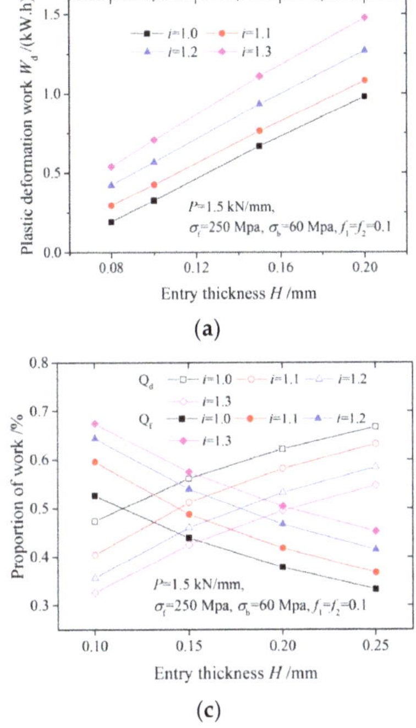

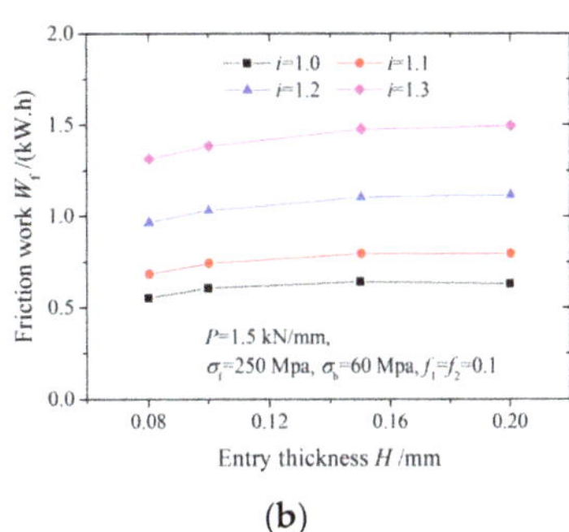

Figure 6. Variation of (**a**) plastic deformation work, (**b**) friction work, and (**c**) proportions of plastic deformation work and friction work with entry thickness.

Figure 7 illustrates the effects of the friction coefficient and speed ratio on total friction work and its proportions in total rolling work. As the speed ratio increases, both the total friction work and the proportion of total friction work increase. The total friction work and the proportion of total friction work also increase with an increased friction coefficient.

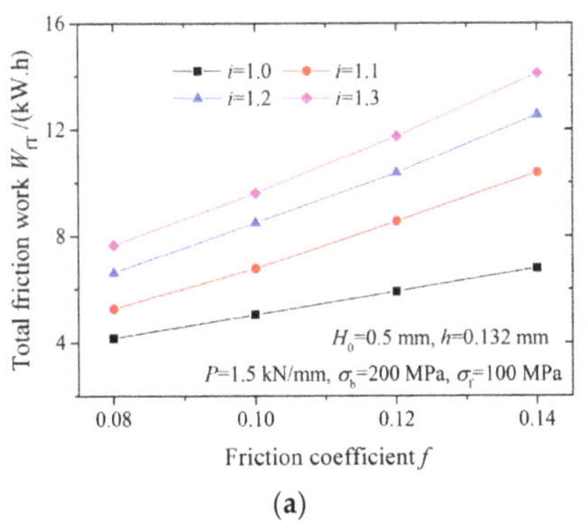

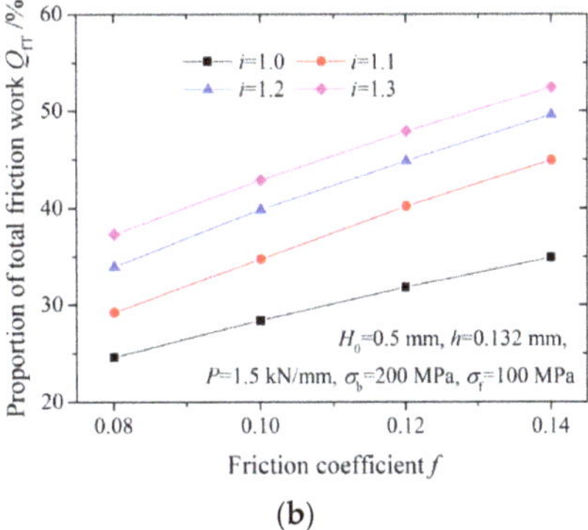

Figure 7. Variation of (**a**) total friction work and (**b**) proportion of total friction work with friction coefficient.

The curves in Figure 8 show the effects of plane deformation resistance and the speed ratio on the total plastic deformation work, total friction work and proportion of total friction work in multi-pass rolling. Given that the deformation force increases with an increase in plane deformation resistance, and the levels of accumulated reduction are the same, the total plastic deformation work increases with an increase in plane deformation resistance. The increase in plane deformation resistance causes the percentage of the cross-shear zone to increase, thus increasing the total friction work. The proportion of total friction work decreases with an increase in plane deformation resistance because the increment rate in the total plastic deformation work is greater than in the total friction work.

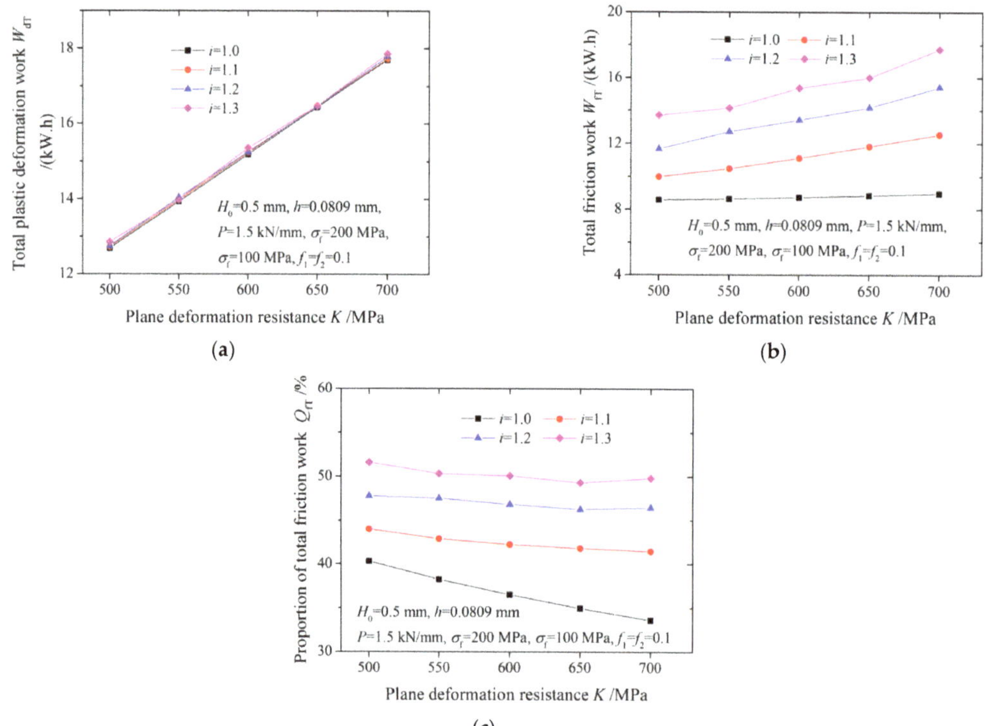

Figure 8. Variation of (**a**) total plastic deformation work, (**b**) total friction work, and (**c**) proportion of total friction work with plane deformation resistance.

Figure 9 shows the effects of the speed ratio on the total friction work and the proportions of total friction work under various back tensions. The percentage of the cross-shear zone decreases with the increase in back tension, resulting in a decrease in the relative slip between the strip and roll. Therefore, the total friction work decreases slightly with an increasing back tension.

The effects of front tension are similar to that of back tension.

The effects of the entry thickness and the friction coefficient on the proportions of friction work are shown in Figure 10. There is a special line on which the proportion of friction work is equal to 50%; in other words, the proportion of friction work is the same as the proportion of plastic deformation work. At points T1, T2, and T3 in Figure 10a, which are called the threshold points, the curve is divided into two parts, and the friction work is greater than the plastic deformation work on the left part. On the contrary, at points S1, S2, and S3 in Figure 10b, the friction work is greater than the plastic deformation work on the right part.

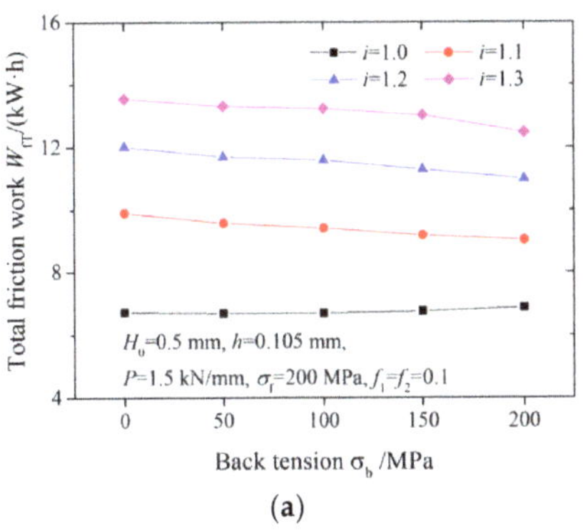

(a)

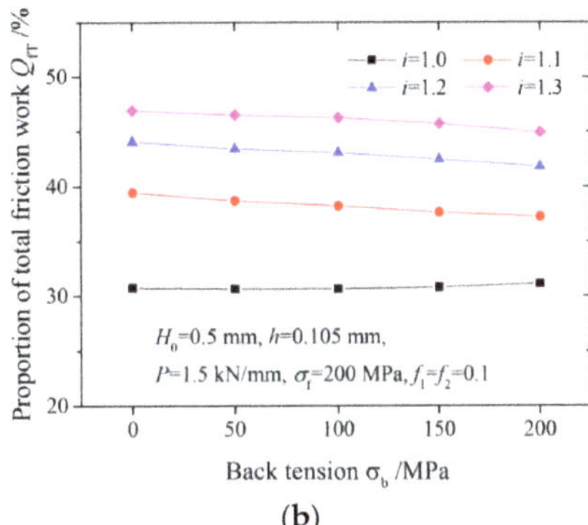

(b)

Figure 9. Variation of (**a**) total friction work and (**b**) proportion of total friction work with back tension.

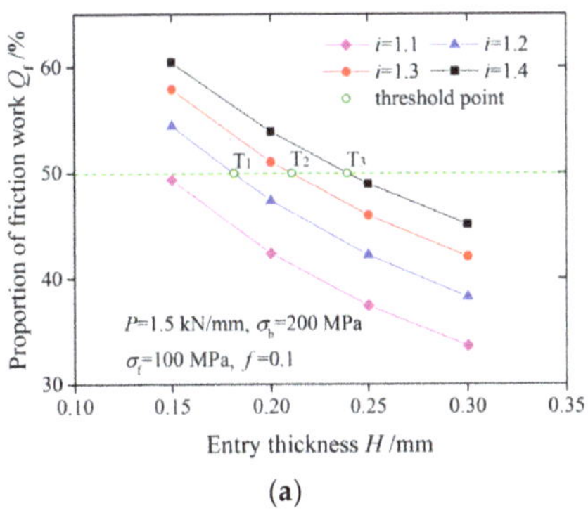

(a)

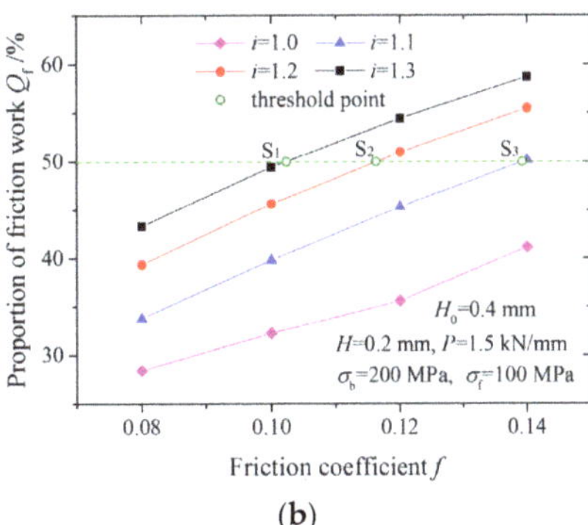

(b)

Figure 10. Variation of threshold point and proportions of friction work with: (**a**) entry thickness; (**b**) friction coefficient.

The threshold point is related to friction conditions in the deformation zone and the deformation degree; it is affected by parameters such as thickness, the friction coefficient, the speed ratio, front and back tension, deformation resistance and the reduction ratio.

4. Conclusions

Mathematic models based on the slab method are built to calculate the rolling pressure, roll force, roll torque, friction power, rolling power and rolling work for the three typical region deformation zone configurations in the asymmetrical rolling of a thin strip. The effects of rolling parameters on energy consumption (friction work and plastic deformation work) are investigated by analyzing computed results. The following conclusions are obtained.

- In all cases, an increase in speed ratio leads to the increase in friction work, thus resulting in the increase in the rolling work, increase in the proportion of friction work and decrease in the proportion of plastic deformation work.
- The increase in entry thickness, deformation resistance of the strip and the friction coefficient cause friction work and rolling work to increase. The increase in entry thickness and deformation resistance also leads to an increase in plastic deformation work.
- The increase in the friction coefficient leads to an increase in the proportion of friction work. The increase in deformation resistance of the strip, front tension and back tension result in a decrease in the proportion of friction work.

- In conditions of a very thin strip being rolled with a large speed ratio, the proportion of friction work exceeds that of plastic deformation work. The concept of the threshold point, at which the friction work is equal to the plastic deformation work, was proposed for this case.
- The comparison between the roll forces obtained from the experimental measurement and calculation results shows a maximum error of 9.7% and an RMS error of 5.9%.

Author Contributions: Conceptualization, X.S.; methodology, X.S.; software, Q.Z. and X.S.; validation, Q.Z., X.S. and X.L.; formal analysis, Q.Z. and X.S.; investigation, Q.Z.; resources, X.S.; data curation, Q.Z.; writing—original draft preparation, Q.Z.; writing—review and editing, X.L.; visualization, Q.Z.; supervision, X.L.; project administration, X.L. All authors have read and agreed to the published version of the manuscript.

Funding: This work was financially supported by the program for the National Natural Science Foundation of China No. 51374069.

Institutional Review Board Statement: Not applicable.

Informed Consent Statement: Not applicable.

Data Availability Statement: Not applicable.

Conflicts of Interest: The authors declare no conflict of interest.

References

1. Siebel, E. Zur Theorie des Walzvorganges bei ungleich angetriebenen Walzen. *Arch. Eisenhüttenw.* **1941**, *15*, 125–128. [CrossRef]
2. Johnson, W.; Needham, G. *An Experimental Study of Asymmetrical Rolling*; Applied Mechanics Convention: Cambridge, UK, 1966.
3. Lu, J.-S.; Harrer, O.-K.; Schwenzfeier, W.; Fischer, F.D. Analysis of the bending of the rolling material in asymmetrical sheet rolling. *Int. J. Mech. Sci.* **2000**, *42*, 49–61. [CrossRef]
4. Salimi, M.; Sassani, F. Modified slab analysis of asymmetrical plate rolling. *Int. J. Mech. Sci.* **2002**, *44*, 1999–2023. [CrossRef]
5. Hao, L.; Di, H.S.; Gong, D.Y. Analysis of Sheet Curvature in Asymmetrical Cold Rolling. *J. Iron Steel Res. Int.* **2013**, *20*, 34–37. [CrossRef]
6. Hwang, Y.M.; Tzou, G.Y. An analytical approach to asymmetrical cold strip rolling using the slab method. *J. Mater. Eng. Perform.* **1993**, *2*, 597–606. [CrossRef]
7. Hwang, Y.M.; Tzou, G.Y. Analytical and experimental study on Asymmetrical sheet rolling. *Int. J. Mech. Sci.* **1997**, *39*, 289–303. [CrossRef]
8. Hwang, Y.M.; Chen, T.H. Analysis of asymmetrical sheet rolling by stream function method. *JSME Int. J. Ser. Mech. Mater. Eng.* **1996**, *39*, 598–605. [CrossRef]
9. Hwang, Y.M.; Chen, D.C.; Tzou, G.Y. Study on Asymmetrical Sheet Rolling by the Finite Element Method. *Chin. J. Mech.* **1999**, *15*, 149–155. [CrossRef]
10. Gao, H.; Ramalingam, S.C.; Barber, G.C. Analysis of asymmetrical cold rolling with varying coefficients of friction. *J. Mate. Process. Technol.* **2002**, *124*, 178–182. [CrossRef]
11. Tian, Y.; Guo, Y.H.; Wang, Z.D.; Wang, G.D. Analysis of rolling pressure in asymmetrical rolling process by slab method. *J. Iron Steel Res. Int.* **2009**, *16*, 22–26, 38. [CrossRef]
12. Zhang, S.H.; Zhao, D.W.; Gao, C.R.; Wang, G.D. Analysis of asymmetrical sheet rolling by slab method. *Int. J. Mech. Sci.* **2012**, *65*, 168–176. [CrossRef]
13. Wang, J.; Liu, X.H.; Guo, W.P. Analysis of mechanical parameters for asymmetrical strip rolling by slab method. *Int. J. Adv. Manuf. Technol.* **2018**, *98*, 2297–2309. [CrossRef]
14. Wang, J.; Liu, X.H.; Sun, X.K. Study on the relationship between asymmetrical rolling deformation zone configuration and rolling parameters. *Int. J. Mech. Sci.* **2020**, *187*, 105905. [CrossRef]
15. Pan, S.C.; Huang, M.N.; Tzou, G.-Y.; Syu, S.W. Analysis of asymmetrical cold and hot bond rolling of unbounded clad sheet under constant shear friction. *J. Mate. Process. Technol.* **2006**, *177*, 114–120. [CrossRef]
16. Skoblik, R.; Rydz, D.; Stradomski, G. Analysis of Asymmetrical Rolling Process of Multilayer Plates. *Solid State Phenom.* **2010**, *165*, 348–352. [CrossRef]
17. Wang, H.Y.; Zhang, D.H.; Zhao, D.W. Analysis of Asymmetrical Rolling of Unbonded Clad Sheet by Slab Method Considering Vertical Shear Stress. *ISIJ Int.* **2015**, *55*, 1058–1066. [CrossRef]
18. Zhu, Q.; Yu, J.M.; Qi, K.M. Problem of reducibility in sheet cold-rolling and gauge suitable to roll. *J. North. Uni. Technol.* **1988**, *4*, 420–426.
19. Tzou, G.Y.; Huang, M.N. Study on the minimum thickness for the asymmetrical PV cold rolling of sheet. *J. Mate. Process. Technol.* **2000**, *105*, 344–351. [CrossRef]

20. Tang, D.L.; Liu, X.H.; Li, X.Y.; Peng, L.G. Permissible Minimum Thickness in Asymmetrical Cold Rolling. *J. Iron Steel Res. Int.* **2013**, *20*, 21–26. [CrossRef]
21. Tang, D.L.; Liu, X.H.; Song, M.; Yu, H.L. Experimental and Theoretical Study on Minimum Achievable Foil Thickness during Asymmetric Rolling. *PLoS ONE* **2014**, *9*, e106637. [CrossRef]
22. Wang, J.; Liu, X.H. Study on minimum rollable thickness in asymmetrical rolling. *Int. J. Adv. Manuf. Technol.* **2021**. [CrossRef]
23. Chen, J.Q.; Hu, X.L.; Liu, X.H. Softening Effect on Fracture Stress of Pure Copper Processed by Asynchronous Foil Rolling. *Materials* **2019**, *12*, 2319. [CrossRef]
24. Song, M.; Liu, X.H.; Liu, L.Z. Size effect on mechanical properties and texture of pure copper foil by cold rolling. *Materials* **2017**, *10*, 538. [CrossRef]
25. Sun, X.K.; Liu, X.H.; Wang, J.; Qi, J.L. Analysis of asymmetrical rolling of strip considering percentages of three regions in deformation zone. *Int. J. Adv. Manuf. Technol.* **2020**, *110*, 763–775. [CrossRef]
26. Sun, X.K.; Liu, X.H.; Wang, J.; Qi, J.L. Analysis of asymmetrical rolling of strip considering two deformation region types. *Int. J. Adv. Manuf. Technol.* **2020**, *110*, 2767–2785. [CrossRef]
27. Alexa, V.; Kiss, I.; Cioată, V.G.; Rațiu, S.A. Modelling and simulation of the asymmetric rolling process—Establishing the optimal technology parameters to asymmetric rolling. *IOP Conf. Ser. Mater. Sci. Eng.* **2019**, *477*, 012025. [CrossRef]
28. Song, M.; Liu, X.H.; Liu, X.; Liu, L.Z. Ultrafine microstructure and texture evolution of aluminum foil by asymmetric rolling. *J. Cent. South Univ.* **2017**, *24*, 2783–2792. [CrossRef]
29. Li, C.S.; Ma, B.; Song, Y.L.; Zheng, J.J.; Wang, J.K. Grain refinement of non-magnetic austenitic steels during asymmetrical hot rolling process. *J. Mater. Sci. Technol.* **2017**, *33*, 1572–1576. [CrossRef]
30. Nam, S.K.; Lee, J.H.; Kim, G.H.; Lee, D.N.; Kim, I. Texture Analysis for Enhancement of R-value in Asymmetrically Rolled Al Alloy Sheet. *J. Mater. Eng. Perform.* **2019**, *28*, 5186–5194. [CrossRef]
31. Dhinwal, S.S.; Toth, L.S.; Lapovok, R.; Hodgson, P.D. Tailoring One-Pass Asymmetric Rolling of Extra Low Carbon Steel for Shear Texture and Recrystallization. *Materials* **2019**, *12*, 1935. [CrossRef] [PubMed]
32. Zhang, T.; Li, L.; Lu, S.H.; Gong, H.; Wu, Y.X. Comparisons of different models on dynamic recrystallization of plate during asymmetrical shear rolling. *Materials* **2018**, *11*, 151. [CrossRef] [PubMed]
33. Amegadzie, M.Y.; Bishop, D.P. Effect of asymmetric rolling on the microstructure and mechanical properties of wrought 6061 aluminum. *Mater. Today Commun.* **2020**, *25*, 101283. [CrossRef]

Article

Effect of CeO$_2$ Size on Microstructure, Synthesis Mechanism and Refining Performance of Al-Ti-C Alloy

Yanli Ma [1], Taili Chen [2], Lumin Gou [2] and Wanwu Ding [1,2,*]

[1] Technology Center, Jiuquan Iron and Steel (Group) Co., Ltd., Jiayuguan 735100, China; mayanli@jiugang.com
[2] School of Materials Science and Engineering, Lanzhou University of Technology, Lanzhou 730050, China; chen_taili@163.com (T.C.); goulumin@163.com (L.G.)
* Correspondence: dingww@lut.cn; Tel.: +86-1511-7183-365

Abstract: The effects of CeO$_2$ size on the microstructure and synthesis mechanism of Al-Ti-C alloy were investigated using a quenching experiment method. A scanning calorimetry experiment was used to investigate the synthesis mechanism of TiC, the aluminum melt in situ reaction was carried out to synthesize master alloys and its refining performance was estimated. The results show that the Al-Ti-C-Ce system is mainly composed of α-Al, Al$_3$Ti, TiC and Ti$_2$Al$_{20}$Ce. The addition of CeO$_2$ obviously speeds up the progress of the reaction, reduces the size of Al$_3$Ti and TiC and lowers the formation temperature of second-phase particles. When the size of CeO$_2$ is 2–4 μm, the promotion effect on the system is most obvious. The smaller the size of CeO$_2$, the smaller the size of Al$_3$Ti and TiC and the lower the formation temperature. Al-Ti-C-Ce master alloy has a significant refinement effect on commercial pure aluminum. When the CeO$_2$ size is 2–4 μm, the grain size of commercial pure aluminum is refined to 227 μm by Al-Ti-C-Ce master alloy.

Keywords: CeO$_2$ size; quenching experiment method; DSC; refining performance

Citation: Ma, Y.; Chen, T.; Gou, L.; Ding, W. Effect of CeO$_2$ Size on Microstructure, Synthesis Mechanism and Refining Performance of Al-Ti-C Alloy. *Materials* **2021**, 14, 6739. https://doi.org/10.3390/ma14226739

Academic Editor: Daniela Kovacheva

Received: 6 September 2021
Accepted: 2 November 2021
Published: 9 November 2021

Publisher's Note: MDPI stays neutral with regard to jurisdictional claims in published maps and institutional affiliations.

1. Introduction

A number of studies show that grain refiner for aluminum and its alloys can significantly improve the mechanical properties, casting properties, deformation processing properties and surface quality of materials [1,2].

Al-Ti-B master alloy has been the most widely used grain refiner and has been applied commercially [3]. However, many defects have been found in the application of Al-Ti-B master alloy thus limiting its development, i.e., a larger aggregation tendency of TiB$_2$ in the melt, large size, and the possibility of Zr and Si poisoning [4,5]. In order to obtain a good refining efficiency, a series of novel master alloys have emerged, such as Al-Ti-C, Al-Ti-B-C and Al-Ti-C/B-RE master alloy [6–9]. Al-Ti-C master alloy especially has been extensively studied by researchers, and it has been found that it can be remedy certain defects of Al-Ti-B master alloy [10,11]. Unfortunately, the refining efficiency of Al-Ti-C is unstable, and the possible reasons for this are that the active of TiC is insufficient, TiC in the melt aggregates easily and the morphology of Al$_3$Ti and TiC is easily affected by preparation conditions. For the most part, the wetting between Al and C is poor, and the formation of TiC is therefore difficult [12].

Ao et al. [13] found that La$_2$O$_3$ added to the Al-Ti-C-CuO system could promote the wettability of C and Al melt. The results of thermodynamic analysis from differential scanning calorimetry (DSC) experiments showed that La$_2$O$_3$ could promote metastable phase transfer into stable phases. Furthermore, the even distribution of TiC can directly influence the refining efficiency of Al-Ti-C. Wang L. D. et al. [14] analyzed the thermodynamics of the Al-Ti-C-Ce system using DSC experiments, and the result showed that rare earth oxide promoted the reaction and made the distribution of TiC even, forming a new Ti$_2$Al$_{20}$Ce phase. Moreover, CeO$_2$ has a catalytic effect and can promote the reaction. Korotcenkov et al. [15] considered that the catalytic activity of CeO$_2$ is closely related to its structure, morphology and size. The research of Auffan et al. [16] and Liu et al. [17] shows that CeO_2 nanoparticles have a high specific surface area, and the smaller the size of CeO_2 nanoparticles, the higher the specific surface area and the greater the activity. Preliminary studies have shown that when the content of CeO_2 was 4 wt.%, it had a significant promoting effect on the Al-Ti-C system [18]. On this basis, it is necessary to further study the effect of CeO_2 particle size on phase transition and microstructure transformation of Al-Ti-C system.

In this study, the aluminum melt in situ reaction was carried out to prepare Al-Ti-C-Ce master alloys. The influence of the size of CeO_2 on the thermodynamic process, reaction products and microstructures of the prepared Al-Ti-C-Ce system were studied in detail through quenching experiments, and the reaction mechanism of the Al-Ti-C-Ce system is summarized in this paper. The goal of the abovementioned studies was to determine the influence mechanism on the microstructure and phase transformation with different sizes of CeO_2 in the Al-Ti-C-Ce system based on thermodynamic analysis and DSC experiments.

2. Experiment

2.1. Quenching Experiments

Al powders (99.0 wt.%, 80–100 μm), Ti powders (99.0 wt.%, 45–65 μm) and C powders (99.0 wt.%, 10–20 μm) were prepared with Al/Ti/C = 5:2:1 (molar ratio) and 4 wt.% CeO_2 powder (99.0 wt.%, with different average sizes, i.e., 30 nm, 50 nm, 1 μm, 2–4 μm). The mixture was evenly mixed in a PULVERISETTE-5 high-speed planetary ball mill with a ball-to-material ratio = 3:1. The rotation speed and the total milling time were 350 r/min and 3 h, respectively. Then, 50 g of the mixture was cold-pressed into a cylindrical prefabricated block of Φ 25 mm $\times$ 50 mm on an AG-10TA universal test stretching machine (Shimadzu Corporation, Kyoto, Japan) and dried in a drying oven at 200 °C for 2 h.

Simultaneously, commercial pure aluminum (99.7 wt.%, A 99.7) was melted in alumina crucible by using a SG-7.5–10 type crucible furnace (Zhonghuan Experimental Furnace Corporation, Tianjin, China). After the temperature was raised to 800 °C, the prefabricated block was added to the melt through the graphite bell jar. According to the previously determined typical sampling time (8, 50, 60, 80 and 90 s), the reaction block was quickly removed at each time point and subjected to high pressure ice brine flow quenching to obtain quenched samples.

The specimens with a size of 10 mm^3 taken from the quenched samples were electrolytically polished by a reagent (10 vol.% $HClO_4$ + 90 vol.% absolute ethanol, voltage is 20 V and the time is about 10~30 s) after mechanical grinding and polishing. The phase constituents of specimens were analyzed using a D8 advance X-ray diffractometer (XRD, Shimadzu Corporation, Kyoto, Japan). The microstructure was examined using a JSM-6700F scanning electron microscope (SEM) equipped with an energy dispersive spectrometer (EDS, Shimadzu Corporation, Kyoto, Japan). In addition, DSC analysis of the Al-Ti-C-Ce system was conducted with a NETZSCH STA 449F3 instrument (NETZSCH, Hanau, Germany) with a heating rate of 20 K/min and samples weighing less than 10 mg.

2.2. Grain Refinement Experiments

To evaluate the grain refining effect of Al-Ti-C master alloy after adding CeO_2, the different master alloys were prepared using the aluminum melt in situ reaction method, and then master alloys were used to refine A 99.7. The specific experimental process is as follows:

Firstly, A 99.7 ingots were melted and heated up to 800 °C in an alumina crucible by using a SG-7.5–10 type crucible furnace (Zhonghuan Experimental Furnace Corporation, Tianjin, China). A proper quality of prefabricated block was then added to the melt after preheating at 200 °C for 2 h. After being held at 800 °C for 15 min, the melt was finally poured into a steel die with an inner diameter of 45 mm, a diameter of 70 mm and a height of 70 mm. By adjusting the size and addition of CeO_2, Al-Ti-C and Al-Ti-C-Ce master alloys were obtained. The specific experimental parameters and numbers of master alloys are shown in Table 1.

Table 1. The experimental parameters and numbers of master alloys.

Sample No.	Composite of Master Alloy	Size of CeO$_2$	Preparation Temperature/°C	Holding Time/Min
#1	Al-Ti-C	None	800	15
#2	Al-Ti-C-Ce	30 nm	800	15
#3	Al-Ti-C-Ce	50 nm	800	15
#4	Al-Ti-C-Ce	1 μm	800	15
#5	Al-Ti-C-Ce	2–4 μm	800	15

Thereafter, an appropriate amount of A 99.7 was melted in a crucible electrical resistance furnace at about 730 °C. The 0.3 wt.% master alloys were sectioned at the position of 10 mm from the bottom on #1~#5 alloys and then added to the melt. After that, the melt was held for 10 min and finally poured into the steel mold. By solidification, the refined samples were obtained. For comparison, standard A 99.7 was treated in the same process. A specimen with a thickness of 10 mm was cut along the bottom of the refined sample. A cross profile of the specimen was etched using Keller's reagent (25 vol.% H$_2$O + 15 vol.% HNO3 + 15 vol.% HF + 45 vol.% HCl). The grain structures of specimens were taken by a camera, and the average grain sizes were calculated using the linear intercept method.

3. Results and Discussion

3.1. The Raw Materials of CeO$_2$ and the Mixed Particles

The raw materials of different CeO$_2$ size are presented in Figure 1. From Figure 1a,b, it can be seen that the morphology of CeO$_2$ with a size of 30 and 50 nm is spherical nanoparticles. In contrast, the 1 and 2–4 μm sized CeO$_2$ gathered together in a short rod shape and small block shape, as shown in Figure 1c,d.

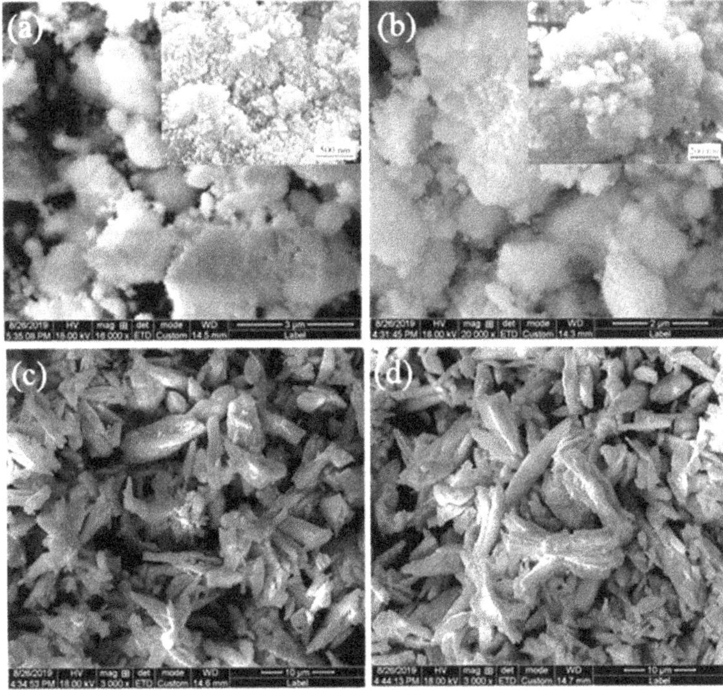

Figure 1. Different sizes of CeO$_2$: (**a**) 30nm; (**b**) 50nm; (**c**) 1 μm; (**d**) 2–4 μm.

Figures 2 and 3 show the microstructure and EDS results of the mixed powders (Al, Ti, C and CeO_2) with different CeO_2 size after ball milling. As shown Figure 2, Al, Ti, C and CeO_2 powders were evenly mixed after ball milling. Combining the EDS mapping analysis of Al, Ti, C, Ce and O elements as shown in Figure 3, it can be seen that the bright white particles with larger size are Ti, gray particles are Al and black particles are C. The bright white particles with small size are mainly enriched with Ce and O elements, which can be judged as CeO_2 particles. As can be observed in Figure 2a,b, for nano-CeO_2 particles, ball milling mainly leads to uniform dispersion. However, for the mixed powder of CeO_2 with a larger size, ball milling not only contributes to dispersion but also destroys the aggregation between CeO_2, breaking the CeO_2 into smaller particles. A statistical analysis of the particle size of CeO_2 found that after ball milling, the size of 1 μm CeO_2 particles was reduced to about 400 nm (see Figure 2c), and the size of 2–4 μm CeO_2 particles was reduced to about 200 nm (see Figure 2d).

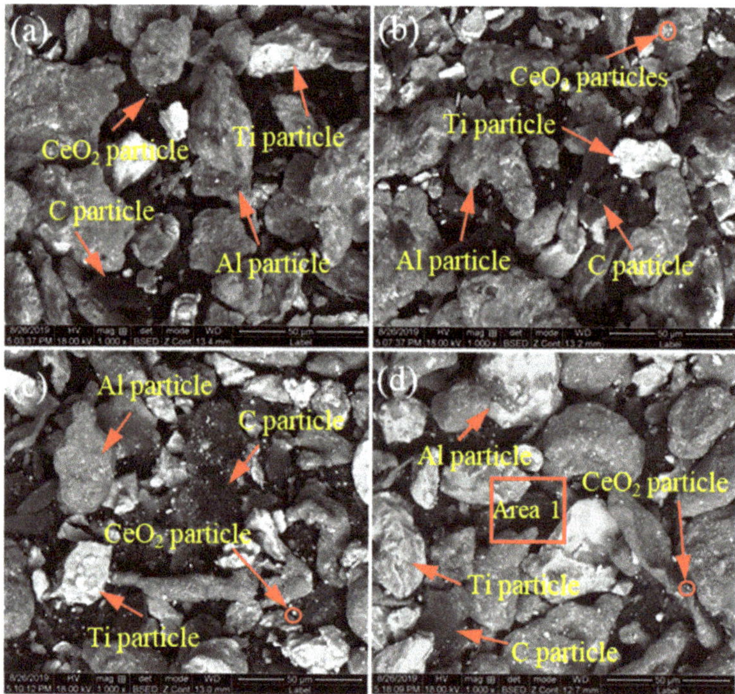

Figure 2. The microstructure of the mixed powders (Al, Ti, C and CeO_2) with different CeO_2 size after ball milling: (**a**) 30 nm; (**b**) 50 nm; (**c**) 1 μm; (**d**) 2–4 μm.

3.2. Effect of CeO_2 Particle Size on Phase Transformation and Microstructure Transformation of Al-Ti-C Master Alloy during Preparation

In order to study the influence of CeO_2 size on the microstructure transformation process of Al-Ti-C system, the microstructure of quenched samples with different CeO_2 sizes at the same reaction time (60 s) was analyzed. SEM images and EDS results are shown in Figure 4. Comparing Figure 4a,b, it is found that when the CeO_2 size is 30 and 50 nm, it has little effect on the reaction process of the system. As can be seen, the molten Al closely connects Ti particles, and incomplete Ti particles and incomplete melting Al particles were reserved in system after solidification. As can be seen from Figure 4c, a variety of compounds with different sizes and morphologies are formed around the Ti particles, which are Al_3Ti, TiC, Al-Ti compounds and CeC_2 particles. It can be seen from

Figure 4d that when the CeO$_2$ size is 1 μm, there is no close connection between the Ti particles and the aluminum matrix. Otherwise, compared to Figure 4a,b the degree of reaction is low, and there are obvious incompletely reacted Al, Ti and C particles in the system. When CeO$_2$ with a size of 2–4 μm is added to the system, as shown in Figure 4e,f, it can be seen that the size of the Ti particles is reduced, the reaction intensity is enhanced, the density of the system is increased and the particles are closely connected. Especially, A large number of TiC, Al$_3$Ti and Ti$_2$Al$_{20}$Ce phases are formed in the Al matrix.

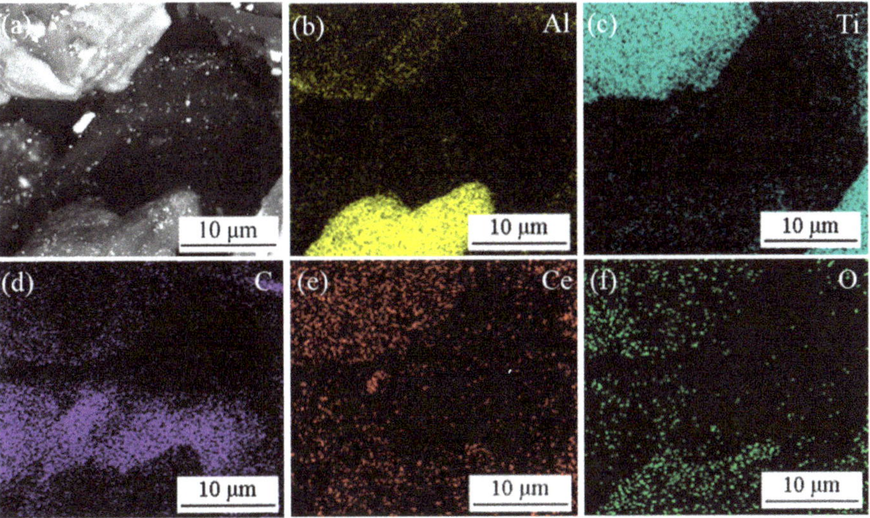

Figure 3. SEM image and map scanning patterns of Area 1 in Figure 2d: (**a**) SEM image; (**b–f**) map scanning patterns of Al, Ti, C, Ce and O elements, respectively.

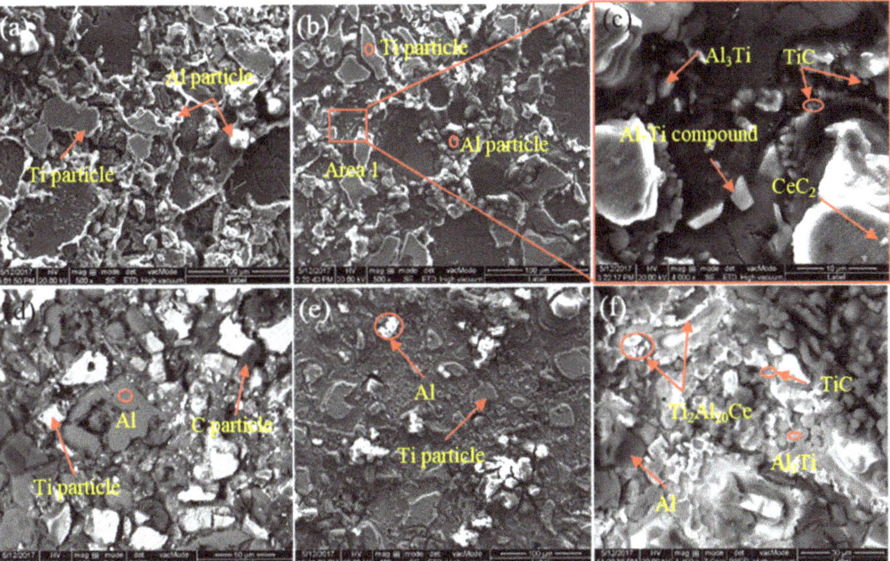

Figure 4. The microstructure of quenching samples with different CeO$_2$ size at 60 s: (**a**) 30 nm; (**b**) 50 nm; (**c**) SEM image of Area 1; (**d**) 1 μm; (**e**) low magnification image and (**f**) high magnification image of samples with CeO$_2$ of the size of 2–4 μm.

This difference in response is because the larger the CeO_2 particles, the larger the specific surface area and the stronger the catalytic effect, which obviously accelerates the progress of the reaction. For micron-sized CeO_2 particles, due to the crushing of CeO_2 particles during ball milling, more active sites are provided, and the larger the particle size, the more severe the degree of crushing, the more active sites and the greater the promotion of the reaction.

The microstructure of the quenched samples with different CeO_2 sizes added under complete reaction was observed. The phase, microstructure characteristics and the average grain size of the second phase particles of the quenched samples were analyzed, as shown in Figures 5–7.

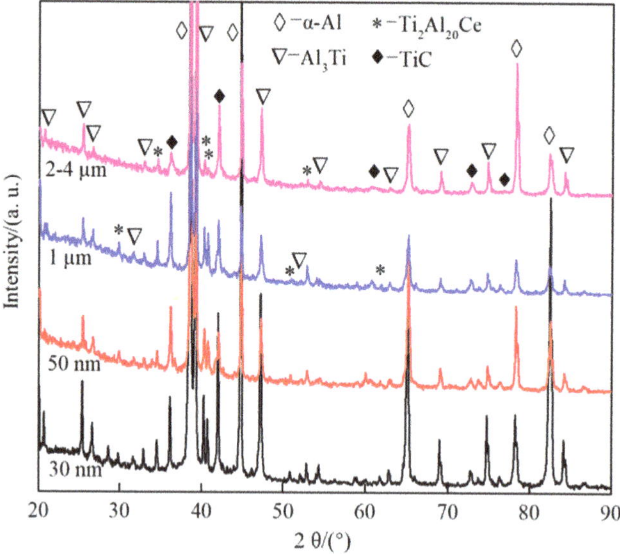

Figure 5. XRD patterns of prefabricated blocks with different CeO_2 size under complete reaction.

Figure 5 shows XRD patterns of prefabricated blocks with different CeO_2 size under complete reaction. As can be seen from Figure 5, under complete reaction, the main phases in the sample are α-Al, Al_3Ti, TiC and $Ti_2Al_{20}Ce$. When the CeO_2 particle size is 30 nm and 2–4 μm, the diffraction peak intensity of TiC ($2\theta = 41.7°$) is higher than 50 nm and 1 μm, and the diffraction peak intensity of Al_3Ti, TiC and $Ti_2Al_{20}Ce$ showed a weakening trend with the increase in CeO_2 size.

Figure 6 shows SEM images of the quenching samples of prefabricated block with different CeO_2 size under complete reaction. Comparing Figure 6a–d, it is found that the size of CeO_2 particles has a great influence on the quantity, morphology and distribution of Al_3Ti and TiC. When the size of CeO_2 is 30 nm and 2–4 μm, the amount of TiC and Al_3Ti is large and the size is relatively uniform, the distribution is dispersed in the matrix. When the size of CeO_2 is 50 nm, TiC and Al_3Ti adhere to each other in an irregular shape, and the size is not uniform. When the size is 1 μm, the number of TiC particles is small, and Al_3Ti is evenly distributed in the matrix.

Figure 7 shows the statistics chart of the second phase particles in prefabricated block with different CeO_2 size under complete reaction. It can be seen from Figure 7 that the particle size of CeO_2 mainly affects the size of TiC particles, and has little effect on the size of Al_3Ti. When the size of CeO_2 is 30 nm, the average size of Al_3Ti and TiC is the smallest at 4.0 and 1.1 μm, respectively. With the increase in CeO_2 particle size, the average size of TiC gradually increases, the average size of Al_3Ti increases first and then tends to be flat and CeO_2 particle size is 2–4 μm. The average size of Al_3Ti in the quenched samples is 6 μm, and the average size of TiC is about 2.2 μm.

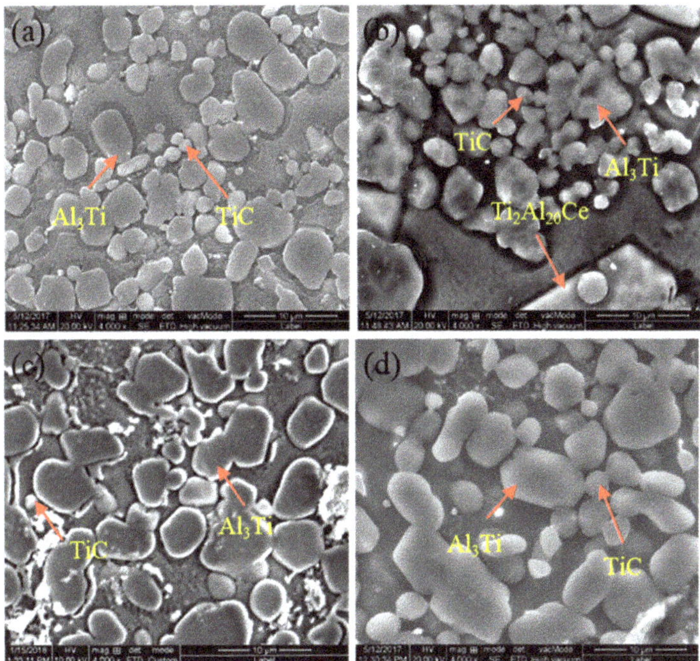

Figure 6. SEM images of the quenching samples of prefabricated block with different CeO_2 size under complete reaction: (**a**) 30 nm; (**b**) 50 nm; (**c**) 1 μm and (**d**) 2–4 μm.

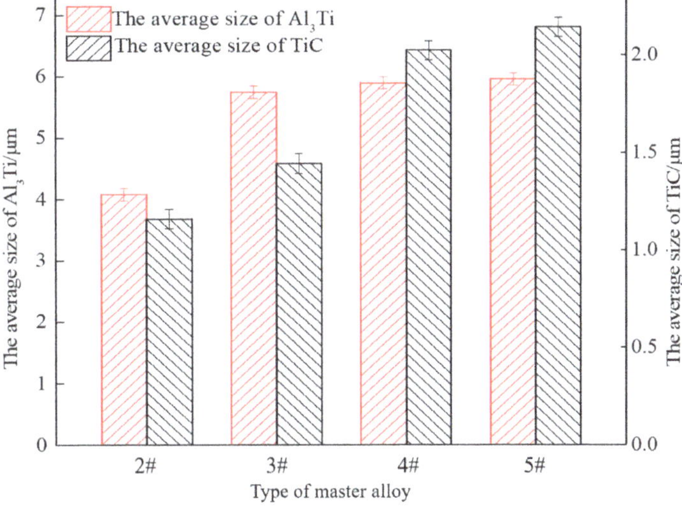

Figure 7. The statistics chart of the second phase particles in prefabricated block with different CeO_2 size under complete reaction.

From the above analysis, it can be seen that the content and size of CeO_2 have a significant effect on the formation, distribution, size, number and morphology of phases in the Al-Ti-C system. In order to further study the formation and transformation process of various phases, the pre-fabricated blocks with CeO_2 size of 2–4 µm and 4 wt.% were taken as the research objects to study the reaction products and microstructures of the Al-Ti-C-Ce system at different stages. The XRD pattern and microstructure are shown in Figures 8–10.

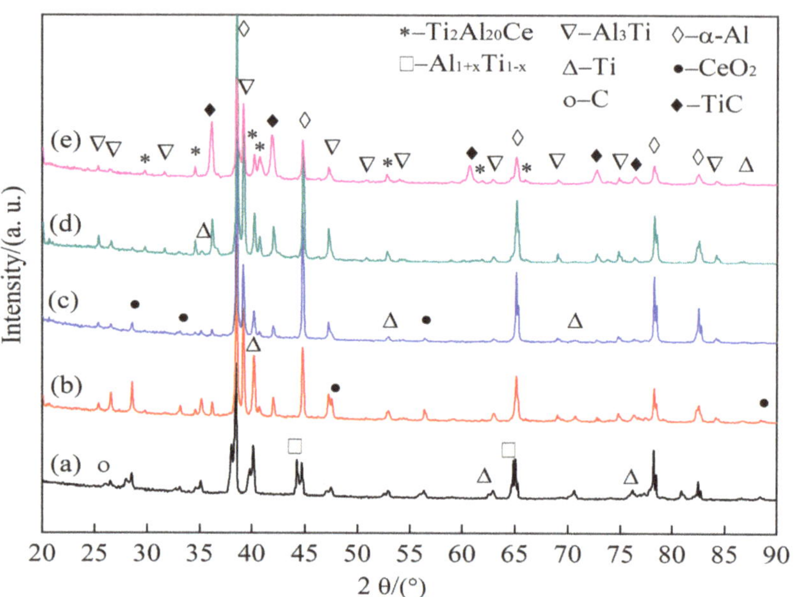

Figure 8. XRD spectra of prefabricated blocks with CeO_2 particle size of 2–4 µm at different reaction times: (**a**) 8, (**b**) 50, (**c**) 60, (**d**) 80, (**e**) 90 s.

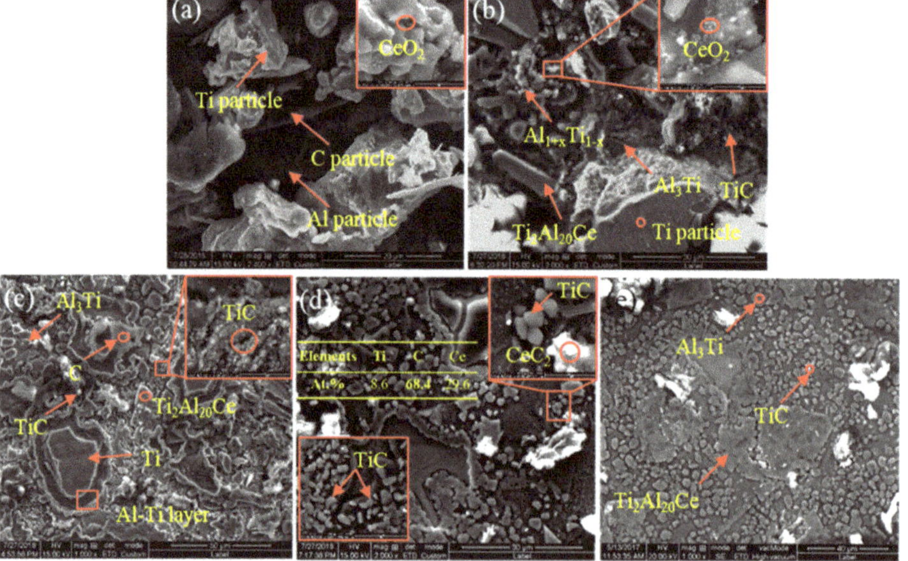

Figure 9. SEM images of #5 quenching samples prepared at different times: (**a**) 8; (**b**) 50; (**c**) 60; (**d**) 80 and (**e**) 90 s.

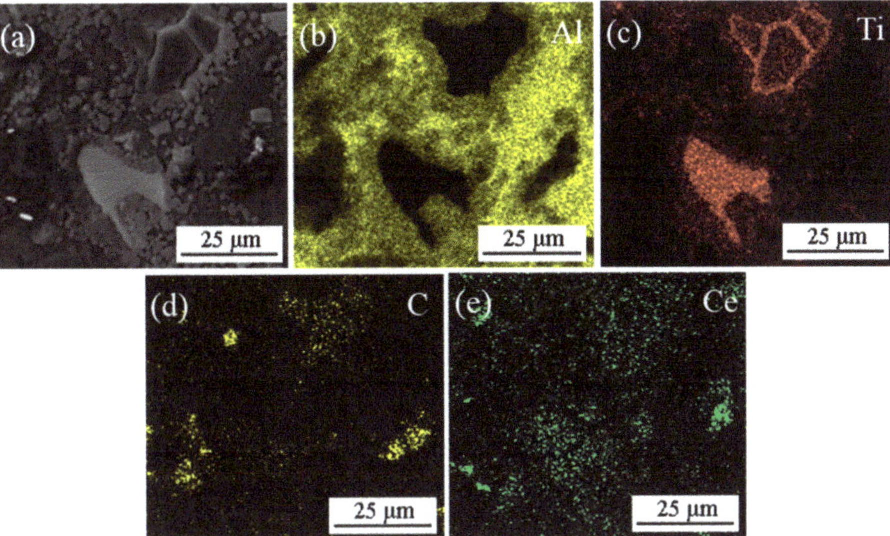

Figure 10. The map scan patterns of Figure 9d: (**a**) backscattered electron image of Figure 9d; (**b–e**) map scanning patterns of Al, Ti, C and Ce elements, respectively.

As can be seen from Figure 8a, in the initial stage of the reaction (8 s), the main detection phases in the system are Ti, C, CeO_2 and a small amount of $Al_{1+x}Ti_{1-x}$, indicating that there is a weak solid–solid diffusion process between solid Al and solid Ti at this stage. When the reaction time reaches 50 s, TiC, Al_3Ti and trace $Ti_2Al_{20}Ce$ are formed, as shown in Figure 8b. As the reaction progresses, the intensity of the diffraction peaks of Al_3Ti and TiC decreases first and then increases. The intensity of the diffraction peaks of $Ti_2Al_{20}Ce$ continues to increase. When the reaction reaches 90 s, a large amount of TiC, Al_3Ti and $Ti_2Al_{20}Ce$ are synthesized in the system; only weak incomplete reaction Ti peaks were detected in Figure 8c–e.

As can be seen from Figure 9a, the system basically maintains the original particle state. Figure 9b shows that when the reaction reaches 50 s, the molten Al in the system spreads on the surface of the Ti particles, and a small amount of Al1+xTi1-x is formed under the solid–liquid diffusion reaction [19,20] and then transformed into $Ti_2Al_{20}Ce$ under the action. In addition, around 3 μm Al_3Ti and 1 μm TiC were formed around the Ti particles, and unreacted Ti particles and CeO_2 also existed in the system. Figure 9c shows that there is a size of about 5 μm Al3Ti, and bulk Ti2Al20Ce are speculated to be caused by the enrichment of the Al element and the Ce element around the Ti element. In addition, solid C reacts with Ti to form a large number of TiC particles, and the whole system reacts to form a wrapped structure of Ti/Al-Ti/Al_3Ti/$Ti_2Al_{20}Ce$ [21,22]. The results of Figure 9d and its surface scanning energy spectrum (see Figure 10) show that rare earth elements are mainly enriched with the surface of C element and around the Ti element, and some rare earth elements react with dissolved Ti atoms and liquid Al to form Al_3Ti and $Ti_2Al_{20}Ce$. The other part reacts with solid C to form CeC_2 and then reacts with dissolved Ti atoms to generate TiC. At the same time, because the affinity between Ti atoms and C atoms is greater than that of Al and C, dissolved Ti atoms directly generate TiC particles with dissolved C atoms. As time goes on, TiC particles grow up, and the encapsulated structure gradually disappears. Al_3Ti particles are free from the encapsulated structure and freely distributed in the matrix, and the rare earth Ce element continues to enrich around Al3Ti, which hinders the growth of Al_3Ti particles. The aspect reacts with it to generate $Ti_2Al_{20}Ce$ and grows up by combining [22,23], as shown in Figure 9e. In Figure 9e, TiC particles with an average size of about 2 μm, small bulk Al_3Ti with an average size of about 6 μm and irregular bulk $Ti_2Al_{20}Ce$ with an average size of about 35 μm were formed in the late stage of the reaction.

In order to further explore the specific reaction process of the Al-Ti-C-Ce system, the influence of CeO_2 on the TiC synthesis method of Al-Ti-C system was studied. Combined with the DSC curve of the Al-Ti-C system in the previous research report [24], differential scanning calorimetry analysis was performed on the preforms with CeO_2 size of 30 nm and 2–4 μm. The DSC curve is shown in Figure 11. The corresponding reaction heat changes are shown in Table 2. From the curve, an obvious endothermic peak and two exothermic peaks can be seen. The first endothermic peaks at temperatures of 670.5 and 671.2 °C represented the melting of aluminum, and the corresponding reaction endotherms were 173.5 and 195.5 J/g. Subsequently, the carbothermal reaction of CeO_2 with C and oxygen produces CeC_2, the reaction of Al with Ti produces Al_3Ti and CeC_2 reacts with dissolved Ti atoms to produce TiC. These reactions release a lot of heat [14,25]. Combined with literature reports [26,27], in the aluminum-rich melt, Al and Ti first form Al_3Ti, and the first exothermic peak can be determined as the formation peak of Al_3Ti. The formation temperatures are 783.2 and 791.7 °C, corresponding to the exothermic heat of 168.8 and 137.7 J/g. The second exothermic peak is the peak formed by TiC, the formation temperatures are 926.1 and 933.6 °C, and the corresponding exotherms of the reaction are 92.99 and 198.6 J/g. Compared with the Al-Ti-C system, it is found that the addition of CeO_2 does not affect the melting point temperature of aluminum, but it reduces the latent heat of aluminum melting and accelerates the melting of aluminum. The smaller the size of CeO_2, the lower the heat required for aluminum melting. Furthermore, the addition of CeO_2 reduces the formation temperature and heat release of Al_3Ti and TiC. The smaller the size of CeO_2, the lower the formation temperature. The lower the heat released, the lower the temperature of the system, which affects the growth rate and growth habit of the second phase particles and is beneficial to obtaining Al_3Ti and TiC with smaller size and regular shape.

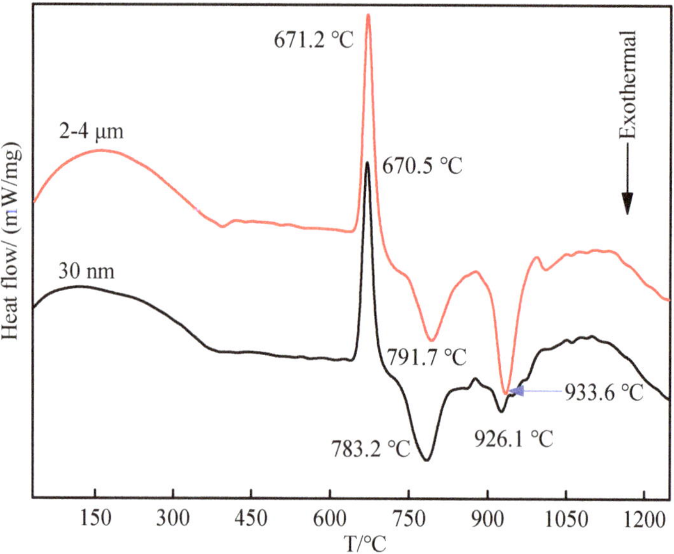

Figure 11. DSC curves of Al-Ti-C-Ce with different CeO_2 size of: 30 nm and 2–4 μm.

Table 2. The corresponding heat change of the reaction.

Type of Alloy	$\triangle H_{Al}/(J/g)$	$\triangle H_{Al3Ti}/(J/g)$	$\triangle H_{TiC}/(J/g)$
Al-Ti-C	214.7	−163.4	−838.8
Al-Ti-C-Ce (30 nm)	173.5	−168.8	−92.99
Al-Ti-C-Ce (2–4 μm)	195.5	−137.7	−198.6

3.3. Refinement Performance Evaluation

Figure 12 shows the macrographs of A 99.7 with different 0.3% master alloys. Figure 12a indicates that the unrefined pure aluminum is mainly composed of coarse equiaxed crystals in the center and columnar crystals in the middle and edges. From Figure 12b, it can be seen that Al-Ti-C master alloy has a significant refinement effect on A 99.7. After adding it into the melt of A 99.7, the coarse equiaxed crystals are refined into fine equiaxed crystals, while the range of columnar crystals is narrowed and the size is reduced. It can be seen from Figure 12c–f that the refining effect of Al-Ti-C-Ce master alloy is higher than Al-Ti-C master alloy, and the #2 alloy and #5 alloy have a better refining effect than #3 alloy and #4 alloy. Among them, the refining effect of #4 alloy with CeO$_2$ size of 1 μm is the worst.

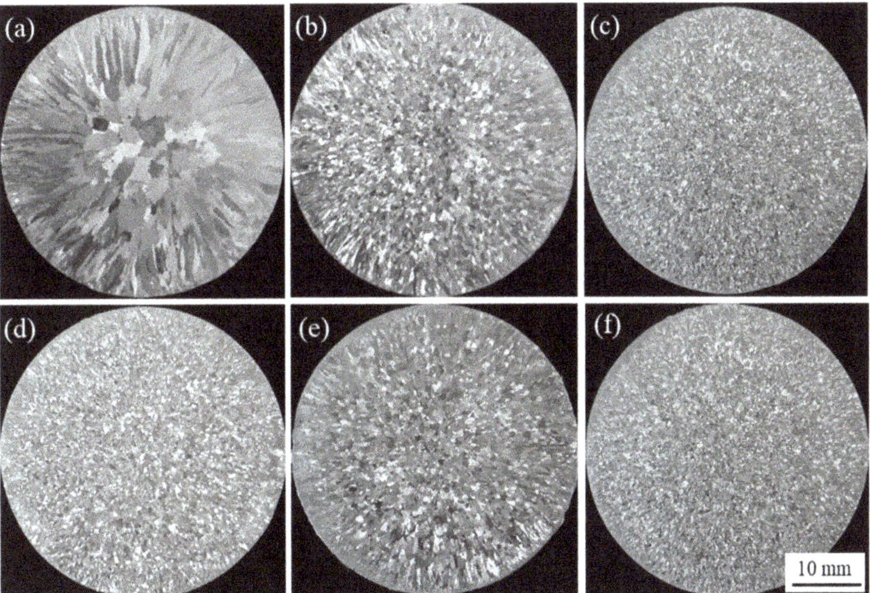

Figure 12. The macrographs of A 99.7 with 0.3% different master alloys: (**a**) unrefined; (**b**) #1; (**c**) #2; (**d**) #3; (**e**) #4; (**f**) #5.

Figure 13 shows the statistical results of the average grain size of the refined sample in Figure 12. It can be seen from Figure 13 that after 0.3% of Al-5Ti-0.62C, master alloys can refine the A 99.7 grain size from 1430 to 790 μm, while 0.3 wt.% #2 alloy refined the grain size of A 99.7 to 248 μm, and with the increased size of CeO$_2$, grain size of A 99.7 tends to increase first and then decrease. The #5 alloy has a significant refinement effect on A 99.7, which can refine the grain size to 227 μm. The above analysis shows that the addition of CeO$_2$ with a size of 2–4 μm to master alloy can promote its refining effect.

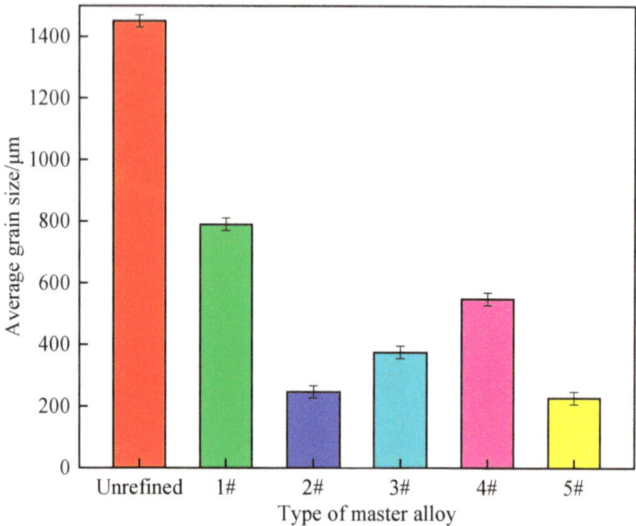

Figure 13. Average grain size of the refining samples in Figure 12.

4. Conclusions

(1)   The Al-Ti-C-Ce system is mainly composed of α-Al, Al_3Ti, TiC and $Ti_2Al_{20}Ce$. The addition of CeO_2 can obviously speed up the progress of the reaction, and the promotion effect of CeO_2 size is 2–4 microns.

(2)   The addition of CeO_2 can promote the uniform distribution of Al_3Ti and TiC, reduce the size of Al_3Ti and TiC, and the smaller the size of CeO_2, the smaller the size of Al_3Ti and TiC synthesized in the system.

(3)   The addition of CeO_2 reduces the latent heat of melting of aluminum, accelerates the melting of aluminum, promotes the reaction process and lowers the formation temperature of second-phase particles. The smaller the size of CeO_2, the lower the formation temperature of Al_3Ti and TiC, and the smaller the heat released by the synthesis of TiC.

(4)   Al-Ti-C-Ce master alloy has a significant refinement effect on A 99.7. With the increase in CeO_2 size, the refinement effect tends to increase first and then decrease. When the CeO_2 size is 2–4 μm, Al-Ti-C-Ce master alloy has the best refining effect.

Author Contributions: W.D. and T.C. conceived and designed the experiments; Y.M. and T.C. carried out the experiments and data collection; W.D., Y.M., T.C. and L.G. analyzed the data, Y.M. and W.D. contributed reagents/materials/analysis tools: Y.M., T.C. and L.G. wrote the paper. All authors have read and agreed to the published version of the manuscript.

Funding: National Natural Science Foundation of China (grant numbers 52161006 and 51661021) Industrial Support Plan Project of Gansu Provincial Department of Education (2021CYZC-23). China Postdoctoral Science Foundation, 2019M653896XB.

Institutional Review Board Statement: Not applicable.

Informed Consent Statement: Not applicable.

Data Availability Statement: Exclude this statement.

Acknowledgments: This work was supported by the National Natural Science Foundation of China (grant numbers 52161006 and 51661021) and Industrial Support Plan Project of Gansu Provincial Department of Education (2021CYZC-23). The authors would like to acknowledge the China Postdoctoral Science Foundation, 2019M653896XB.

Conflicts of Interest: The authors declare no conflict of interest.

References

1. Birol, Y. Grain refining efficiency of Al-Ti-C alloys. *J. Alloys Compd.* **2006**, *422*, 128–131. [CrossRef]
2. Kumar, G.; Murty, B.S.; Chakraborty, M. Development of A-Ti-C grain refiners and study of their grain refining efficiency on Al and Al-7Si alloy. *J. Alloys Compd.* **2005**, *396*, 143–150. [CrossRef]
3. Xxa, B.; Yfa, B.; Hui, F.C.; Qwa, B.; Gda, B.; Gla, B. The grain refinement of 1070 alloy by different Al-Ti-B mater alloys and its influence on the electrical conductivity. *Results Phys.* **2019**, *14*, 102482.
4. Wang, Y.; Fang, C.M.; Zhou, L. Mechanism for Zr poisoning of Al-Ti-B based grain refiners. *Acta Mater.* **2018**, *164*, 428–439. [CrossRef]
5. Li, Y.; Hu, B.; Liu, B. Insight into Si poisoning on grain refinement of Al-Si/Al-5Ti-B system. *Acta Mater.* **2020**, *187*, 51–65. [CrossRef]
6. Peng, T.; Li, W.F.; Wang, K. Effect of Al-Ti-C master alloy addition on microstructures and mechanical properties of cast eutectic Al-Si-Fe-Cu alloy. *Mater. Des.* **2017**, *115*, 147–157.
7. Li, P.T.; Liu, S.D.; Zhang, L.L. Grain refinement of A356 alloy by Al-Ti-B-C master alloy and its effect on mechanical properties. *Mater. Des.* **2013**, *47*, 522–528. [CrossRef]
8. Wang, H.J.; Xu, J.; Kang, Y.L. Effect of Al-5Ti-1B-1Re on the microstructure and hot crack of as-cast Al-Zn-Mg-Cu alloy. *J. Mater. Eng. Perform.* **2014**, *23*, 1165–1172. [CrossRef]
9. Zhou, X.F.; He, Y.D.; Zhang, Y.Q. The Microstructure and Solidification Mechanism of Al-Ti-C-Ce Prepared by Petroleum Coke. *Mater. Sci. Forum* **2019**, *960*, 14–20. [CrossRef]
10. Ding, H.M.; Liu, X.F.; Yu, L. Influence of zirconium on grain refining efficiency of Al-Ti-C master alloys. *J. Mater. Sci.* **2007**, *42*, 9817–9821. [CrossRef]
11. Doheim, M.A.; Omran, A.M.; Abdel, G.A. Evaluation of Al-Ti-C master alloys as grain refiner for aluminum and its alloys. *Metall. Mater. Trans. A* **2011**, *42*, 2862–2867. [CrossRef]
12. Jiang, H.; Sun, Q.; Zhang, L. Al-Ti-C master alloy with nano-sized TiC particles dispersed in the matrix prepared by using carbon nanotubes as C source. *J. Alloys Compd.* **2018**, *748*, 774–782. [CrossRef]
13. Ao, M.; Liu, H.; Dong, C. The effect of La_2O_3 addition on intermetallic-free aluminium matrix composites reinforced with TiC and Al_2O_3 ceramic particles. *Ceram. Int.* **2019**, *45*, 12001–12009. [CrossRef]
14. Wang, L.D.; Wei, Z.L.; Yang, X.B. Thermodynamics analysis of Al-Ti-C-RE prepared by rare earth oxide Ce_2O_3. *Chin. J. Nonferrous Met.* **2013**, *23*, 2928–2935.
15. Korotcenkov, G.; Al-Douri, Y. (Eds.) *Metal Oxides: Powder Technologies*; Elsevier: Cambridge, MA, USA, 2020.
16. Auffan, M.; Rose, J.; Bottero, J.Y. Towards a definition of inorganic nanoparticles from an environmental, health and safety perspective. *Nat. Nanotechnol.* **2009**, *4*, 634. [CrossRef] [PubMed]
17. Liu, H.; Xing, X.; Rao, L. Interface relation between CeO_2/TiC in hypereutectic Fe-Cr-C-Ti-CeO_2 hardfacing alloy. *Mater. Chem. Phys.* **2019**, *222*, 181–192. [CrossRef]
18. Ding, W.W.; Chen, T.L.; Zhao, X.Y. Effect of CeO_2 on Microstructure and Synthesis Mechanism of Al-Ti-C Alloy. *Materials* **2018**, *11*, 2508. [CrossRef]
19. Liu, X.F.; Bian, X.F. *Master Alloys for the Structure Refinement of Aluminium Alloys*, 1st ed.; Central South University: Changsha, China, 2012; pp. 7–10, 57–64, ISBN 978-7-5487-0450-8.
20. Liu, X.F.; Wang, Z.Q.; Zhang, Z.G. The relationship between microstructures and refining performances of Al-Ti-C master alloys. *Mater. Sci. Eng. A* **2002**, *332*, 70–74.
21. Xu, C.; Xiao, W.L.; Zhao, W.T. Microstructure and formation mechanism of grain-refining particles in Al-Ti-C-RE grain refiners. *J. Rare Earths* **2015**, *33*, 553–560. [CrossRef]
22. Wang, K.; Cui, C.X.; Wang, Q. The microstructure and formation mechanism of core-shell-like $TiAl_3$/Ti_2Al_{20}Ce in melt-spun Al-Ti-B-Re grain refiner. *Mater. Lett.* **2012**, *85*, 153–156. [CrossRef]
23. Jarfors, A.; Fredriksson, H.; Froyen, L. On the thermodynamics and kinetics of carbides in the aluminum-rich corner of the Al-Ti-C system. *Mater. Sci. Eng. A* **1991**, *135*, 119–123. [CrossRef]
24. Ding, W.W.; Chen, T.L.; Zhao, X.Y. Investigation of Microstructure of Al-5Ti-0.62C System and Synthesis Mechanism of TiC. *Materials* **2020**, *13*, 310. [CrossRef] [PubMed]
25. Nie, B.X.; Liu, H.M.; Qu, Y.; H, Q.L. Effect of La_2O_3 on the in-situ reaction products in Al-CuO-Ti-C system. *Mater. Res. Express* **2019**, *6*, 046555. [CrossRef]
26. Dong, K. (Al-) Study on Reaction Mechanism and Mechanical Properties of Copper Matrix Composites Synthesized by In-Situ Reaction of Ti-C-Cu System. Master's Thesis, Nanjing University of Technology, Nanjing, China, 2016.
27. Wang, P. Thermodynamic Analysis and Kinetic Mechanism of Ti-Al-C System. Ph.D. Thesis, Wuhan University of Technology, Wuhan, China, 2008.

Article

Effect of Grain Refiner on Fracture Toughness of 7050 Ingot and Plate

Fang Yu [1,*], Xiangjie Wang [1], Tongjian Huang [2] and Daiyi Chao [2]

[1] School of Material Science and Engineering, Northeastern University, Shenyang 110044, China; wangxj@epm.neu.edu.cn

[2] Shandong Nanshan Aluminum Co., Ltd., Longkou 265713, China; htj20141119@icloud.com (T.H); cdy19861226@163.com (D.C.)

* Correspondence: yufang1989@163.com; Tel.: +86-166-5355-8810

Citation: Yu, F.; Wang, X.; Huang, T.; Chao, D. Effect of Grain Refiner on Fracture Toughness of 7050 Ingot and Plate. *Materials* **2021**, *14*, 6705. https://doi.org/10.3390/ma14216705

Academic Editor: Filippo Berto

Received: 11 September 2021
Accepted: 2 November 2021
Published: 7 November 2021

Publisher's Note: MDPI stays neutral with regard to jurisdictional claims in published maps and institutional affiliations.

Abstract: In this paper, two types of grain refining alloys, Al-3Ti-0.15C and Al-5Ti-0.2B, were used to cast two types of 7050 rolling ingots. The effect of Al-3Ti-0.15C and Al-5Ti-0.2B grain refiners on fracture toughness in different directions for 7050 ingots after heat treatment and 7050-T7651 plates was investigated using optical electron microscopy (OEM) and scanning electron microscopy (SEM). Mechanical properties testing included both tensile and plane strain fracture toughness (K_{IC}). The grain size was measured from the surface to the center of the 7050 ingots with two different grain refiners. The fracture surface was analyzed by SEM and energy dispersive spectrometer (EDS). The experiments showed the grain size from edge to center was reduced in 7050 ingots with both the TiC and TiB refiners, and the grain size was larger for ingots with the Al-3Ti-0.15C grain refiner at the same position. The tensile properties of 7050 ingots after heat treatment with Al-3Ti-0.15C grain refiner were 1–2 MPa lower than the ingot with the Al-5Ti-0.2B grain refiner. For the 7050-T7651 100 mm thick plate with the Al-3Ti-0.15C grain refiner, for the L direction, the tensile properties were lower by about 10~15 MPa; for the plate with the Al-3Ti-0.15C refiner than plate with Al-5Ti-0.2B refiner, for the LT direction, the tensile properties were lower by about 13–18 MPa; and for the ST direction, they were lower by about 8–10 MPa compared to that of Al-5Ti-0.2B refiner. The fracture toughness of the 7050-T7651 plate produced using the Al-3Ti-0.15C ingot was approximately 2–6 MPa $\cdot \sqrt{m}$ higher than the plate produced from the Al-5Ti-0.2B ingot. Fractography of the failed fracture toughness specimens revealed that the path of crack propagation of the 7050 ingot after heat treatment produced from the Al-3Ti-0.15C grain refiner was more tortuous than in the ingot produced from the Al-5Ti-0.2B, which resulted in higher fracture toughness.

Keywords: Al-5Ti-0.2B; Al-3Ti-0.15C; 7050 ingot; 7050-T7651 plate; fracture toughness; grain size

1. Introduction

7050 Aluminum alloy have been widely used in the aerospace industry due to their low quench sensitivity, high fracture toughness, high fatigue crack growth rate (FCGR), and excellent stress corrosion resistance [1]. With the development of the aviation industry, the requirement for comprehensive performance of materials is getting higher, which means the raw material needs to have higher fracture toughness, higher FCGR, and high fatigue properties [2–4]. Normally, the tensile property levels decrease as the toughness level increases by alloying and heat treatment [5]. Therefore, the goal that most aluminum manufacturers want to pursue is to improve the fracture toughness of 7050 plates without the loss of the tensile property.

The effect of the microstructure on aluminum fracture toughness has been studied by many researchers. The main factors affecting fracture toughness [5,6] are grain size, recrystallization ratio, coarse insoluble phase, and strengthening precipitates. The relationship between recrystallization fraction (large angle grain boundary volume) and fracture toughness [7,8] was quantitatively analyzed. The higher the recrystallization fraction, the lower the fracture toughness. Hahn et al. [5] found that reducing the Fe, Si content and increasing the amount of smaller typical Cr, Mn, Zr-bearing particles could improve facture toughness, and the ratio of Zn:Mg [9] could also influence the fracture toughness. Zhang, Dumont et al. [10–12] investigated the effect of heat solution treatment process and quench parameters on the fracture toughness of 7050 and 7085 thick plate. Chen et al. [13] observed that the deformation texture can improve the fracture toughness and tensile properties of 7B50-T7751 plate effectively, and increase different directions of anisotropy. Other researchers also reported the effect of homogenization process [14] and aging parameter [15] on fracture toughness.

Although many experiments have been conducted to observe the factors that affect fracture toughness, there are still certain issues that have not been clarified. The grain size is known to be one of the major factors that affects fracture toughness [5], and grain refiners decide the final grain size and the quality of the ingot that is used for the manufacture of aviation products. A large number of studies have focused on the effect of grain refiners on the grain size and mechanical properties of ingots or products [16–21]. The effect of Al-5Ti-0.25C and Al-5Ti-0.2B additions on the grain size of different wrought alloys (AA1050, AA3004, AA5182, AA6063, AA7050, AA7475, AA8079) under different melting temperatures and addition rates was investigated [16] and it was found that the grain size decreased when increasing the rate of the refiner, and in all alloys, with the exception of AA3004, fine grain sizes (130 μm) were achieved with the addition of 1 kg/t. Nagaumi et al. [17] studied the effect of Al-3Ti-0.2B and Al-5Ti-1B on the microstructure of 7050 ingots and found that the refining fading of the Al-3Ti-0.2C refiner was more evident than that of Al-5Ti-1B grain refiner when the soaking time was increased. Huang et al. [18] noticed that compared with the 7050 alloy using the Al-5Ti-1B grain refiner, the distribution of the second phases present in 7050 aluminum alloy was more dispersed and uniform than the alloy using Al-5Ti-0.2C refiner. They also found that increasing the amount of Al-5Ti-0.2C grain refiner was more effective at easing the "Zr poisoning" phenomenon of 7050 alloy and could improve the strength and hardness while keeping good elongation. There are few reports [22] that discuss the relationship between grain refiner and fracture toughness of aviation aluminum products.

To examine the relationship between grain refiner and fracture toughness, a systematic study was conducted in 7050 ingots and 7050-T7651 100 mm thick plate with the grain refiner of Al-3Ti-0.15C and Al-5Ti-0.2B, respectively. The prepared plate was subjected to the same heat solution treatment and aging process. Microstructure characterization was then performed to evaluate the difference of the 7050 ingot and plate with two different grain refiners, and this study has guiding significance for industrial production of the 7050-T7651 plate, which needs higher fracture toughness.

2. Materials and Methods

The materials used for this study were commercial 7050 ingots with Al-3Ti-0.15C and Al-5Ti-0.2B grain refiner provided by Shan Dong Nanshan Aluminum Co., Ltd. at Longkou, China. The feeding amount for both Al-3Ti-0.15C and Al-5Ti-0.2B grain refiner was 2.5 kg/t. The ingot size was 440 mm × 1420 mm × 5000 mm and the composition is shown in Table 1.

Table 1. Chemical Composition of 7050 Aluminum Alloy (wt%).

Si	Fe	Cu	Mn	Mg	Cr	Zn	Ti	Zr	Al
0.08	0.10	2.2	0.10	2.15	0.02	6.2	0.06	0.10	remainder

After casting was completed, slices of the two different grain refined ingots were taken to evaluate the microstructure, grain size, and mechanical properties of the as-cast material according to the sampling diagram in Figure 1. The ingot samples were solution heat treated and artificially aged (heat solution treatment and age process were 479 °C × 50 min and 121 °C × 4 h + 163 °C × 20 h, respectively) prior to mechanical testing to eliminate the effects of the hot rolling process. The tensile properties and fracture toughness of ingot after heat treatment were then measured by an Instron 5985 tensile machine (INSTRON, Boston, MA, USA) and Instron 8802fatigue test machine (INSTRON, Boston, MA, USA), respectively. The fracture surface for ingots with different grain refiners were analyzed by scanning electron microscopy (SEM, FEI, Hillsboro, OR, USA).

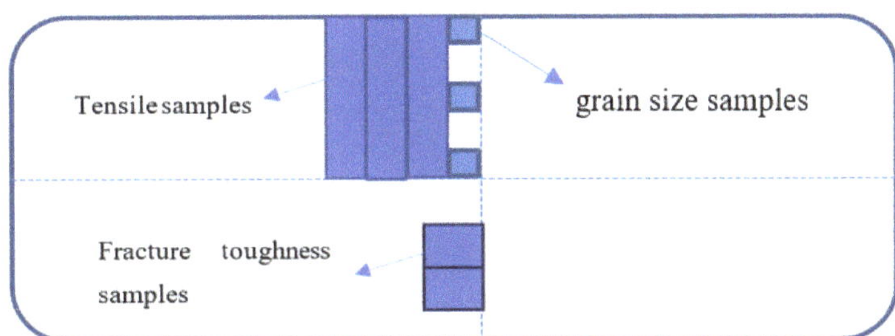

Figure 1. Samples locations for 7050 ingot.

A 100 mm plate for the 7050 alloy was hot rolled from ingots that were cast with two different types of grain refining alloy. The first set of 7050 plates were produced using Al-3Ti-0.15C grain refined ingots, and the second set of 7050 plates were produced using Al-5Ti-0.2B grain refined ingots according to the preparation process shown in Figure 2. Ensuring the consistency of thermomechanical processing enables the impact of grain refining on the wrought microstructure and mechanical properties to be isolated. All of the TiC and TiB_2 AA7050 ingots were homogenized together in the same furnace using a two-step process (step I: 465 °C soaking for 8 h, step II: 478 °C soaking for 20 h). After the homogenization process was finished, the ingots with Al-3Ti-0.15C and Al-5Ti-0.2B refiner were hot rolled to a 100 mm plate, then solution heat treated and artificially aged together to ensure consistent thermomechanical processes and to get the 7050 plate with T7651 temper (7050-T7651 plate).The heat solution heat treatment process was soaking 240 min at 479 °C, and the aging was a two-step process: Step I was soaking 4 h at 121 °C and Step II was holding 18 h at 163 °C.

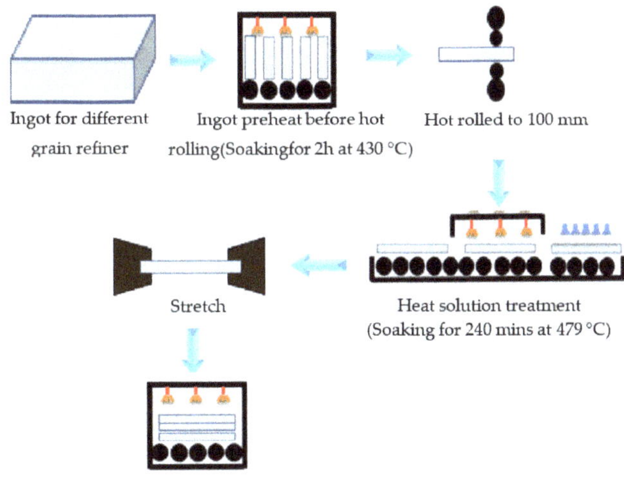

Figure 2. Scheme of the preparation process of the studied 7050-T7651 plate.

Samples for the 100 mm 7050-T7651 plate were taken according to the requirement of the AMS2355 standard for tensile property and fracture toughness test, and the tensile properties and fracture toughness were tested on T/4 of plate by an Instron 5985 tensile machine and Instron 8802 fatigue test machine, respectively. Specimen orientations for fracture toughness are shown in Figure 3.

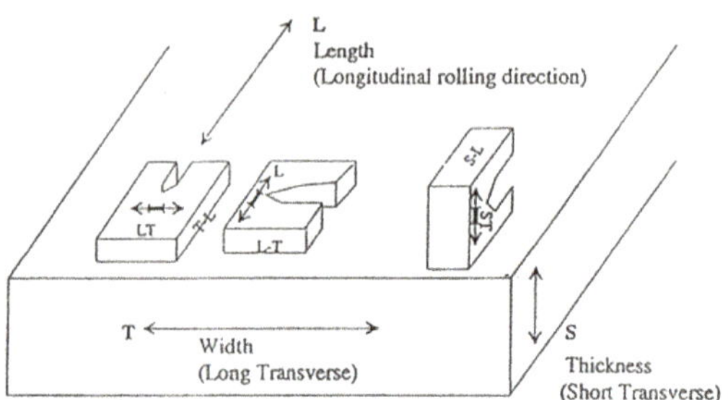

Figure 3. Specimen orientations used for fracture toughness test.

3. Results

3.1. The Effect of Grain Refiner on 7050 Ingot Grain Size

The grain size from the edge to the center of the 7050 ingot along with the casting direction was observed, five pictures were selected for each position, and the average grain size was analyzed by the intersection method according to ASTM E112 standard [23]. The microstructure and calculated grain size of the ingot with Al-5Ti-0.2B and Al-3Ti-0.15C grain refiner are shown in Figures 4 and 5, respectively. From Figure 4, the grain size from edge to center is increased, with a grain size of 95.99 μm, 101.04 μm, and 121.96 μm from edge to center for the ingot with Al-5Ti-0.2B grain refiner and 127.31 μm, 140.50 μm, and 165.32 μm with Al-3Ti-0.15C grain refiner, respectively. According to the study [17] by Hiromi, the average grain size of the 7050 ingot without grain refiner was 210 μm, and both the Al-5Ti-0.2B and Al-3Ti-0.15C grain refiner have refined the grain size.

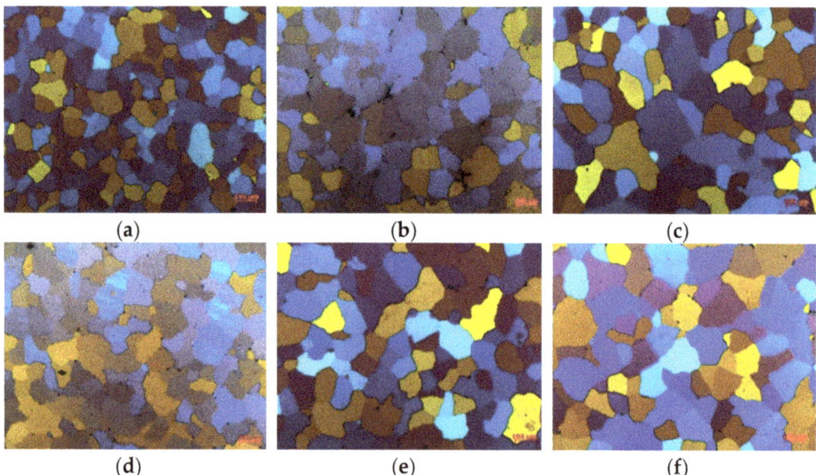

Figure 4. Grain structure of 7050 ingot: (**a**) Al-5Ti-0.2B surface, (**b**) Al-5Ti-0.2B 1/4 thickness, (**c**) Al-5Ti-0.2B 1/2 thickness, (**d**) Al-3Ti-0.15C surface, (**e**) Al-3Ti-0.15C 1/4 thickness, and (**f**) Al-3Ti-0.15C 1/2 thickness.

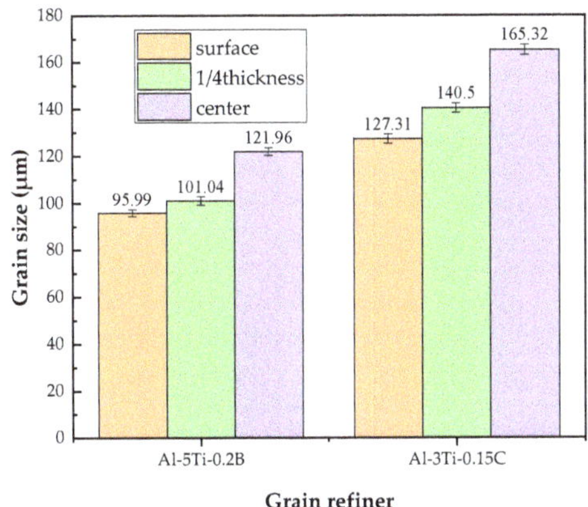

Figure 5. Average grain size for ingot using different grain refiners.

3.2. The Effect of Grain Refiner on Tensile Property of 7050 Ingot and 7050-T7651 100 mm Plate

Table 2 shows the tensile properties of the 7050 ingot along with casting direction after heat treatment with Al-5Ti-0.2B and Al-3Ti-0.15C grain refiners. The tensile property of the 7050 ingot with Al-3Ti-0.15C grain refiner is a little lower than that of the ingot with Al-5Ti-0.2B grain refiner.

Table 2. Tensile Properties for 7050 Ingot after Treatment with Different Grain Refiners.

Grain Refiner	Ultimate Strength (MPa)	Yield Strength (MPa)	Elongation (%)
Al-5Ti-0.2B	531.0 ± 1.04	474.5 ± 0.50	6.0 ± 0.50
Al-3Ti-0.15C	529.5 ± 0.50	473.6 ± 0.76	7.5 ± 0.50

Table 3 depicts the tensile properties of 7050-T7651 100 mm thick plate with different grain refiners. According to the requirements of AMS 2355 standard [24], the mechanical properties at 1/4 thickness should be tested for plates with thickness of more than 38.1 mm. Therefore, for 7050-T7651 100 mm thick plate, the mechanical properties were tested at the location of 1/4 thickness to investigate the influence of different grain refiners on the mechanical properties. As can be seen from Table 3, compared with the addition of Al-5Ti-0.2B refiner, the tensile properties of 7050-T7651 100 mm thick plate with the addition of Al-3Ti-0.15C grain refiner in all directions are slightly reduced, and the elongation is not much different. For the L direction, the tensile property is lower by about 10–15 MPa for the plate with Al-3Ti-0.15C refiner than for the plate with Al-5Ti-0.2B refiner; for the LT direction, the tensile property is lower by about 13–18 MPa than that of Al-5Ti-0.2B refiner; and for the ST direction, the yield strength is lower by about 8–10 MPa than that of Al-5Ti-0.2B refiner.

Table 3. Tensile Properties for 7050-T7651 100 mm Plate with Different Grain Refiners.

Direction	Al-5Ti-0.2B			Al-3Ti-0.15C		
	Ultimate Strength (MPa)	Yield Strength (MPa)	Elongation %	Ultimate Strength (MPa)	Yield Strength (MPa)	Elongation %
L	532.0 ± 1.44	485.8 ± 2.70	11.3 ± 0.55	522.8 ± 2.44	470.5 ± 2.81	12.8 ± 0.34
LT	543.3 ± 1.91	482.6 ± 3.01	9.6 ± 0.45	529.7 ± 1.37	464.0 ± 2.06	10.5 ± 0.34
ST	531.2 ± 0.47	449.7 ± 0.24	7.6 ± 0.24	523.5 ± 0.63	438.9 ± 1.86	8.8 ± 0.69

3.3. Effect of Different Grain Refiners on Fracture Toughness of 7050 Ingot and Plate

Figure 6 shows fracture toughness test results for the 7050 ingot along with casting direction and 7050-T7651 100 mm plate at 1/4 thickness with different grain refiners. The fracture toughness of the ingot is increased by the addition of Al-3Ti-0.15C grain refiner. Comparing fracture toughness for different directions of the 7050-T7651 100 mm plate with Al-5Ti-0.2B grain refiner, plates with Al-3Ti-0.15C grain refiner have higher fracture toughness at L-T, T-L, and S-L directions, and the difference of the fracture toughness for the L-T direction for the 7050 product with different grain refiners is more obvious. The fracture toughness of the 7050-T7651 plate with Al-3Ti-0.15C grain refiner for the L-T direction is about 5 MPa $\cdot \sqrt{m}$ higher than for the plate with Al-5Ti-0.2B grain refiner.

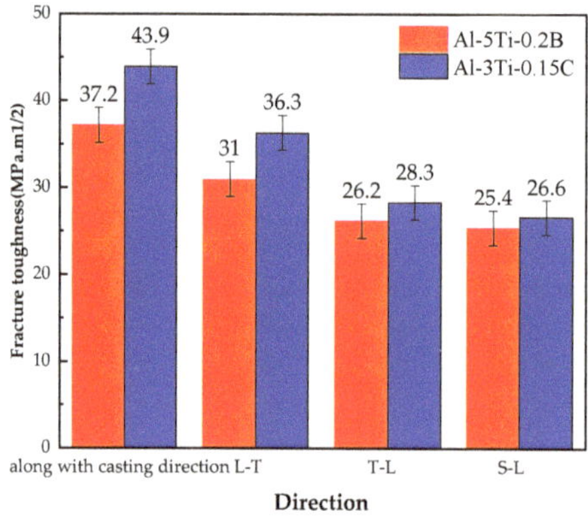

Figure 6. Fracture toughness of 7050 ingot after heat treatment and 7050-T7651 100 mm plate for different grain refiners in different directions.

The fracture surfaces of the 7050 ingot fracture toughness specimens after heat treatment were analyzed by SEM to observe the effect of the grain refiners. Figure 7 shows fractography for different grain refiners of the ingots after heat treatment. The fracture surface exhibits different character for ingots with different grain refiners. Intergranular failure is observed linking shear planes from adjacent grains in ingots with Al-5Ti-0.2B grain refiner, whereas ingots with Al-3Ti-0.15C grain refiner exhibits transgranular failure via sheared planes covered with fine dimples. The crack path of the ingot with Al-3Ti-0.15C grain refiner is more tortuous.

The EBSD analysis results of different refiners for ingots after heat treatment at the crack tip are shown in Figure 8. As can be seen from Figure 8, the ingot containing Al-3Ti-0.15C refiner has intergranular and transgranular fracture, and the ingot produced by Al-5Ti-0.2B refiner presents intergranular fracture. Moreover, the ingot microstructure has a genetic effect: the grain size of the ingot with Al-3Ti-0.15C grain refiner is larger under the same refiner adding amount and casting process compared with Al-5Ti-0.2B grain refiner, which leads to increase in fracture toughness of the final product.

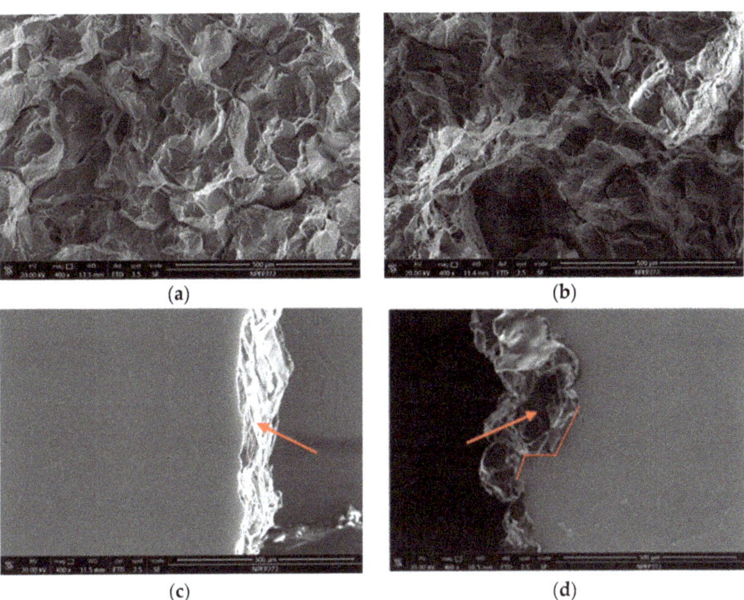

Figure 7. Fracture surfaces of ingot for different grain refiners: (**a**,**c**) Al-5Ti-0.2B grain refiner, (**b**,**d**) Al-3Ti-0.15C grain refiner.

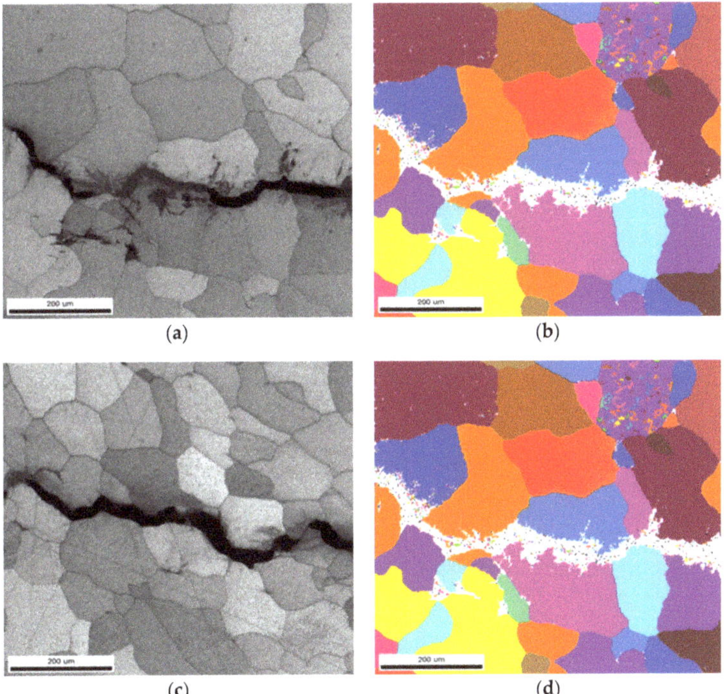

Figure 8. EBSD analysis for 7050 ingot fracture surface at crack tip: (**a**,**b**) ingot with Al-3Ti-0.15C grain refiner, (**c**,**d**) ingot with Al-5Ti-0.2B grain refiner.

SEM analysis of the fracture surface for 7050-T7651 100 mm thick plate with different refiners were carried out. The samples in the T-L direction have both intergranular and transgranular fractures, and a large amount of second phases are distributed at the same time. The enlarged SEM photos of the fracture surfaces with different grain refiners are shown in Figure 9. The energy dispersive spectrometer (EDS) analysis shows both TiB_2 and TiC particles have a granular shape, and the size of TiB_2 particles is larger than that of TiC particles; the TiC particles were difficult to find under the scope of SEM. In addition, TiB_2 particles are more aggregated than TiC particles, whereas TiC particles are more dispersed and have stronger resistance to crack propagation. The basic characteristics of fracture are basically consistent with the paper [25] researched by Zhang Xinming from Central South University.

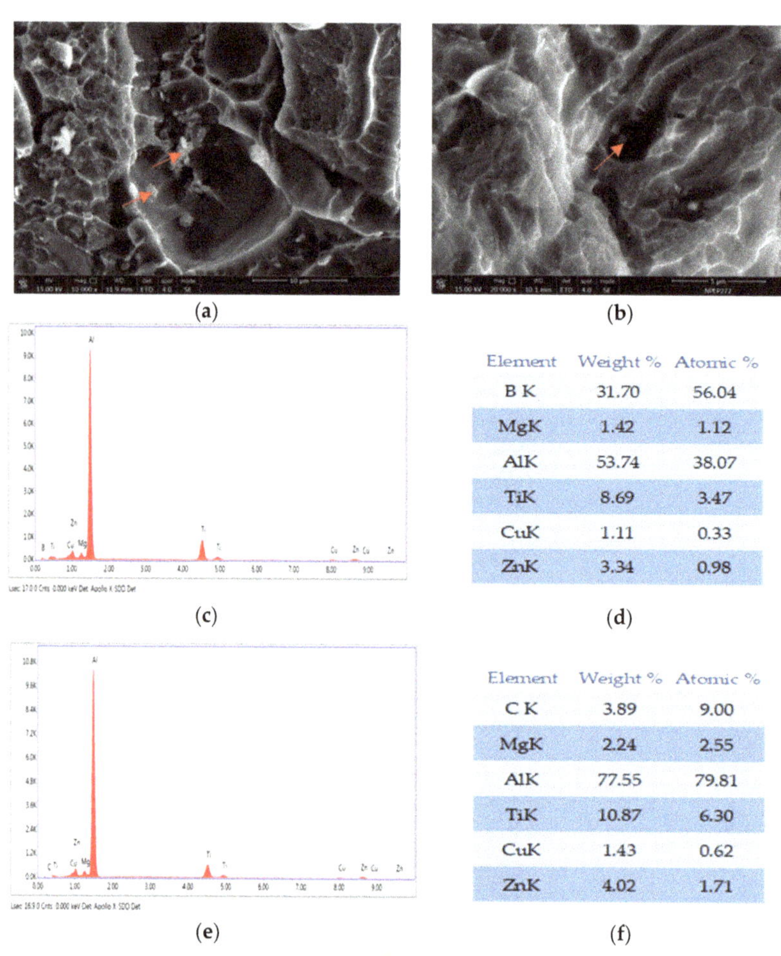

Element	Weight %	Atomic %
B K	31.70	56.04
MgK	1.42	1.12
AlK	53.74	38.07
TiK	8.69	3.47
CuK	1.11	0.33
ZnK	3.34	0.98

(d)

Element	Weight %	Atomic %
C K	3.89	9.00
MgK	2.24	2.55
AlK	77.55	79.81
TiK	10.87	6.30
CuK	1.43	0.62
ZnK	4.02	1.71

(f)

Figure 9. T-L Fracture surface for 7050 plate: (**a**) Al-5Ti-0.2B grain refiner, (**b**) Al-3Ti-0.15C grain refiner (**c,d**) EDS result of particles from (**a**), (**e,f**) EDS result of particles from (**b**).

4. Discussion

The ingot is solidified from the surface to the center in turn. The surface layer has a high degree of supercooling, high nucleation rate, and relatively small grains because of the faster heat dissipation. Moreover, heat dissipation of the center is slow, the degree of undercooling is small, and the nucleation rate is low, resulting in larger grains in the center. Literature [18] shows that the refining effect of Al-5Ti-0.2B is better than that of Al-3Ti-0.15C, primarily because of the microstructure of the refiner and the morphology and distribution of the refined particles. As shown in Figure 10, TiB_2 crystal in the Al-5Ti-0.2B grain refiner has a closed hexagonal structure, which cannot provide a nucleation core for the matrix. The thin Al_3Ti layer is enriched on the {0001} plane, which is used as the basis to promote α-Al nucleation. The nucleation directions of Al_3Ti are <110>{112} and <210>{112}, respectively. During solidification, Al_3Ti can nucleate in multiple directions and form a "halo" structure with "circle" shape; TiC crystal has face centered cubic (FCC) structure, which can be used as grain refining core for nucleation at the {111} plane and form "petalous" shape, and the number of nucleation is lower than that of the Al-Ti-B refiner. It is also related to the microstructure of the grain refiner and the morphology and distribution of the particles. The literature [17] shows that TiC particles in Al-3Ti-0.15C are unstable. When held in a soaking furnace for a long time, TiC particles will aggregate and react to form more stable Al_4C_3, whose refining effect is far less than that of TiC particles. Although for Al-Ti-B grain refiner, long soaking time also causes agglomeration and deposition of TiB_2 particles, studies have shown [21] that Al-Ti-B is still effective in refining pure aluminum, even if the holding time in the furnace is as long as 24 h. Combined with the above factors, the grain refinement effect of Al-3Ti-0.15C is not as good as that of Al-5Ti-0.2B. Therefore, with the same amount of addition, the grain size of the ingot with Al-5Ti-0.2B grain refiner is smaller.

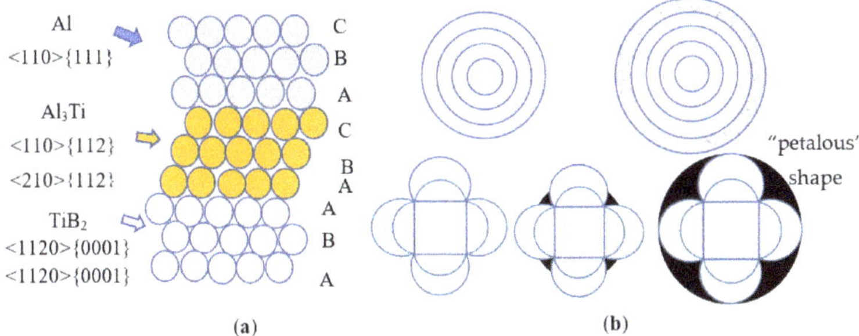

Figure 10. Grain refiner nucleation schematic diagram: (**a**) Orientation relationship between Al, Al_3Ti, and TiB_2; (**b**) Nucleation diagram of Al-5Ti-0.2B and Al-3Ti-0.15C refiner during solidification.

If the grain size is small enough, plasticity will be enhanced without detrimental, low energy, intergranular fracture for thin products under plane stress. However, for thick products under plane strain, fracture is usually controlled by coarse particles, and a recrystallized grain structure is preferable. Teleshov [26] studied the effect of grain size on fracture toughness and strength of AK4-1CH extrusion products. It was shown that the larger grain size products have higher fracture toughness than the smaller grain size products in L-T, T-L, and S-L directions, especially in the T-L direction. By comparing the microstructure, properties, and the distance between the vertical crack plane and the second phase in the propagation direction, it was found that the main reason for the different fracture toughness of AK4-1CH products may be the smaller extent of grain boundary in the path of plastic deformation. To overcome this deformation, the deformation in neighboring grains is needed. However, the stress generated far from the crack tip is not enough, which leads to the increase in the size of the plastic deformation zone, resulting in the increase in KIC. According to the study [27] of Krasovskii, the depth of the plastic deformation zone is related to the fracture toughness and yield strength of the product. According to the equation $\lambda = 0.128 * (KIC/\delta_{0.2})^2$, the value for the ingot with different grain refiners is 1.080 mm and 0.787 mm, respectively, which are larger than the average grain size. Fracture path for the 7050 ingot with different grain refiners is shown in Figure 11. T range of the plastic deformation zone with large grain size was large, which affected the fracture mode at the crack tip.

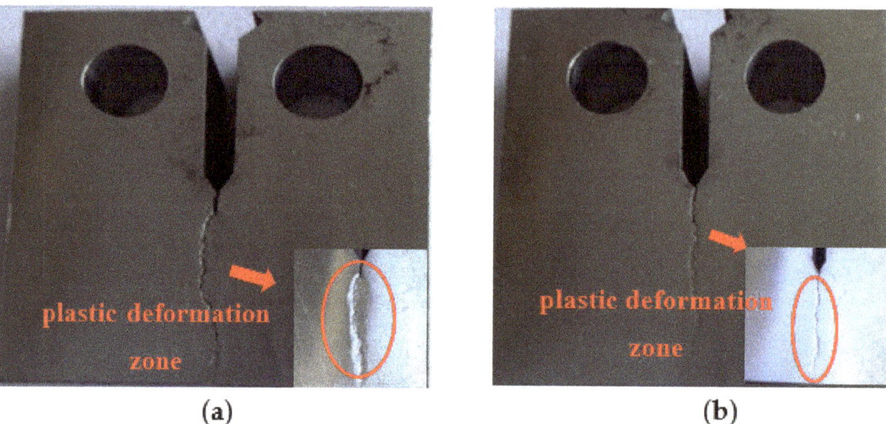

(a)　　　　　　　　　　　　　　　**(b)**

Figure 11. Fracture path for 7050 ingot with different grain refiners: (**a**) Al-3Ti-0.15C (**b**) Al-5Ti-0.2B.

As discussed in Section 2, the recrystallization ratio is one of the factors affecting fracture toughness: the higher the recrystallization fraction, the lower the fracture toughness. The recrystallization ratio of the 7050-T7651 plate in different thickness positions with Al-5Ti-0.2B and Al-3Ti-0.15C grain refiners are shown in Figure 12. From surface to 1/2 thickness, the recrystallization ratio for the 7050 plate with Al-5Ti-0.2B grain refiner is 11.2%, 10.3%, and 2.4%, respectively, and 12.3%, 8.1%, and 1.5% with Al-5Ti-0.2B grain refiner. There are no obvious differences in the recrystallization ratio for the 7050-T7651 plate with different grain refiners.

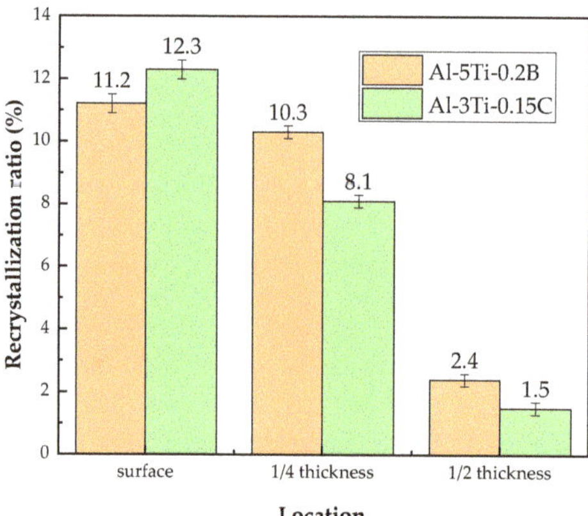

Figure 12. Recrystallization ratio of 7050-T7651 plate with different grain refiners.

The Fe and Si bearing phase formed during solidification is brittle and not coherent with the matrix, easily forms pores, and becomes a source of crack propagation, thus strongly reducing the fracture toughness of the alloy. The size of the precipitate and the interface binding force between the precipitate and the matrix have a great influence on the fracture toughness. The larger the size of the precipitate, the weaker the binding force between the phase and the matrix, which leads to the crack occurrence and the inferior fracture toughness. Gokhale, Deshpande et al. [7,28] quantitatively analyzed the relationship between microstructure and fracture toughness, and found that the larger the surface area of the second phase particle per unit volume, the lower the fracture toughness. Therefore, in order to improve the fracture toughness of 7050 products, the content of Fe and Si and the size of the dispersion phase must be strictly controlled. In the 7050-T7651 plate with Al-5Ti-0.2B grain refiner, TiB_2 particles are easy to gather in the process of casting and have a genetic effect; TiB_2 particles act as heterogeneous nucleation to precipitate in the aging process. The large precipitate phase will be become the source of cracks, which may reduce the local plastic deformation ability and also the fracture toughness.

5. Conclusions

A study has been performed on the effect of grain refiner on the tensile property and fracture toughness of 7050 ingots and 7050-T7651 100 mm thick plate. The following conclusions may be drawn from this work.

The experiments showed that the grain size from surface to center of the ingot with the addition of Al-5Ti-0.2B grain refiner was 95.99 µm,101.04 µm, and 121.96 µm, respectively, and 127.31 µm, 140.50 µm, and 165.32 µm with the addition of Al-3Ti-0.15C grain refiner. The refining ability of Al-5Ti-0.2B grain refiner was better than that of Al-3Ti-0.15C.

For both 7050 ingots and 7050-T7651 100 mm thick plate, the tensile property with Al-5Ti-0.2B grain refiner was higher than that of Al-3Ti-0.15C grain refiner. However, the fracture toughness of 7050 ingots and 7050-T7651 plates made with Al-3Ti-0.15C grain refiner was higher than that made with Al-5Ti-0.2B grain refiner, and fracture path for the 7050 ingot with Al-3Ti-0.15C after heat treatment was more tortuous than for the ingot with Al-5Ti-0.2B.

The fractured surface for the 7050-T7651 plate showed both TiB_2 and TiC particles have a granular shape, and the size of the TiB_2 particles was larger than that of the TiC particles. TiC particles were hard to find under the scope of SEM, whereas TiB_2 particles were easy to find and easy to aggregate. The combination of grain size and uniform distribution of the second phase results in higher fracture toughness for 7050 ingots and 7050-T7651 plate with Al-3Ti-0.15C grain refiner.

Author Contributions: Conceptualization, F.Y. and X.W.; methodology, F.Y. and T.H.; software, F.Y.; validation, X.W., D.C. and T.H.; formal analysis, F.Y.; investigation, F.Y.; resources, X.W.; data curation, D.C.; writing—F.Y.; writing—review and editing, F.Y., D.C. and X.W. All authors have read and agreed to the published version of the manuscript.

Funding: This research was funded by the Fundamental Research Funds for the National Natural Science Foundation of China (U1708251); the Key Research and Development Program of Liaoning (2020JH2/10700003).

Institutional Review Board Statement: The study was conducted according to the guidelines of the Northeastern University, and approved by the School of Material Science and Engineering.

Informed Consent Statement: Informed Consent was obtained from all subjects involved in the study.

Data Availability Statement: The data presented in this study are available upon request.

Conflicts of Interest: The authors declare no conflict of interest.

References

1. Staley, J.T.; Liu, J.; Hunt, H.W. Aluminum alloys for aerostructures. *Adv. Mater. Process.* **1997**, *152*, 10–17.
2. Yang, S.J.; Dai, S.L. A glimpse at the development and application of aluminum alloys in aviation industry. *Mater. Rev.* **2005**, *2*, 76–80.
3. Deng, Y.L.; Zhang, X.M. Development of Aluminum and Aluminum alloy. *Chin. J. Nonferr. Met.* **2019**, *9*, 2115–2141.
4. Ji, H. Development and application of 700 high strength aluminum alloys on airplane. *Aeronaut. Sci. Technol.* **2015**, *6*, 75–78.
5. Hahn, G.T.; Rosenfield, A.R. Metallurgical Factors Affecting Fracture Toughness of Aluminum Alloys. *Metall. Trans. A* **1975**, *6A*, 653–668. [CrossRef]
6. Starke, E.A.; Staley, J.T. Application of modern aluminum alloys to aircraft. *Prog. Aerospace Sci.* **1996**, 747–783. [CrossRef]
7. Deshpande, N.U.; Gokhale, A.M.; Denzer, D.K.; Liu, J. Relationship Between Fracture Toughness, Fracture Path and Microstructure of 7050 Aluminum Alloy: Part 1. Quantitative Characterization. *Metall. Mater. Trans. A* **1998**, *29*, 1191–1201. [CrossRef]

8. Morere, B.; Ehrstrom, J.C.; Gregson, P.J.; Sinclair, I. Microstructural Effects on Fracture Toughness in AA7010 Plate. *Metall. Mater. Trans. A* **2000**, *31*, 2503–2515. [CrossRef]

9. Qin, C.; Gou, G.Q.; Chen, X.L.; Chen, H. Effect of alloying elements on mechanical property and fracture of A7N01S-T5 aluminum alloy. *Chin. J. Mater. Res.* **2015**, *7*, 535–541.

10. Zhang, X.M.; He, D.G.; Liu, S.D.; Han, N.M.; Zhang, R. Effects of multi-stage promotively-solutionizing treatment on strength and fracture toughness of 7050 aluminum alloy thick plate. *Chin. J. Nonferr. Met.* **2012**, *22*, 1546–1552.

11. Dumont, D.; Deschamps, A.; Brechet, Y. On the relationship between microstructure, strength and toughness in AA7050 aluminum alloy. *Mater. Sci. Eng. A* **2003**, *356*, 326–336. [CrossRef]

12. Chen, S.Y.; Chen, K.H.; Dong, P.X.; Ye, S.P.; Huang, L.P. Effect of heat treatment on stress corrosion cracking, fracture toughness and strength of 7085 aluminum alloy. *Trans. Nonferr. Met. Soc. Chin.* **2014**, *24*, 2320–2325. [CrossRef]

13. Chen, G.H.; Li, G.A.; Chen, J.Z.; Gang, R.J.; Wei, W.H. Effect of rolling microstructure characteristics on fracture toughness of 7B50-T7751 aluminum alloy thick plate. *Light Alloy Fabr. Technol.* **2018**, *6*, 29–33.

14. Wu, L.M.; Wang, W.H.; Hsu, Y.F.; Shan, T. Effects of homogenization treatment on recrystallization behavior and dispersoid distribution in an Al-Zn-Mg-Sc-Zr alloy. *J. Alloys Compd.* **2008**, *456*, 163–169. [CrossRef]

15. Han, N.M.; Zhang, X.M.; Liu, S.D.; Huang, L.Y.; Xin, X.; He, D.G. Effects of retrogression and re-aging on strength and fracture toughness of aluminum alloy 7050. *Chin. J. Nonferr. Met.* **2012**, *7*, 1871–1882.

16. Schneider, W.; Keaens, M.A.; Mcgarry, M.J.; Whitehead, A.J. A Comparison of the Behaviour of AlTiB and AlTiC Grain Refiners. *Light Metals* **1998**, *3*, 400–408.

17. Hiromi, N.; Guo, S.J.; Xue, G.X.; Ke, M.H. Effect of grain refiner on microstructure of 7050 aluminum alloy with casting temper. *Light Alloy Fabr. Technol.* **2012**, *20*, 27–43.

18. Huang, Y.C.; Du, Z.Y.; Xiao, Z.B.; Yan, X.X. Effect of Al-Ti-C and Al-Ti-B on Microstructure and Mechanical performance of 7050 Aluminum alloy. *J. Mater. Eng.* **2015**, *43*, 75–80.

19. Liu, M. Development overview of grain refinement for aluminum alloys. *Foundry Tech.* **2018**, *39*, 2429–2432.

20. Liu, L.Y.; Sun, B.H.; Han, Y. Study on Morphology of Al-5Ti-1B Master Alloy and Its Effect on Grain Refinement of Aluminum Alloy. *Alum. Processing.* **2019**, *2*, 15–18.

21. Hu, Y.J.; Zhong, J.Y.; Zhang, Y.L. Effect of Al-Ti-C and Al-Ti-B wire rod on grain of aluminum alloy flat ingot. *Light Alloy. Fabr. Technol.* **2007**, *10*, 7–8.

22. Mostafa, A.; Adaileh, W.; Awad, A.; Kilani, A. Mechanical Properties of Commercial Purity Aluminum Modified by Zirconium Micro-Additives. *Crystals* **2021**, *11*, 270. [CrossRef]

23. ASTM International. ASTM E112-13. Standard Test Methods for Determining Average Grain Size. In *ASTM Book of Standards*; ASTM International: West Conshohocken, PA, USA, 2013.

24. SAE International AMS2355. Quality Assurance, Sampling and Testing Aluminum Alloys and Magnesium Alloy Wrought Products (Except Forging Stock), and Rolled, Forged, or Flash Welded Rings. In *Aerospace Material Specification*; SAE International: Warrendale, PA, USA, 2017.

25. Zhang, X.M.; Han, N.M.; Liu, S.D. Inhomogeneity of texture, tensile property and fracture toughness of 7050 aluminum alloy thick plate. *Chin. J. Nonferr. Met.* **2020**, *20*, 202–207.

26. Teleshov, V.V.; Shtovba, Y.K.; Smolentsev, V.I.; Sirotkina, O.M. Effect of grain size on the fracture toughness and strength of alloy AK4-1ch. *Met. Sci. Heat Treat.* **1993**, *25*, 29–34. [CrossRef]

27. Krasovskii, A.Y. Brittleness of Metals at Low Temperature. *Naukova Dumka* **1980**, *13*, 340.

28. Gokhale, A.M.; Deshpande, N.U.; Denzer, D.K. Relationship Between Fracture Toughness, Fracture Path, and Microstructure of 7050 Aluminum Alloy Part II. Multiple Micro-mechanisms-Based Fracture Toughness Model. *Metall. Mater. Trans. A* **1998**, *29*, 1203–1210. [CrossRef]

Article

In Situ Observation of the Tensile Deformation and Fracture Behavior of Ti–5Al–5Mo–5V–1Cr–1Fe Alloy with Different Microstructures

Suping Pan [1,2], **Mingzhu Fu** [1], **Huiqun Liu** [1,3,*], **Yuqiang Chen** [4,*] **and Danqing Yi** [1,3]

1   School of Materials Science and Engineering, Central South University, Changsha 410083, China; pan-su-ping@163.com (S.P.); fumingzhu@csu.edu.cn (M.F.); yioffice@csu.edu.cn (D.Y.)
2   Advanced Research Center, Central South University, Changsha 410083, China
3   State Key Laboratory of Powder Metallurgy, Central South University, Changsha 410083, China
4   Hunan Engineering Research Center of Forming Technology and Damage Resistance Evaluation for High Efficiency Light Alloy Components, Hunan University of Science and Technology, Xiangtan 411201, China
*   Correspondence: liuhuiqun@csu.edu.cn (H.L.); yqchen1984@163.com (Y.C.)

Abstract: The plastic deformation processes and fracture behavior of a Ti–5Al–5Mo–5V–1Cr–1Fe alloy with bimodal and lamellar microstructures were studied by room-temperature tensile tests with in situ scanning electron microscopy (SEM) observations. The results indicate that a bimodal microstructure has a lower strength but higher ductility than a lamellar microstructure. For the bimodal microstructure, parallel, deep slip bands (SBs) are first noticed in the primary α (α_p) phase lying at an angle of about 45° to the direction of the applied tension, while they are first observed in the coarse lath α (α_L) phase or its interface at grain boundaries (GBs) for the lamellar microstructure. The β matrix undergoes larger plastic deformation than the α_L phase in the bimodal microstructure before fracture. Microcracks are prone to nucleate at the α_p/β interface and interconnect, finally causing the fracture of the bimodal microstructure. The plastic deformation is mainly restricted to within the coarse α_L phase at GBs, which promotes the formation of microcracks and the intergranular fracture of the lamellar microstructure.

Keywords: Ti–5Al–5Mo–5V–1Cr–1Fe alloy; in situ observation; slip band; microcrack; fracture mechanism

Citation: Pan, S.; Fu, M.; Liu, H.; Chen, Y.; Yi, D. In Situ Observation of the Tensile Deformation and Fracture Behavior of Ti–5Al–5Mo–5V–1Cr–1Fe Alloy with Different Microstructures. *Materials* **2021**, *14*, 5794. https://doi.org/10.3390/ma14195794

Academic Editors: Andrey Belyakov and Carmine Maletta

Received: 21 July 2021
Accepted: 29 September 2021
Published: 3 October 2021

Publisher's Note: MDPI stays neutral with regard to jurisdictional claims in published maps and institutional affiliations.

1. Introduction

Due to their strength, corrosion resistance, and heat resistance, near-β-titanium alloys have been widely used in aerospace and automotive engineering as structural materials [1]. With the continuous progress of aerospace and automotive technology, more challenging requirements are imposed on the mechanical properties of titanium alloys [2]. The design of titanium alloys with better mechanical properties is therefore the focus of many studies [3–6].

Previous studies found that the mechanical properties of titanium alloys are closely related to the morphology and distribution of the α phase [7–9]. Two types of typical microstructure (bimodal and lamellar microstructures) can be obtained in titanium alloys by different heat treatments [10]. Many studies concerning the microstructure–property relationship of titanium alloys showed that bimodal variants have a higher ductility and lower strength than their lamellar counterparts [11–13]. Qin et al. [14] found that the yield strength ($\sigma_{0.2}$) of lamellar variants reaches 1900 MPa while that of bimodal alloys is around 1600 MPa. Zheng et al. [15] found that the elongation of bimodal variants is 11.5–14.5% and that of lamellar alloys is 4.5–9.5%. Therefore, the microstructure plays an essential role in determining the mechanical properties of titanium alloys.

Many studies have been performed to reveal the essential mechanism underpinning the different mechanical properties of titanium alloys induced by bimodal and lamellar microstructures. Wu et al. [16] found that the higher ductility of bimodal alloys can be mainly attributed to the more numerous deformation mechanisms available to the equiaxed primary α phase (α_p) in bimodal alloys (including dislocation slips, twins, and shear bands) than those of the α lath (α_L) in lamellar structures. Some believed that the tensile fracture of near-β-titanium alloys is sensitive to micro-void nucleation. Qin et al. [17] found that the tensile fracture of lamellar alloys is a process of the initiation of nano-scale voids, followed by their growth and coalescence in the deformation band, and transgranular shearing. The reason for the formation of micro-voids is the stress concentration caused by the difference in the strength of the α phase and β matrix. Stress concentration at the GBs is derived from the precipitation of the α_L phase along the α/β interface, which results in micro-void nucleation. This behavior is mainly responsible for the low ductility of lamellar alloys. Prior research into the deformation and properties of materials is based on ex situ test methods, which make it difficult to provide direct evidence of the deformation and fracture behavior; therefore, detailed information concerning the evolution of a microstructure, and its corresponding effect on the fracture process under tension, remains unclear.

Recently, in situ tensile SEM observation has become a powerful and effective tool to evaluate the deformation behavior from a microstructural perspective by capturing the microstructural evolution dynamically [18,19]. Huang et al. [20] indicated that localized stress concentration at the GBs derived from the geometric incompatibility between neighboring α grains was mainly responsible for microcrack formation in a Ti–6Al alloy. Zhang et al. [21] studied the deformation mechanism of a Ti–5Al–2Sn–2Zr–4Cr–4Mo alloy with a bimodal microstructure, implying that the α_p had high compatibility of deformation and the slip line in the α_p phase was the primary deformation mechanism. Shao et al. [22] found that microcracks were primarily initiated along the α_L phase at the edges of the sample. To date, however, the essential mechanism of the effect of microstructure on the mechanical properties of titanium alloys remains unclear, which hinders any further attempts to improve the mechanical properties of alloys.

In this study, Ti–5Al–5Mo–5V–1Cr–1Fe alloy samples with bimodal and lamellar microstructures were prepared through heat treatments. Their room-temperature tensile deformation process and fracture behavior were monitored in real time by in situ SEM observations. On this basis, the aim of this work is to ascertain the essential effects of microstructure on the mechanism of tensile deformation and fracture of titanium alloys. The results of this work are expected to provide a basis for future improvement in the mechanical properties of such titanium alloys.

2. Materials and Methods

2.1. Materials

The as-received material in this study was a forged Ti–5Al–5Mo–5V–1Cr–1Fe alloy provided by Baoti Group Ltd (Baoji, China) as cuboidal specimens (98.0 mm in length, 20.0 mm in width, and 7.0 mm in height) with a nominal composition of 5.07% Al, 4.81% Mo, 4.74% V, 1.06% Fe, 0.95% Cr, and the rest Ti (all in wt%). The $\alpha + \beta/\beta$ transition temperature ($T_{\alpha + \beta \to \beta}$) of this alloy is about 865 °C.

To obtain the desired microstructure, two as-received cuboids were, respectively, solution-treated at 830 °C (below $T_{\alpha + \beta \to \beta}$) and 895 °C (above $T_{\alpha + \beta \to \beta}$) for 2 h. Then, these cuboids were cooled in a furnace (FC) to 750 °C and held for 2 h before air cooling (AC). Thereafter, they were aged at 600 °C for 8 h, before air cooling (AC) to room temperature (Figure 1).

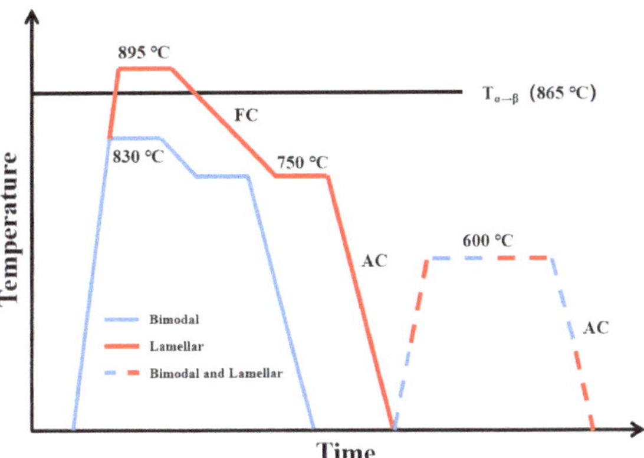

Figure 1. Heat treatment routes for bimodal and lamellar samples.

2.2. In Situ Tensile Test

The sample for the in situ tensile test was cut by wire-electrode cutting with a gauge length of 1.5 mm, a gauge width of 1.5 mm, and a thickness of 1.0 mm (Figure 2). It was mechanically polished using emery papers with SiC (5 μm, 3 μm, 1 μm, and 0.25 μm). Then, it was chemo-mechanically polished using Al_2O_3 (0.04 μm) suspension to remove the work-hardening resulting from previous mechanical polishing. A micro-stage (Figure 2a) and Mini MTS (Liweiauto Ltd., Hangzhou, China) controller (with a maximum load capacity of 2500 N) were employed to clamp the sample and control the tensile strain rate, respectively. The Mini MTS system was adopted to measure the mechanical properties of in situ samples. Before in situ testing, the sample was mechanically polished and then stretched at a speed of 1.5 μm/s at room temperature in vacuum.

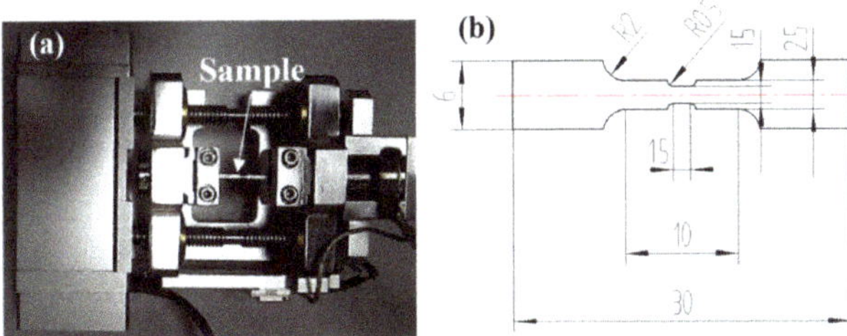

Figure 2. (**a**) Mechanical test bench; (**b**) geometry and dimensions of in situ tensile test specimens at room temperature (units: mm).

Processes during the in situ tensile deformation in bimodal and lamellar microstructures were observed using a JSM-5600 field-emission gun SEM (JEOL Ltd., Tokyo, Japan) with an accelerating voltage of 15 kV. Several interruptions were allowed by the loading system during tensile testing: this allowed the load to be held while capturing the SEM micrographs, after which the tensile test was resumed from the same applied load and displacement at the same rate. Three repeated in situ tensile tests were carried out under each set of experimental conditions, and typical results were provided. The tensile direction for the corresponding SEM images printed herein was parallel to the vertical direction.

Electron backscattered diffraction (EBSD) measurements were conducted before and after in situ tensile tests using an AZtec system (Oxford Instruments Group, Oxford, UK) coupled to a Hitachi-Regulus 8230 cold field emission SEM (Hitachi High-Technologies Corporation, Tokyo, Japan). The operating voltage used was 20 kV to optimize the quality of the diffraction patterns. The EBSD samples were electropolished using a solution of 8% perchloric acid ($HClO_4$) and 92% CH_4O at $-25\,^{\circ}C$.

3. Results

3.1. Microstructure of As-Heat-Treated Samples

Figure 3a shows backscattered electron (BSE)-SEM images of bimodal samples before in situ tensile testing. Since this sample was solution-treated at a temperature below $T_{\alpha+\beta\to\beta}$, some coarse globular α_p phases were maintained, which were evenly distributed in the β matrix. For the lamellar microstructure seen in Figure 3b, α_p was completely dissolved during solution treatment at a temperature above $T_{\alpha+\beta\to\beta}$ (i.e., 895 $^{\circ}C$ for 2 h), while coarse α_L, formed during subsequent low-temperature aging treatment, was distributed along GBs of the β matrix.

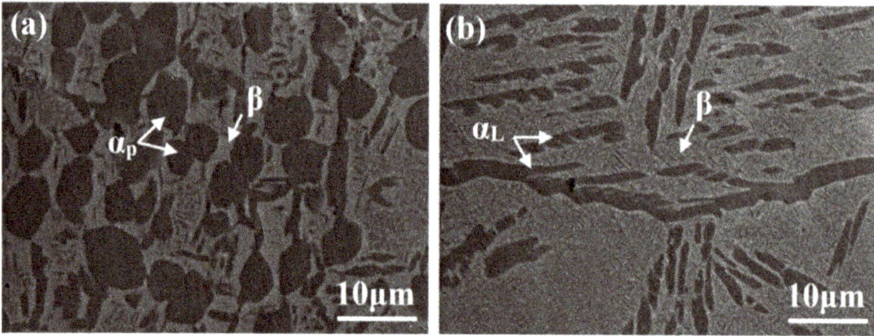

Figure 3. BSE-SEM images of samples before stretching in situ; (**a**) bimodal microstructure; (**b**) lamellar microstructure.

3.2. The Stress–Displacement Curves during In Situ Testing

The stress–displacement curves of bimodal and lamellar microstructures under in situ stretching are demonstrated in Figure 4. The drops in the curves are caused by slight stress relaxation during the pauses for SEM imaging, in which three typical drops for each sample are marked, respectively (A, B, and C for the bimodal microstructure and A′, B′, and C′ for the lamellar microstructure).

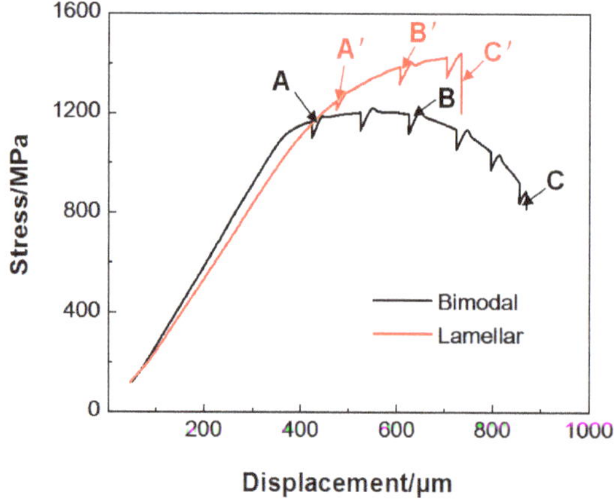

Figure 4. The stress–displacement curves of samples with bimodal and lamellar microstructures stretched in situ at room temperature.

As can be seen from Figure 4, $\sigma_{0.2}$ and the ultimate tensile stress (σ_{th}) of the bimodal microstructure are 1041.8 and 1220.5 MPa, which are much lower than those of specimens with a lamellar microstructure ($\sigma_{0.2}$ = 1102.7 MPa, σ_{th} = 1441.3 MPa). The maximum tensile displacement of the bimodal microstructure is about 872 μm, while that of the lamellar microstructure is about 736 μm. This result indicates that the lamellar microstructure is stronger, but less ductile than the bimodal microstructure, which is consistent with findings from previous studies [11–15].

3.3. Microstructure Evolution during In Situ Stretching

3.3.1. Microstructure Evolution of Bimodal Microstructure

Figure 5a illustrates the in situ SEM images of the bimodal microstructure at position A. At this stage, some parallel and deep SBs can be seen in a small number of α_p phase regions. These SBs generally lie at an angle of 41° to 49° with the tensile direction. As shown in Figure 5c, most of them are quite short (only several microns) and strictly confined within a single α_p phase, while some of them not only cross the whole α_p phase grain, but also pass through the α_p/β interface and enter the region containing the β matrix (Figure 5b). In addition, there is a certain region of distortion arising at the α_p/β interface (Figure 5d), which may be attributed to the deformation incompatibility between α_p and β due to their different crystal structures.

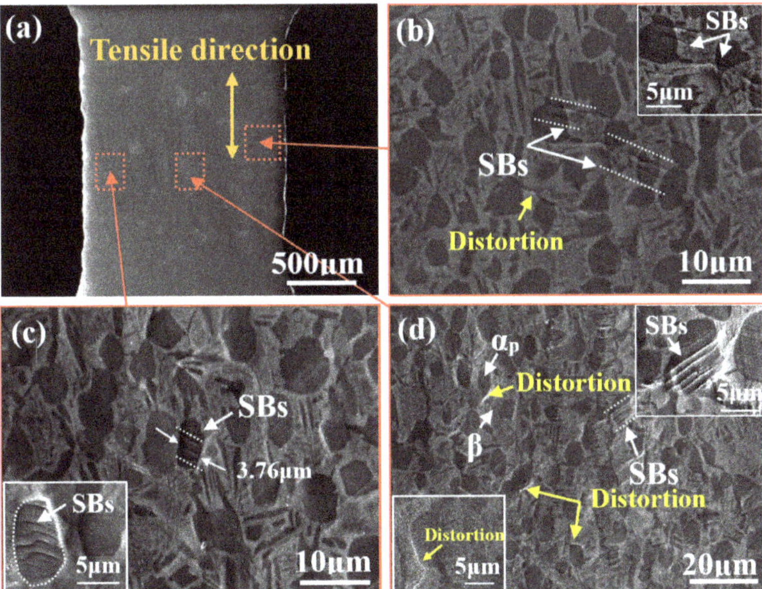

Figure 5. (**a**) In situ SEM images of the bimodal microstructure at position A; (**b**) magnified image showing SBs passing through the α_p/β interface; (**c**) SBs formed in the region containing the α_p phase; (**d**) distortion at the α_p/β interface region.

Figure 6 illustrates BSE-SEM images of bimodal samples stretched at position B. Obvious necking appears (Figure 6a) and some microcracks form on the edge of the sample (Figure 6b) caused by the increasing strain. The microcrack tends to propagate along the α_p/β interface and gradually grows to the center of the sample, as shown in Figure 6b. Noticeable SBs can be found in the β matrix adjacent to regions of α_p phase and the distortion at the α_p/β interface increases in severity (Figure 6d). This indicates that a higher stress concentration arises at this region near the α_p/β interface. In addition,

Figure 6c shows that the SBs in the α_p phase become deepened and some tiny microcracks are also initiated in the β matrix close to regions containing the α_p phase.

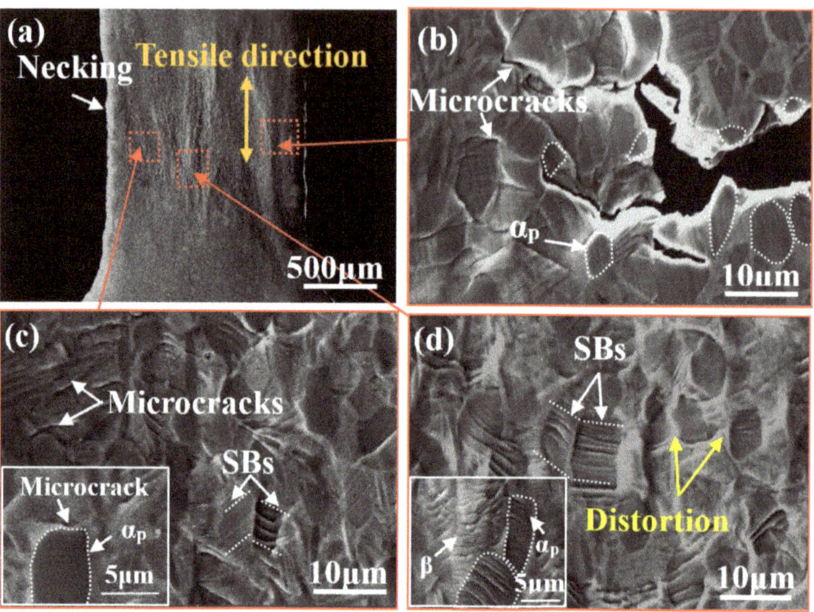

Figure 6. (**a**) In situ SEM images of the bimodal microstructure at position B; (**b**) magnified image showing microcracks formed on the edge of the sample; (**c**) microcracks initiated from the α_p/β interface; (**d**) distortion was aggravated around the α_p/β interface.

Figure 7 illustrates in situ SEM images of the bimodal microstructure at position C (after fracture). The fracture surface is relatively rough (Figure 7e) and significant necking can be observed at the region close to the fracture surface (Figure 7c,f). As shown in Figure 7b,d, many microcracks form in the region close to the fracture surface: these nucleate at and generally propagate along the α_p/β interface (the crack path is tortuous). Although the number of SBs increases, the α_p phase still maintains a granular shape (Figure 7d) and grains are wrapped by the significantly distorted β matrix (Figure 7d). As can be seen in Figure 7a, the morphology of the β matrix near the fractured zone becomes streamlined in shape, implying that it undergoes significant plastic deformation before fracture. Therefore, the β matrix undergoes greater deformation than the α_p phase during in situ stretching; in addition, this deformation is relatively uniformly distributed in the β matrix, probably due to the excellent deformation compatibility.

To assess the deformation behavior of the bimodal microstructure during in situ tensile loading, an area of the sample was selected for tracking, and the SEM images thereof at different stages were recorded. Figure 8a,c show the SEM images of the same area on a specimen with a bimodal microstructure surface at the stage of positions A and B, respectively: with increasing strain, the sample surface became significantly rougher and took on an undulate appearance (Figure 8c). To determine the strain in this local area, the changes in distance between two α_p phase grains on the sample surface were measured. Stretched from position A to position B, the distance increased from 135.71 μm (position A) to 167.85 μm (position B), indicating some 23.68% of the plastic deformation appears along the tensile direction (ε_L) in this region.

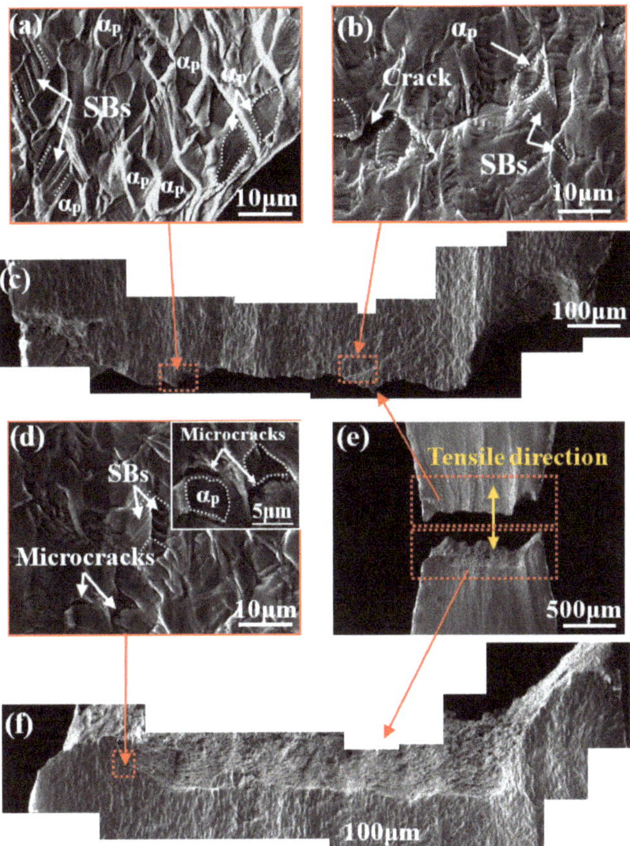

Figure 7. (**a**) The streamlined shape of the β matrix near the fractured zone at position C; (**b**) a crack propagated along the α_p/β interface; (**c**,**f**) SEM images showing the rough fracture surface of the sample; (**d**) magnified image showing intensified SB formation in the α_p phase and microcracking at the α_p/β interface; (**e**) macroscopical SEM images of the bimodal microstructure at position C.

Figure 8b,d are magnified images of Figure 8a,c, respectively. Although the sample underwent severe plastic deformation, the distribution of SBs in the β matrix remained relatively uniform. For further analysis, 25 α_p phase grains were selected for calibration (and assigned serial numbers 1 to 25). The size changes of these 25 α_p phase grains along the tensile direction from positions A to B were statistically studied (Table 1); most α_p phase regions have ε_L values of less than 10%, and their average ε_L value is 9.72%, which is only 41% of the average ε_L value over the region. This result demonstrates that during tensile deformation, the β matrix was subject to larger plastic deformation than the α_p phase.

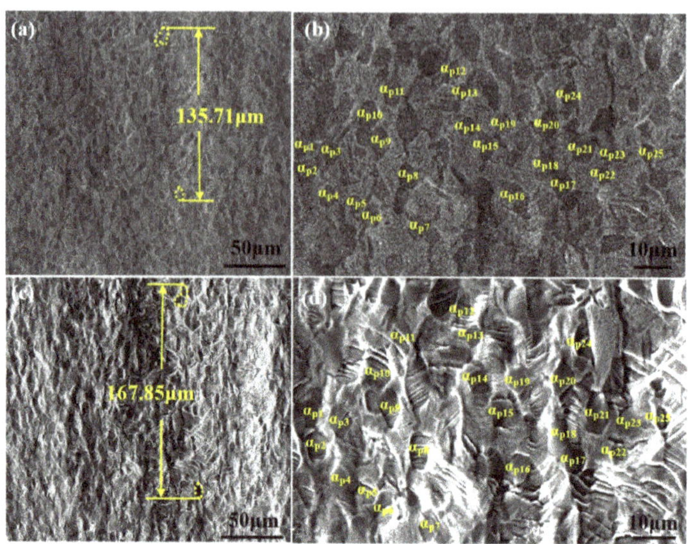

Figure 8. In situ SEM images of bimodal samples; (**a**) a selected area at position A; (**b**) magnified image showing the morphologies of α_p phase at position A; (**c**) the area in (**a**) stretched to position B; (**d**) magnified image illustrating the deformation within α_p phase at position B.

Table 1. Calculation of ε_L values for αp phase (Figure 8) from positions A to B.

No.	Position A (μm)	Position B (μm)	Δ (Position A→B) (μm)	ε_L (%)
α_{p1}	6.54	7.15	0.61	9.33
α_{p2}	6.78	7.30	0.52	7.67
α_{p3}	6.84	7.52	0.68	9.94
α_{p4}	5.67	6.13	0.46	8.11
α_{p5}	5.34	5.77	0.43	8.05
α_{p6}	6.78	7.44	0.66	9.73
α_{p7}	7.12	7.73	0.61	8.57
α_{p8}	7.46	8.32	0.86	11.53
α_{p9}	8.64	9.43	0.79	9.14
α_{p10}	7.54	8.61	1.07	14.19
α_{p11}	8.13	9.27	1.14	14.02
α_{p12}	5.49	5.98	0.49	8.93
α_{p13}	4.91	5.39	0.48	9.78
α_{p14}	5.17	5.62	0.45	8.70
α_{p15}	7.80	8.25	0.45	5.77
α_{p16}	6.19	6.86	0.67	10.82
α_{p17}	6.89	7.29	0.40	5.81
α_{p18}	8.22	9.12	0.90	10.95
α_{p19}	5.94	6.43	0.49	8.25
α_{p20}	9.83	10.57	0.74	7.53
α_{p21}	8.14	9.36	1.22	14.99
α_{p22}	4.87	5.21	0.34	6.98
α_{p23}	5.59	6.28	0.69	12.34
α_{p24}	8.64	9.78	1.14	13.19
α_{p25}	6.19	6.72	0.53	8.56
			Average	9.72

3.3.2. Evolution of the Lamellar Microstructure

Figure 9 shows BSE-SEM images of specimens with a lamellar microstructure during in situ stretching at position A′ on the stress–displacement curve (Figure 4). As shown in Figure 9d, some parallel SBs form within or at the boundaries of the α_L phase at an angle of about 45° to the tensile direction. Differing from those in specimens with a bimodal microstructure, in which such SBs measure only several microns, these SBs aligned along the length of the α_L phase are significantly longer and can always grow to several tens of microns.

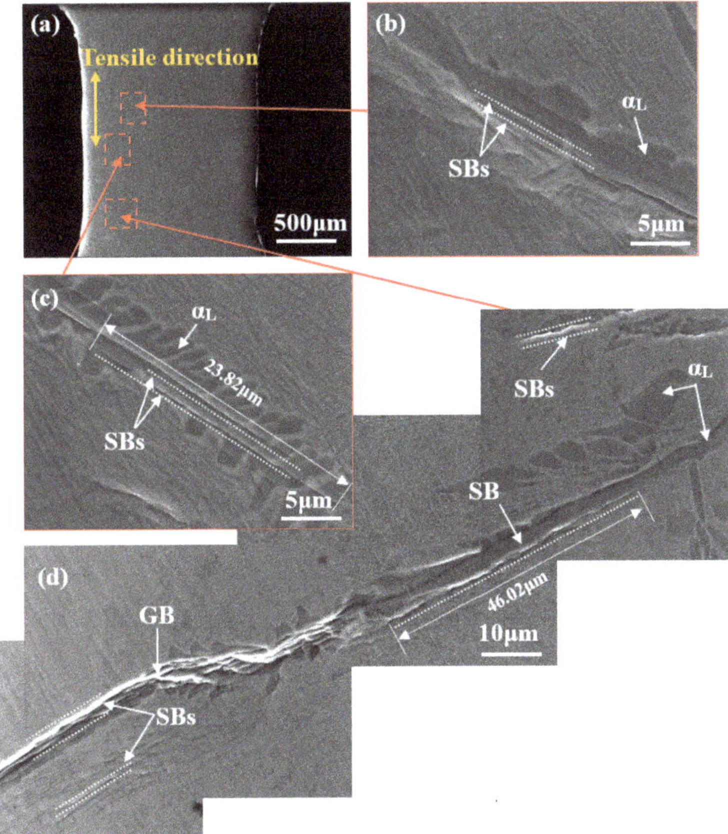

Figure 9. (a) In situ SEM images of lamellar microstructure at position A′; (b,c) magnified image showing several SBs in an α_L phase or at its interface; (d) magnified image showing plentiful SBs in a long coarse α_L phase at a GB at an angle of about 45° to the tensile direction.

Figure 10 illustrates BSE-SEM images of lamellar samples during in situ stretching at position B′. As the strain increases, SBs gradually extend and connect with each other along the length of α_L at which some extremely long SBs are formed. As shown in Figure 10d, the connected SBs in the α_L at the grain boundary are over one hundred microns in length, besides which certain microcracks evolving from SBs can also be found in the α_L grains or at their interfaces.

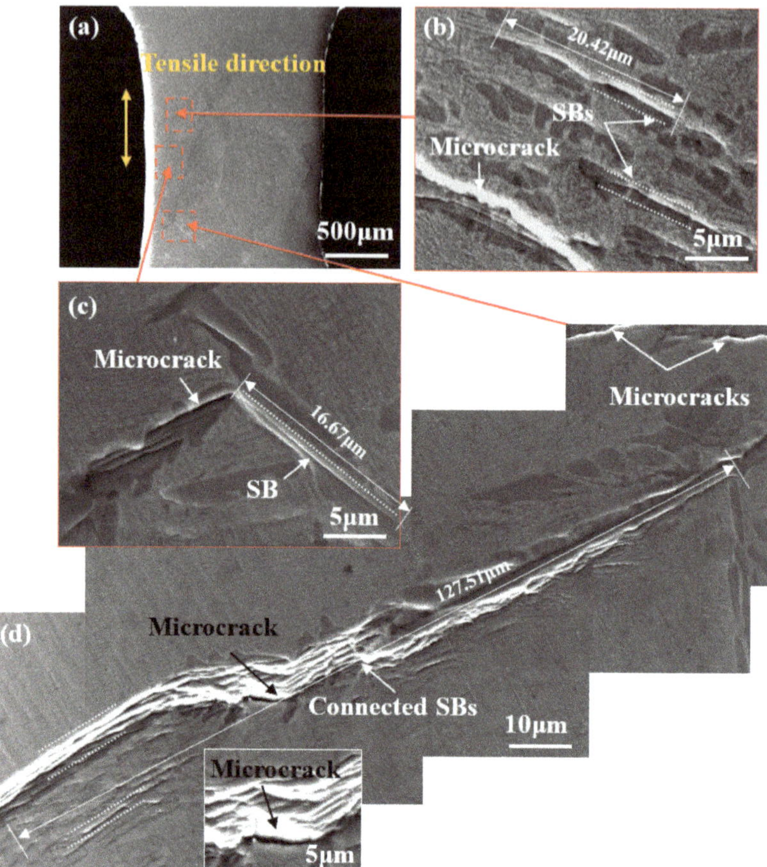

Figure 10. (**a**) In situ SEM images of lamellar microstructure at position B′; (**b**,**c**) magnified image showing several microcracks formed in α_L phase or at their interfaces; (**d**) magnified image showing SBs connected with each other and forming an SB with a length of 127.51 μm.

Figure 11 exhibits in situ SEM images of lamellar samples during stretching at position C′. No obvious necking occurs in the sample until it fractures, which indicates that the lamellar microstructure undergoes less plastic deformation than the bimodal microstructure. Although there are slight SBs in the β matrix adjacent to the fracture surface, they are fewer in number than in specimens with a bimodal microstructure. Judging from Figure 11c,d, specimens with a lamellar microstructure generally fracture along the GBs, which results in a sharp fracture surface.

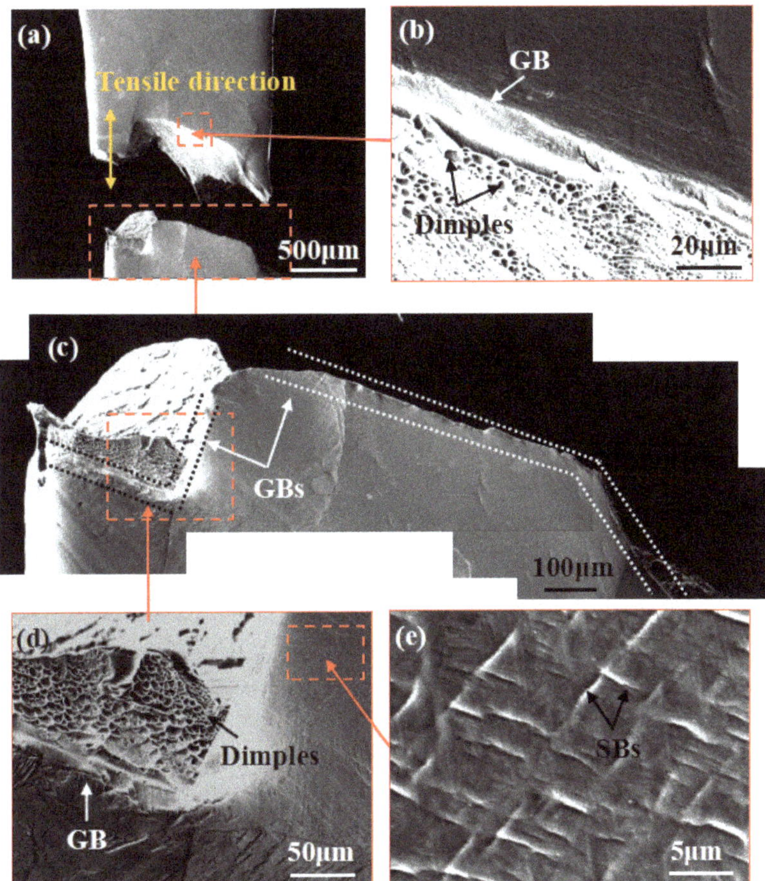

Figure 11. (**a**) In situ SEM images of the lamellar microstructure at position C′; (**b**,**d**) magnified images showing many small, shallow dimples on the fracture surface; (**c**) magnified image showing the sample with a shape fracture surface generally breaking along the GB; (**e**) some short crossed SBs are shown in the β matrix adjacent to the fracture surface.

As shown in Figure 12a,c, the changes in a selected area of the lamellar microstructure from position A′ to position B′ are tracked. The changes in roughness of the sample surface are small and there are fewer SBs compared with those in specimens with a bimodal microstructure (Figure 11). The ε_L value in this local area from position A′ to position B′ is found to be 1.33% (less than that in specimens with a bimodal microstructure).

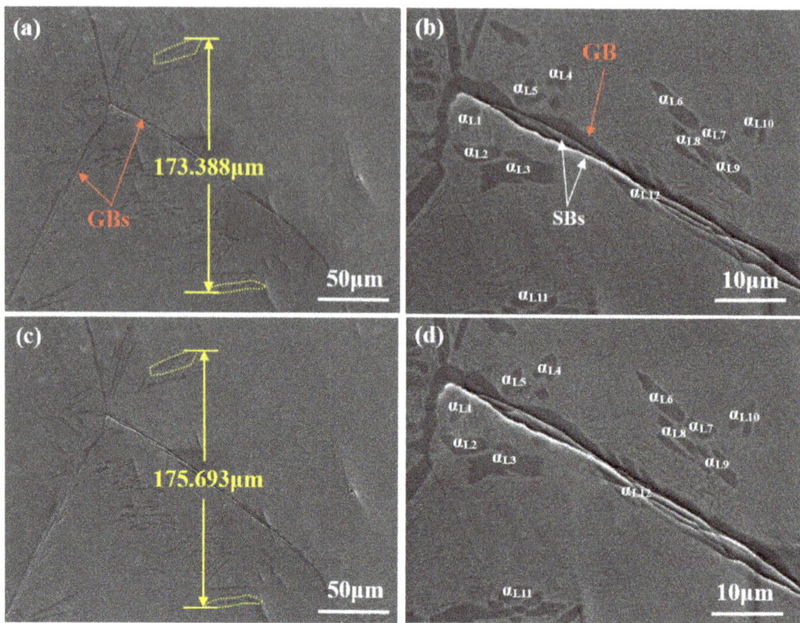

Figure 12. In situ SEM images of the lamellar microstructure; (**a**) a selected area at position A′; (**b**) magnified image showing the morphologies of α_L phase at position A′; (**c**) the area in (**a**) stretched to position B′; (**d**) magnified image illustrating the deformation within α_L phase at position B′.

Figure 12b,d demonstrate magnified versions of Figure 12a,c, respectively. To quantify the deformation behavior, 12 α_L phase grains were selected for calibration and assigned serial numbers 1 to 12. As shown in Figure 12b, SBs are mainly found in the large α_L phase (α_{L12}), which can extend along the GB, while they are scarcely found in the β matrix or small α_L phase regions. With the increase in strain from position A′ to position B′, SBs in α_{L12} deepen, while the microstructure in the β matrix and small α_L phase changes little (Figure 12d). The ε_L values of the selected 12 α_L phase grains deformed from position A′ to position B′ were statistically analyzed (Table 2). According to the data in Table 2, the relatively small α_L phase (α_{L1}-α_{L11}) has quite small ε_L values (less than 1.0%), while the ε_L value of the long α_L phase at the GB (α_{L12}) is found to be 5.79%, which is 3.35 times greater than the ε_L value across this area; the deformation in the lamellar microstructure is mainly concentrated in the large α_L phase found at the GB (i.e., α_{L12}).

Table 2. Calculation of ε_L values of α_L phase (Figure 12) from positions A′ to B′.

No.	Position A′ (μm)	Position B′ (μm)	Δ (Position A′→B′) (μm)	ε_L (%)
α_{L1}	3.441	3.459	0.018	0.52
α_{L2}	2.957	2.973	0.016	0.54
α_{L3}	6.068	6.108	0.040	0.66
α_{L4}	3.925	3.934	0.009	0.23
α_{L5}	4.084	4.108	0.024	0.59
α_{L6}	7.189	7.257	0.068	0.95
α_{L7}	2.996	3.012	0.016	0.53

Table 2. *Cont.*

No.	Position A′ (μm)	Position B′ (μm)	Δ (Position A′→B′) (μm)	ε_L (%)
α_{L8}	5.839	5.874	0.035	0.60
α_{L9}	6.226	6.283	0.057	0.92
α_{L10}	5.144	5.181	0.037	0.72
α_{L11}	2.527	2.545	0.018	0.71
α_{L12}	26.774	28.324	1.550	5.79
-	-	-	Average	1.06

4. Discussion

Recently, investigators have found that mechanical behavior and its related mechanism of action are sensitive to the initial microstructure of titanium alloys. Specimens with a bimodal microstructure are found to have a lower strength but higher ductility than those with a lamellar microstructure [11–15], while the essential reason for the differences in mechanical behavior between bimodal and lamellar microstructures is still hotly debated. Huang et al. [23] found that the α_p phase in specimens with a bimodal microstructure plays a major role in accommodating the plastic strain of titanium alloys due to its good compatibility during deformation, while Tan et al. [24] stated that cracks are readily initiated at SBs in the α_p phase, which lies at the crux of the tensile deformation. Liu et al. [25,26] considered that a high stress concentration at the α_L/β grain boundary results in intergranular fracture and low ductility of specimens with a lamellar microstructure, whereas Qin et al. [27] found that the crack nucleates inside the β grains and will spread under high tensile stress without hindrance in the larger β grains, leading to the low plasticity of specimens with a lamellar microstructure. In this study, their quite different mechanical properties were found to be essentially attributed to different microstructural evolutions during tensile loading.

4.1. Deformation Mechanisms and Microstructural Evolution of the Bimodal Microstructure

Based on the results of in situ SEM observation, the deformation mechanisms and microstructural evolution of the bimodal microstructure are shown schematically in Figure 13.

The bimodal microstructure contains coarse globular α_p grains distributed in the β matrix (Figure 13a). Under a relatively small strain (Figure 13b), many parallel, deep SBs are formed inside some α_p phase regions due to the limited slip systems of α_p and the stress concentration caused by their relatively large size. As previously proved by Semiatin et al. [28], the ratio of critical resolved shear stress in the α phase at room temperature was 1:0.7:3.2 for basal ({0001}<11−20>), prismatic ({1−100}<11−20>), and pyramidal slip ({1−101}<11−20>), respectively. This indicates that pyramidal slip is difficult to take place and the α_p phase is more likely to slip along its basal or prismatic plane at room temperature. Figure 14 demonstrates four examples of SB identification for a bimodal microstructure after in situ stretching. These SBs essentially correspond to prismatic or basal slip systems and to single slip behavior with a relatively large Schmid factor (SF) (SF > 3.7), which agrees well with the results of Semiatin et al. [28].

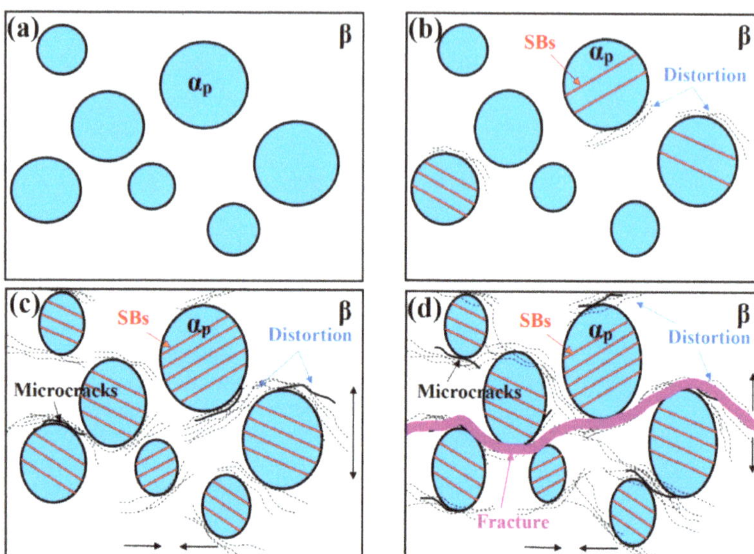

Figure 13. Schematic mechanism showing the evolution of bimodal microstructure during in situ stretching; (**a**) the initial microstructure; (**b**) at a relatively low strain; (**c**) at a relatively high strain; (**d**) schematic mechanism of deformation of bimodal samples after fracture.

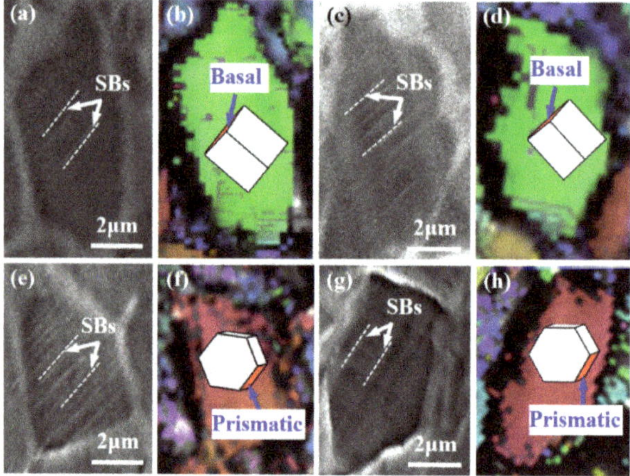

Figure 14. Examples of identification of activated slip systems on the α_p phase in bimodal microstructure; (**a**,**b**) a basal slip activates (SF = 0.38) in the α_p phase; (**c**,**d**) a basal slip activates (SF = 0.37) in the α_p phase; (**e**,**f**) a prismatic slip activates (SF = 0.45) in the α_p phase; (**g**,**h**) a prismatic slip activates (SF = 0.46) in the α_p phase.

In addition, as shown in Figure 15a, the orientations of α_p phases are randomly distributed in the bimodal microstructure. For the same α_p phase, there is a large difference in SF values for basal and prismatic slips (Figure 15b,c); the one with the maximum SF is supposed to be the easiest to activate [21]. Therefore, it is difficult for the α_p phase to activate both basal and prismatic slips simultaneously. SBs are prone to occur along a single basal or prismatic plane of the α_p phase, which is oriented solely along the maximum shear stress direction, i.e., at an angle of 45° to the applied tension. Differing from the α_p

phase, there are significantly more slip systems in the β phase, among which {1−10}<111>, {11−2}<111>, and {12−3}<111> slips were thought to be the three typical cases [28,29]. As presented in Figure 15d,f, there is no significant difference in the SF values for these three types of slips in the β phase. Thus, it is difficult to observe β phase regions' slip strictly along one single plane especially under a small strain.

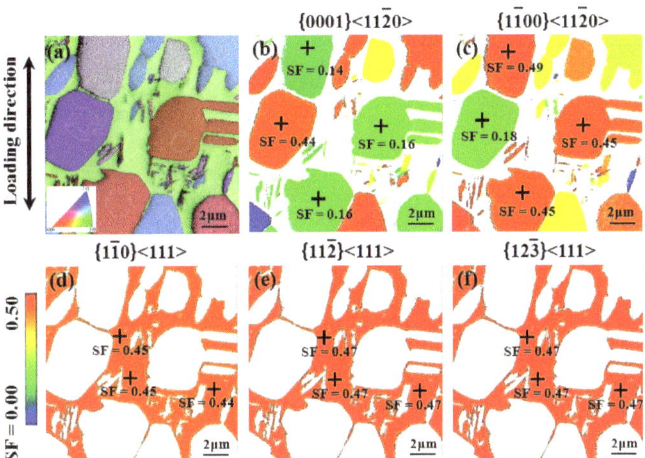

Figure 15. EBSD images of bimodal microstructure before stretching in situ; (**a**) inverse pole-figure (IPF) map; (**b**) SF map of α for basal slip; (**c**) SF map of α for prismatic slip; (**d**) SF map of β for {1−10}<111> slip; (**e**) SF map of β for {11−2}<111> slip; (**f**) SF map of β for {12−3}<111> slip.

In the bimodal microstructure, although a small number of SBs in the α_p phase can pass through the α_p/β interface, most are restricted at the interface between β and α_p and confined to within a single α_p phase grain. This will lead to an increased stress concentration at the α_p/β interface, therefore distorting the interface. As the strain increases (Figure 13c), the α_p phase is elongated slightly along the tensile direction (Figure 8).

β is softer than α_p due to lower concentrations of solute Al in β [30], thus β should bear more significant deformation than the α_p phase at the same stress, while due to the more numerous slip systems and greater deformation coordination in BCC β, parallel, deep SBs may be less likely to form in the β phase when the strain is relatively small (Figures 6 and 7). Caused by the different deformation behaviors between α_p and β, stress concentration at the interface gradually increases with increasing strain, finally generating microcracks. Then, with the further increase in strain (Figure 13d), the number of SBs in the α_p phase increases slightly due to its more limited slip system, while that in the β matrix increases to a much greater extent. This further aggravates the stress concentration at the α_p/β interface, making the microcrack propagate along the α_p/β interface and into the β matrix. As the microcracks grow, they gradually come closer together, whereupon they tend to interconnect to form a main crack. Additionally, as the main crack grows, it is likely to bridge these microcracks formed at the α_p/β interfaces. This finally leads to the zig-zag crack path and rough fracture surface of the sample since the α_p grains are randomly distributed in the β matrix (Figure 13d).

It should also be pointed out that, because of the connected distribution of the β phase and isolated distribution of the α_p phase in specimens with a bimodal microstructure, localized deformation readily propagates into the surrounding area by way of the soft β phase, with only a minor role played by the harder α_p phase. This ensures relatively uniform deformation, giving rise to the excellent ductility of the bimodal microstructure.

4.2. Deformation Mechanisms and Microstructural Evolution of the Lamellar Microstructure

Figure 16 schematically presents the deformation mechanisms and microstructural evolution of the lamellar microstructure based on the results of in situ SEM observation.

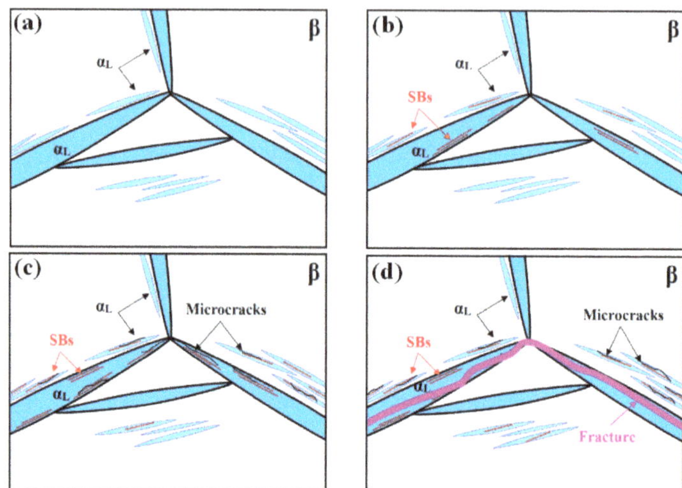

Figure 16. Schematic mechanism showing the evolution of lamellar samples during in situ stretching; (**a**) initial microstructure; (**b**) at a relatively low tensile displacement; (**c**) at a relatively high tensile displacement; (**d**) schematic mechanism of the deformation of lamellar samples after fracture.

Differing from the bimodal microstructure, many long and coarse lamellar α_L grains, rather than globular α_p regions, are seen in the lamellar microstructure: these are mainly distributed at or near GBs (Figure 16a). Since these α_L phase regions are much harder than the β phase [31], they can exert a strong fencing effect and therefore separate each β grain into a relatively isolated region between which dislocation cannot easily pass. This in turn leads to a great stress concentration at the α_L phase. Therefore, deep SBs are first observed in the coarse α_L phase at GBs under a relatively low tensile strain (Figure 16b). As shown in Figure 17, these SBs correspond to prismatic slip behavior (SF > 4.2) and generally lie along the plane at an angle of approximately 45° to the direction of the applied tension (i.e., the direction of maximum shear stress).

Meanwhile, the deformation in the β matrix is much smaller due to the strong fencing effect of the α_L (Figure 9). With increasing strain, the number of SBs in the large α_L phase increases while that in the β matrix remains low (Figure 10), suggesting that the deformation in lamellar microstructures is non-uniform and mainly concentrated in the large α_L grains at GBs. As a result of this inhomogeneous deformation, microcracks are readily initiated from, and propagate along, the SBs (Figure 16c). The formation of microcracks in these α_L regions in turn produces a greater stress concentration, releasing stress accumulation in the β matrix. This aggravates the inhomogeneity of the deformation of specimens with a lamellar microstructure. Finally, the sample fractures along the large α_L grains at GBs, leading to the low plasticity of the lamellar microstructure (Figure 16d).

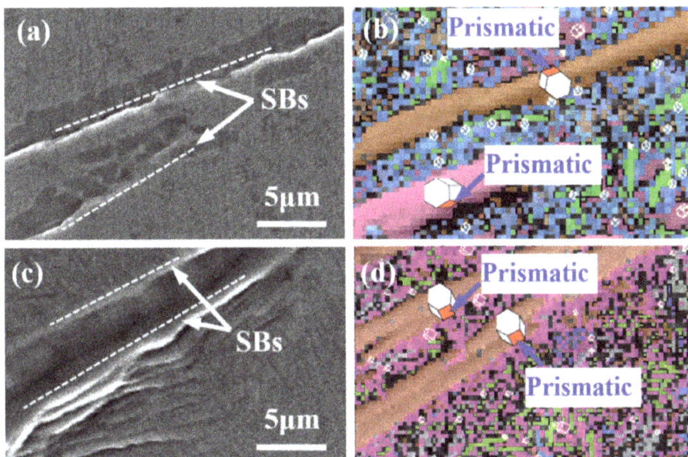

Figure 17. Examples of identification of activated slip systems on α_L phase in lamellar microstructure; (**a,b**) prismatic slips activate in two α_L phase with SF values of 0.44 and 0.42, respectively; (**c,d**) prismatic slips activate in two α_L phases with SF values of 0.43 and 0.45, respectively.

5. Conclusions

In this study, the microstructural evolution and fracture mechanisms of a Ti–5Al–5Mo–5V–1Cr–1Fe alloy with bimodal and lamellar microstructures were investigated through in situ tensile SEM observation. The following conclusions can be obtained:

1. For the bimodal microstructure, parallel and deep SBs, at around 45° to the tensile direction, are first observed in the α_p phase due to the limited slip systems therein and the stress concentration caused by its large size. These SBs mainly correspond to prismatic or basal slip systems. With increasing strain, distortion at the α_p/β interface arises, leading to the evolution of microcracks: the interconnection of the microcracks at the α_p/β interface finally leads to the fracturing of the bimodal microstructure.
2. For the lamellar microstructure, parallel and deep SBs are first observed in coarse α_L phase grains or its interfaces at GBs, making an angle of around 45° to the tensile direction. With increasing strain, these SBs grow along the length of the α_L phase and gradually interconnect, thus forming microcracks. The lamellar microstructure finally fractures along the α_L phase at GBs through the interconnection of those microcracks in the α_L phase.
3. Due to the connected distribution of β and isolated distribution of the α_p phase in the bimodal microstructure, the localized deformation readily propagates into the surrounding area through β, which is softer and has a better ability to undergo plastic deformation. This results in the more uniform deformation and higher ductility of specimens with a bimodal microstructure.
4. The coarse α_L phase at the GBs greatly restricts the deformation in the β matrix during tensile loading, which in turn produces a significant stress concentration and local deformation in the coarse α_L phase. This finally leads to intergranular fracture and contributes to the higher strength and lower ductility of those specimens with a lamellar microstructure.

Author Contributions: Investigation, S.P. and Y.C.; Data acquisition Y.C., S.P. and M.F.; Formal analysis, M.F., H.L., D.Y. and S.P.; Resources, H.L. and Y.C.; Writing—original draft, S.P. and M.F.; Conceptualization, D.Y. H.L. and Y.C. All authors have read and agreed to the published version of the manuscript.

Funding: This research was funded by the State Key Laboratory of Powder Metallurgy (Grant No. 10500-621022001), Central South University, Changsha, China.

Institutional Review Board Statement: Not applicable.

Informed Consent Statement: Informed consent was obtained from all subjects involved in the study.

Data Availability Statement: The data presented in this study are available on request from the corresponding author.

Conflicts of Interest: The authors declare no conflict of interest.

References

1. Gao, F.Y.; Guo, Y.F.; Yang, S.L.; Yu, Y.; Yu, W. Fatigue properties of friction stir welded joint of titanium alloy. *Mater. Sci. Eng. A* **2020**, *793*, 139819. [CrossRef]
2. Vitus, M.T.; Li, C.; Wang, S.F.; Li, J.W.; Xu, X.J. Mechanical properties of near alpha titanium alloys for high–temperature applications—A review. *Aircr. Eng. Aerosp. Technol.* **2020**, *92*, 521–540.
3. Chen, W.; Li, C.; Zhang, X.Y.; Chen, C.; Lin, Y.C.; Zhou, K.C. Deformation−induced variations in microstructure evolution and mechanical properties of bi−modal Ti−55511 titanium alloy. *J. Alloys Compd.* **2019**, *783*, 709–717. [CrossRef]
4. Zhang, D.D.; Liu, N.; Chen, Y.Y.; Zhang, G.Q.; Tian, J.; Kong, F.T.; Xiao, S.L.; Sun, J.F. Microstructure Evolution and Mechanical Properties of PM-Ti43Al9V0.3Y Alloy. *Materials* **2020**, *13*, 198. [CrossRef]
5. Di, W.; Tian, Y.Y.; Zhang, L.G.; Wang, Z.Y.; Sheng, J.W.; Wang, W.L.; Zhou, K.C.; Liu, L.B. Optimal Design of High–Strength Ti–Al–V–Zr Alloys through a Combinatorial Approach. *Materials* **2018**, *11*, 1603.
6. Shao, H.; Shan, D.; Wang, K.X.; Zhang, G.J.; Zhao, Y.Q. Massive α precipitation selectivity and tensile fracture behavior of TC18 alloy. *J. Alloys Compd.* **2019**, *797*, 10–17. [CrossRef]
7. Li, C.; Chen, J.; Li, W.; He, J.J.; Qin, W.; Ren, Y.J.; Chen, J.L.; Chen, J.H. Study on the relationship between microstructure and mechanical property in a metastable β titanium alloy. *J. Alloys Compd.* **2015**, *627*, 222–230. [CrossRef]
8. Cui, N.; Wu, Q.Q.; Bi, K.X.; Xu, T.W.; Kong, F.T. Effect of Heat Treatment on Microstructures and Mechanical Properties of a Novel β-Solidifying TiAl Alloy. *Materials* **2019**, *12*, 1672. [CrossRef] [PubMed]
9. Yadav, P.; Saxena, K.K. Effect of heat−treatment on microstructure and mechanical properties of Ti alloys: An overview. *Mater. Today Proc.* **2020**, *26*, 2546–2557. [CrossRef]
10. Zhu, W.G.; Sun, Q.Y.; Tan, C.S.; Li, P.; Xiao, L.; Sun, J. Tensile brittleness and ductility improvement in a novel metastable β titanium alloy with lamella structure. *J. Alloys Compd.* **2020**, *827*, 154311. [CrossRef]
11. Srinivasu, G.; Natraj, Y.; Bhattacharjee, A.; Nandy, T.K.; Rao, G.V.S.N. Tensile and fracture toughness of high strength β Titanium alloy, Ti–10V–2Fe–3Al, as a function of rolling and solution treatment temperatures–ScienceDirect. *Mater. Des.* **2013**, *47*, 323–330. [CrossRef]
12. Kar, S.K.; Suman, S.; Shivaprasad, S.; Chaudhuri, A.; Bhattacharjee, A. Processing–microstructure–yield strength correlation in a near β Ti alloy, Ti–5Al–5Mo–5V–3Cr. *Mater. Sci. Eng. A* **2014**, *610*, 171–180. [CrossRef]
13. Wu, G.Q.; Shi, C.L.; Sha, W.; Sha, A.X.; Jiang, H.R. Microstructure and high cycle fatigue fracture surface of a Ti–5Al–5Mo–5V–1Cr–1Fe titanium alloy. *Mater. Sci. Eng. A* **2013**, *575*, 111–118. [CrossRef]
14. Qin, D.Y.; Li, Y.L. The role of microstructure and stress state in dynamic mechanical behavior of Ti−5Al−5V−5Mo−3Cr alloy. *Mater. Charact.* **2018**, *147*, 421–433. [CrossRef]
15. Zheng, C.; Wang, F.C.; Cheng, X.W.; Fu, K.Q.; Liu, J.X.; Wang, Y.F.; Liu, T.T.; Zhu, Z.X. Effect of microstructures on ballistic impact property of Ti-6Al-4V targets. *Mater. Sci. Eng. A* **2014**, *608*, 53–62. [CrossRef]
16. Wu, D.; Liu, L.B.; Zhang, L.G.; Wang, W.L.; Zhou, K.C. Tensile deformation mechanism and micro−void nucleation of Ti−55531 alloy with bimodal microstructure. *J. Mater. Res. Technol.* **2020**, *9*, 15442–15453. [CrossRef]
17. Qin, D.Y.; Li, Y.L.; Zhang, S.Y.; Zhou, L. On the tensile embrittlement of lamellar Ti–5Al–5V–5Mo–3Cr alloy. *J. Alloys Compd.* **2016**, *663*, 581–593. [CrossRef]
18. Zhao, Q.Y.; Yang, F.; Torrens, R.; Bolzoni, L. In-situ observation of the tensile deformation and fracture behaviour of powder-consolidated and as–cast metastable beta titanium alloys. *Mater. Sci. Eng. A* **2019**, *750*, 45–59. [CrossRef]
19. Wan, X.; Zhu, K.; Xu, Y.J.; Han, B.S.; Jing, T. In-Situ Observation of Fracture Behavior of Ti–Aluminide Multi–Layered Composites Produced by a Hybrid Sintering Process. *Materials* **2019**, *12*, 1568. [CrossRef] [PubMed]
20. Huang, S.X.; Zhao, Q.Y.; Lin, C.; Wu, C.; Zhao, Y.Q.; Jia, W.J.; Mao, C.L. In-situ investigation of tensile behaviors of Ti–6Al alloy with extra low interstitial. *Mater. Sci. Eng. A* **2021**, *809*, 140958. [CrossRef]
21. Zhang, S.F.; Zeng, W.D.; Zhao, Q.Y.; Ge, L.L.; Zhang, M. In situ SEM study of tensile deformation of a near–β titanium alloy. *Mater. Sci. Eng. A* **2017**, *708*, 574–581. [CrossRef]
22. Shao, H.; Zhao, Y.Q.; Ge, P.; Zeng, W.D. In-situ SEM observations of tensile deformation of the lamellar microstructure in TC21 titanium alloy. *Mater. Sci. Eng. A* **2013**, *559*, 515–519. [CrossRef]
23. Huang, J.; Wang, Z.R.; Xue, K.M. Cyclic deformation response and micromechanisms of Ti alloy Ti–5Al–5V–5Mo–3Cr–0.5Fe. *Mater. Sci. Eng. A* **2011**, *528*, 8723–8732. [CrossRef]

24. Tan, C.S.; Sun, Q.Y.; Xiao, L.; Zhao, Y.Q.; Sun, J. Characterization of deformation in primary α phase and crack initiation and propagation of TC21 alloy using in-situ SEM experiments. *Mater. Sci. Eng. A* **2018**, *725*, 33–42. [CrossRef]
25. Liu, R.; Hui, S.X.; Ye, W.J.; Xiong, B.Q.; Yu, Y.; Fu, Y.Y. Dynamic fracture toughness of TA15ELI alloy studied by instrumented impact test. *Rare Metals* **2010**, *29*, 608–612. [CrossRef]
26. Liu, R.; Hui, S.X.; Ye, W.J.; Yu, Y.; Fu, Y.Y.; Song, X.Y.; Deng, X.G. Tensile and fracture properties of Ti–62A alloy plate with different microstructures. *Rare Metals* **2012**, *31*, 420–423. [CrossRef]
27. Qin, D.Y.; Lu, Y.F.; Liu, Q.; Zheng, L.; Zhou, L. Transgranular shearing introduced brittlement of Ti–5Al–5V–5Mo–3Cr alloy with full lamellar structure at room temperature. *Mater. Sci. Eng. A* **2013**, *572*, 19–24. [CrossRef]
28. Semiatin, S.L.; Bieler, T.R. Effect of texture and slip mode on the anisotropy of plastic flow and flow softening during hot working of Ti–6Al–4V. *Metall. Mater. Trans. A* **2001**, *32*, 1787–1799. [CrossRef]
29. Banerjee, D.; Williams, J.C. Microstructure and slip character in titanium alloys. *Def. Sci. J.* **2014**, *36*, 191–206. [CrossRef]
30. Chong, Y.; Bhattacharjee, T.; Park, M.H.; Shibata, A.; Tsuji, N. Factors determining room temperature mechanical properties of bimodal microstructures in Ti–6Al–4V alloy. *Mater. Sci. Eng. A* **2018**, *730*, 217–222. [CrossRef]
31. Chong, Y.; Deng, G.Y.; Yi, J.H.; Shibata, A.; Tsuji, N. On the strain hardening abilities of α+β titanium alloys: The roles of strain partitioning and interface length density. *J. Alloys Compd.* **2019**, *811*, 152040. [CrossRef]

materials

Article

Cobalt Content Effect on the Magnetic Properties of $Ni_{50-x}Co_xMn_{35.5}In_{14.5}$ Annealed Ribbons

Łukasz Dubiel [1,*], Andrzej Wal [2], Ireneusz Stefaniuk [3], Antoni Żywczak [4], Piotr Potera [2] and Wojciech Maziarz [5]

[1]   Department of Physics and Medical Engineering, Faculty of Mathematics and Applied Physics, Rzeszow University of Technology, Powstancow Warszawy 12, 35-959 Rzeszow, Poland

[2]   Institute of Physics, College of Natural Sciences, University of Rzeszow, Pigonia 1, 35-959 Rzeszow, Poland; wal@ur.edu.pl (A.W.); ppotera@ur.edu.pl (P.P.)

[3]   Center of Teaching Technical and Natural Sciences, University of Rzeszow, Pigonia 1, 35-959 Rzeszow, Poland; istef@ur.edu.pl

[4]   Academic Centre for Materials and Nanotechnology, AGH University of Science and Technology, A. Mickiewicza 30, 30-059 Krakow, Poland; zywczak@agh.edu.pl

[5]   Institute of Metallurgy and Materials Science, Polish Academy of Sciences, Reymonta 25, 30-059 Krakow, Poland; w.maziarz@imim.pl

*   Correspondence: l.dubiel@prz.edu.pl

Abstract: We present a study of the annealing effect and its influence on magnetic and structural properties for a series of Heusler alloys $Ni_{50-x}Co_xMn_{35.5}In_{14.5}$ ($x = 0, 3, 5$) prepared in ribbon form. We studied the morphology and composition using scanning electron microscopy (SEM) equipped with an X-ray microanalyzer (EDX). The magnetic properties were determined by two methods: electron magnetic resonance (EMR) and vibrating sample magetometer (VSM). We found that cobalt content in the annealed samples reveals an additional magnetic phase transition at lower temperatures.

Keywords: Ni-Co-Mn-In ribbons; Heusler alloys; EMR; transformation; annealing

Citation: Dubiel, Ł.; Wal, A.; Stefaniuk, I.; Żywczak, A.; Potera, P.; Maziarz, W. Cobalt Content Effect on the Magnetic Properties of $Ni_{50-x}Co_xMn_{35.5}In_{14.5}$ Annealed Ribbons. *Materials* **2021**, *14*, 5497. https://doi.org/10.3390/ma14195497

Academic Editor: Hailiang Yu

Received: 26 August 2021
Accepted: 17 September 2021
Published: 23 September 2021

1. Introduction

The Ni-Mn-based alloys are a very attractive group of multifunctional materials due to their unusual properties and potential application in different areas of life, e.g., in medicine, spintronics, or environmentally friendly refrigeration. Moreover, the physical properties of these materials, such as their magnetocaloric properties or magnetic-shape memory, strongly depend on the atomic ordering and the location of atoms in the crystal structure. Even slight changes in lattice site occupation result in noticeable changes in both magnetic and structural properties of these materials. Such modification can be done in two ways, firstly by substitution of appropriate atoms, as in the case of Ni-Co-Mn-In, which was obtained by substitution of Ni atoms in the ternary alloy with Co [1–4]. The second, simpler method is by applying well-defined heat treatment to the base alloy [5,6]. In the literature, one can find various examples of heat-treating, using Ni-Mn-based alloys, such as annealing with slow cooling or quenching [1,5,7,8]. The heat-treating duration and temperature also vary, but generally, in the ribbons, this time is relatively short compared to bulk samples [9].

The fabrication of ribbons by the melt-spinning method leads to stresses as a result of the rapid solidification process. The annealing process leads to relaxation [10,11], i.e., the removal of such stresses, which are associated with a change in the distance between manganese atoms. This, in turn, may cause changes in the magnetic interactions between these atoms [12], and thus changes in their magnetic properties.

The annealing parameters were chosen on the basis of experiments described in literature [1,5,7,11,13]. The order–disorder transition (ODT), which for the Ni-Mn-In alloy type takes place at 950 K [13], plays a very important role in the selection of the annealing temperature. Typically, annealing is carried out at a temperature greater than 950 K. For example, Recarte et al. [7] report that annealing at such temperatures followed by a slow cooling favors the production of a larger fraction of the $L2_1$ structure than rapid cooling. Although, there are also reports [1] that long-term annealing below the ODT temperature also promotes the production of the $L2_1$ structure.

Furthermore, there are other annealing methods used in off-stoichiometric Ni-Mn-In and Ni-Mn-Sn alloys [14,15], including annealing in a magnetic field. This procedure leads to the decomposition of the off-stoichiometric alloy into the cubic $L2_1$ phase and a second phase with lower symmetry [14–17], i.e., ferromagnetic $Ni_{50}Mn_{25}X_{25}$, $(X = In, Sn)$, precipitates are embedded in a $Ni_{50}Mn_{50}$ non-magnetic matrix.

In this paper, we investigated annealed off-stoichiometric Ni-Co-Mn-In ribbons with three different compositions. Based on electron magnetic resonance (EMR) spectra and magnetization data, we discussed in detail the magnetic properties of these alloys and their dependence on the cobalt content in the alloy.

2. Materials and Methods

Three alloys $Ni_{50-x}Co_xMn_{35.5}In_{14.5}$ $(x = 0, 3, 5)$ were prepared by the melt-spinning method according to the procedure, which was described in references [2]. Before measurements the samples were annealed at 1173 K for 30 min and then slowly cooled in a furnace for 12 h. Samples with different cobalt content were labeled NC0MI, NC3MI, and NC5MI for 0, 3, and 5 at.% of cobalt, respectively.

Morphology and composition of samples were inexamined using Tescan Vega 3 scanning electron microscope (TESCAN ORSAY HOLDING, Brno, Czech Republic) equipped with the EDX Bruker Quantax microanalysis system.

The electron magnetic resonance measurements were performed using a Bruker ELEXYS E580 spectrometer (Karlsruhe, Germany) equipped with the Bruker liquid N gas flow cryostat with the 41131 VT digital controller (Bruker Analytische Messtechnik, Rheinstetten, Germany), within the temperature range 110 K $\leq T \leq$ 450 K. The EMR signal was registered with the following parameters, a 100 kHz modulation and amplitude of 0.1 mT. The microwave power was 23.77 mW for NC0MI, NC5MI and 15.00 mW for NC3MI, respectively. All EMR measurements were performed in the X-band (9.44 GHz). The resonance position was calibrated by DPPH.

The magnetization results were obtained by the use of LakeShore 7407 vibrating sample magnetometer (VSM, Westerville, OH, USA). The field cooled (FC) and the field heated (FH) curves were registered at $\mu_0 H = 0.01$ T and $\mu_0 H = 0.1$ T in the temperature range of 85 K $\leq T \leq$ 450 K.

3. Results and Discussion

3.1. Morphology and Composition

Figure 1 shows the SEM images of surfaces for melt-spun and annealed (1173 K) NC3MI and NC5MI at room temperature (RT). The all melt-spun ribbons (Figure 1a,c) exhibit a granular surface topology. The grain size is not uniform and grains are randomly distributed throughout the area of ribbons. Although the ribbon is obtained from polycrystalline ingots, the columnar crystal directions in the ribbon are preferentially oriented perpendicular to the ribbon plane [18,19].

The surface morphology of the ribbons after annealing has changed significantly. Additionally there is also a difference in the morphology of the annealed ribbons with different cobalt content.

As shown in Figure 1b, the grains in NC3MI become more uniform after annealing and the largest grains have disappeared. In addition to grains, on the surface of the ribbons one can see the second phase in stripes form. EDS analysis showed that the second phase composition differs from the matrix composition by a significant increase in cobalt concentration (11.35 at.%), at the expense of manganese (29.65 at.%) and indium (11.76 at.%). Figure 1d presents the SEM image of the surface of NC5MI. The morphology of this sample is very rich and significantly different from the two previous alloys. First of all, the increase in grain size after annealing is clearly visible. In addition to the second phase in the striped form, one can see the eutectoid form on the surface of this ribbon. Similarly to the NC3MI ribbon, the second phase composition is different from the matrix composition.

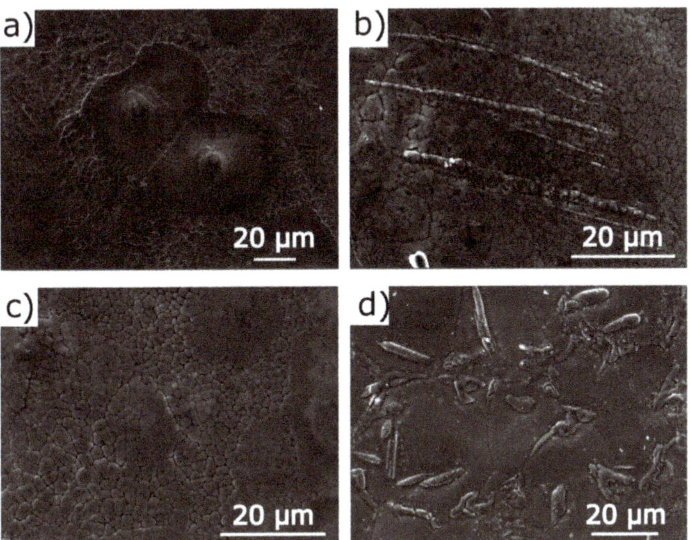

Figure 1. Microstructure of $Ni_{50-x}Co_xMn_{35.5}In_{14.5}$ ribbons at RT: $Ni_{47}Co_3Mn_{35.5}In_{14.5}$ as-spun (**a**) and annealed (**b**) ribbons; $Ni_{45}Co_5Mn_{35.5}In_{14.5}$ as-spun (**c**) and annealed (**d**) ribbons.

3.2. Magnetic Properties

The magnetic properties of the $Ni_{50-x}Co_xMn_{35.5}In_{14.5}$ ribbons were studied as a function of temperature. The magnetization curves $M(T)$ obtained for the samples with different content of Co are presented in Figure 2. The FH and FC magnetization curves for NC3MI and NC5MI were obtained in a field 0.01 T. For NC0MI, magnetization cures were registered at 0.1 T due to a very weak signal.

The nature of the $M(T)$ curves for annealed ribbons differs significantly from non-annealed ribbons [20–22]. The $M(T)$ of the NC5MI sample displayed in Figure 2f is similar to magnetization of the ribbon before annealing [21]. This magnetization curve does not exhibit any thermal hysteresis and has a clear shift toward higher temperatures relative to non-annealed alloy [20,21].

Also, for the two remaining annealed samples, NC0MI and NC3MI, no clear temperature hysteresis is visible, and the behavior of $M(T)$, especially for NC0MI, is unusual. In order to better present phase transitions and to determine precisely the temperatures of these transitions, one should consider the temperature dependence of the derivative of magnetization dM/dT (Figure 2a–c). At high temperature ranges, each sample exhibits a magnetic phase transition from paramagnetism to ferromagnetism with Curie temperatures equal to 300 K, 363 K, and 392 K for NC0MI, NC3MI, and NC5MI, respectively. However, the magnetization below T_C suggests that these materials are multi-phased. Especially for NC0MI, there is a visible narrow region when magnetization is saturated, but with the further reduction of temperature, the magnetization starts to increase again around 225 K. Similar magnetization curves were observed for $Ni_{50.3}Mn_{35.3}In_{14.4}$ alloy produced by annealing under a magnetic field [15] and for $Ni_{2.14}Mn_{0.55}Sb_{1.3}$ after homogenization [23]. Considering the derivative of M versus T (Figure 2a), one can see that at around 150 K the next magnetic ordering temperature occurs. The character of this magnetic phase transition strongly depends on Co content. With increasing of cobalt concentration, the intensity of the low temperature peak in the dM/dT value gradually decreases. For a sample containing 3 at.% of cobalt this peak still exits, but its minimum is much less visible than for NC0MI. In the Figure 2c corresponding to the NC5MI sample the minimum in the low temperature region almost disappears.

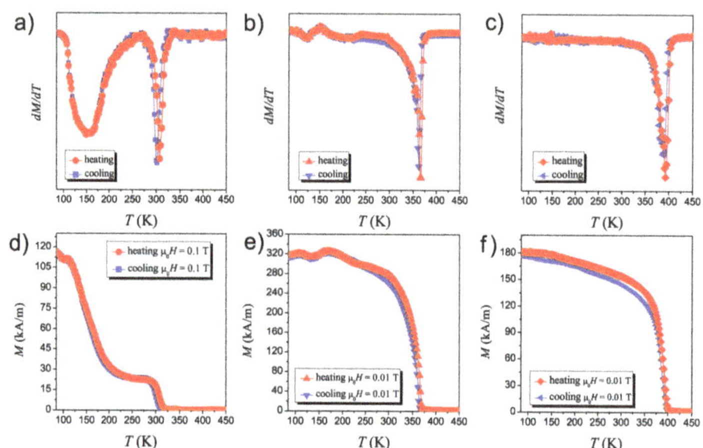

Figure 2. FH and FC magnetization curves as a function of temperature of $Ni_{50-x}Co_xMn_{35.5}In_{14.5}$ annealed ribbons: $Ni_{50}Mn_{35.5}In_{14.5}$ (**d**), $Ni_{47}Co_3Mn_{35.5}In_{14.5}$ (**e**) and $Ni_{45}Co_5Mn_{35.5}In_{14.5}$ (**f**). The upper panel contains the corresponding dM/dT curves (**a–c**).

Such behavior of the $M(T)$ curve may suggest that there are regions of different magnetic properties in the sample, which are revealed during the temperature change. Most likely, these are antiferromagnetic and ferromagnetic regions [14], the mutual volume ratio of which changes during temperature changes. Figure 2a shows that around 150 K, the material undergoes a phase change followed by a further increase in magnetization, meaning that the ferromagnetic region grows at the expense of another one.

Comparing this nature of the change in magnetization as a function of temperature with the results of measurement of the magnetization of a sample with the same composition but not subjected to annealing, it can be concluded that the annealing causes a change in the structure of the Heusler alloy. The authors of [14] report on the phase separation phenomenon in the $Ni_{50}Mn_{45}In_5$ alloy during annealing in the presence of a magnetic field. In our case, annealing took place without an external field and the composition of the sample was different, containing more indium and less manganese. It seems, however, that the annealing process could also in our case cause a change in the structure, i.e., precipitation of various phases of the alloy. Indirect evidence for this may be the composition measurements carried out for the surfaces of the annealed samples, which reveal a different percentage of elements in relation to the unheated samples [20]. The comparison of Figure 2a–c shows that increasing the cobalt concentration causes homogeneity of these areas, i.e., the structural heterogeneity in the volume of the alloy decreases.

Another explanation of the observed properties can be based on [23], the authors of which report a similar relationship $M(T)$ of $Ni_{2.14}Mn_{0.55}Sb_{1.31}$ samples subjected to annealing at the temperature of 850 °C. They assume that the crystal structure of the samples is the same in the whole volume, but the elements that compose it may be in different places in different areas. This can lead to magnetic interactions of the FM and AFM type, which compete with each other at different temperatures, resulting in the observed magnetization.

Figure 3 displays some selected EMR spectra for $Ni_{50-x}Co_xMn_{35.5}In_{14.5}$ ribbons with different concentrations of Co atoms. The spectra were registered within the temperature range of $110\,K \leq T \leq 450\,K$. In general, the annealing process does not drastically change the character of the EMR signal registered for $Ni_{50-x}Co_xMn_{35.5}In_{14.5}$ [20,21]. At the highest temperatures, the main line has an unchanged asymmetric character. Such an observation confirms that the ribbons before annealing were already partially ordered materials.

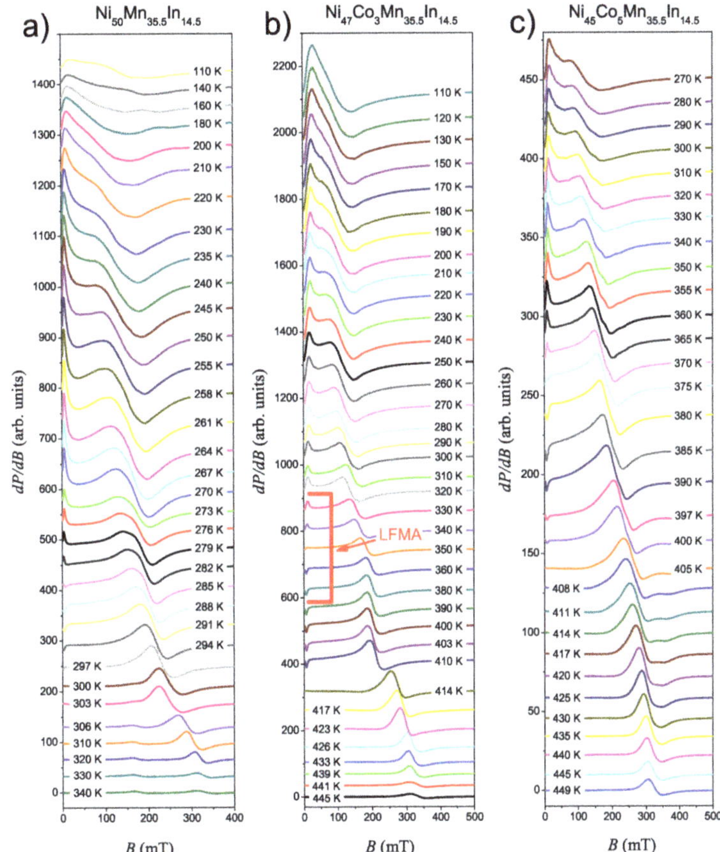

Figure 3. Selected EMR spectra of derivative of the resonance absorption (dP/dB) for $Ni_{50-x}Co_x$ $Mn_{35.5}In_{14.5}$ ribbons: $Ni_{50}Mn_{35.5}In_{14.5}$r (**a**), $Ni_{47}Co_3Mn_{35.5}In_{14.5}$ (**b**) and $Ni_{45}Co_5Mn_{35.5}In_{14.5}$ (**c**) registered during a cooling process. Spectra are vertically shifted for clarity. The example of the LFMA signal is marked in red.

At the highest temperatures the EMR spectra of NC0MI ribbon include two separated Dyson lines: very weak line 1 ($B_R \sim 330$ mT) and line 2 ($B_R \sim 200$ mT). Within the temperature range $400\,K \leq T \leq 450\,K$, the amplitude of the line 1 is so weak that its contribution to the EMR signal can be neglected. Below 400 K, the amplitude of line 1 gradually increases, and at a temperature of about 345 K, line 1 reaches a height equal line 2. With the further reduction in temperature, the intensity of line 1 increases and this line shifts towards weaker fields, and in 300 K it overlaps with line 2. For this sample, the mainline (line 1) is similar in nature to the paramagnetic lines observed for the Ni-Mn-In ribbons in previous work [20,24].

The spectra of the sample NC3MI are presented in Figure 3b. In the temperature range $426\,K \leq T \leq 445\,K$, a single strong, asymmetric line is observed. This is a typical paramagnetic line. Below 426 K EMR line starts broadening and shifts toward a low-field region. Figure 3c shows the EMR spectra for NC5MI registered within the temperature range of

270 K $\leq T \leq$ 450 K. At the highest temperatures in the EMR spectrum, a single asymmetrical Dysonian line (line 1) is visible. In the range of temperatures 430 K $\leq T \leq$ 450 K, the position of the line slightly changes. During cooling of the sample, below 425 K the main line shifts towards the low field region, and, additionally, in the EMR spectrum around 340 mT a much weaker symmetrical line appears (line 2). As the temperature decreases these two lines become better separated.

The resonance conditions below the magnetic phase transition temperature in ferromagnetic materials can be considered within the classical theory of ferromagnetic resonance proposed by Kittel [25]. The thickness of the ribbon (parallel to x-axis) is far smaller than other sample dimensions and we can treat it, in the first approximation, as a plane surface, where the magnetic field is parallel to it. This model corresponds to a special case of the Kittel equation, where the demagnetization factors are equal to $N_x = 1$ and $N_y = N_z = 0$, and the Kittel equation will be simplified considerably (SI units):

$$\omega_0 = \gamma[(\mu_0 M + B)B]^{1/2}, \tag{1}$$

where ω_0 is the microwave angular frequency ($\omega_0 = 2\pi f_0$), γ is the gyromagnetic ratio, B means the static magnetic field, μ_0 stands for the permeability, and M denotes the saturation magnetization. A Typical continuous wave EMR experiment has a fixed microwave frequency, solving the Equation (1) for M, assuming B as the value of the magnetic field corresponding to the resonance condition.

Table 1 lists the M values calculated based on Equation (1) and determined from the hysteresis loops for particular temperatures. The latter value of magnetization M for each temperature was obtained from $M(H)$ dependence presented in Figure 4. As can be seen for the NC3MI ribbon, the magnetization value M determined from Equation (1) is slightly higher than that from the figure. This difference can be related to the fact that EMR probes only the surface of a sample ($\sim$1 μm), while the VSM measurements provide information about bulk magnetization. Thus, in the EMR results, the outer material layer is taken into account, where the interaction of a ferromagnetic nature is dominant, while VSM measures the magnetization of the entire sample volume, thus the total magnetization may be reduced as a result of the presence of a different magnetic ordering.

Table 1. The magnetization value determined from the solution of the Kittel equation (M) and from hysteresis loops (M^*) for NC0MI and NC3MI samples.

Label	T (K)	M (kA/m)	M^* (kA/m)
NC0MI	300	147	26
	100	608	157
NC3MI	350	330	242
	300	553	416
	100	879	694

At temperatures below T_C, in the EMR spectra registered for each sample, an additional signal appears, known in the literature as low-field microwave absorption (LFMA). Generally, the LFMA is observed for soft ferromagnetic materials, such as manganites, amorphous ribbons, magnetic thin films, magnetic nanoparticles [26–30]. In addition, LFMA is also used to detect ferromagnetism and to determine T_C. Furthermore, the absence of LFMA in the soft ferromagnetic sample is a good indicaton of existence of superparamagnetic state [29,30].

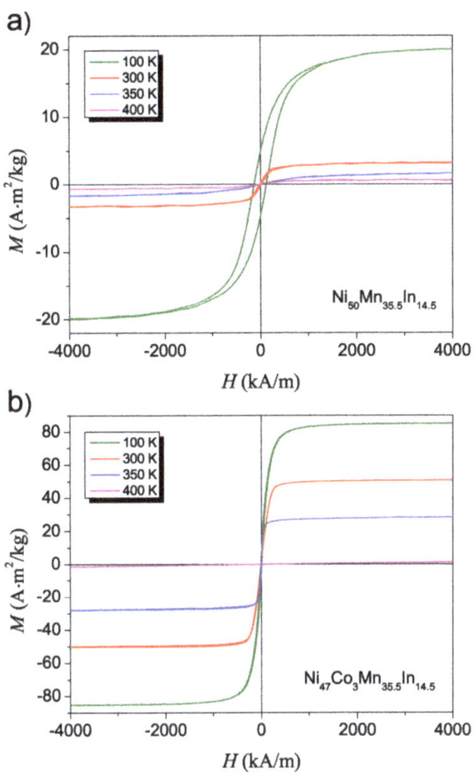

Figure 4. The field dependence of magnetization for $Ni_{50}Mn_{35.5}In_{14.5}$ (**a**) and $Ni_{47}Co_3Mn_{35.5}In_{14.5}$ (**b**) ribbons at the selected temperatures.

In our prior works, we observed LFMA in the Ni-Mn-In and Ni-Co-Mn-In ribbons [22,24] but the behaviour of LFMA in annealed ribbons is more complex. Below T_C we observe only the fragment of the LFMA signal, but this signal has the same phase as the main line. With the further reduction in temperature, the LFMA signal changes their phase. The change in the signal phase for the sample without cobalt occurs at a temperature of approximately 291 K, for the sample with the cobalt content of 3 at.%, this temperature is higher, approx. 351 K. For the sample with the highest cobalt content (5 at.%), the phase change temperature is approximately 370 K. An especially interesting behavior of the LFMA can be observed for the NC3MI ribbon. The Curie temperature determined from VSM data for sample NC3MI is equal to 364 K (see Figure 2b), while the value of T_C determined from the temperature dependence of the integrated intensity of EMR line is equal 421 K (see Table 2). The difference in the value of Curie temperature obtained from VSM and EMR is a consequence of the different natures of both techniques. The EMR is sensitive to changes in the environment of the magnetic ions, so it is a good method for detecting local changes in properties of materials [21]. For conducting samples, due to an existing skin depth phenomenon, and the local character of the EMR method, one can notice a discrepancy between results determined from EMR data and VSM technique. In our prior measurements, however, for as-spun $Ni_{45}Co_5Mn_{35.5}In_{14.5}$ [21] and annealed $Ni_{50}Mn_{35.5}In_{14.5}$ [24], only a slight difference in T_C values determined by using the EMR and VSM technique was observed, although for those samples the skin depth effect also exists. The difference in the Curie temperature values determined in both methods for samples described in this report may be explained by the annealing process, which caused the inhomogeneity of the samples. Ferromagnetic structures may have formed on the sample surface, which is manifested by the Curie 421 K temperature determined by the EMR method and the differences in the magnetization values calculated using the EMR and VSM techniques and those listed in Table 1. The sample taken as a whole, neglecting the thin layer near the surface, remains paramagnetic at this temperature, as indicated by $M(T)$ and $M(H)$ dependences. Ferromagnetism in the macro scale is revealed at a lower temperature, about 360 K, which is visible both in the magnetization data and in the anomalous behaviour of the EMR line, suggesting the occurrence of a phase transition.

Table 2. The corresponding EMR results of NC0MI, NC3MI, and NC5MI samples and comparison with literature data. T_C are determined from the temperature dependence of integral intensity, a and b parameters are determined from temperature dependence of the EMR linewidth.

Label	Composition	T_C (K)	a (mT)	b (mT/K)
NC0MI	$Ni_{50}Mn_{35.5}In_{14.5}$	300	−260	0.87
[24] *	$Ni_{50}Mn_{35.5}In_{14.5}$	298	−220	0.8
NC3MI	$Ni_{47}Co_3Mn_{35.5}In_{14.5}$	421	-	-
[22] **	$Ni_{47}Co_3Mn_{35.5}In_{14.5}$	340	−244	0.77
NC5MI	$Ni_{45}Co_5Mn_{35.5}In_{14.5}$	414	−9	0.1

* annealed at 600 K; ** non-annealed.

Figure 5 exhibits the temperature dependence of linewidth $\Delta B(T)$ for NC0MI and NC5MI. In the temperature range of 280 K $\leq T \leq$ 450 K the EMR spectra for NC0MI (Figure 5a) consists of two lines. In the whole temperature range, the ΔB of the line 2 varies slightly between 30–40 mT. In contrast, ΔB of line 1 in the temperature range of 330 K $\leq T \leq$ 400 K decreases linearly, and then, after approaching Curie temperature, the linewidth starts to increase. This behaviour of $\Delta B(T)$ around T_C is typical for ferromagnetic metals [31]. The high temperature EMR signal for NC5MI within temperature range 395 K $\leq T \leq$ 450 K can be also divided into two lines, but the linewidth for NC5MI (Figure 5b) has a different character than the temperature dependence for NC0MI. The main line, which is labelled 1, shows linear behaviour of $\Delta B(T)$ only in a narrow range of temperature (430 K $\leq T \leq$ 450 K) and its value is about 30 mT. Below these temperatures, the ΔB of both line increases. For NC3MI, no such relationship was determined due to the unregularly nature of changes in linewidth as a function of temperature.

The linear region of $\Delta B(T)$ observed in the temperature dependence for both samples, can be expressed by the Korringa relaxation model [32]:

$$\Delta B(T) = a + b \cdot T, \tag{2}$$

where $a = \Delta B_0$ is the residual width and b is the Korringa rate.

The value of the last parameter b, describing the slope of $\Delta B(T)$, depends mainly on concentration of $3d$ elements, i.e., the b value increases when the number of $3d$ electrons decreases [33]. For NC0MI, the value of b is about 10 G/K (see Tabel 2) and this value is typically observed for transition metal EMR [34]. A similar value of this parameter was determined for the in-situ ribbon [20]. For the linear part of line 1 of NC5MI, the value of b is smaller and equal to 1 G/K. A similar value was observed in $YBaMn_2O_6$ [35], where the bottleneck scenario for the spin relaxation of the manganese ion via mobile e_g electron was noticed.

The main values determined from the EMR results, such as T_C, a, and b are collected in Table 2. Moreover, the table contains the EMR parameter values for in-situ [21,22] and another annealed ribbon [24] presented in our earlier works. The smallest changes in the EMR parameter values were observed for the sample without cobalt. However, in the paramagnetic regime, a weak additional line can be observed, which can be considered a singularity for Ni-Mn-In ribbons. Other parameters, including the Curie temperature, differ only by 2 K, which shows that the annealing parameters are not native, and the annealing itself significantly shifts the Curie temperature, i.e., for the in-situ ribbon, this temperature is 267 K [20] and it overlaps with the martensitic transformation temperature. It looks different in the case of NC3MI and NC5MI ribbons. For the sample with 3 at.% of cobalt the temperature dependence of $\Delta(B)$ has an irregular character and cannot be described using the Korringa relaxation model, while for as-cast ribbon with the same chemical composition, one can observe a typical Korringa relaxation process [22]. Furthermore, the Curie temperature, determined from the temperature dependence of the integral intensity, increases significantly relative to the non-annealed sample ($T_C = 340$ K) [24] and differs from the Curie temperature for the NC3MI sample, determined from the temperature dependence of magnetization. In the case of the NC5MI ribbon, unlike the NC0MI ribbon and the non-annealed NC3MI ribbon, the width of the paramagnetic line (line 1) does not change significantly with temperature changes (above 430 K) and oscillates around the value $\sim$30 mT, and the parameters of the linear fit of the fragment $\Delta B(T)$ above T_C, equals $a \approx 0$ mT, and $b = 1$ G/K. Such parameter values and the shape of $\Delta B(T)$ indicate that the temperature dependence of the line width for the annealed NC5MI ribbon cannot be described by the Korring relaxation model.

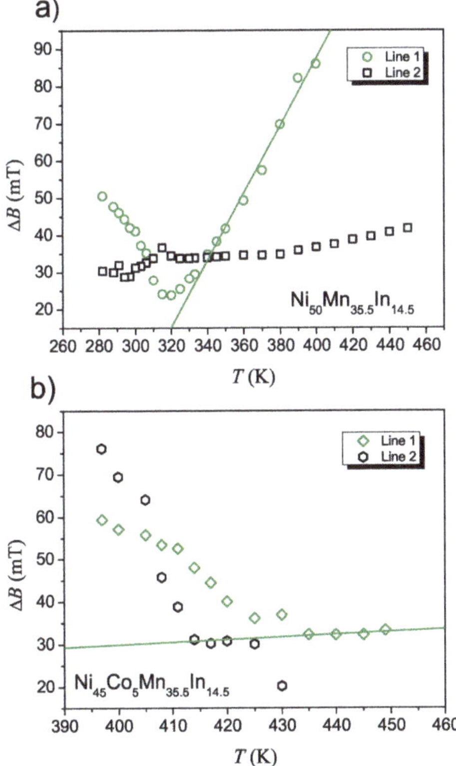

Figure 5. Temperature dependence of the EMR linewidth ΔB for Ni$_{50}$Mn$_{35.5}$In$_{14.5}$r (**a**) and Ni$_{45}$Co$_5$Mn$_{35.5}$In$_{14.5}$ (**b**) ribbons. The solid lines represent the best fit of the linewidth above T_C, using the expression (1).

4. Conclusions

The effect of cobalt content of annealed Ni$_{50-x}$Co$_x$Mn$_{35.5}$In$_{14.5}$ for ($x = 0, 3, 5$) is investigated through scanning electron microscopy, electron magnetic resonance, and vibrating sample magnetometer. The main conclusions can be listed as follows:

- The EMR and VSM results confirm that the annealing process shifts the magnetic phase transition characteristic temperatures. Furthermore, VSM measurement shows the existence of an additional magnetic phase transition below T_C.
- Comparing the EMR spectra for as-cast ribbons, published in our prior works, with results for annealed ribbons presented in this work we observe changes. The annealing caused the existence of an addition line in the paramagnetic regime for NC0MI and NC5MI samples.
- The differences in the Curie temperature determined by two methods (VSM and EMR) for the annealed NC3MI sample suggest that it is magnetically inhomogeneous, i.e., the subsurface layers have different magnetic properties than the interior of the sample.
- The scanning electron microscopy results show that the annealing process modifies the morphology of the samples with cobalt content. For ribbon without cobalt, the changes are not observed.
- The EMR spectra for all ribbons, registered below Curie temperature, contain the LFMA signal. During the cooling process, the LFMA signal changes its phase to the opposite of the main EMR signal.

Author Contributions: Conceptualization, Ł.D.; methodology, Ł.D. and A.W.; validation, I.S. and A.Ż.; formal analysis, Ł.D. and A.W.; investigation, Ł.D., I.S., A.Ż., P.P. and W.M.; writing—original draft preparation, Ł.D.; writing—review and editing, Ł.D. and A.W. All authors have read and agreed to the published version of the manuscript.

Funding: This research received no external funding.

Institutional Review Board Statement: Not applicable.

Informed Consent Statement: Not applicable.

Data Availability Statement: Not applicable.

Acknowledgments: We would like to thank Wojciech Bochnowski from the College of Natural Sciences for a fruitful discussion. This research was undertaken in the Center for Innovation and Transfer of Natural Sciences and Engineering Knowledge at the University of Rzeszow and in the Academic Centre for Materials and Nanotechnology at the AGH University of Science and Technology.

Conflicts of Interest: The authors declare no conflict of interest.

References

1. Ito, W.; Nagasako, M.; Umetsu, R.Y.; Kainuma, R.; Kanomata, T.; Ishida, K. Atomic ordering and magnetic properties in the $Ni_{45}Co_5Mn_{36.7}In_{13.3}$ metamagnetic shape memory alloy. *Appl. Phys. Lett.* **2008**, *93*, 232503. [CrossRef]
2. Maziarz, W. SEM and TEM studies of magnetic shape memory NiCoMnIn melt spun ribbons. *Solid State Phenom.* **2012**, *186*, 251–254. [CrossRef]
3. Ju, J.; Hu, L.; Bao, C.; Shuai, L.; Yan, C.; Wang, Z. Microstructure and Magnetic Field-Induced Strain of a Ni-Mn-Ga-Co-Gd High-Entropy Alloy. *Materials* **2021**, *14*, 2514. [CrossRef]
4. Umetsu, R.; Yasumura, H.; Narumi, Y.; Kotani, Y.; Nakamura, T.; Nojiri, H.; Kainuma, R. Soft X-ray absorption spectroscopy and magnetic circular dichroism under pulsed high magnetic field of Ni-Co-Mn-In metamagnetic shape memory alloy. *J. Alloys Compd.* **2022**, *890*, 161590. [CrossRef]
5. González-Legarreta, L.; Sánchez, T.; Rosa, W.O.; García, J.; Serantes, D.; Caballero-Flores, R.; Prida, V.M.; Escoda, L.; Suñol, J.J.; Koledov, V.; et al. Annealing Influence on the Microstructure and Magnetic Properties of Ni–Mn–In Alloys Ribbons. *J. Supercond. Nov. Magn.* **2012**, *25*, 2431–2436. [CrossRef]
6. Guo, J.; Zhong, M.; Zhou, W.; Zhang, Y.; Wu, Z.; Li, Y.; Zhang, J.; Liu, Y.; Yang, H. Grain Size Effect of the γ Phase Precipitation on Martensitic Transformation and Mechanical Properties of Ni–Mn–Sn–Fe Heusler Alloys. *Materials* **2021**, *14*, 2339. [CrossRef] [PubMed]
7. Recarte, V.; Pérez-Landazábal, J.; Sánchez-Alarcos, V.; Rodríguez-Velamazán, J. Dependence of the martensitic transformation and magnetic transition on the atomic order in Ni–Mn–In metamagnetic shape memory alloys. *Acta Mater.* **2012**, *60*, 1937–1945. [CrossRef]
8. Sánchez-Alarcos, V.; Recarte, V.; Pérez-Landazábal, J.; Gómez-Polo, C.; Rodríguez-Velamazán, J. Role of magnetism on the martensitic transformation in Ni–Mn-based magnetic shape memory alloys. *Acta Mater.* **2012**, *60*, 459–468. [CrossRef]

9. Sánchez, T.; Turtelli, R.S.; Grössinger, R.; Sánchez, M.; Santos, J.; Rosa, W.D.O.D.; Prida, V.; Escoda, L.; Suñol, J.; Koledov, V.; et al. Exchange bias behavior in $Ni_{50.0}Mn_{35.5}In_{14.5}$ ribbons annealed at different temperatures. *J. Magn. Magn. Mater.* **2012**, *324*, 3535–3537. [CrossRef]

10. Wang, D.; Peng, K.; Gu, B.; Han, Z.; Tang, S.; Qin, W.; Du, Y. Influence of annealing on the magnetic entropy changes in $Fe_{81.6}Mo_4Zr_{3.3}Nb_{3.3}B_{6.8}Cu_1$ amorphous ribbons. *J. Alloys Compd.* **2003**, *358*, 312–315. [CrossRef]

11. Xuan, H.C.; Xie, K.X.; Wang, D.H.; Han, Z.D.; Zhang, C.L.; Gu, B.X.; Du, Y.W. Effect of annealing on the martensitic transformation and magnetocaloric effect in $Ni_{44.1}Mn_{44.2}Sn_{11.7}$ ribbons. *Appl. Phys. Lett.* **2008**, *92*, 242506. [CrossRef]

12. Şaşıoğlu, E.; Sandratskii, L.M.; Bruno, P. First-principles calculation of the intersublattice exchange interactions and Curie temperatures of the full Heusler alloys Ni_2MnX (X = Ga, In, Sn, Sb). *Phys. Rev. B* **2004**, *70*, 024427. [CrossRef]

13. Miyamoto, T.; Ito, W.; Umetsu, R.Y.; Kainuma, R.; Kanomata, T.; Ishida, K. Phase stability and magnetic properties of $Ni_{50}Mn_{50-x}In_x$ Heusler-type alloys. *Scr. Mater.* **2010**, *62*, 151–154. [CrossRef]

14. Çakır, A.; Acet, M.; Farle, M. Shell-ferromagnetism of nano-Heuslers generated by segregation under magnetic field. *Sci. Rep.* **2016**, *6*, 28931. [CrossRef] [PubMed]

15. Çakır, A.; Acet, M.; Wiedwald, U.; Krenke, T.; Farle, M. Shell-ferromagnetic precipitation in martensitic off-stoichiometric Ni-Mn-In Heusler alloys produced by temper-annealing under magnetic field. *Acta Mater.* **2017**, *127*, 117–123. [CrossRef]

16. Schlagel, D.; McCallum, R.; Lograsso, T. Influence of solidification microstructure on the magnetic properties of Ni–Mn–Sn Heusler alloys. *J. Alloys Compd.* **2008**, *463*, 38–46. [CrossRef]

17. Yuhasz, W.; Schlagel, D.; Xing, Q.; McCallum, R.; Lograsso, T. Metastability of ferromagnetic Ni–Mn–Sn Heusler alloys. *J. Alloys Compd.* **2010**, *492*, 681–684. [CrossRef]

18. Ma, S.C.; Wang, D.H.; Zhong, Z.C.; Luo, J.M.; Xu, J.L.; Du, Y.W. Peculiarity of magnetoresistance in high pressure annealed $Ni_{43}Mn_{41}Co_5Sn_{11}$ alloy. *Appl. Phys. Lett.* **2013**, *102*, 032407. [CrossRef]

19. Chen, F.; Huang, Q.; Jiang, Z.; Xuan, H.; Zhang, M.; Xu, X.; Zhao, J. Large magnetoresistance in highly textured $Mn_{44.7}Ni_{43.5}Sn_{11.8}$ melt spun ribbons. *Smart Mater. Struct.* **2016**, *25*, 055031. [CrossRef]

20. Dubiel, Ł.; Stefaniuk, I.; Wal, A.; Żywczak, A.; Dziedzic, A.; Maziarz, W. Magnetic and structural phase transition in $Ni_{50}Mn_{35.5}In_{14.5}$ ribbon. *J. Magn. Magn. Mater.* **2019**, *485*, 21–26. [CrossRef]

21. Dubiel, Ł.; Żywczak, A.; Maziarz, W.; Stefaniuk, I.; Wal, A. Magnetic phase transition and exchange bias in $Ni_{45}Co_5Mn_{35.5}In_{14.5}$ Heusler alloy. *Appl. Magn. Reson.* **2019**, *50*, 809–818. [CrossRef]

22. Łukasz, D.; Wal, A.; Stefaniuk, I.; Żywczak, A.; Maziarz, W. Electron magnetic resonance study of the $Ni_{47}Co_3Mn_{35.5}In_{14.5}$ ribbons. *J. Magn. Magn. Mater.* **2021**, *530*, 167930. [CrossRef]

23. Guha, S.; Datta, S.; Panda, S.K.; Kar, M. Room temperature magneto-caloric effect and electron transport properties study on $Ni_{2.14}Mn_{0.55}Sb_{1.31}$ alloy. *J. Alloys Compd.* **2020**, *843*, 156033. [CrossRef]

24. Łukasz, D.; Stefaniuk, I.; Wal, A.; Kuźma, M. Effect of annealing on the magnetic ordering and electron magnetic resonance of melt-spun Ni-Mn-In ribbons. *J. Magn. Magn. Mater.* **2020**, *504*, 166638. [CrossRef]

25. Kittel, C. On the Theory of Ferromagnetic Resonance Absorption. *Phys. Rev.* **1948**, *73*, 155–161. [CrossRef]

26. Valenzuela, R.; Alvarez, G.; Montiel, H.; Gutiérrez, M.P.; Mata-Zamora, M.E.; Barrón, F.; Sánchez, A.Y.; Betancourt, I.; Zamorano, R. Characterization of magnetic materials by low-field microwave absorption techniques. *J. Magn. Magn. Mater.* **2008**, *320*, 1961–1965. [CrossRef]

27. Alvarez, G.; Montiel, H.; Barron, J.F.; Gutierrez, M.P.; Zamorano, R. Yafet-Kittel-type magnetic ordering in $Ni_{0.35}Zn_{0.65}Fe_2O_4$ ferrite detected by magnetosensitive microwave absorption measurements. *J. Magn. Magn. Mater.* **2010**, *322*, 348–352. [CrossRef]

28. Lee, J.; Kim, J.; Kim, K.H. Effect of the magnetization process on low-field microwave absorption by FeBN magnetic thin film. *Phys. Status Solidi A Appl. Mater. Sci.* **2014**, *211*, 1900–1902. [CrossRef]

29. Montiel, H.; Alvarez, G.; Conde-Gallardo, A.; Zamorano, R. Microwave absorption behavior in Cr_2O_3 nanopowders. *J. Alloys Compd.* **2015**, *628*, 272–276. [CrossRef]

30. Yahya, M.; Hosni, F.; M'Nif, A.; Hamzaoui, A.H. ESR studies of transition from ferromagnetism to superparamagnetism in nano-ferromagnet $La_{0.8}Sr_{0.2}MnO_3$. *J. Magn. Magn. Mater.* **2018**, *466*, 341–350. [CrossRef]

31. Demishev, S. Electron Spin Resonance in Strongly Correlated Metals. *Appl. Magn. Reson.* **2020**, *51*, 473–522. [CrossRef]

32. Barnes, S. Theory of electron spin resonance of magnetic ions in metals. *Adv. Phys.* **1981**, *30*, 801–938. [CrossRef]

33. Kaczmarska, K.; Pierre, J. Influence of transition metals on magnetic properties of GdT_2X_2. *J. Alloys Compd.* **1997**, *262–263*, 248–252. [CrossRef]

34. Deisenhofer, J.; von Nidda, H.A.K.; Loidl, A.; Sampathkumaran, E. ESR investigation of the spin dynamics in $(Gd_{1-x}Y_x)_2PdSi_3$. *Solid State Commun.* **2003**, *125*, 327–331. [CrossRef]

35. Schaile, S.; von Nidda, H.A.K.; Deisenhofer, J.; Loidl, A.; Nakajima, T.; Ueda, Y. Korringa-like relaxation in the high-temperature phase of *A*-site ordered $YBaMn_2O_6$. *Phys. Rev. B* **2012**, *85*, 205121. [CrossRef]

 *materials*

Article

Deformation Behavior and Properties of 7075 Aluminum Alloy under Electromagnetic Hot Forming

Zhihao Du [1], **Zanshi Deng** [1], **Xiaohui Cui** [1,2,3,*] **and Ang Xiao** [1]

[1] College of Mechanical and Electrical Engineering, Central South University, Changsha 410083, China; dzh7695782@126.com (Z.D.); dzscsu@csu.edu.cn (Z.D.); xiaoang@csu.edu.cn (A.X.)

[2] Light Alloy Research Institute, Central South University, Changsha 410083, China

[3] State Key Laboratory of High Performance Complex Manufacturing, Central South University, Changsha 410083, China

* Correspondence: cuixh622@csu.edu.cn; Tel.: +86-153-8802-8791

Abstract: High-strength 7075 aluminum alloy is widely used in the aerospace industry. The forming performance of 7075 aluminum alloy is poor at room temperature. Therefore, hot forming is mainly adopted. Electromagnetic forming is a high-speed forming technology that can significantly improve the forming limit of difficult-to-deform materials. However, there are few studies on electromagnetic hot forming of 7075-T6 aluminum alloy. In this study, the deformation behavior of 7075-T6 aluminum alloy in the temperature range of 25 °C to 400 °C was investigated. As the temperature increased, the sheet forming height first decreased, then increased. When the forming temperature is between 200 °C and 300 °C, η phase coarsening leads to a decrease in stress and hardness of the material. When the forming temperature is between 300 °C and 400 °C, continuous dynamic recrystallization of 7075 aluminum alloy occurs, resulting in grain refinement and an increase in stress and hardness. The results of numerical simulations and experiments all show that the forming height and deformation uniformity of the sheet metal are optimal at 400 °C, compared to 200 °C.

Keywords: 7075-T6 aluminum alloy; electromagnetic forming; hot forming; numerical simulation; mechanical property

Citation: Du, Z.; Deng, Z.; Cui, X.; Xiao, A. Deformation Behavior and Properties of 7075 Aluminum Alloy under Electromagnetic Hot Forming. *Materials* **2021**, *14*, 4954. https://doi.org/10.3390/ma14174954

Academic Editor: Frank Czerwinski

Received: 19 July 2021
Accepted: 24 August 2021
Published: 30 August 2021

Publisher's Note: MDPI stays neutral with regard to jurisdictional claims in published maps and institutional affiliations.

1. Introduction

With rapid developments in aerospace engineering, high-speed trains, and large aircraft, the demand for high-strength, lightweight structural parts is increasing. High-strength and lightweight alloy structural parts can improve material properties. High-strength aluminum alloys offer the advantages of low density and high strength [1,2] and are widely used in the aerospace industry. However, poor plasticity of high-strength aluminum alloys at room temperature limits their application [3,4].

Under hot forming conditions, the yield stress of high-strength aluminum alloys decreases dramatically, and the plastic deformation ability increases. Xiao et al. [5] studied the hot compression test of an Al-Zn-Mg-Cu alloy in the deformation temperature range of 350–450 °C and the strain rate range of 0.001–1 s^{-1}. Based on the microstructure characteristics, the optimum processing parameters were obtained in the temperature range of 380–405 °C and the strain rate range of 0.006–0.035 s^{-1}. Wang et al. [6] studied the formability of AA2024 aluminum alloy at temperatures ranging from 350 °C to 493 °C using tensile tests. The highest plastic forming performance was observed at 450 °C. However, the ductility of the material sharply declined when the temperature exceeded 450 °C. This is due to precipitation of solute atoms on grain boundaries, resulting in grain boundary separation, which reduces the ductility of the material. Kumar et al. [7] studied the tensile properties of 7020-T6 aluminum alloy in a temperature range of 150 °C to 250 °C. The plastic deformation capacity increased with increasing temperature. However, at temperatures higher than 150 °C, the material exhibited dynamic recovery and η′

phase coarsening, resulting in a decrease in tensile and yield stress. Tomoyoshi et al. [8] studied the influence of temperature on the stamping of 2024 aluminum alloy. When the temperature was 400 °C, springback in the stamping part was reduced to zero. However, when the temperature rose to 500 °C, the sheet fractured as the temperature was close to the solidus temperature. Behrens et al. [9] studied the formability of 7022 and 7075 aluminum alloys at temperatures ranging from 150 °C to 300 °C. At 300 °C, the maximum strain increased. Xiao et al. [10] studied the formability of 7075 through hot uniaxial tensile tests. The forming limit was higher at 400 °C than at 20 °C, 450 °C, and 500 °C. Sławomi et al. [11] studied the formability of 7075-T6 aluminum alloy vehicle supports between 20 °C and 240 °C. The parts broke at 20 °C. When the forming temperature was 240 °C, parts were formed accurately, and the tensile stress of the material was reduced by 11% compared with the original state. Huo et al. [12] studied the formability and the microstructural evolution of 7075-T6 sheets at temperatures ranging from 20 °C to 250 °C through sheet bulging experiments. The results showed fine η′ phase and GP zones at 200 °C, leading to the best formability.

Electromagnetic forming (EMF) is a high-speed forming technology that utilizes pulse magnetic field force to process metal [13–15]. In their classic review paper, Psyk et al. [16] pointed out that EMF can greatly reduce springback and improve the forming limit of materials compared with traditional quasi-static forming. Liu et al. [17] formed V-shaped parts with flanges through electromagnetic-assisted stamping (EMAS) using 5052 aluminum alloy. A uniform pressure coil was used to apply pulsed magnetic pressure to the whole part, which effectively reduced tensile stress on the outer surface and compressive stress on the inner surface. Cui et al. [18] proposed EMAS with magnetic force reverse loading. When the die structure did not change, springback in the V-shaped 5052 aluminum alloy part after stamping was reduced. The results show that the equivalent plastic strain and plastic dissipated energy increase after EMF, whereas the stress and elastic strain energy decrease. Cui et al. [19] proposed electromagnetic increment forming (EMIF), which was used to form large 3003 aluminum alloy parts with a moving discharge of small coils. Cui et al. [20] proposed an electromagnetic partitioning forming technology that can precisely manufacture and achieve springback control of large bending radius parts using 3003 aluminum alloy. During the process, elastic deformation is transformed into plastic deformation, and the residual stress is very small after EMF.

A great deal of research has been carried out on the formability of materials. Feng et al. [21] used a V-shaped die and tapered die to perform EMF of 5052 aluminum alloy plates. Compared with quasi-static stamping, maximum strain was increased by about 30% in the side wall of the V-shaped specimens and by about 100% in the tapering specimen. Fang et al. [22] performed drawing experiments using EMIF and 5052 aluminum alloy. Compared with quasi-static stamping, the drawing height was improved by 140% and the dislocation density was increased. Su et al. [23] compared the forming limits of 2219-O aluminum sheets under quasi-static, uniaxial electromagnetic force tensile, and dynamic tensile loads through unidirectional tensile experiments. Under the uniaxial electromagnetic force tensile condition, the forming limit of 2219-O sheets increased by 45.4% compared with the quasi-static tensile condition, and 3.7% to 4.3% compared with the dynamic tensile loading. Electromagnetic hot forming has also been proposed for difficult-to-deform materials. Xu et al. [24] studied the formability of AZ31 magnesium alloy sheets at different temperatures. When the temperature increased from 25 °C to 200 °C, the forming limit increased. However, when the temperature rose to 250 °C, the forming limit decreased. In contrast to 25 °C and 250 °C, grain refinement was clearly observed at 200 °C. Therefore, electromagnetic hot forming is an effective way to improve the formability of difficult-to-deform materials. However, previous research has mainly focused on aluminum alloy materials with good room-temperature plasticity. There are fewer studies on electromagnetic hot forming of high-strength aluminum alloys such as 7075 aluminum alloy.

In this paper, electromagnetic hot forming of 7075-T6 aluminum alloy is proposed. The deformation behavior and mechanical properties of sheet metal were studied at different temperatures both macroscopically and microscopically via experimentation, and the forming process was analyzed using numerical simulations.

2. Experiment

2.1. Experimental Apparatus

Figure 1a shows the 200 kJ Electromagnetic Forming (EMF) machine (Central South University, Changsha, Hunan, China)and 100-ton hydraulic press (Zhongyou, Tengzhou, Shandong, China) used in the experiment. The rated voltage and capacitance of the EMF machine were 25 kV and 640 μF, respectively. A Rogowski coil and oscilloscope were used to measure the discharge current flowing through the coil. Figure 1b shows the experimental tool used for electromagnetic hot forming, including the coil, sheet, and forming die. During the hot forming process, the sheet is first heated in a heating stove, and then the mold-contained heating rods are used to press the sheet. Figure 1c shows the material resistance tester (Anbai, Changzhou, Jiangsu, China). In this study, the forming temperatures were set to 25 °C, 200 °C, 300 °C, and 400 °C. The resistivity of aluminum alloy at different temperatures was obtained using a resistance tester.

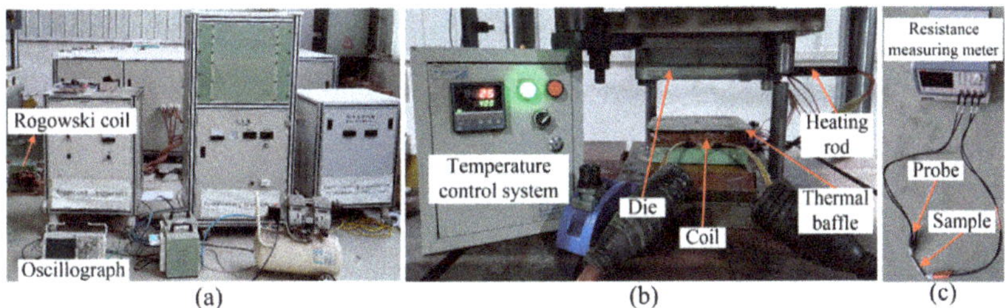

Figure 1. Experimental apparatus: (**a**) electromagnetic forming machine; (**b**) forming apparatus; (c) resistance measuring instruments.

Figure 2a shows the 2D structure of electromagnetic hot forming. The coil is a spiral structure. The cross section (length × width) of the coil wire was 3 mm × 10 mm, and the distance between each turn of the coil was 3 mm. Figure 2a shows a three-dimensional (3D) drawing of the forming die. The diameter of the center opening of the forming die was 100 mm, and the radius of the fillet was 10 mm. Four holes were arranged on the side of the forming die to install the heating rod. The material used in the experiment was 7075-T6 aluminum alloy with the following dimensions: 200 mm × 200 mm × 1 mm. To facilitate subsequent analysis, Path 1 on the sheet was selected. Two special nodes on Path 1 were selected. Node A is located at the sheet center, which is 100 mm away from the sheet edge. Node B is 70 mm away from the sheet edge, as shown in Figure 2c.

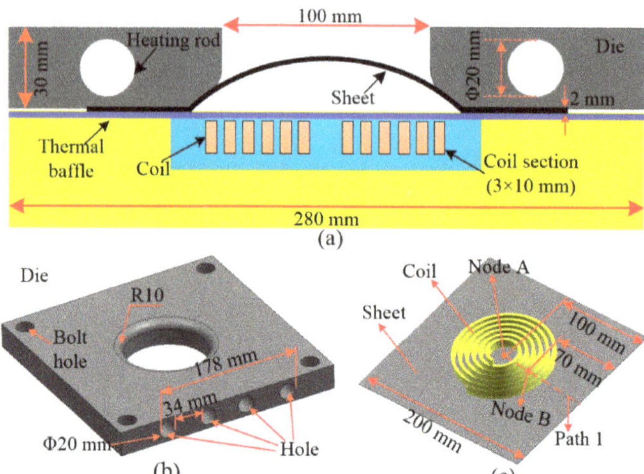

Figure 2. Structure of electromagnetic hot forming: (**a**) forming diagram; (**b**) forming die; (**c**) coil and sheet.

Figure 3 shows the current curves passing through the coil at different temperatures, when the discharge voltage is 5 kV. When the sheet temperature increases, the sheet resistance increases. The mutual inductance between the coil and the sheet decreases, while the total inductance of the discharge system increases with the sheet temperature increase. Thus, the peak current values decrease slightly at 400 °C compared with 200 °C.

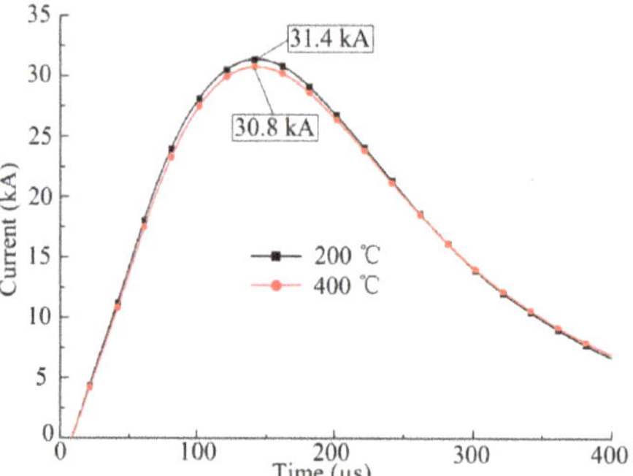

Figure 3. Current curve at 5 KV.

2.2. Analysis of Forming Results

Figure 4 shows the relationship between maximum forming height of sheet metal and voltage at different temperatures. The forming height increases with increasing voltage. The same trend was observed at each temperature. The forming height of sheet metal first decreased, then increased with increasing temperature. The forming height was lowest at 200 °C and highest at 400 °C.

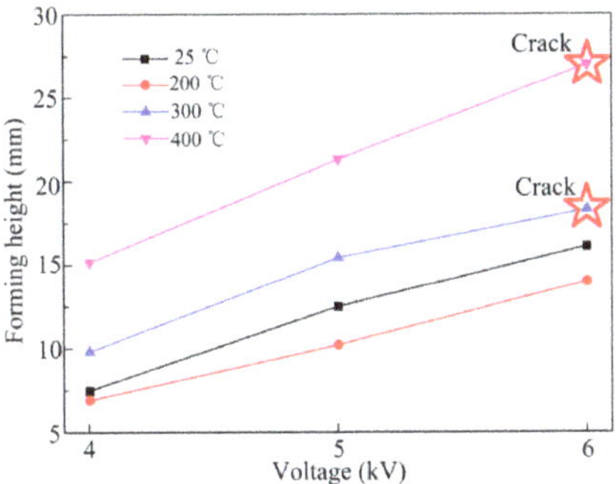

Figure 4. Change in sheet metal deformation with voltage at different forming temperatures.

Figure 5 shows the fracture types of the sheet metal for different voltages and temperatures. When the forming temperature was 25 °C and the discharge voltage was 8 kV, necking occurred at the top of the sheet metal. When the forming temperature was 300 °C and 400 °C, the edge of the sheet metal cracked under a discharge voltage of 6 kV.

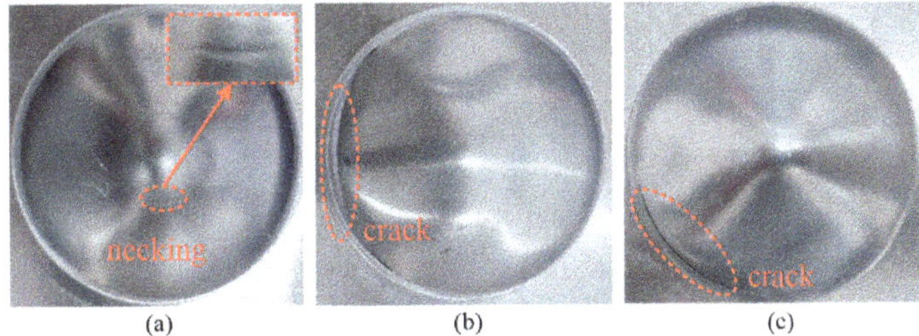

Figure 5. Fracture types at different voltages and temperatures: (**a**) 25 °C, 8 kV; (**b**) 300 °C, 6 kV; (**c**) 400 °C, 6 kV.

The forming results of sheet metal at different temperatures with a discharge voltage of 5 kV are shown in Figure 6. The forming heights at 25 °C, 200 °C, 300 °C and 400 °C were 12.5 mm, 10.2 mm, 15.4 mm and 21.3 mm, respectively. As the temperature increased, the forming height of the sheet metal first decreased, then increased. Compared with 25 °C, the forming height of sheet metal decreased by 18.5% at 200 °C, while the forming heights at 300 °C and 400 °C increased by 23% and 70.4%, respectively.

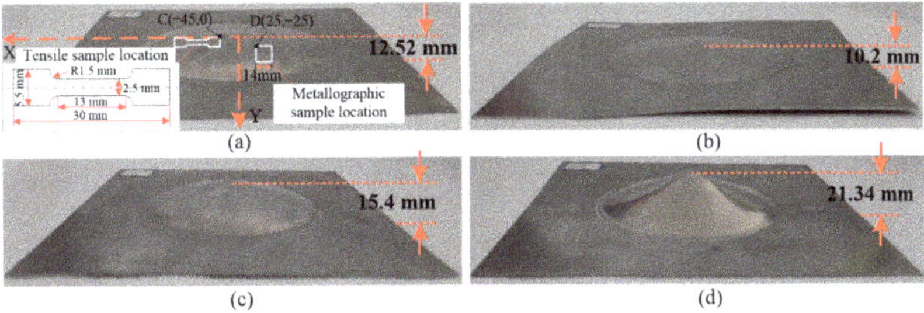

Figure 6. Experimental results for different forming temperatures at a voltage of 5 kV: (**a**) 25 °C; (**b**) 200 °C; (**c**) 300 °C; (**d**) 400 °C.

After forming at 5 kV, mechanical properties of 7075-T6 sheets at different temperatures were examined. The tensile sample was cut with point C (45, 0) as the reference point, and the hardness sample was cut with point D $(-25, -25)$ as the reference point, as shown in Figure 6a. Figure 7 shows the tensile curves and deformed samples obtained with different forming temperatures. At 25 °C, the yield stress, tensile stress, and hardness of the deformed material were 503 MPa, 560 MPa, and 194 HV, respectively. At 200 °C, the values were 475 MPa, 521 MPa, and 184 HV, respectively. At 200 °C, the yield stress, tensile stress, and hardness of the material decreased by 5.5%, 7%, and 5.2%, respectively, compared with 25 °C. At 300 °C, the yield stress, tensile stress, and hardness of the deformed material were 231 MPa, 379 MPa, and 108.3 HV, respectively. At 400 °C, the values were 256 MPa, 450 MPa, and 142.7 HV, respectively. The mechanical properties of the material were lowest when the forming temperature was 300 °C. In addition, the forming height was lower at 200 °C compared with 25 °C. Therefore, the maximum forming height of the sheet metal and good mechanical properties were obtained with a forming temperature of 400 °C.

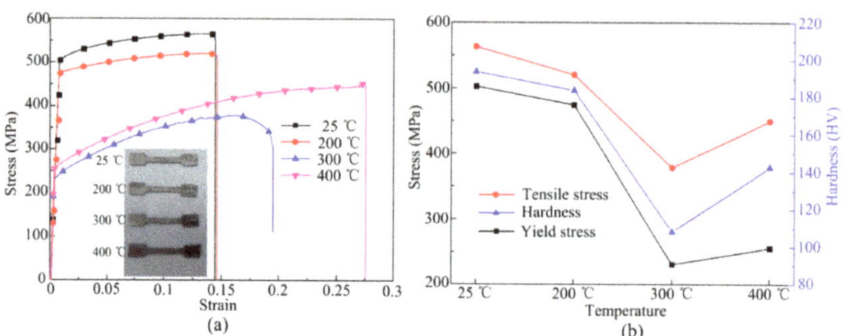

Figure 7. Mechanical properties of sheet metal after forming: (**a**) tensile curve; (**b**) stress and hardness curves.

3. Finite Element Simulation

A finite element model for electromagnetic field analysis was established in an ANSYS/EMAG (14.0) module, as shown in Figure 8a. The model includes the coil, sheet material, air field, and far field air. Figure 8b shows the electromagnetic force distribution on the sheet at 400 °C. Figure 8c shows the magnetic force distribution along Path 1. The magnetic force generated at node A is close to 0, while the largest magnetic force occurs at node B. A finite element model of the deformation field was established in ABAQUS/EXCRITE (6.13), as shown in Figure 8d. The model consists of a die, a blank holder, and a sheet. To shorten the computation time, the die and blank holder were set as rigid bodies. In the simulation process, the electromagnetic force calculated in ANSYS (14.0) was imported into ABAQUS (6.13) to analyze the deformation of the sheet material. Then, the deformation results were imported into ANSYS/EMAG (14.0) to calculate the electromagnetic force for the next step.

The forming results at different temperatures (200 °C and 400 °C) were numerically simulated. The resistivity of the materials at 200 °C and 400 °C is 9.2×10^{-8} Ω·m and 12.9×10^{-8} Ω·m, respectively. Figure 9 shows the stress–strain curves of 7075-T6 aluminum alloy at different temperatures. To investigate the influence of strain rate on the mechanical properties of the materials, the Johnson–Cook constitutive model was adopted, expressed as

$$\sigma = \sigma_s \lfloor 1 + C \ln \varepsilon^* \rfloor \tag{1}$$

where σ_s is the stress curve of different temperatures under quasi-static conditions (see Figure 9), and C is the strain rate coefficient. For 7075-T6 aluminum alloy, C = 0.034. ε^* is the strain rate of the material.

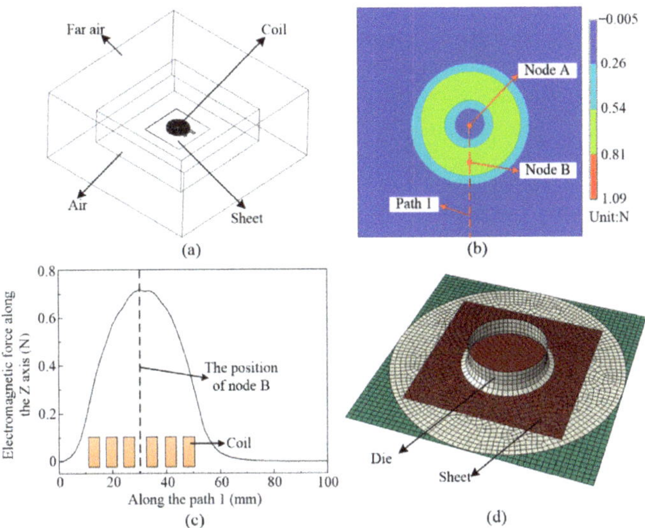

Figure 8. Finite element model of electromagnetic forming: (**a**) electromagnetic field model; (**b**) contours of magnetic force distribution; (**c**) magnetic force along Path 1; (**d**) structure field.

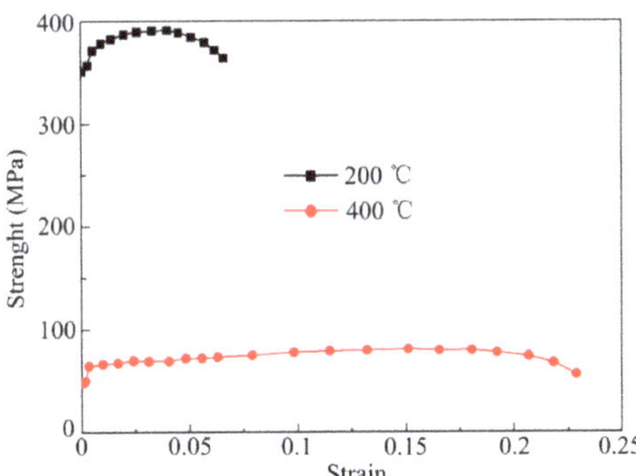

Figure 9. Stress–strain of materials at different temperatures.

Figure 10 shows the sheet deformation profiles obtained by experimentation and numerical simulation. At 200 °C and 400 °C, the experimental forming heights are 10.2 mm and 21.34 mm, respectively. The forming heights by simulation are 10.7 mm and 22.14 mm, respectively. The maximum errors between the simulation and experiment are less than 6% at four temperatures. The simulation results are consistent with the experimental results. Moreover, the deformation profile at 200 °C is wavy, whereas the deformation profile at 400 °C is smooth.

From Figures 4–7 and 10, the forming height of the sheet at 200 °C is the lowest, and the sheet deformation uniformity is poor. At 400 °C, the deformation uniformity and forming height of the parts are the best. Therefore, the deformation process was compared between 200 °C and 400 °C. Figure 11a shows the sheet deformation profile at different time points during the forming process at 200 °C. The sheet forming height is highest at 180 μs, and then vibration appears in the sheet, resulting in a wavy profile. Figure 11b shows the deformed profile at different time points during the forming process at 400 °C. Yield stress in the material is greatly reduced, leading to a significant reduction in the vibration of the material during deformation, compared with 200 °C. At the forming temperature of 400 °C, when the deformation is terminated, the deformation profile is uniform.

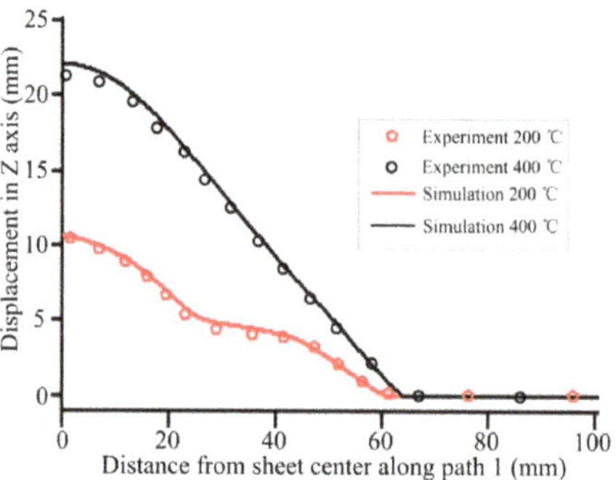

Figure 10. Comparison of simulated and experimental deformed profiles.

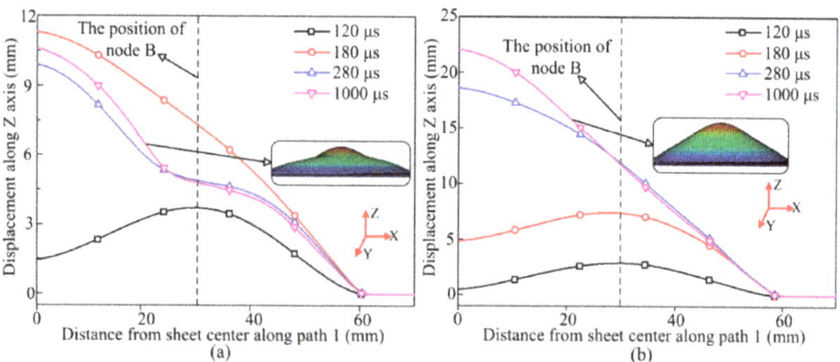

Figure 11. Deformation profile of sheet material at different time points during the forming process at different temperatures: (**a**) 200 °C; (**b**) 400 °C.

Figure 12a shows the equivalent plastic strain on the sheet at 200 °C at different times. After 180 μs, the equivalent plastic strain on the sheet increases. The equivalent plastic strain appears wavy at 1000 μs. Figure 12b shows the equivalent plastic strain on the sheet at 400 °C at different times. After 180 μs, the equivalent plastic strain increases sharply over time. The strain on the sheet is uniformly distributed. At 1000 μs, the greater the distance away from the sheet center, the smaller the plastic strain appears.

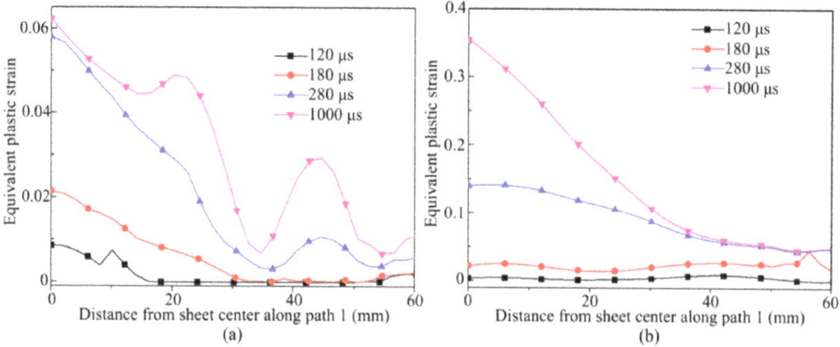

Figure 12. Equivalent plastic strain of sheet material at different time points during the forming process at different temperatures: (**a**) 200 °C; (**b**) 400 °C.

Figure 13 shows three principal stresses with time of node A at a voltage of 5 kV and forming temperatures of 200 °C and 400 °C. σ_1 is along the tangential direction of the sheet, σ_2 is along the width direction of the sheet, and σ_3 is along the thickness direction of the sheet. At 200 °C, σ_1 reaches its first peak of 571 MPa at 120 μs, and σ_2 and σ_3 reach 515 MPa and −35MPa, respectively. At 400 °C, σ_1 reaches its first peak value of 110 MPa at 160 μs, and σ_2 and σ_3 are 91 MPa and −1 MPa, respectively. The stress state of node A is approximately bidirectional isotension. After σ_1 reaches its first peak, the stress oscillates significantly.

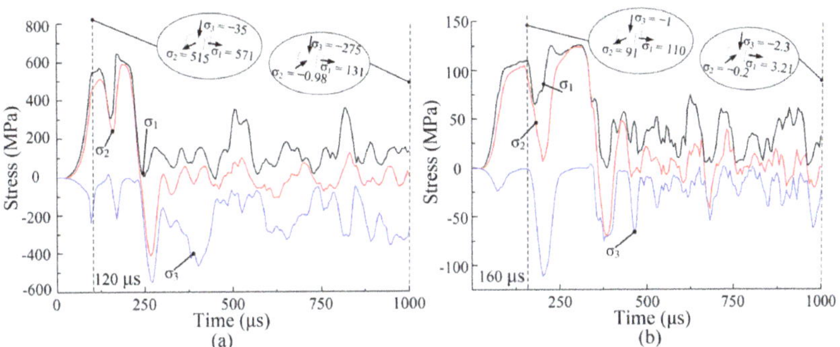

Figure 13. Variation of three principal stresses with time at node A during the forming process at different temperatures: (**a**) 200 °C; (**b**) 400 °C.

Figure 14 shows three principal stresses and time curve of node B at a voltage of 5 kV and forming temperatures of 200 °C and 400 °C. At 200 °C, σ_1 reaches its first peak value of 587 MPa at 185 μs, and σ_2 and σ_3 are 386 MPa and −30 MPa, respectively. At 400 °C, σ_1 reaches its first peak value of 104 MPa at 155 μs, and σ_2 and σ_3 are 87.3 MPa and −19.4 MPa, respectively. Node B is subjected to bidirectional tensile stress. Similar to node A, the stress oscillates and gradually decreases after the first peak. From Figures 13 and 14, the three principal stresses first increase and then oscillate and decay. Moreover, smaller stresses appear on the sheet after the sheet deformation at 400 °C, compared with 200 °C. Therefore, the sheet forming at a higher temperature can obviously reduce the internal stress of sheet metal. Many scholars have found that stress oscillation leads to a large decrease in the internal stress of the sheet during the EMF process, which can reduce the springback of the sheet [25–27].

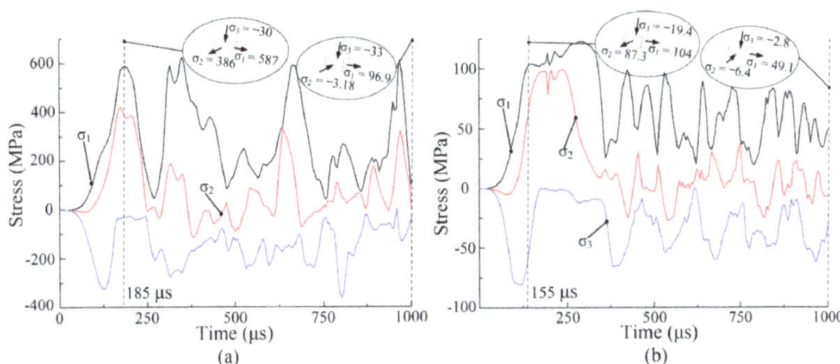

Figure 14. Variation of three principal stresses with time at node B: (**a**) 200 °C; (**b**) 400 °C.

Figure 15a,b show the equivalent plastic strain and plastic strain rate of node A when the discharge voltage is 5 kV. At 200 °C, the maximum equivalent plastic strain and plastic strain rate of node A are 0.058 and 1692 s^{-1}, respectively. At 400 °C, the maximum equivalent plastic strain and plastic strain rate of node A are 0.322 and 3981 s^{-1}, respectively. The equivalent plastic strain increases at first and then remains stable. Compared with a forming temperature of 200 °C, a larger plastic strain and strain rate in the sheet metal were obtained at 400 °C. This is because yield stress decreases with increasing temperature, leading to greater plastic deformation of the sheet metal at 400 °C.

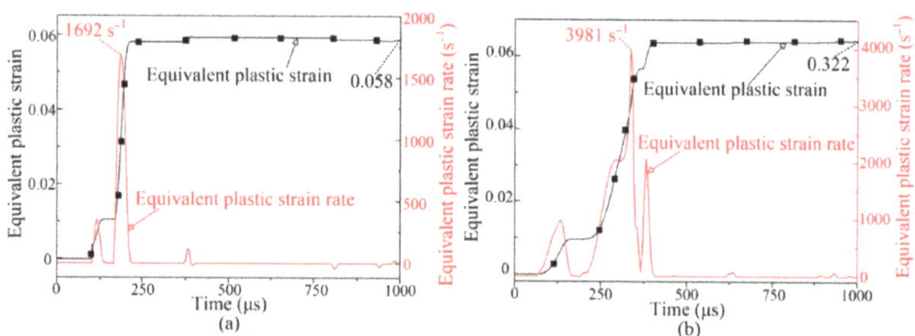

Figure 15. Equivalent plastic strain and strain rate: (**a**) 200 °C, node A; (**b**) 400 °C, node A.

4. Discussion of Results

The microstructure of the deformed sheet was observed and analyzed through a transmission electron microscope (TEM). The TEM samples were prepared by the electrolytic double-spray method. The electrolytic double-spray solution was a mixture of 30% nitric acid and 70% methanol. The temperature was controlled by liquid nitrogen at −35 °C−−25 °C. Figure 16 shows the changes in size of the second phase of the material at 200 °C, 300 °C, and 400 °C. In the forming process of the 7075-T6 aluminum alloy, the main second phase is the η phase. With the increase of temperature, the η phase will be coarser. The increase of η phase size will weaken the strengthening effect and decrease the mechanical properties, as shown in Equation (2) [28]:

$$\sigma = C \bullet f_t^{1/2} \bullet r^{-1} \tag{2}$$

where σ represents precipitate hardening, C is material constants, f_t is the volume fraction of precipitates, and r is the particle size.

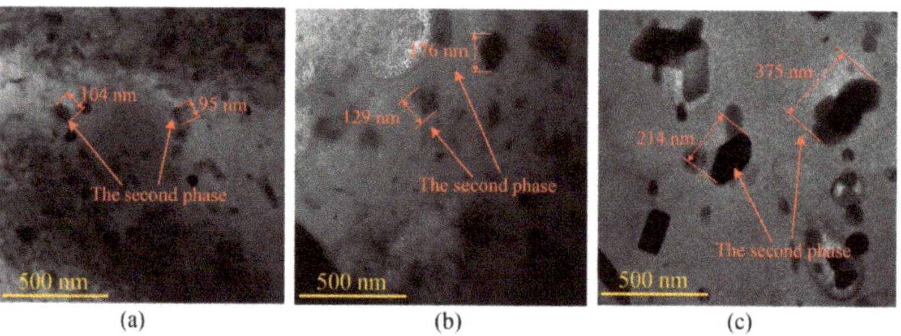

Figure 16. The distribution of the second phase particles: (**a**) 200 °C; (**b**) 300 °C; (**c**) 400 °C.

Compared with 25 °C, 200 °C, and 300 °C, the second phase was further coarsened at 400 °C. Thus, the change of the second phase at 400 °C was not the main factor affecting mechanical properties of the material. In order to analyze the reason for the improvement of mechanical properties at 400 °C, the changes of grain size with temperature were analyzed. The metallographic experiments were performed with an optical digital microscope. Figure 17 shows the grain distributions after sheet forming. The average grain size was about 32 μm, 33 μm, 34 μm, and 16 μm at 25 °C, 200 °C, 300 °C, and 400 °C, respectively. The grain refinement occurs at 400 °C compared with other forming temperatures. This is because the dynamic recovery effect appears at 400 °C for 7075 aluminum alloy [29]. Based on the Hall–Petch relationship (Equation (3)) [30], the mechanical properties of the material improve if the grain size decreases. Therefore, the material stress and hardness increase as the forming temperature increases from 300 °C to 400 °C.

$$\sigma = \sigma_0 + k \bullet d^{-1/2} \tag{3}$$

where σ and σ_0 represent the yield stress and the friction stress when dislocations glide on the slip plane, respectively, k represents material constants, and d is the grain size.

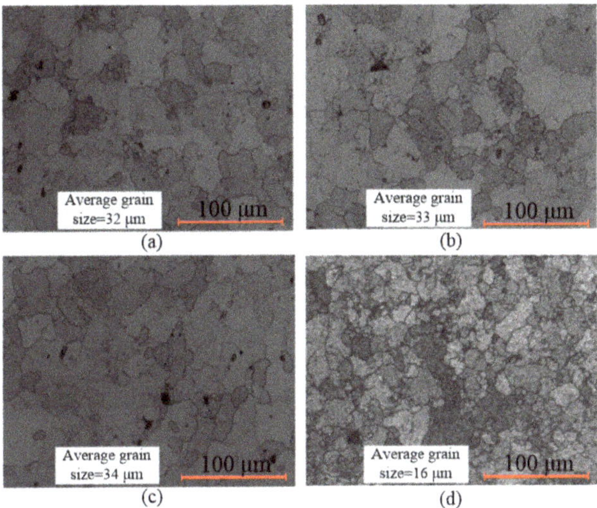

Figure 17. Grain distribution in sheet metal with different forming temperatures: (**a**) 25 °C; (**b**) 200 °C; (**c**) 300 °C; (**d**) 400 °C.

5. Conclusions

7075 aluminum alloy is easy to break at room temperature. In order to increase the forming height of 7075 aluminum alloy, electromagnetic hot forming is proposed. The simulation and experimental results at 25 °C, 200 °C, 300 °C, and 400 °C were compared to prove the accuracy of the FEM model. The main conclusions can be summarized as follows:

(1) Compared with a forming temperature of 200 °C, a more uniform deformation profile and larger plastic strain and strain rate, as well as smaller stresses in the sheet metal, were obtained at 400 °C.

(2) When the temperature was increased from 25 °C to 400 °C, the deformation height first decreased, then increased. The lowest deformation height occurred at 200 °C.

(3) When the forming temperature was increased from 25 °C to 400 °C, the stress and hardness first decreased, then increased. The least favorable mechanical properties occurred at 300 °C. This is because of obvious coarsening in the second phase at 300 °C. Continuous dynamic recrystallization of 7075 aluminum alloy occurs at 400 °C, which leads to grain refinement and improved mechanical properties.

Author Contributions: The research was conceived by X.C. and Z.D. (Zhihao Du); Z.D. (Zhihao Du). and Z.D. (Zanshi Deng) planned and performed all the experiments; A.X. collected all data; theoretical and experimental analysis were performed by Z.D. (Zhihao Du), X.C. and A.X.; the manuscript was reviewed by X.C. and Z.D. (Zhihao Du); the manuscript was written by Z.D. (Zhihao Du). All authors have read and agreed to the published version of the manuscript.

Funding: This work was supported by the National Natural Science Foundation of China (Grant Number: 51775563 and 51405173), Innovation Driven Program of Central South University (Grant number: 2019CX006), and the Project of State Key Laboratory of High Performance Complex Manufacturing, Central South University (ZZYJKT2020-02).

Institutional Review Board Statement: Not applicable.

Informed Consent Statement: Not applicable.

Data Availability Statement: Data is contained within the article.

Conflicts of Interest: The authors declare that they have no conflict of interest in this work. We declare that we do not have any commercial or associative interest that represents a conflict of interest in connection with the work.

References

1. Rasoul, J.N.; Ali, F.; Mahdi, S. Electromagnetically activated high-speed hydroforming process: A novel process to overcome the limitations of the electromagnetic forming process. *CIRP J. Mater. Sci. Technol.* **2019**, *27*, 21–30. [CrossRef]
2. Yu, H.P.; Sun, L.C.; Zhang, X.; Wang, S.L.; Li, C.F. Experiments on electrohydraulic forming and electromagnetic forming of aluminum tube. *Int. J. Adv. Manuf. Technol.* **2017**, *89*, 3169–3176.
3. Zhang, Z.Q.; Zhang, X.K.; He, D.Y. Forming and warm die quenching process for AA7075 aluminum alloy and its application. *J. Mater. Eng. Perform.* **2020**, *29*, 620–625.
4. Rong, H.; Hu, P.; Ying, L.; Hou, W.B.; Zhang, J.H. Thermal forming limit diagram (TFLD) of AA7075 aluminum alloy based on a modifified continuum damage model: Experimental and theoretical investigations. *Int. J. Mech. Sci.* **2019**, *156*, 59–73.
5. Xiao, D.; Peng, X.Y.; Liang, X.P.; Deng, Y.; Xu, G.F.; Yin, M.Z. Research on Constitutive Models and Hot Workability of As-Homogenized Al-Zn-Mg-Cu Alloy During Isothermal Compression. *Met. Mater. Int.* **2017**, *23*, 591–602.
6. Wang, L.; Strangwood, M.; Balint, D.; Lin, J.; Dean, T.A. Formability and failure mechanisms of AA2024 under hot forming conditions. *Mater. Sci. Eng. A* **2011**, *528*, 2648–2656.
7. Kumar, M.; Sotirov, N.; Chimani, C.M. Investigations on warm forming of AW-7020-T6 alloy sheet. *J. Mater. Proc. Technol.* **2014**, *214*, 1769–1776.
8. Tomoyoshi, M.; Ken-ichiro, M.; Ryosuke, Y. Hot stamping of high-strength aluminium alloy aircraft parts using quick heating. *CIRP Ann. Manuf. Technol.* **2017**, *66*, 269–272.
9. Behrens, B.; Nürnberger, F.; Bonk, C. Inflfluences on the formability and mechanical properties of 7000-aluminum alloys in hot and warm forming. *J. Phys. Conf. Ser.* **2017**, *896*, 12004.
10. Xiao, W.C.; Wang, B.Y.; Zheng, K.L. An experimental and numerical investigation on the formability of AA7075 sheet in hot stamping condition. *Int. J. Adv. Manuf. Technol.* **2017**, *92*, 3299–3309.
11. Sławomir, P.; Paweł, K.; Zbigniew, G. Warm forming of 7075 aluminum alloys. *Procedia Eng.* **2017**, *207*, 2399–2404.
12. Huo, W.T.; Hou, L.G.; Zhang, Y.S. Warm formability and post-forming microstructure/property of high-strength AA 7075-T6 Al alloy. *Mater. Sci. Eng. A* **2016**, *675*, 44–54.
13. Hyeonil, P.; Daeyong, K.; Jinwoo, L.; Se-Jong, K.; Youngseon, L. Effect of an aluminum driver sheet on the electromagnetic forming of DP780 steel sheet. *J. Mater. Proc. Technol.* **2016**, *235*, 158–170.
14. Zhang, S.Y.; Ali, N.; Brad, K. Numerical Model and Experimental Investigation of Electromagnetic Tube Compression with Field Shaper. *Procedia Manuf.* **2018**, *26*, 537–542.
15. Xiong, W.R.; Wang, W.P.; Wan, M.; Li, X.J. Geometric issues in V-bending electromagnetic forming process of 2024-T3 aluminum alloy. *J. Manuf. Process.* **2015**, *19*, 171–182.
16. Psyk, V.; Risch, D.; Kinsey, B.L.; Tekkaya, A.E.; Kleiner, M. Electromagnetic forming—A review. *J. Mater. Proc. Technol.* **2011**, *211*, 787–829.
17. Liu, W.; Zou, X.F.; Huang, S.Y.; Lei, Y. Electromagnetic-assisted calibration for surface part of aluminum alloy with a dedicated uniform pressure coil. *Int. J. Adv. Manuf. Technol.* **2019**, *100*, 721–727.
18. Cui, X.H.; Zhang, Z.W.; Du, Z.H.; Yu, H.L. Inverse bending and springback-control using magnetic pulse forming. *J. Mater. Proc. Technol.* **2020**, *275*, 116374.
19. Cui, X.H.; Mo, J.H.; Li, J.J. Electromagnetic incremental forming (EMIF): A novel aluminum alloy sheet and tube forming technology. *J. Mater. Proc. Technol.* **2014**, *214*, 409–427.
20. Cui, X.H.; Du, Z.H.; Xiao, A.; Yan, Z.Q. Electromagnetic partitioning forming and springback control in the fabrication of curved parts. *J. Mater. Proc. Technol.* **2021**, *288*, 116889.
21. Feng, F.; Li, J.J.; Huang, L.; Su, H.L.; Li, H.Z.; Zhang, Y.J.; Cao, S.J. Formability enhancement of 5052 aluminium alloy sheet in electromagnetic impaction forming. *Int. J. Adv. Manuf. Technol.* **2021**, *112*, 2639–2655.
22. Fang, J.X.; Mo, J.H.; Li, J.J. Microstructure difference of 5052 aluminum alloys under conventional drawing and electromagnetic pulse assisted incremental drawing. *Mater. Charact.* **2017**, *129*, 88–97.
23. Su, H.L.; Huang, L.; Li, J.J.; Zhu, H.; Feng, F.; Li, H.W.; Yan, S.L. Formability of AA 2219-O sheet under quasi-static, electromagnetic dynamic, and mechanical dynamic tensile loadings. *J. Mater. Sci. Technol.* **2021**, *70*, 125–135.

24. Xu, J.R.; Xie, X.Y.; Wen, Z.S.; Cui, J.J.; Zhang, X.; Zhu, D.B.; Liu, L. Deformation behaviour of AZ31 magnesium alloy sheet hybrid actuating with Al driver sheet and temperature in magnetic pulse forming. *J. Manuf. Process.* **2019**, *37*, 402–412.
25. Golovashchenko, S.F. Springback calibration using pulsed electromagnetic field. *AIP Conf. Proc.* **2005**, *778*, 284–285.
26. Iriondo, E.; Alcaraz, J.L.; Daehn, G.S.; Gutiérrez, M.A.; Jimbert, P. Shape calibration of high strength metal sheets by electromagnetic forming. *J. Manuf. Process.* **2013**, *15*, 183–193.
27. Wang, X.; Huang, L.; Li, J.J.; Su, H.L.; Feng, F.; Ma, F. Investigation of springback during electromagnetic-assisted bending of aluminium alloy sheet. *Int. J. Adv. Manuf. Technol.* **2019**, *105*, 375–394.
28. Zhang, Q.L.; Luan, X.; Du, J.Q.; Fu, M.W.; Wang, K.H. Development of the post-form strength prediction model for a high strength 6xxx aluminium alloy with pre-existing precipitates and residual dislocations. *Int. J. Plast.* **2019**, *119*, 230–248.
29. Gourdet, S.; Montheillet, F. A model of continuous dynamic recrystallization. *Acta Mater.* **2003**, *51*, 2685–2699.
30. Yu, H.H.; Xin, Y.C.; Wang, M.Y.; Liu, Q. Hall-Petch relationship in Mg alloys: A review. *J. Mater. Sci. Technol.* **2018**, *34*, 248–256.

materials

Article

An Efficient Approach to the Five-Axis Flank Milling of Non-Ferrous Spiral Bevel Gears

Hao Xu [1,†], Yuansheng Zhou [2,3,†], Yuhui He [2,3,*,†] and Jinyuan Tang [2,3]

1 Changsha Zhongchuan Gear & Transmission Driveline Co., Ltd., Changsha 410200, China; xuhao518@sina.cn
2 State Key Laboratory of High Performance Complex Manufacturing, Central South University, Changsha 410083, China; zyszby@csu.edu.cn (Y.Z.); jytangcsu_312@163.com (J.T.)
3 College of Mechanical and Electrical Engineering, Central South University, Changsha 410083, China
* Correspondence: csuhyh@csu.edu.cn
† These authors contributed equally to this work.

Abstract: Five-axis flank milling has been applied in industry as a relatively new method to cut spiral bevel gears (SBGs) for its flexibility, especially for the applications of small batches and repairs. However, it still has critical inferior aspects compared to the traditional manufacturing ways of SBGs: the efficiency is low, and the machining accuracy may not ensure the qualified meshing performances. To improve the efficiency, especially for cutting non-ferrous metals, this work proposes an approach to simultaneously cut the tooth surface and tooth bottom by a filleted cutter with only one pass. Meanwhile, the machining accuracy of the contact area is considered beforehand for the tool path optimization to ensure the meshing performances, which is further confirmed by FEM (finite element method). For the convenience of the FEM, the tooth surface points are calculated with an even distribution, and the calculation process is efficiently implemented with a closed-form solution. Based on the proposed method, the number (or total length) of the tool path is reduced, and the contact area is qualified. Both the simulation and cutting experiment are implemented to validate the proposed method.

Citation: Xu, H.; Zhou, Y.; He, Y.; Tang, J. An Efficient Approach to the Five-Axis Flank Milling of Non-Ferrous Spiral Bevel Gears. *Materials* **2021**, *14*, 4848. https://doi.org/10.3390/ma14174848

Keywords: five-axis flank milling; spiral bevel gears; non-ferrous; CNC machining; tool path planning

Academic Editor: Tomasz Trzepieciński

Received: 1 July 2021
Accepted: 21 August 2021
Published: 26 August 2021

Publisher's Note: MDPI stays neutral with regard to jurisdictional claims in published maps and institutional affiliations.

1. Introduction

Spiral bevel and hypoid gears are significant components of power transmission systems used in automobiles, helicopters, etc. Spiral bevel gears perform rotation about intersecting axes, while hypoid gears rotate about crossed axes. Both spiral bevel gears and hypoid gears can be manufactured in the same way. In the following article, we only mention SBGs for the convenience. SBGs are mainly cut by three conventional approaches, face milling [1,2], face hobbing [3,4], and hobbing [5]. In some cases, the tooth surfaces should be modified by changing the machine setting or cutter motion. Achtmann and Bär [6] applied modified helical motion and roll to produce optimally fitted bearing ellipse. Fan [7] used higher-order polynomials to represent the cradle increment angle of the machine setting than the conventional way and developed TCA (tooth contact analysis) programs in the Gleason commercial software CAGE. Simon [8] reduced the transmission errors by defining the cradle radial setting and the cutting ratio with fifth-order polynomial functions and optimizing them. All of these approaches are equipped with special gear manufacturing machine tools, which rely on corresponding manufacturers.

Unlike the conventional approaches, computer numerical control (CNC) milling has also been introduced as a new technology to cut SBGs on general CNC machine tools. Although CNC milling has a lower production rate than the conventional approaches, it takes advantage of cutting SBGs for small batches, prototypes and repairs (When the gear is worn, it can be repaired by CNC milling). In addition, since the rigidity of CNC milling machine tools is poorer than the conventional gear machine tools, the cutting efficiency is limited to the application of CNC milling to the gears with the hard material. This limitation is not a serious problem for the non-ferrous metals. Hence, it would be beneficial to find the efficient way for CNC milling of non-ferrous gears.

With the recent development of CNC milling technologies, CNC milling of SBGs becomes an advanced technology applied in some gear manufacturers. Many researchers have studied the manufacturing of SBGs with CNC milling. Tsiafis et al. [9] studied design and manufacturing of SBGs using CNC milling machines. Li et al. [10] proposed a novel integrated design and machining mode of SBGs based on universal CNC machine tools. In order to solve the problem that the tooth surface points cannot cover the whole tooth surface, Wang et al. [11] proposed a new adaptive geometric meshing theory with an advanced study of cutter geometry and meshing theory. Shih et al. proposed a method to manufacture face-milled SBGs and the cutter head on a five-axis CNC machine [12] and then carried out real cutting experiments [13]. Álvarez et al. [14] studied different machining strategies of five-axis CNC milling to improve the machining quality of SBGs. Based on the updated Kriging model, Deng et al. [15] proposed a method to improve the accuracy of tooth surface reconstruction. Gleason's UpGear method [16] is intended for CoSIMT-type tools on Gleason-Heller 5-axis CNC machines.

End milling is a mode of CNC milling to remove the material around the midst of the cutter's flat end. DMG-Mori's gearMILL [17] offers SBG modules cutting with end mill and ball mill tools. The alternative mode, flank milling, removes the material along the cutter flank. Comparing end milling and flank milling, flank milling takes advantage of quality enhancement, manufacturing time and cost reduction [18,19], and it has been widely applied to manufacturing a ruled surface [20]. For some of the existing practical models of SBGs, their working parts are ruled surfaces, or some are close to ruled surfaces. Taking the example of the gears cut by Gleason's face-milled methods, which include non-generated and generated methods: the working part of the non-generated gear is a conical cutter surface, which is a ruled surface; the working part of the generated gear is the envelope surface of the conical cutter surface, and it is close to a ruled surface [21]. Hence, it makes sense to apply the flank milling to cut SBGs.

Flank milling and the manufacturing of SBGs are two different disciplines. Exksting research on flank milling is mainly about the minimization of machining errors. The corresponding issues include envelope surface [22–25], cutter-workpiece engagements [26–30], geometrical deviations [24,31,32], tool path planning strategies [33–39], tool path optimization with: particle swarm optimization method [40], dynamic programming method [41], local method [42], global method [43], constraints [44], generic cutters [45], the consideration of tool path smoothness [46], the influences of tool axis [47] and cutter runout [48,49], cutter optimization of shape [50] and size [51], etc. In contrast, the manufacturing of SBGs values the working performances [21], such as contact path and transmission errors. By considering the difference, Zhou et al. [52,53] focused on the machining of tooth surface area, and an extra pass (tool path) was needed to machine the bottom of a tooth slot of SBGs, so the machining efficiency can be further improved. Meanwhile, they did not give the solution of TCA for the flank milling of SBGs, and it is very difficult, due to the complicated tooth surface geometry, but important since, it directly evaluates the meshing performances of the machined SBGs.

In this work, a filleted end mill cutter is used to simultaneously cut the tooth surface and tooth bottom with only one pass, and this method improves the machining efficiency by avoiding the extra pass to cut the tooth bottom, which is necessary for the methods in [52,53]. In addition, compared with the traditional flank milling method considering the whole tooth surface errors, the tool path planning strategy and optimization model proposed in this paper focus on the machining accuracy near the tooth surface contact area, which can ensure that the tooth surface has better meshing performances. In order to verify the effectiveness and authenticity of the proposed method, the TCA with FEM is applied with a novel tooth surface modeling approach to check the meshing performances of the machined SBGs, and actual machining experiments are carried out. The analysis results show that the contact area is qualified. In Section 2, the tooth surface and meshing of SBGs are introduced. The optimization model of tool path planning is established in Section 3. With the planned tool path, a new closed-form representation is proposed to efficiently generate tooth surface points with an even distribution for the convenience of building the FEM model. Subsequently, the TCA of the flank milling of SBGs is carried out based on the FEM, as described in Section 4. The cutting simulation and experiments are explained in Section 5.

2. Tooth Surface and Meshing of SBGs

The 3D model of an SBG with 33 teeth is shown in Figure 1. Each tooth slot has both a convex side and a concave side. The depth of the tooth slot gradually decreases from heel to toe. Here we assume that the tooth surface is a given designed surface and represented as $\mathbf{g}(h, \phi)$. The details of the tooth surface model can be referred to as [2,53].

When a pair of SBGs are meshing to transfer power during the process of two tooth surfaces engaging with each other, they contact at different points when they are treated as rigid bodies. Practically, they contact at small ellipses around the contact points due to the deformation of the tooth surface. The contact paths are formed by connecting these points, and the contact area is generated by joining these contact ellipse. Once both tooth surfaces of a pair of gear drive are obtained, the contact path and area can be calculated by TCA [21]. For a gear drive, edge contact should be avoided since it decreases work performance and serves life. A good design of tooth surface should be capable of avoiding the edge contact while considering practical circumstances, including manufacturing errors, load, and errors of alignment. Subsequently, an ideal contact path is usually chosen as the middle of a pair of tooth surfaces of a gear drive [21], as shown in Figure 1.

Since the tooth surface is a complex 3D surface, it is difficult to define the ideal contact path directly. Alternatively, it can be defined according to the gear blank. As shown in Figure 1, the ideal contact path is coincident to the middle of a pair of tooth surface. A point $\mathbf{q}'$ on the middle of a pair of tooth surface can be mathematically represented according to the gear blank parameters. When $\mathbf{q}'$ is defined by the gear blank parameters, the contact points on the tooth surface can also calculated according to a mapping approach as shown in Figure 1. When point $\mathbf{q}'$ is rotated along with z_g, a circular curve is formed and intersected with the convex side of tooth surface as $\mathbf{q}$. Subsequently, the mapping relationship can be used to calculate the contact point on the tooth surface [52]. We have a system of two equations in two unknowns [52]:

$$\begin{cases} Z(h^*, \phi^*) = z_q \\ X^2(h^*, \phi^*) + Y^2(h^*, \phi^*) = r_q^2. \end{cases} \tag{1}$$

where $X(h^*, \phi^*)$, $Y(h^*, \phi^*)$, $Z(h^*, \phi^*)$ are the coordinates of the tooth surface in S_g. Once both unknowns h^* and ϕ^* are calculated according to Equation (1), the contact point on the contact path is obtained by submitting into the tooth surface $\mathbf{g}(h, \phi)$.

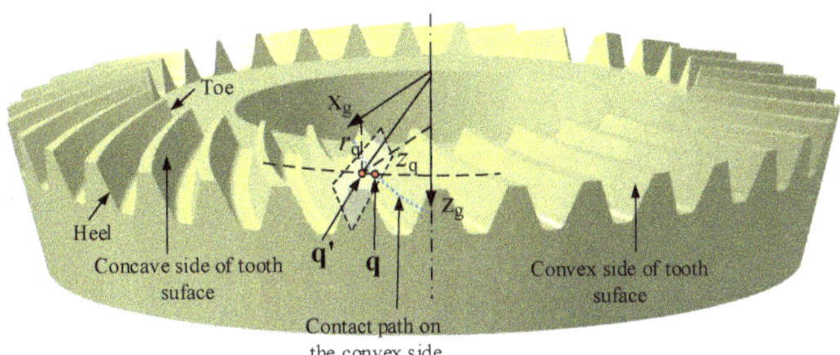

Figure 1. Tooth surface and contact path.

3. Tool Path Planning Strategy and Optimization Model for Flank Milling of SBGs

3.1. Tool Path Planning Strategy

As shown in Figure 2, a process is considered to generate a cutter envelope surface tangent to the tooth surface along the contact path. At each instant of the process, the cutter surface is tangent to the tooth surface at a point, which is represented on both surfaces as **p** and **q**, respectively. The trajectories of the tangent point on both surfaces are cutter contact (CC) line **p**(ϕ) and contact path **q**(ϕ), respectively. Once **p** contributes as a point on the cutter envelope surface, it means the cutter envelope surface is also tangent to the designed surface at **p**. The necessary conditions of the cutter envelope surface tangent with the tooth surface along **q**(ϕ) are summarized in terms of two aspects [53].

- The cutter surface is tangent to the tooth surface along **q**(ϕ).
- For **p**(ϕ), it must satisfy

$$\frac{d\rho(h)}{dh} \cdot \frac{dh(\phi)}{d\phi} = 0. \tag{2}$$

The left side of Equation (2) is directly determined by the CC line on the cutter surface. For a cylinder, $\rho(h)$ is a constant and $h(\phi)$ can vary.

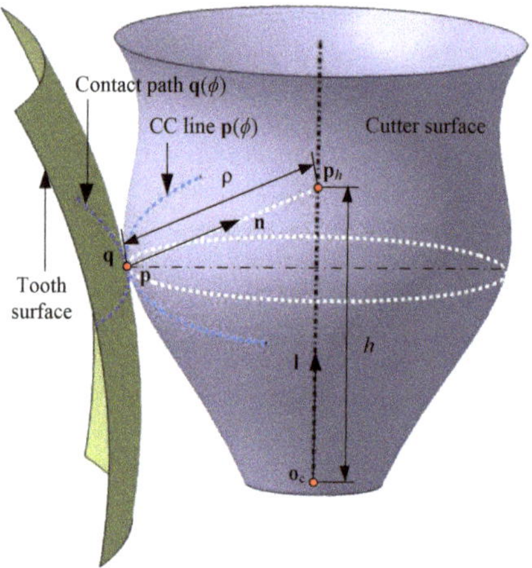

Figure 2. The tangency condition between the cutter envelope surface and tooth surface.

With both models of tooth surface and contact path, the tool path planning for five-axis flank milling is implemented with two steps. First, the tool path strategy based on the necessary conditions is used to make the cutter envelope surface tangent to the designed surface along the contact path. Second, cutter orientations are optimized to obtain the minimal geometric deviations of the contact area.

In order to improve the efficiency, the fillet end mill cutter is used to process the tooth slot once to ensure the one-time machining of tooth surface and fillet part. As shown in Figure 3, a filleted end mill cutter is used only one pass to machine the convex side of the tooth surface. According to the necessary conditions, the cutter is planned to be tangent to the tooth surface along the contact path. For a point **q** on the contact path, a local coordinate system $S_q(\mathbf{q}; \mathbf{n}, \mathbf{t}, \mathbf{d})$ is established. **n** is the unit normal vector of both cutter envelope surface and tooth surface. **t** is the unit tangent vector of the contact path at **q**. **d** is obtained as $\mathbf{d} = \mathbf{n} \times \mathbf{t}$. Furthermore, the tool path planning strategy is stated with two aspects: the cutter axis **l** is a unit vector in the tangent plane **tqd**; cutter tip point $\mathbf{o}_c$ is determined according to the tangent between the cutter and root cone.

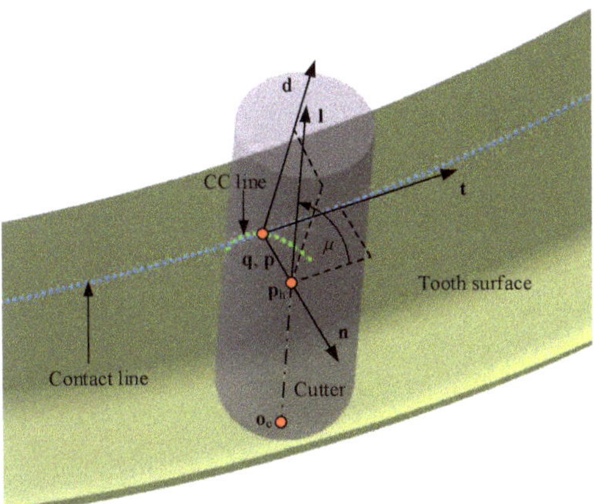

Figure 3. Tool path planning strategy.

As shown in Figure 3, the tool axis l could be a varied unit vector in the tangent plane, and it can be defined with an angle μ. $\mathbf{o}_c$ is the tip point. With a given l, $\mathbf{o}_c$ can be determined with a solution to h calculated according to the condition that the cutter is tangent with root cone, which will be explained later. We have

$$\mathbf{l}(\phi) = \cos(\mu(\phi)) \cdot \mathbf{t}(\phi) + \sin(\mu(\phi)) \cdot \mathbf{d}(\phi)$$
$$\mathbf{o}_c(\phi) = \mathbf{p}_h(\phi) - h(\phi) \cdot \mathbf{l}(\phi) \tag{3}$$
$$= \mathbf{q}(\phi) + R \cdot \mathbf{n}(\phi) - h(\phi) \cdot \mathbf{l}(\phi)$$

where R is the radius of the cutter. It should be noted that because the fillet part is directly machined by the bottom of the cutter, it is necessary to ensure that the toroidal surface of the cutter is tangent to the root cone. As shown in Figure 4, the distance between $\mathbf{p}_h$ and the tip point is h_m, the distance between $\mathbf{p}_h$ and the root cone is $(h_q+c)/2$ and marked as M. Therefore, the following equation holds

$$\begin{cases} h_m = M\sin\mu + r(1 - \sqrt{1 - (\frac{M\cos\mu - d}{r})^2}) \\ M = \frac{h_q+c}{2} \end{cases} \tag{4}$$

where r is the fillet radius of cutter and d is the cutter constant. When the toroidal surface is tangent to the root cone, h_m can be approximately regarded as h in Equation (3). Therefore, h in Equation (3) can be replaced by h_m in Equation (4) to obtain the coordinates of tip point $\mathbf{o}_c$.

Because $\mathbf{n}$, $\mathbf{t}$, and $\mathbf{d}$ can be calculated in coordinate system S_g, l and $\mathbf{o}_c$ can also be obtained in S_g by transforming from S_q. Since μ is the variable to define l, a polynomial function in terms of the motion parameter ϕ is applied to define the μ as

$$\mu(\phi) = u_1 + u_2 \cdot \phi + u_3 \cdot \phi^2 + u_4 \cdot \phi^3 \tag{5}$$

where $u_i(i = 1 \sim 4)$ are the coefficients of the polynomials. Assuming $\mathbf{x} = [u_1, u_2, u_3, u_4]$, the optimization problem with respect to the variable $\mathbf{x}$ is proposed as follows to minimize the geometric deviations of the contact area.

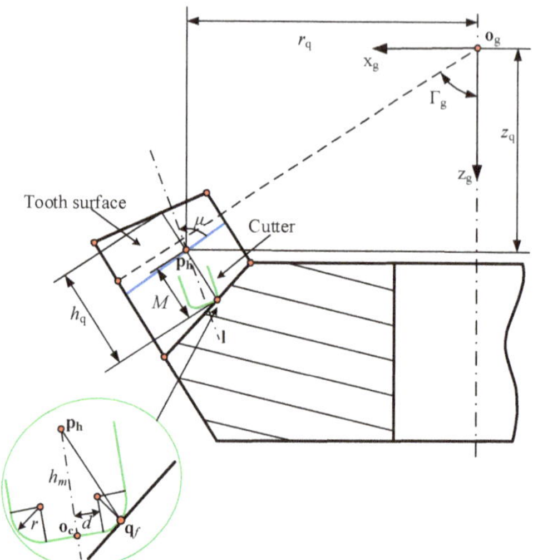

Figure 4. Cutter constraints in machining.

In summary, the tool path planning strategy generally meets the following conditions.

- The necessary condition of the cutter envelope surface tangent to the designed tooth surface is satisfied.
- The constraint condition of the toroidal surface of the cutter tangent to root cone is satisfied.
- The constraint condition that the swing angle of tool axis μ is a function of motion parameter ϕ is satisfied.

3.2. Optimization Model to Minimize the Geometric Deviations of the Contact Area

Once the cutter envelope surface and tooth surface are tangent along the contact path, both of them have the same normal curvature at every point on the tangent direction of the contact path. An effective way to reduce the geometric deviations around this point is to minimize the relative normal curvature of direction **d** [53]. Furthermore, for the geometric deviations of the contact area, it is an effective way to minimize the overall relative normal curvatures of direction **d** along the contact path. By inserting Equation (3) into Equations (10) and (11), the cutter envelop surface can be obtained. When the representations of theoretical tooth surface and cutter envelope surface are known, the normal curvature of any point on the surface along any direction can be calculated according to the theory of differential geometry. For a point $\mathbf{q}(u,v)$ on the contact path, the principal curvature and principal direction at this point can be calculated by Equation (6).

$$\mu_{1,2} = \frac{-(LG - NE) \pm \sqrt{(LG - NE)^2 - 4(LF - ME)(MG - NF)}}{2(LF - ME)}$$

$$L\mu_{1,2} + M = k_{1,2}(E\mu_{1,2} + F) \tag{6}$$

$$e_{1,2} = \frac{r_u\mu_{1,2} + r_v}{|r_u\mu_{1,2} + r_v|}$$

where L, M, and N are the first fundamental homogeneous coefficients, and E, F, and G are the second fundamental homogeneous coefficients. $k_{1,2}$ and $e_{1,2}$ are the principal curvature and the principal direction at $\mathbf{q}$, respectively. Suppose that the angle between the tangent direction **d** and the principal direction e_1 at point **q** is θ, as shown in Figure 5. Then, according to Euler formula, the normal curvature of point **q** along the tangent direction **d** is

$$k_d = k_1 \cos^2\theta + k_2 \sin^2\theta \tag{7}$$

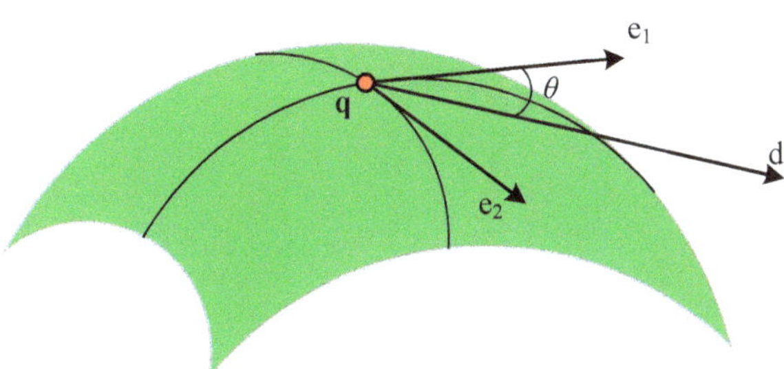

Figure 5. Principal direction and tangent direction.

In conclusion, for a set of data points $\mathbf{q}_i (i = 1 \sim N)$ sampled from the contact path, the following model can be used to describe the error between the cutter envelope surface and the designed tooth surface.

$$\min \sum_{1}^{N} (k_d - k_e(\mathbf{x}))^2$$

$$s.t.\ \mu_{min} \le \mu_i \le \mu_{max}, i = 1 \sim N \tag{8}$$

$$h_m = M \sin\mu + r(1 - \sqrt{1 - (\frac{M \cos\mu - d}{r})^2}), M = \frac{h_q + c}{2}.$$

where **x** is optimal variable; μ_{min} and μ_{max} are used to define the range of μ_i; k_e and k_d are the normal curvatures along the direction **d** of designed surface and cutter envelope surface, respectively. In some cases, the absolute value of relative normal curvature is close to 0, so the minimum relative curvature radius can be used to describe the closeness between two surfaces, as shown in Figure 6.

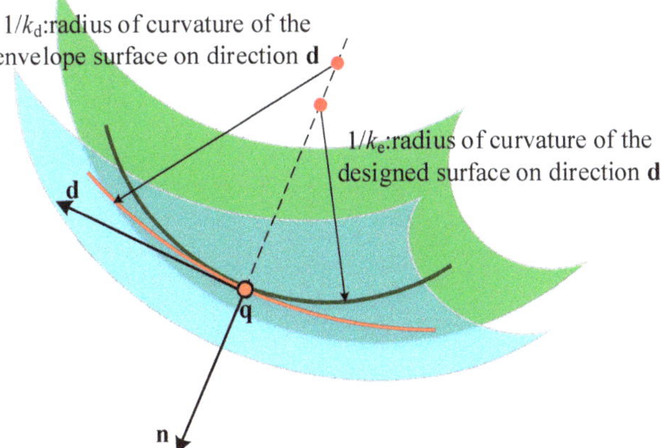

Figure 6. Description deviation of relative radius of curvature.

The corresponding optimization model is replaced by

$$\min \sum_{1}^{N} (\frac{1}{k_d} - \frac{1}{k_e(\mathbf{x})})^2$$

$$s.t. \ \mu_{min} \le \mu_i \le \mu_{max}, i = 1 \sim N \tag{9}$$

$$h_m = M\sin\mu + r(1 - \sqrt{1 - (\frac{M\cos\mu - d}{r})^2}), M = \frac{h_q + c}{2}.$$

4. Modeling and Contact Analysis of SBGs

In order to ensure the good contact performances of the tooth surface after flank milling, the contact analysis of the tooth surface based on the FEM can be carried out, so the three-dimensional model of the tooth surface is needed. In order to facilitate the modeling in an efficient way, tooth surface points should be generated with an approximately even distribution [54]. Although [54] gave an effective way to achieve this goal, a complicated computation algorithm was applied to solve a global optimization problem. In contrast, we here give a new way, which directly applies closed-form calculations to efficiently generate the tooth surface points.

Assume that the cutter surface is expressed as $\mathbf{r}(h, \theta)$, where h and θ are two independent parameters. When the cutter surface moves continuously along a tool path defined with parameter ϕ, a family of surfaces of the cutter surface is generated. The envelope surface is the boundary of the family surfaces. The envelope surface can be calculated as a closed-form result according to the geometric meshing theory (or geometric envelope approach) [2,31] as

$$\mathbf{r}(h, \phi) = \mathbf{o}_c(\phi) + h(\phi) \cdot \mathbf{l}(\phi) + \rho(h) \cdot \mathbf{n}(h, \phi) \tag{10}$$

where, as shown in Figure 2, the cutter surface is represented as a surface of revolution; $\mathbf{o}_c$ is the cutter tip point; $\mathbf{n}$ is the unit normal of the cutter surface at a point $\mathbf{p}$; ρ is the distance between $\mathbf{p}$ and $\mathbf{p}_h$, which is the intersected point of $\mathbf{n}$ and $\mathbf{l}$, the cutter axis; α is the angle formed by $\mathbf{n}$ and $\mathbf{l}$. $\mathbf{n}$ in Equation (10) can be calculated as as [2,31]

$$\mathbf{n}(h, \phi) = \frac{\cos\alpha \cdot \mathbf{v}_h^2}{(\mathbf{l} \times \mathbf{v}_h)^2} \cdot \mathbf{1} - \frac{(\mathbf{l} \cdot \mathbf{v}_h) \cdot \cos\alpha}{(\mathbf{l} \times \mathbf{v}_h)^2} \cdot \mathbf{v}_h \pm \frac{\sqrt{(\mathbf{l} \times \mathbf{v}_h)^2 - \cos^2\alpha \cdot \mathbf{v}_h^2}}{(\mathbf{l} \times \mathbf{v}_h)^2} \cdot (\mathbf{l} \times \mathbf{v}_h) \tag{11}$$

where $\mathbf{v}_h$ is the velocity of $\mathbf{p}_h$.

However, for some special cases, Equations (10) and (11) are not appropriate to obtain the contact points covering the whole tooth surface, which will be mentioned later. Now, another method of calculating the envelope surface is introduced by taking the filleted end mill cutter as an example.

As shown in Figure 7, assume that $\mathbf{p}_2$ is a contact point on the toroidal surface, $\mathbf{n}_2$ denotes the unit normal of the cutter surface at $\mathbf{p}_2$, $\mathbf{t}$ is a unit vector on the plane determined by $\mathbf{p}_2$ and $\mathbf{l}$, and $\mathbf{t}$ is orthogonal to the tool axis $\mathbf{l}$. The geometric characteristic can be expressed as

$$\mathbf{n}_2 \cdot (\mathbf{e} \times \mathbf{l}) = 0 \tag{12}$$

Combining with this geometric characteristic and envelope condition, we can obtain the contact points on the toroidal part, which will be described next.

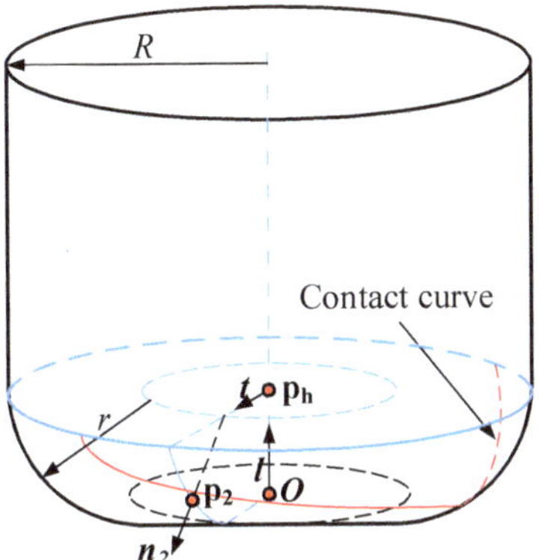

Figure 7. Calculating the contact point on the filleted end mill cutter.

As shown in Figure 7, $\mathbf{p_2}$ can be expressed as

$$\mathbf{p_2} = \mathbf{o} + r\mathbf{l} + (R - r)\mathbf{e} + r\mathbf{n_2} \tag{13}$$

and the velocity $\mathbf{v_{p2}}$ on point $\mathbf{p_2}$ can be obtained as

$$\mathbf{v}_{p2} = \mathbf{v}_o + \mathbf{w} \times (\mathbf{p_2} - \mathbf{O}) = \mathbf{v}_o + \mathbf{w} \times (r\mathbf{l} + (R - r)\mathbf{t} + r\mathbf{n_2}) \tag{14}$$

Considering the envelope theory, the envelope condition can be written as

$$\mathbf{n_2} \cdot \mathbf{v}_{p2} = \mathbf{n_2} \cdot (\mathbf{v}_O + \mathbf{w} \times (r\mathbf{l} + (R - r)\mathbf{t})) = 0 \tag{15}$$

Assume that $\mathbf{p_h}$ is the center point of the circle, which is the intersection of cylindrical part and toroidal part, the velocity of point $\mathbf{p_h}$ is $\mathbf{v_h}$. Hence, Equation (15) can be rewritten as

$$\mathbf{n_2} \cdot (\mathbf{v}_h + (R - r)(\mathbf{w} \times \mathbf{t})) = 0 \tag{16}$$

According to the Equations (12) and (16) , $\mathbf{n_2}$ can be obtained as

$$\mathbf{n_2} = \pm \frac{(\mathbf{l} \times \mathbf{t}) \times (\mathbf{v}_h + (R - r)(\mathbf{w} \times \mathbf{t}))}{|(\mathbf{l} \times \mathbf{t}) \times (\mathbf{v}_h + (R - r)(\mathbf{w} \times \mathbf{t}))|} \tag{17}$$

By substituting Equation (17) into Equation (13), corresponding contact points can be calculated. By sampling parameter $\mathbf{t}$ with all directions on the plane orthogonal to the tool axis $\mathbf{l}$, the contact curve on the toroidal part can be obtained.

The envelope surface can be calculated by using the above two methods, but in some cases, one method may be more suitable than the other.

In Figure 8, the contact line on the cylindrical surface is mainly in the height direction. If the discrete h method is used, more contact points (blue points) will be obtained. If the discrete θ method is used, fewer contact points (green points) will be obtained. The contact line on the toroidal surface is mainly in the width direction. If the discrete h method is used, fewer contact points (blue points) will be obtained. If the discrete θ method is used, more contact points (green points) will be obtained. Therefore, in order to obtain contact points that cover the whole tooth surface uniformly, we need to select the appropriate method to calculate the envelope surface according to different situations.

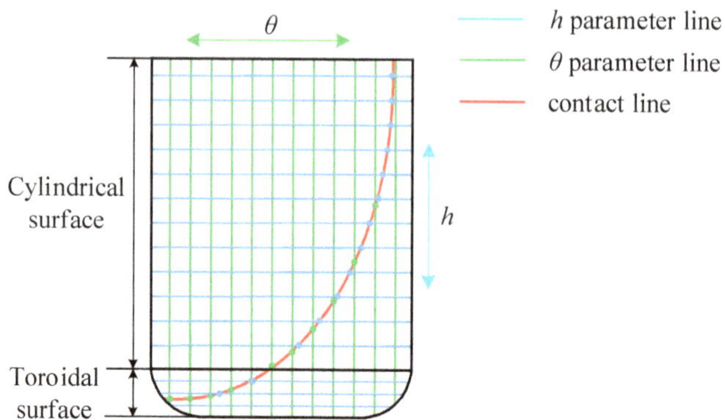

Figure 8. Calculation of contact line.

According to the data in Table 1, combined with the tool path planning method and closed-form modeling method proposed in this paper, the optimization model is established according to Equation (9). For this model, the range of μ_i is chosen as $\left[\frac{\pi}{6}, \frac{5\pi}{6}\right]$, and $N = 21$. With the optimal result, the machined surface is calculated according to Equations (10) and (11) and modeled in CATIA V5R20, as shown in Figure 9. The 3D model was imported into ABAQUS to establish the FEM. The material properties are Young's modulus 206,800 and Poisson's ratio 0.3. Using the five tooth calculation model, both the gear and the pinion contain 131,600 elements, and the element type is solid element C3D8R. Checking the mesh quality indicates that both the error mesh and the warning mesh are 0. The torque is 2625 N·m, which is applied to the gear. Set the number of CPU cores to 30. After submitting the job, the analysis lasted for 6 h 20 min, and the analysis results are shown in Figure 10. There is a slight edge contact between the gear and the pinion, but the overall contact area is basically in the tooth direction.

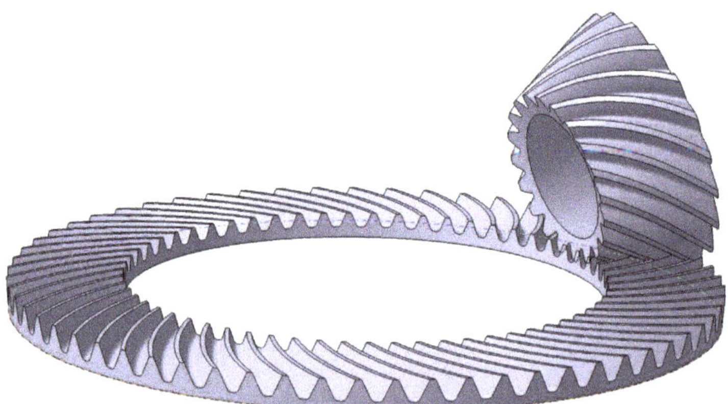

Figure 9. 3D model of SBGs.

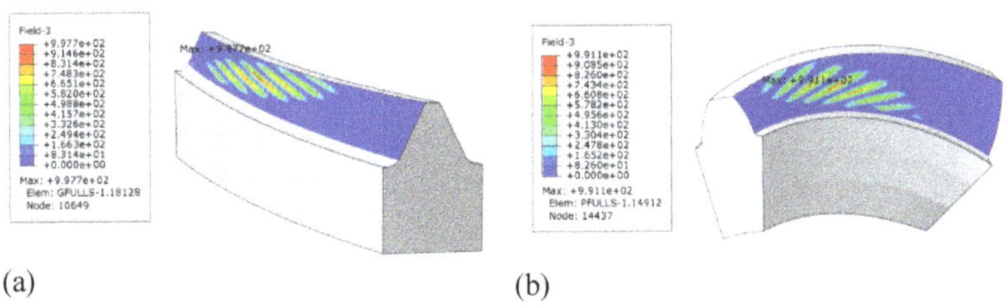

Figure 10. (**a**) Contact area on the gear; (**b**) Contact area on the pinion.

5. Machining Simulation and Experiment

An SBG represented by the parameters in Table 1 was machined. The radius and fillet radius of the cutter are 2 mm and 1.5 mm, respectively. The cutter has been checked without interference on both sides of the working part of the tooth surface.

Table 1. Main data of a face-milled generated SBG.

Blank Data			
Parameter	**Value**	**Parameter**	**Value**
Gear tooth number	61	Pinion tooth number	20
Module	4.8338	Shaft angle	90.0000°
Pinion handle	Right hand	Mean spiral angle	32°
Face width	27.5000 mm	Clearance	1.0300 mm
Outer addendum	1.7600 mm	Outer dedendum	7.6700 mm
Face angle	76.1167°	Root angle	69.5833°
Blade Data			
Parameter	**Value**	**Parameter**	**Value**
Average radius	63.5000 mm	Point width	2.5400 mm
Pressure angle	22.0000°	Fillet radius	1.5000 mm
Machine-Settings			
Parameter	**Value**	**Parameter**	**Value**
Radial setting	64.3718 mm	Cradle angle	−56.7800°
Sliding base	0.0000 mm	Machine center to back	0.0000 mm
Blank offset	−0.2071 mm	Machine root angle	69.5900°
Roll ratio	1.0323	Modified roll coefficients	0.0000

In the experiment, five slots were machined, the calculated tool path was imported into CATIA, and the reasonable advance and retreat tool path was set. Finally, the complete tool path of this experiment was obtained. The simulation processing was verified in Mastercam, as shown in Figure 11a,b. Finally, the five tooth slots model of SBG was obtained and compared with the theoretical tooth surface, as shown in Figure 11c.

The simulation results show that the tooth surface errors in the meshing area are between −0.01 and 0.01 mm, while the tooth surface errors in the top and root area are relatively large.

The corresponding NC code is calculated by using the simulation software, and the actual processing experiment is carried out on the five-axis NC precision engraving machine GR200V A15SH manufactured by Beijing Jingdiao Technology Co., Ltd. (Beijing, China). The gear material is 7075 aluminum alloy and the five tooth slots model of SBG is obtained, as shown in Figure 12.

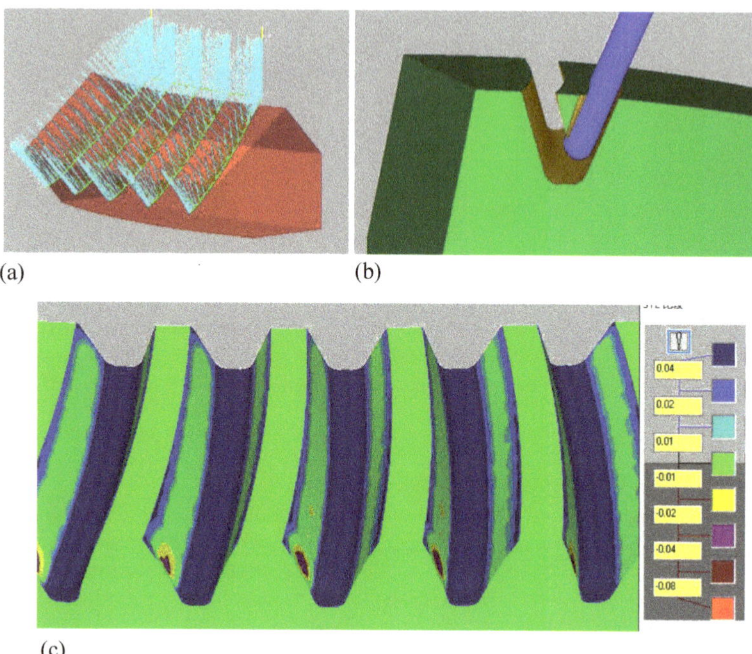

Figure 11. (a) Tool path generation by simulation; (b) simulation machining of five-axis flank milling; (c) tooth surface errors between simulated machining tooth surface and theoretical tooth surface.

Figure 12. Five-axis flank milling of SBG.

The tooth profile errors of the machined tooth surface were measured by CMM (coordinate measuring machine). According to Gleason's standard, 5×9 grid points are planned on the gear shaft section, and the tooth profile errors are calculated with the grid midpoint as the reference point, as shown in Figure 13. According to the tooth surface equation of SBG, the theoretical coordinates $\mathbf{p}$ and normal vector $\mathbf{n}$ of 5×9 grid points can be easily obtained. After the theoretical data are input into the CMM, the machine will automatically touch the tooth surface to obtain the actual measured coordinate values $\mathbf{p}_t$, and the tooth profile errors can be calculated according to Equation (18).

$$\delta = (\mathbf{p} - \mathbf{p}_t) \cdot \mathbf{n} \tag{18}$$

The tooth profile errors of concave and convex sides of tooth surfaces are shown in Figure 14a,b, respectively. The measurement results show that the errors of the middle part of the tooth height direction are relatively small, and the errors of the upper and lower sides gradually increase, which is consistent with the simulation results, and the maximum error is less than 20 μm. The errors of meshing area are larger than the simulation result, which is mainly due to the existence of certain machine tool errors and measurement errors. Referring to the five-axis machining center in Hermle, Germany, which is a world-class brand, the accuracy of machined parts is generally between 0.005 and 0.01 mm. The errors between the actual machined surface and the simulated surface are less than this order of magnitude. Therefore, the error fluctuation value is in a reasonable range, which shows the correctness of the given tool path.

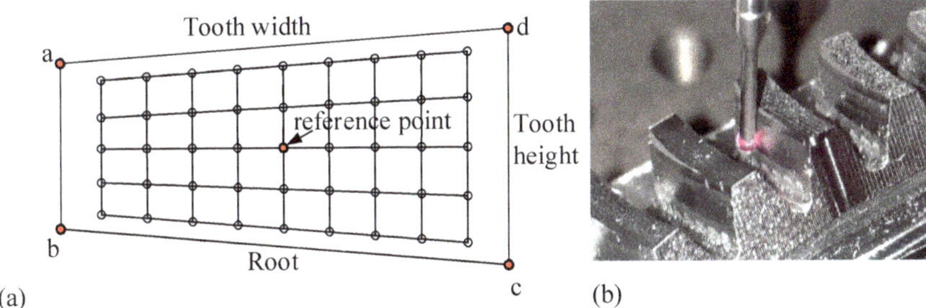

Figure 13. (**a**) Measurement mesh; (**b**) measuring tooth profile errors.

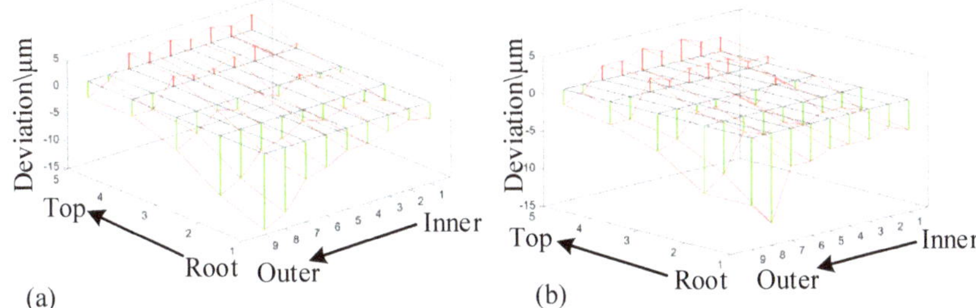

Figure 14. (**a**) Convex side; (**b**) Concave side.

6. Discussion

According to the tooth surface geometry and its meshing utilization, each side of the tooth surface of SBGs is cut by the five-axis flank milling with only one pass. A filleted end mill cutter is applied to implement the tool path planning. Due to the special requirements of gears in meshing performances, the contact area is accurately machined by the cutter flank side with a tangency condition between the cutter envelop surface and tooth surface along the contact path. The fillet part is directly machined by the cuter bottom. The tooth surface of flank milling can be obtained by two different explicit calculation methods. In some cases, one method may be more suitable than the other. By optimizing the tool path, the accuracy of tooth surface and contact performance can be improved. The machining efficiency is improved because each side of the tooth surface is machined by one pass rather than multiple passes.

In order to verify the effectiveness and authenticity of the method proposed in this paper, contact simulation analysis and flank milling experiments were carried out, respectively. The simulation results show that there is a slight edge contact between the gear and the pinion, and the contact trace is slightly closer to the inner diameter of the gear, but the overall contact area is basically towards the tooth direction. Theoretically, the contact trace should be located in the middle of the tooth surface. However, for the following two reasons, the results shown in Figure 10 are the best results that the proposed method can achieve at present.

- The contact trace of SBGs are sensitive to the topological morphology of the tooth surface. Small changes in the tooth surface may cause changes in the position and orientation of the contact trace.
- The research content of this paper aims to adopt an efficient flank milling method to ensure low tooth surface error near the contact area, which is not a real research method to fully realize the pre-control of meshing performances.

Further research is needed to realize the pre-control of meshing performances of SBG based on five-axis flank milling. The experimental results show that the tooth surface error of the meshing area obtained by NC simulation is less than 10 μm, while the maximum error of the actual machining is less than 20 μm. According to the five-axis machining center in Hermle , Germany, which is a world-class brand, the accuracy of machined parts is generally between 0.005 and 0.01 mm. The errors between the actual machined surface and the simulated surface is less than this order of magnitude. Therefore, the error fluctuation value is in a reasonable range, which shows the correctness of the given tool path.

For some cases, the high machining accuracy is required for not only the contact area, but also the other areas. For those cases, multiple passes are needed to machine the convex or concave sides of the tooth surface, and some further machining processes might also be needed to accurately machine the fillet part of tooth surface. Furthermore, future work about practical machining, measurement, and error control would also be valuable.

Author Contributions: Conceptualization, H.X. and Y.H.; Formal analysis, H.X. and Y.H.; Funding acquisition, Y.H. and Y.Z.; Investigation, H.X.; Methodology, Y.Z.; Project administration, H.X.; Software, Y.Z.; Supervision, J.T.; Validation, Y.Z. and J.T.; Visualization, J.T.; Writing—original draft, H.X.; Writing—review and editing, Y.Z. and J.T. All authors have read and agreed to the published version of the manuscript.

Funding: This research was funded by National Key R&D Project of China through Grant No. 2018YFB2001700, the National Natural Science Foundation of China through Grant No. 51705541 and State Key Laboratory of High Performance Complex Manufacturing through Grant No. Kfkt2019-08.

Institutional Review Board Statement: Not applicable.

Informed Consent Statement: Not applicable.

Data Availability Statement: The data presented in this study are available on request the corresponding author.

Conflicts of Interest: The authors declare no conflict of interest.

References

1. Tsay, C.B.; Lin, J.Y. A mathematical model for the tooth geometry of hypoid gears. *Math. Comput. Model.* **1993**, *18*, 23–34. [CrossRef]
2. Zhou, Y.; Chen, Z.C. A new geometric meshing theory for a closed-form vector representation of the face-milled generated gear tooth surface and its curvature analysis. *Mech. Mach. Theory* **2015**, *83*, 91–108. [CrossRef]
3. Shih, Y.P.; Fong, Z.H.; Lin, G. Mathematical Model for a Universal Face Hobbing Hypoid Gear Generator. *J. Mech. Des.* **2007**, *129*, 38–47. [CrossRef]
4. Vimercati, M.; Piazza, A. Applications of a mathematical model for representation of face hobbed hypoid and spiral bevel gear geometry. In Proceedings of the International Conference on Gears, Garching Germany, 14–16 September 2005; VDI-Berichte 2005; pp. 467–483.
5. Lelkes, M.; Marialigeti, J.; Play, D. Numerical Determination of Cutting Parameters for the Control of Klingelnberg Spiral Bevel Gear Geometry. *J. Mech. Des.* **2002**, *124*, 761–771. [CrossRef]
6. Achtmann, J.; Baer, G. Optimized Bearing Ellipses of Hypoid Gears. *J. Mech. Des.* **2003**, *125*, 739–745. [CrossRef]
7. Qi, F. Computerized Modeling and Simulation of Spiral Bevel and Hypoid Gears Manufactured by Gleason Face Hobbing Process. *J. Mech. Des.* **2006**, *128*, 1315–1327.
8. Simon, V.V. Design and Manufacture of Spiral Bevel Gears With Reduced Transmission Errors. *J. Mech. Des.* **2015**, *131*, 041007–041017. [CrossRef]

9. Tsiafis, I.; Mamouri, P.; Kompogiannis, S. Design and manufacturing of spiral bevel gears using CNC milling machines. In *IOP Conference Series: Materials Science and Engineering*; IOP Publishing: Bristol, UK; 2018; Volume 393, p. 012066.

10. Li, J.; Gao, Z.; Zhang, P.; Feng, L.; Yin, G.; Wang, H.; Su, J.; Ma, W. An integration method of design and machining for spiral bevel gears based on universal CNC machine tools. *J. Braz. Soc. Mech. Sci. Eng.* **2020**, *42*, 1–14. [CrossRef]

11. Wang, S.; Zhou, Y.; Tang, J.; Xiao, Z. An adaptive geometric meshing theory for the face-milled generated spiral bevel gears. *Forsch. Ing.* **2019**, *83*, 775–780. [CrossRef]

12. Shih, Y.P.; Lai, K.L.; Sun, Z.H.; Yan, X.L. Manufacture of Face-Milled Spiral Bevel Gears on a Five-Axis CNC Machine. In *Proceedings of the 14th IFToMM World Congress*; National Taiwan University: Taipei, Taiwan, 2015; pp. 328–335.

13. Shih, Y.P.; Sun, Z.H.; Lai, K.L. A flank correction face-milling method for bevel gears using a five-axis CNC machine. *Int. J. Adv. Manuf. Technol.* **2017**, *91*, 3635–3652. [CrossRef]

14. Álvarez, Á.; Calleja, A.; Ortega, N.; de Lacalle, L.N.L. Five-Axis Milling of Large Spiral Bevel Gears: Toolpath Definition, Finishing, and Shape Errors. *Metals* **2018**, *8*, 353. [CrossRef]

15. Deng, C.; Yan, H.; Chen, Y.; Yi, W. Tooth surface reconstructing method of spiral bevel gear generated by duplex helical method based on renewal Kriging model (in chinese). *J. Cent. South Univ. (Sci. Technol.)* **2019**, *50*, 1351–1356.

16. Gaiser, U. 5-Axis Gear Manufacturing Gets Practical. In *Gear Technology*; American Gear Manufacturers Association: Chicago, IL, USA, 2017.

17. Lochbihler, T. Complete machining of gears on 5X multitasking machines. In *The Proceedings of the JSME International Conference on Motion and Power Transmissions*; The Japan Society of Mechanical Engineers: Tokyo, Japan, 2017; Volume 2017, pp. 2–4.

18. Li, Z.L.; Tuysuz, O.; Zhu, L.M.; Altintas, Y. Surface form error prediction in five-axis flank milling of thin-walled parts. *Int. J. Mach. Tools Manuf.* **2018**, *128*, 21–32. [CrossRef]

19. Li, Z.L.; Zhu, L.M. Compensation of deformation errors in five-axis flank milling of thin-walled parts via tool path optimization. *Precis. Eng.* **2019**, *55*, 77–87. [CrossRef]

20. Harik, R.F.; Gong, H.; Bernard, A. 5-axis flank milling: A state-of-the-art review. *Comput. Aided Des.* **2013**, *45*, 796–808. [CrossRef]

21. Litvin, F.L.; Fuentes, A. *Gear Geometry and Applied Theory*; Cambridge University Press: Cambridge, MA, USA, 2004.

22. Bo, P.; Bartoň, M.; Plakhotnik, D.; Pottmann, H. Towards efficient 5-axis flank CNC machining of free-form surfaces via fitting envelopes of surfaces of revolution. *Comput. Aided Des.* **2016**, *79*, 1–11. [CrossRef]

23. Lu, Y.A.; Wang, C.Y.; Zhou, L. Geometric deviation evaluation for a five-axis flank milling tool path using the tool swept envelope. *Int. J. Adv. Manuf. Technol.* **2019**, *105*, 1811–1821. [CrossRef]

24. Liu, X.; Zhou, Y.; Tang, J. A comprehensive adaptive approach to calculating the envelope surface of the digital models in CNC machining. *J. Manuf. Syst.* **2020**, *57*, 119–132. [CrossRef]

25. Zhu, L.; Lu, Y. Geometric conditions for tangent continuity of swept tool envelopes with application to multi-pass flank milling. *Comput. Aided Des.* **2015**, *59*, 43–49. [CrossRef]

26. Li, Z.L.; Wang, X.Z.; Zhu, L.M. Arc–surface intersection method to calculate cutter–workpiece engagements for generic cutter in five-axis milling. *Comput. Aided Des.* **2016**, *73*, 1–10. [CrossRef]

27. Li, Z.L.; Zhu, L.M. Mechanistic modeling of five-axis machining with a flat end mill considering bottom edge cutting effect. *J. Manuf. Sci. Eng.* **2016**, *138*, 111012. [CrossRef]

28. Li, Z.L.; Zhu, L.M. An Accurate Method for Determining Cutter-Workpiece Engagements in Five-Axis Milling With a General Tool Considering Cutter Runout. *J. Manuf. Sci. Eng.* **2018**, *140*, 021001. [CrossRef]

29. Chang, Z.; Chen, Z.C.; Mo, R.; Zhang, D.; Deng, Q. An accurate and efficient approach to geometric modeling of undeformed chips in five-axis CNC milling. *Comput. Aided Des.* **2017**, *88*, 42–59. [CrossRef]

30. Chang, Z.; Chen, Z.C.; Zhao, J.; Zhang, D. A Generic Approach to Modeling Geometry of Un-Deformed Chip by Mathematical Representing Envelopes of Swept Cutter in Five-Axis CNC Milling. In Proceedings of the ASME 2014 International Design Engineering Technical Conferences and Computers and Information in Engineering Conference, Buffalo, NY, USA, 17–20 August 2014.

31. Zhou, Y.; Chen, Z.C.; Yang, X. An accurate, efficient envelope approach to modeling the geometric deviation of the machined surface for a specific five-axis CNC machine tool. *Int. J. Mach. Tools Manuf.* **2015**, *95*, 67–77. [CrossRef]

32. Wang, L.; Li, W.; Si, H.; Yuan, X.; Liu, Y. Geometric deviation reduction method for interpolated toolpath in five-axis flank milling of the S-shaped test piece. *Proc. Inst. Mech. Eng. Part B J. Eng. Manuf.* **2020**, *234*, 910–919. [CrossRef]

33. Yuan, C.; Mi, Z.; Jia, X.; Lin, F.; Shen, L. Tool orientation optimization and path planning for 5-axis machining. *J. Syst. Sci. Complex.* **2021**, *34*, 83–106. [CrossRef]

34. Mane, H.; Pande, S. Adaptive tool path planning strategy for 5-axis CNC machining of free form surfaces. In Proceedings of the ASME 2019 14th International Manufacturing Science and Engineering Conference, Erie, PA, USA, 10–14 June 2019.

35. Yan, G.; Chen, H.; Zhang, X.; Qu, C.; Ju, Z. A dimension-driven adaptive programming for tool-path planning and post-processing in 5-axis form milling of hyperboloidal-type normal circular-arc gears. *Int. J. Adv. Manuf. Technol.* **2020**, *106*, 2735–2746. [CrossRef]

36. Li, Y.; Zeng, L.; Tang, K.; Xie, C. Orientation-point relation based inspection path planning method for 5-axis OMI system. *Robot. Comput. Integr. Manuf.* **2020**, *61*, 101827. [CrossRef]

37. Tunc, L.T. Smart tool path generation for 5-axis ball-end milling of sculptured surfaces using process models. *Robot. Comput. Integr. Manuf.* **2019**, *56*, 212–221. [CrossRef]

38. Vu, D.D.; Monies, F.; Rubio, W. A new optimization tool path planning for 3-axis end milling of free-form surfaces based on efficient machining intervals. In *AIP Conference Proceedings*; AIP Publishing LLC: Melville, NY, USA, 2018; Volume 1960, p. 070011.
39. Yi, J.; Chu, C.H.; Kuo, C.L.; Li, X.; Gao, L. Optimized tool path planning for five-axis flank milling of ruled surfaces using geometric decomposition strategy and multi-population harmony search algorithm. *Appl. Soft Comput.* **2018**, *73*, 547–561. [CrossRef]
40. Hsieh, H.T.; Chu, C.H. Particle swarm optimisation (PSO)-based tool path planning for 5-axis flank milling accelerated by graphics processing unit (GPU). *Int. J. Comput. Integr. Manuf.* **2011**, *24*, 676–687. [CrossRef]
41. Wu, P.H.; Li, Y.W.; Chu, C.H. Optimized tool path generation based on dynamic programming for five-axis flank milling of rule surface. *Int. J. Mach. Tools Manuf.* **2008**, *48*, 1224–1233. [CrossRef]
42. Gong, H.; Fang, F.; Hu, X.; Cao, L.X.; Liu, J. Optimization of tool positions locally based on the BCELTP for 5-axis machining of free-form surfaces. *Comput. Aided Des.* **2010**, *42*, 558–570. [CrossRef]
43. Zhu, L.; Zheng, G.; Ding, H.; Xiong, Y. Global optimization of tool path for five-axis flank milling with a conical cutter. *Comput. Aided Des.* **2010**, *42*, 903–910. [CrossRef]
44. Gong, H.; Wang, N. 5-axis flank milling free-form surfaces considering constraints. *Comput. Aided Des.* **2011**, *43*, 563–572. [CrossRef]
45. Gong, H.; Wang, N. Optimize tool paths of flank milling with generic cutters based on approximation using the tool envelope surface. *Comput. Aided Des.* **2009**, *41*, 981–989. [CrossRef]
46. Fountas, N.A.; Vaxevanidis, N.M.; Stergiou, C.I.; Benhadj-Djilali, R. Globally optimal tool paths for sculptured surfaces with emphasis to machining error and cutting posture smoothness. *Int. J. Prod. Res.* **2019**, *57*, 5478–5498. [CrossRef]
47. Ibaraki, S.; Yoshida, I. A five-axis machining error simulator for rotary-axis geometric errors using commercial machining simulation software. *Int. J. Autom. Technol.* **2017**, *11*, 179–187. [CrossRef]
48. Li, Z.L.; Zhu, L.M. Envelope Surface Modeling and Tool Path Optimization for Five-Axis Flank Milling Considering Cutter Runout. *J. Manuf. Sci. Eng.* **2014**, *136*, 041021. [CrossRef]
49. Li, Z.L.; Niu, J.B.; Wang, X.Z.; Zhu, L.M. Mechanistic modeling of five-axis machining with a general end mill considering cutter runout. *Int. J. Mach. Tools Manuf.* **2015**, *96*, 67–79. [CrossRef]
50. Wu, B.; Liang, M.; Zhang, Y.; Luo, M.; Tang, K. Optimization of machining strip width using effective cutting shape of flat-end cutter for five-axis free-form surface machining. *Int. J. Adv. Manuf. Technol.* **2018**, *94*, 2623–2633. [CrossRef]
51. Zhang, W.; Ren, J.; Liang, Y. Optimization of the cutter size and tool orientation for reaching the critical state with multi-constraints in deep and narrow channel parts machining. *Int. J. Adv. Manuf. Technol.* **2020**, *110*, 1969–2001. [CrossRef]
52. Zhou, Y.; Chen, Z.C.; Tang, J. A new method of designing the tooth surfaces of spiral bevel gears with ruled surface for their accurate five-axis flank milling. *J. Manuf. Sci. Eng.* **2017**, *139*, 061004. [CrossRef]
53. Zhou, Y.; Chen, Z.C.; Tang, J.; Liu, S. An innovative approach to NC programming for accurate five-axis flank milling of spiral bevel or hypoid gears. *Comput. Aided Des.* **2017**, *84*, 15–24. [CrossRef]
54. Wang, S.; Zhou, Y.; Liu, X.; Liu, S.; Tang, J. An Advanced Comprehensive Approach to Accurately Modeling the Face-Milled Generated Spiral Bevel Gears. *J. Comput. Inf. Sci. Eng.* **2020**, *21*, 1–36. [CrossRef]

Article

The Correlation Analysis of Microstructure and Tribological Characteristics of In Situ VCp Reinforced Iron-Based Composite

Yun Zhang [1], Richen Lai [2], Qiang Chen [2], Zhen Liu [2], Ruiqing Li [1], Jufei Chen [1] and Pinghu Chen [2,*]

[1] State Key Laboratory of High Performance Complex Manufacturing, Light Alloys Research Institute, College of Mechanical and Electrical Engineering, Central South University, Changsha 410083, China; yun_zhang66@163.com (Y.Z.); liruiqing@csu.edu.cn (R.L.); chenjufei1988@163.com (J.C.)

[2] College of Mechatronics & Control Engineering, Shenzhen University, Shenzhen 518060, China; 1910293034@email.szu.edu.cn (R.L.); 1910293020@email.szu.edu.cn (Q.C.); liuzhen598352368@126.com (Z.L.)

* Correspondence: chenpinghu1986@163.com; Tel.: +86-137-8616-1244

Citation: Zhang, Y.; Lai, R.; Chen, Q.; Liu, Z.; Li, R.; Chen, J.; Chen, P. The Correlation Analysis of Microstructure and Tribological Characteristics of In Situ VCp Reinforced Iron-Based Composite. *Materials* **2021**, *14*, 4343. https://doi.org/10.3390/ma14154343

Academic Editor: Itzhak Green

Received: 15 July 2021
Accepted: 1 August 2021
Published: 3 August 2021

Publisher's Note: MDPI stays neutral with regard to jurisdictional claims in published maps and institutional affiliations.

Abstract: In this study, four kinds of heat treatments were performed to obtain a certain amount of retained austenite, which can result in good toughness and low brittleness accompanied with wear resistance of an in situ VC particle reinforced iron-based composite (VCFC). Microstructure, mechanical properties and wear resistance of the samples under heat treatment of QP, QPT, MQP and MQPT were compared. The experimental results indicated that there is a huge difference in microstructure between MQPT and the other heat treatments. High-proportion retained austenite and white net-like precipitates of M_7C_3 carbide existed in the MQPT-treated sample, but thick M_7C_3 carbide with brittleness was discovered in the other sample. Thereby, high-proportion retained austenite contributed to its low hardness of 634 HV and high tensile strength of 267 MPa, while a maximum hardness of 705.5 HV and a minimum tensile strength of 205 MPa were exhibited in the QPT-treated sample with a V-rich carbide of high hardness, a Cr-rich carbide of brittleness and a high-proportion martensite. Meanwhile, a phase transformation from retained austenite to martensite could increase the hardness and enhance wear resistance based on the transformation-induced plasticity (TRIP) effect; its wear rate was only 1.83×10^{-6} mm^{-3}/(N·m). However, the wear rates of the samples under QP, QPT and MQP heat treatments increased by 16.4%, 44.3% and 41.0%, respectively. The wear mechanism was a synergistic effect of the adhesive wear mechanism and the abrasive wear mechanism. The adhesive wear mechanism was mainly considered in the MQPT-treated sample to reduce the wear rate attributed to high-proportion retained austenite and the existence of wear debris with a W element on the surface of the wear track. However, the abrasive wear mechanism could exist in the other samples because of a lot of thick, brittle M_7C_3, thereby resulting in a higher wear rate due to immediate contact between the designed material and the counterpart.

Keywords: in situ vanadium carbide; wear-resistant composite; heat treatment; phase transformation; mechanical properties; tribological behavior

1. Introduction

The iron-based composites reinforced with hard ceramic particles (carbides, nitrides, metallic oxides, intermetallic compounds, etc.), which possess an excellent hardness and wear resistance, have been applied in the automobile, aerospace, mining, metallurgy and civil engineering industries [1–5]. In situ VC particle reinforced iron-based composites were developed due to spontaneous growth of vanadium carbide during solidification [6,7]. In situ VC particles are uniformly distributed in the iron matrix accompanied with their good metallurgical bonding, which results in good hardness and wear resistance [8]. In order to obtain more excellent wear resistance, a large amount of VC particles should be provided by adding a large amount of V and C elements [9,10], resulting in low toughness, high fatigue damage or an initial failure [11]; this limits the wide application of our designed material for serious impact and high wear components (wear-resistant elbow, milling cutter, washer, etc.) in engineering equipment. To solve this problem, the corresponding heat treatments were performed to improve the toughness because of the formation of retained austenite with a certain ratio [12,13], but the trade-offs of wear resistance and toughness were not realized, thus resulting in the limits of its bright applied future.

Transformation-induced plasticity (TRIP) aided steel has a comprehensive performance with high strength and excellent toughness because of an amount of metastable austenite [14,15], and it can fulfill the purpose of wear resistance-toughness trade-offs by the addition of ceramic particles and post heat treatment [16,17]. The role of retained austenite on wear resistance of nanostructured carbide-free bainitic steel was discussed by P. V. Moghaddam; the results indicated that better wear resistance was obtained because a higher work hardening capacity was increased with a high TRIP effect [18]. In addition, Q&P heat treatment was carried out to obtain the retained austenite, which contributes to high strength properties, and excellent wear resistance was caused by phase transformation of retained austenite [19]. In our previous works, a novel in situ VCp reinforced iron-based composite (VCFC) was designed based on the transformation-induced plasticity (TRIP) effect. Simultaneously, the corresponding heat treatment was performed to obtain the retained austenite (FCC) of metastable structure, which was then transformated from an FCC structure of good toughness to a BCC structure of high hardness under a certain load [20–22]. Previous results indicate that Mn-partition and C-partition can stabilize the retained austenite at room temperature accompanied with excellent wear resistance. However, the wear resistance of the samples with different heat treatments was not studied systematically in our previous works.

In this work, tempering treatment was carried out to tailor the microstructure, mechanical properties and wear resistance based on heat treatment of quenching–partitioning (QP) and Manganese partitioning–quenching–partitioning (MQP) to assess the influence of heat treatment on the microstructure, mechanical properties and wear resistance. In particular, the wear mechanism was explored in detail for different friction conditions.

2. Experimental Procedure

A VCFC was prepared by a 200 kg medium frequency electromagnetic induction furnace (HJZ-300KW, Hengjia, Dongguan, China). The chemical composition mainly included 8.1 V, 3.0 Mn, 2.8 C, 2.5 Cr, 1.5 Mo, 1.5 Ti, 1.5 Si, bal. Fe (wt.%). In order to obtain uniform distribution of all solute elements in the molten iron: Firstly, pig iron was melted in an electromagnetic induction furnace, and the slag floating on the surface of the molten iron was removed. Secondly, ferromanganese iron, ferrochromium, ferromolybdenum and ferrovanadium were gradually added into the molten iron, and then deoxidization was carried out by adding a certain amount of aluminum element. Thirdly, a titanium element was added and considered as a heterogeneous nucleating element. The temperature of the molten iron was adjusted at 1500 °C for a period of time. The molten iron with 1450 °C was poured into the sand system with 200 mm × 100 mm × 20 mm; finally, all ingots were cooled to 200 °C in the casting sand and then the air was cooled to room temperature. Afterwards, a wire-cut electric discharge machine was adopted to cut heat-treated specimens of 50 mm × 30 mm × 10 mm. Four kinds of heat treatment techniques were performed and the detailed process is presented in Table 1. After heat treatment, the samples were cut for microstructural characterization, microhardness testing, tensile measurement and friction-wear testing.

Table 1. Heat treatment schedules of the VCFC.

Heat Treatment Process	Procedure Name			
	QP	QPT	MQP	MQPT
Mn partitioning T/t	-	-	700 °C/15 min	700 °C/15 min
Austenization T/t	1000 °C/30 min	1000 °C/30 min	1000 °C/30 min	1000 °C/30 min
Quenching T/t	100 °C/5 min	100 °C/5 min	100 °C/5 min	100 °C/5 min
C partitioning T/t	360 °C/30 min	360 °C/30 min	360 °C/30 min	360 °C/30 min
Tempering T/t	-	280 °C/120 min	-	280 °C/120 min
Cooling method	in air	in air	in air	in air

All samples for microhardness, tensile strength and wear were prepared by mechanical grinding using 400, 800, 1200 and 1500 grit SiC papers. Firstly, the digital micro Vickers hardness tester (Micro Vickers HV-1000Z, MEGA INSTRUMENTS, Shanghai, China) was employed to measure the microhardness of samples with different heat treatments under a load of 1000 gf with a loading duration time of 15 s; each hardness value was obtained by on average at least five points, resulting in reaching the final value. Secondly, tensile strength was measured using universal testing machines (PWS-E100, Shenggong, Jinan, China) the detailed sizes of the samples and testing parameters have been reported in our previous work [22]. Thirdly, the friction and wear tests were carried out using a ball-on-disk machine (MFT-5000, Rtec Instrument, San Jose, CA, USA) with a φ 4 mm tungsten carbide ceramic ball as the counterpart under different wear loads of 5, 10, 20 N and different wear times of 1, 3, 5 h; meanwhile, the turning speed and radius were set to 200 r/min and 5 mm, respectively. Afterwards, 3D laser confocal scanning microscopy (VK-X200 series, Keyence Corporation of America, Itasca, IL, USA) was performed to characterize the morphology of the wear tracks and to calculate wear rate according to the equation $w = V/(L \cdot F)$; w is wear rate, $mm^3/(m \cdot N)$; V is wear volume, which is equal to $2\pi r \cdot S_{wear}$, mm^3; L represents total travel, which is $2\pi r \cdot 200$ r/min$\cdot$t, m; F is wear load, N. The samples for the friction-wear test were ground by 1500 grit sand paper, but the roughness (Ra) of the surface was still about 3 μm because of the existence of a multi-phase with an iron matrix (relatively soft) and hard carbides such as VC and M_7C_3.

Scanning electron microscope (SEM, TESCAN MIRA 3 LMH/LMU, Brno-Kohoutovice, Czech Republic) operated at 20 kV was employed to characterize the shape, size and distribution of the precipitate using the synergetic characterization of back scattered electron (BSE) and secondary electron (SE); the samples were etched with nitrohydrochloric acid (Huihong, Changsha, China) (the ratio of concentrated nitric acid and hydrochloric acid was 3:1) for 20 s. Meanwhile, an energy dispersive spectroscopy (EDS) detector was used to examine elemental distribution. In addition, phase analysis was performed by high-energy X-ray diffraction (HEXRD, Bruker, D8 discover, Karlsruhe, Germany) with a Cu target ($\lambda = 0.15418$ nm), which was operated at 45 kV and 200 mA, and the data were collected at $2\theta = 20°{\sim}100°$ using a 0.02 step size.

3. Results

The back scattered electron image and the corresponding secondary electron image of the samples with different heat treatments are shown in Figures 1 and 2. In the QP-treated sample, the morphologies mainly include globular-like black particles, irregular thick precipitates with few net-like structure and tiny white precipitates of uniform distribution in the matrix. The uneven distribution of globular-like black particles with the size 10~20 μm is presented in Figure 1a. However, there is a relatively large difference between the QP-treated sample and the other treated samples. For example, black particles are gathered to form flower-like particles when irregular thick precipitates become the thinner and more even distribution in the QPT-treated sample, as shown in Figure 1b. However, uniform distribution of black particles existed in the MQP-treated sample, and the width of irregular thick precipitates decreased accompanied with more consistency, as shown in Figure 1c. Even more specifically, irregular net-like white precipitates were detected without irregular thick precipitates in the MQPT-treated sample, as shown in Figure 1d. Furthermore, according to EDS mapping, black particles are V-rich carbide, and irregular thick precipitates and net-like white precipitates mainly include Fe, Cr and C elements, which can be considered as containing Fe, Cr carbide, as shown in Figure 2.

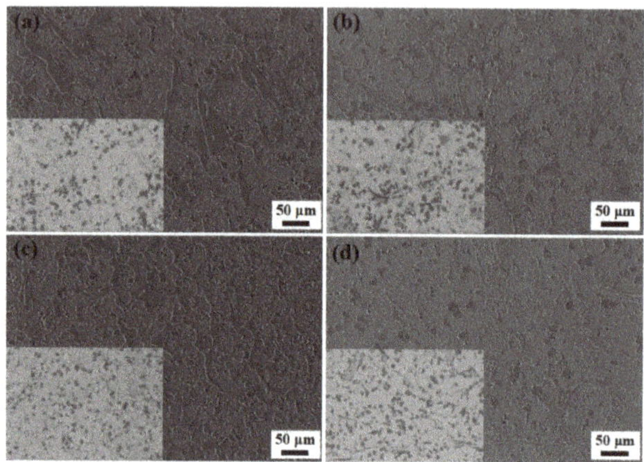

Figure 1. Secondary electron images andbBackscattered electron images of the VCFC with different heat-treated processes. (**a**) QP, (**b**) QPT, (**c**) MQP, (**d**) MQPT.

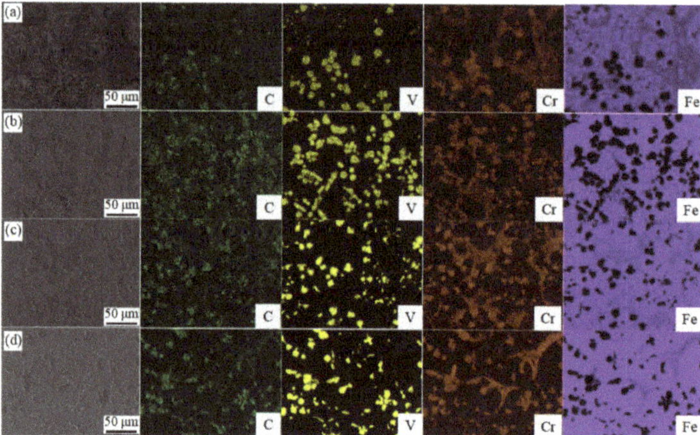

Figure 2. Main alloying elements' distribution of the VCFC with different heat-treated processes. (**a**) QP, (**b**) QPT, (**c**) MQP, (**d**) MQPT.

Combinations with the XRD patterns of the samples after different heat treatments are shown in Figure 3. It can be seen that all the samples were composed of VC, retained austenite of γ-fcc structure, martensite of α-bcc structure, M_7C_3 and M_3C. In this study, the spectra peaks of VC appeared approximately at 2theta = 37.4°, 43.4°, 63.1°, 75.7° and 79.7° for the crystallographic planes of (111), (200), (220), (311) and (222), respectively. In addition, the three strongest peaks of retained austenite appeared at 43.3°, 50.4° and 74.1° for the crystallographic planes of (111), (200) and (220), while the strongest peak of martensite was (110) at 45° approximately. It is also notable that the (110) peak intensity of the MQPT-treated sample decreased while the intensity of four peaks increased relative to the samples under other heat treatments. In addition, the peak intensity of M_7C_3 and M_3C decreased or disappeared in the MQPT-treated sample.

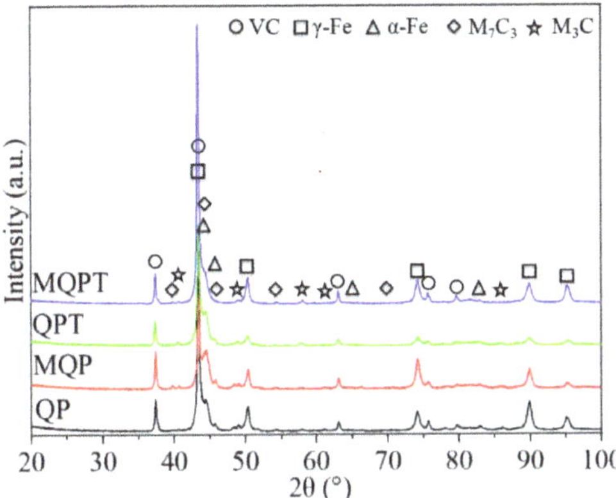

Figure 3. XRD patterns of the VCFC with different heat-treated process.

Figure 4 shows the microhardness and tensile strength of the samples with different heat treatments. The microhardness values of QPT and MQP-treated samples were larger than those under QP and MQPT heat treatments. The minimum value was discovered in the sample under MQPT heat treatment, whose value is only 634 HV and 10% lower than that under the QPT-treated sample. However, there was a relatively contradictory trend that emerged in the tensile strength, as shown in Figure 4b. The minimum value emerged in the sample under QPT heat treatment, while the tensile strength of the QP-treated sample reached a maximum value of above 300 MPa. In general, the wear resistance of the sample with larger hardness was better than that with low hardness. However, according to the results shown in Figure 5, the experimental results of friction-wear testing are not in accordance with the common practice. Figure 5a shows the friction coefficient of the samples under different heat treatments; it is obvious that the friction coefficient of the QPT-treated sample is the minimum value, which is about 0.45, while the maximum value of ~0.6 can appear in the sample under QP heat treatment. Meanwhile, the wear rates of four samples with different heat treatments were compared under the conditions of wear load 20 N and wear time 5 h, as shown in Figure 5b; the wear rate of QP, QPT, MQP and MQPT-treated samples first rose then descended; the maximum value of 2.64×10^{-6} mm^3/(N·m) appeared in the sample under QPT heat treatment; at the same time, the wear rate of the MQPT-treated sample had a minimum value of 1.83×10^{-6} mm^3/(N·m). The tribological behaviors of the MQPT-treated sample were emphasized in the analysis under different wear loads and different wear times. Figure 5c indicates that the wear rate gradually decreased with the increase of the wear load—the wear rate under wear load of 5 N is 5.01×10^{-6} mm^3/(N·m)—but the wear rates of 10 and 20 N decreased by 32% and 63%. In addition, the same tendency existed in the wear rate with the increase of wear time; the wear rate reaches 6.88×10^{-6} mm^3/(N·m) when the wear time is 1 h, but the wear rate can be decreased about 3.7 times when the wear time is 5 h.

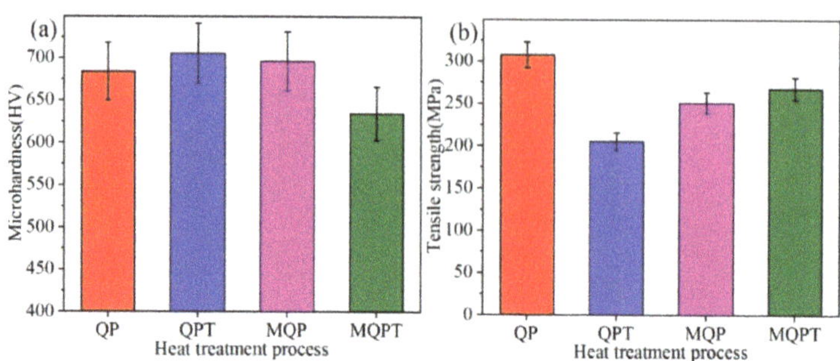

Figure 4. Microhardness (**a**) and tensile strength (**b**) of the VCFC with different heat-treated processes.

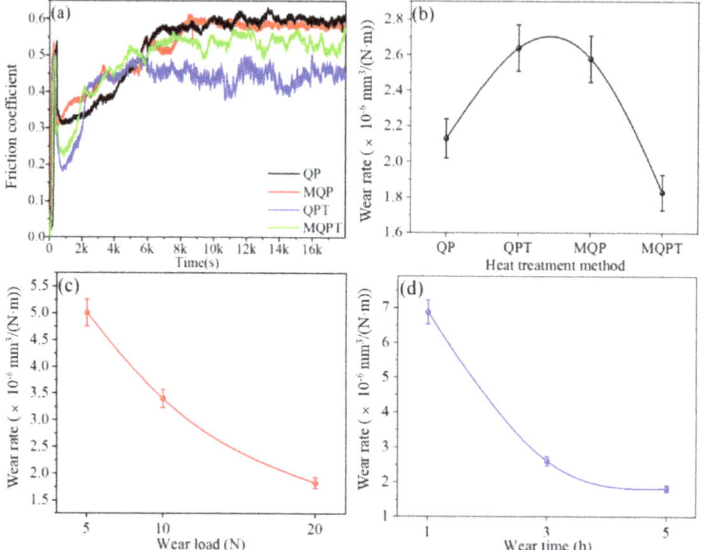

Figure 5. Friction coefficient, microhardness and tensile strength of the VCFC with different heat-treated processes. (**a**) The friction coefficient; (**b**) Wear rate of four specimens under wear load 20 N and wear time 5 h; (**c**) Wear rate of MQPT specimen under wear time 5 h and different wear load; (**d**) Wear rate of MQPT specimen under wear load 20 N and different wear time.

Combined with 3D laser confocal scanning microscopy of the samples under different heat treatments, outline heights were magnified 10 times, as shown in Figures 6 and 7. Figure 6 shows that, for 3D morphology of wear tracks with different heat treatments, the depth of the wear track with the QPT-treated sample is the maximum, and its value can reach about 15 μm under wear load of 20 N and wear time of 5 h, and the deepest region is at the middle of wear track; there is a same phenomenon that appears in the wear track of the MQPT-treated sample, but the depth is only 8 μm and the width is relatively narrow. In the samples of QP and MQP heat treatments, the deepest region is not in the middle of wear tracks but at the edge of wear tracks, as shown in Figure 6a,c; their maximum depth is approximately 10 μm, but the deeper region of the MQP-treated sample is larger than that of the QP-treated sample. Furthermore, the 3D morphology of the wear track for the MQPT-treated sample is discussed emphatically under different wear loads and times, as shown in Figure 7. Due to the low wear load of 5 N, the rough surface still existed at the region of wear track, as shown in Figure 7a, but the rough surface was removed by the counterpart under a wear load of 10 N and multiple deep channels were discovered in the wear track. With the increase of wear load, there was a single channel at the middle of wear track and its depth was larger, as shown in Figure 7c. However, the deep channel was the first to be discovered at the wear time of 1 h for a wear load of 20 N. and the channel became deeper when accompanied with a narrow channel, as shown in Figure 7e, when the wear time was increased to 3 h. Finally, two channels merged to form the single wider channel.

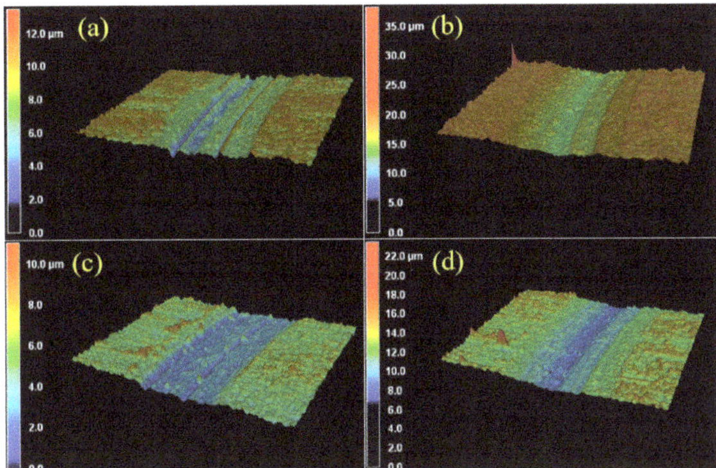

Figure 6. Three-dimensional laser confocal images of the wear tracks with different heat treatment processes under the conditions of wear load 20 N and wear time 5 h. (**a**) QP; (**b**) QPT; (**c**) MQP; (**d**) MQPT.

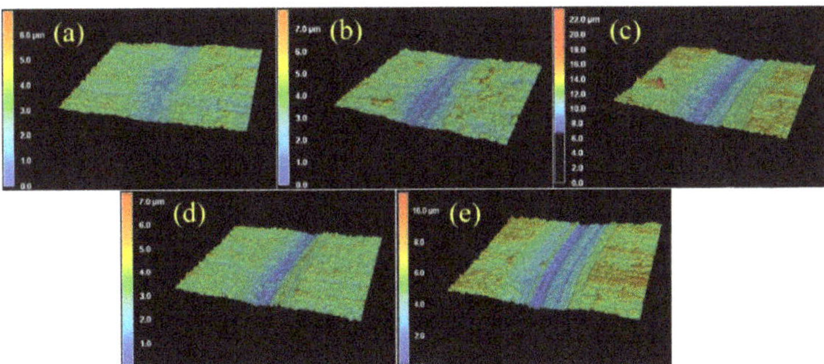

Figure 7. Three-dimensional laser confocal images of the wear tracks of MQPT specimen under different wear loads and wear times. (**a**) 5N, 5h; (**b**) 10N, 5h; (**c**) 20N, 5h; (**d**) 20N, 1h; (**e**) 20N, 3h.

4. Discussion

VC carbide was in situ formed spontaneously above 1400 °C because of negative Gibbs free energy with $\triangle G_0(kJ/mol) = -102.090 + 0.00958T$ (298 K < T < 2273 K) [23–25], but the low cooling rate of the designed material in clay sand can result in multiple phases, such as M_7C_3, $M_{23}C_6$, M_3C carbides, iron matrix, etc. [26,27]. Thereby, the microstructure was mainly compared V-rich carbide, Cr-rich carbide and iron matrix of pearlite and martensite structure, but austenite was never discovered in an as-cast sample [20]. However, according to the calculated results of JMatPro software (Sente, Guildford, UK, version 4.0), the austenite ratio has a maximum value at 1261 °C and starts to rapidly decrease until it disappears completely at 697.2 °C. Therefore, the austenitizing temperature was designed and optimized at 1000 °C, followed by quenching at 100 °C, thus resulting in a certain amount of retained austenite at room temperature [21]; Z.C. Li reported that retained austenite was more stabilized when Mn and C elements dissolved into an austenite structure [28]. Therefore, an Mn-partitioning treatment made retained austenite more stable because it suppresses rapid diffusion of alloying elements in the post heat treatment [22]. However, unfortunately, some brittle precipitates were formed and resulted in weakening mechanical properties and wear resistance. In this work, tempering treatment was performed after QP or MQP treatment, and it was noted that tempering treatment with a long time has an important role in the redistribution of M_7C_3 or M_3C, thus resulting in their width being reduced or disappearing. Typical microstructures for QPT and MQPT-treated samples are shown in Figure 8. The same microstructures of QP and MQP-treated samples exist in the QPT-treated sample, and are composed of VC, M_7C_3, M_3C, a large amount of martensite and a certain retained austenite. The existence of VC, M_7C_3 and a large amount of martensite contributes to high hardness, as shown in Figure 4a. However, the brittle M_7C_3 was easily broken in the process of polishing with glass paper; weakening tensile strength is presented in Figure 4b. By contrast, thick M_7C_3 and tiny M_3C particles were never discovered and net-like M_3C appeared in the MQPT-treated sample accompanied with a large amount of retained austenite, which can result in low hardness [29], and strain-induced transformation from retained austenite to martensite can occur during the tensile process to improve tensile strength [30], as shown in Figures 4, 5 and 8.

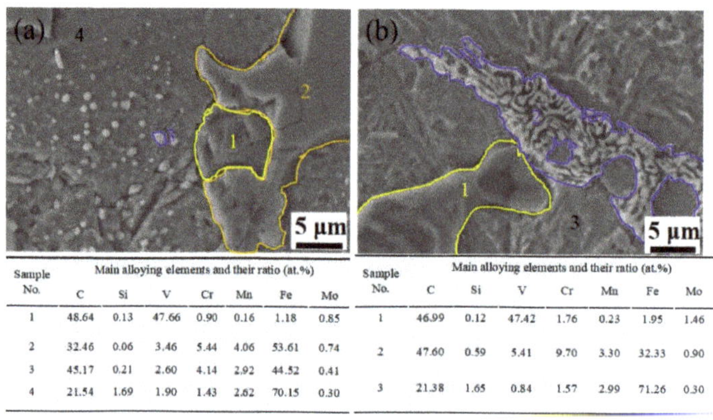

Sample No.	Main alloying elements and their ratio (at.%)						
	C	Si	V	Cr	Mn	Fe	Mo
1	48.64	0.13	47.66	0.90	0.16	1.18	0.85
2	32.46	0.06	3.46	5.44	4.06	53.61	0.74
3	45.17	0.21	2.60	4.14	2.92	44.52	0.41
4	21.54	1.69	1.90	1.43	2.62	70.15	0.30

Sample No.	Main alloying elements and their ratio (at.%)						
	C	Si	V	Cr	Mn	Fe	Mo
1	46.99	0.12	47.42	1.76	0.23	1.95	1.46
2	47.60	0.59	5.41	9.70	3.30	32.33	0.90
3	21.38	1.65	0.84	1.57	2.99	71.26	0.30

Figure 8. Typical microstructures and their elements ratio under different heat treatment processes. (**a**) QPT; (**b**) MQPT.

For the results of microhardness and tensile strength, an unusual phenomenon with high hardness and lower tensile strength appeared in the QPT sample. This can be caused by two reasons. On the one hand, a more brittle non-uniform M_7C_3 carbide existed in the QPT sample, as shown in Figure 1b. Moreover, as shown in Figure 8a, these brittle carbides were broken under a certain load. On the other hand, the ratio of retained austenite was less than that of other samples, the probability of crack growth increased with a decrease in retained austenite. Therefore, the phenomenon of high hardness and lower tensile strength can occur in the QPT sample.

In addition, wear resistance was enhanced by a synergistic effect between microstructure and the TRIP effect. Generally speaking, excellent wear resistance of the materials was attributed to hardness and friction coefficient, but in this work the hardness of the material was a very important factor. S.Z. Wei et al. noted that morphology features (shape, size and distribution) of the carbides played a dominant role in tribological behaviors when hardness was larger than 58 HRC (655 HV); in contrast, the average hardness of the material was the most significant factor when hardness was less than 58 HRC (655 HV) [31,32]. Concerning tribological features of the samples under different heat treatments in this work, the hardness of QP, MQP and QPT-treated samples were higher than 58 HRC (655 HV) through the conversion between Vickers hardness (HV) and Rockwell hardness (HRC) when the hardness of the MQPT-treated sample was lower than 58 HRC (655 HV), as shown in Figure 4a. Combined with microstructure in Figures 1–3, a large amount of VC, M_7C_3 and M_3C carbide existed in the QP, MQP and QPT-treated samples, but brittle M_7C_3 carbide was easily broken by pressure from a relatively large load, thus resulting in weak wear resistance and high wear rate. Besides that, an amount of M_7C_3 with brittle and high hardness was uniformly distributed in QP and MQP samples. During the process of friction and wear, a serious wear also occurred in counterparts when the tested samples were worn seriously; hard debris existed into wear tracks and was pushed to the edge, followed by what is considered as abrasive particles taking part in the friction and wear, so much so that two channels appeared at the edge of wear tracks for QP and QPT samples. Nevertheless, brittle M_7C_3 carbide was never detected in the MQPT-treated sample and Cr, V, Fe-rich net-like precipitates were discovered; while the hardness decreased, wear resistance was enhanced because of the occurrence of martensitic phase transformation.

Figure 5 shows that there was excellent wear resistance in the MQPT-treated sample while serious wear appeared in the QPT-treated sample. The wear mechanism of VCFC has a synergistic effect concerning the adhesive wear mechanism and the abrasive wear mechanism. However, the adhesive wear mechanism played a dominant role in the MQPT-treated sample; W-rich material was peeled from the counterpart to cover the surface of the wear track, but serious wear occurred in several channels in contact with the counterpart, as shown in Figure 9. However, the abrasive wear mechanism played a dominant role in the QPT-treated sample because of the existence of brittle M_7C_3 carbide, as shown in Figure 10. Some VC carbides were enclosed in brittle M_7C_3 carbide, as shown in Figure 8a; brittle M_7C_3 carbide was easily broken and peeled to form a crater around the VC particle, and the debris of the brittle M_7C_3 carbide was served as abrasive particles to flatten out on the surface of the wear track; therefore, serious wear occurred in the QPT-treated sample. On the other hand, metastable retained austenite played an important role in strengthening wear resistance. According to XRD patterns, a higher ratio of retained austenite existed in the MQPT-treated sample, and retained austenite transformed into martensite could reduce the stress concentration, and retained austenite could be considered as a soft phase, which could possess a higher absorbing energy to coalesce with the defects from other phases, thus hindering crack growth during the loading process [33,34]. Meanwhile, the phase transformation from retained austenite to martensite occurred and increased the hardness, thus resulting in excellent wear resistance and low wear rate [35–37].

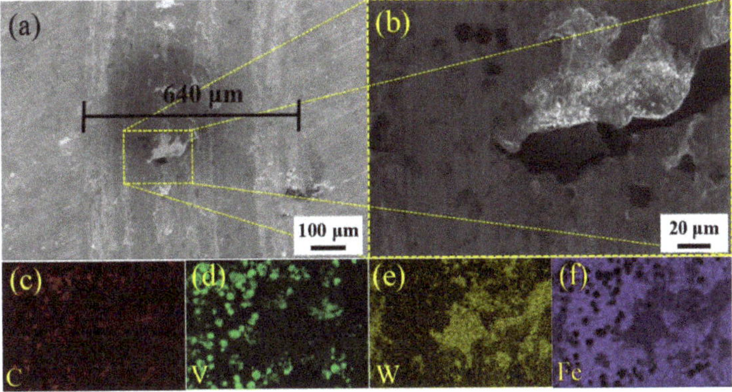

Figure 9. SEM morphologies and the corresponding EDS elemental distribution of wear tracks in the MQPT-treated sample. (**a**) SEM image of wear track; (**b**) Magnified morphology of (**a**) in the yellow box with the dotted line; (**c**) Distribution of C element; (**d**) Distribution of V element; (**e**) Distribution of W element; (**f**) Distribution of Fe element.

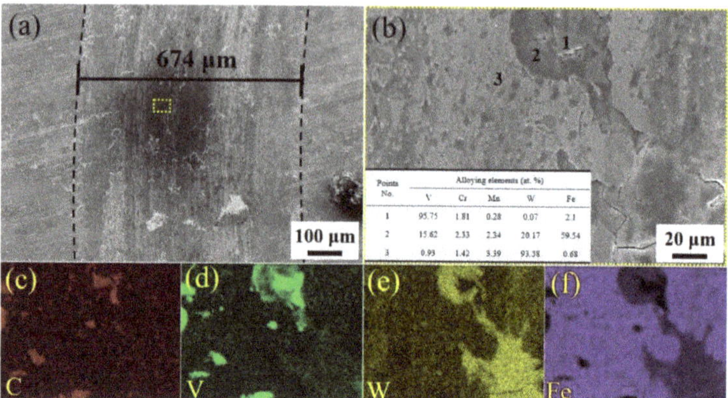

Points No.	Alloying elements (at. %)				
	V	Cr	Mn	W	Fe
1	95.75	1.81	0.28	0.07	2.1
2	15.62	2.33	2.34	20.17	59.54
3	0.93	1.42	3.39	93.38	0.68

Figure 10. SEM morphologies and the corresponding EDS elemental distribution of wear tracks in the QPT-treated sample. (**a**) SEM image of wear track; (**b**) Magnified morphology of (**a**) in the yellow box with the dotted line; (**c**) Distribution of C element; (**d**) Distribution of V element; (**e**) Distribution of W element; (**f**) Distribution of Fe element.

For tribological behaviors of MQPT-treated sample, wear rate under wear load and wear time exhibited a downward trend in a parabola, as shown in Figure 5c,d. This could be attributed to two factors: Hard VC particles and TRIP effect. Firstly, a serious wear occurred in the initial stage because of the rough surface and low hardness. In terms of time, soft retained austenite was firstly ground out from the convex surface, and then martensite was also pushed out, thus resulting in the exposure of VC particles on the surface of the wear track. Serious wear occurred in measured material and the counterpart. Consequently, debris contained a large amount of W element from the counterpart apart from Fe and V elements of VCFC. In this stage, the retained austenite played little or no role in enhancing wear resistance because there was no time for the retained austenite to transform to bcc-structure martensite of high hardness, thus resulting in a higher wear rate and entering the stable stage. Secondly, the slope of the wear rate decreased gradually because of strain-induced work hardening and complete contact between the counterpart and measured material at the stable stage, thus resulting in excellent wear resistance because the high ratio of the bcc structure with a high hardness existed in the matrix, which was attributed to phase transformation of the retained austenite with the increase of wear load and wear time [38].

5. Conclusions

In this work, different heat treatments were employed to tailor the microstructure, mechanical properties and wear resistance of VCFC; meanwhile, SEM, EDS, XRD, EBSD and TEM were employed to evaluate synthetically the relationship between microstructure and mechanical properties and wear resistance of the samples under different heat treatments. The main conclusions are as follows:

(1) Metastable retained austenite with fcc structure could be stabilized at room temperature with an appropriate process of post heat treatments; the volume content of retained austenite under MQPT-treated sample had a maximum value of 25.7%, which is 56.7% higher than that of the QPT-treated sample.

(2) Morphology features of VC and M_7C_3 carbide changed using different heat-treated processes. Phase types did not change while the shape, size and distribution of carbide changed in QP, MQP and QPT-treated samples. However, in the MQPT-treated sample, M_7C_3 carbide and M_3C disappeared, accompanied with the formation of net-like precipitates.

(3) The maximum hardness of 705 HV appeared in the QPT-treated sample accompanied with the minimum tensile strength; this was attributed to a large amount of hard VC and brittle M_7C_3 carbide. However, abundant retained austenite contributed to a minimum hardness of 634 HV and relatively higher tensile strength because the retained austenite possessed low hardness and good toughness to reduce the possibility of cracking during the loading process.

(4) The average hardness increased because phase transformation from retained austenite to martensite occurred during the friction-wear process, thus resulting in excellent wear resistance and lower wear rate with the increase of wear load and wear time; the wear rate of the MQPT-treated sample was only 1.83×10^{-6} mm^{-3}/(N × m) under wear load of 20 N and wear time of 5 h when the wear rate of the QPT-treated sample was 2.64×10^{-6} mm^{-3}/(N × m).

Author Contributions: Conceptualization, P.C. and Y.Z.; methodology, P.C.; software, Z.L.; validation, P.C., Y.Z., R.L. (Richen Lai), Q.C., Z.L. and R.L. (Ruiqing Li); formal analysis, Y.Z. and Z.L.; investigation, R.L. (Richen Lai) and Q.C.; resources, P.C.; data curation, Y.Z. and P.C.; writing—original draft preparation, P.C. and Y.Z.; writing—review and editing, R.L. (Ruiqing Li) and J.C.; visualization, Y.Z.; supervision, J.C.; project administration, P.C.; funding acquisition, P.C., Y.Z. and R.L. (Ruiqing Li).All authors have read and agreed to the published version of the manuscript.

Funding: This work was funded by Open Research Fund of State Key Laboratory of High-Performance Complex Manufacturing, Central South University (Grant No. Kfkt2020-01), Guangdong Basic and Applied Basic Research Foundation (Grant No. 2019A1515011858). Meanwhile, the work was also supported by the Science and Technology Innovation Program of Hunan Province (Grant No. 2020RC2002), Hunan Provincial Natural Science Foundation of China (Grant No. 2019JJ50807) and the Project of State Key Laboratory of High Performance Complex Manufacturing, Central South University (Grant No. ZZYJKT2021-01).

Institutional Review Board Statement: Not applicable.

Informed Consent Statement: Not applicable.

Data Availability Statement: The data presented in this study are available on request from the corresponding author.

Acknowledgments: The authors would like to thank Y.F. (Yecheng Fan) and X.Z. (Xiaobin Zhou) from Shiyanjia Lab (www.shiyanjia.com) for the XRD intensive training.

Conflicts of Interest: The authors declare no conflict of interest.

References

1. Sharma, D.; Mahant, D.; Upadhyay, G. Manufacturing of metal matrix composites: A state of review. *Mater. Today Proc.* **2020**, *26*, 506–519. [CrossRef]

2. Qiu, B.; Xing, S.; Dong, Q. Fabrication and wear behavior of ZTA particles reinforced iron matrix composite produced by flow mixing and pressure compositing. *Wear* **2019**, *428–429*, 167–177. [CrossRef]

3. Song, B.; Wang, Z.; Yan, Q.; Zhang, Y.; Zhang, J.; Cai, C.; Wei, Q.; Shi, Y. Integral method of preparation and fabrication of metal matrix composite: Selective laser melting of in-situ nano/submicro-sized carbides reinforced iron matrix composites. *Mater. Sci. Eng. A* **2017**, *707*, 478–487. [CrossRef]

4. Chen, H.; Lu, Y.; Sun, Y.; Wei, Y.; Wang, X.; Liu, D. Coarse TiC particles reinforced H13 steel matrix composites produced by laser cladding. *Surf. Coat. Technol.* **2020**, *395*, 125867. [CrossRef]

5. Xiang, S.; Ren, S.; Liang, Y.; Zhang, X. Fabrication of titanium carbide-reinforced iron matrix composites using electropulsing-assisted flash sintering. *Mater. Sci. Eng. A* **2019**, *768*, 138459. [CrossRef]

6. Paraye, N.; Ghosh, P.; Das, S. A novel approach to synthesize surface composite by in-situ grown VC reinforcement in steel matrix via TIG arcing. *Surf. Coat. Technol.* **2020**, *399*, 126129. [CrossRef]

7. Wang, Y.; Ding, Y.; Wang, J.; Cheng, F.; Shi, J. In situ production of vanadium carbide particulates reinforced iron matrix surface composite by cast-sintering. *Mater. Des.* **2007**, *28*, 2202–2206. [CrossRef]

8. Moghaddam, E.; Karimzadeh, N.; Varahram, N.; Davami, P. Impact-abrasion wear characteristics of in-situ VC-reinforced austenitic steel matrix composite. *Mater. Sci. Eng. A* **2013**, *585*, 422–429. [CrossRef]

9. Zhong, L.; Ye, F.; Xu, Y.; Li, J. Microstructure and abrasive wear characteristics of in situ vanadium carbide particulate-reinforced iron matrix composites. *Mater. Des.* **2014**, *54*, 564–569. [CrossRef]

10. Zhong, L.; Hojamberdiev, M.; Ye, F.; Wu, H.; Xu, Y. Fabrication and microstructure of in situ vanadium carbide ceramic particulates-reinforced iron matrix composites. *Ceram. Int.* **2013**, *39*, 731–736. [CrossRef]
11. Xu, L.; Wei, S.; Xing, J.; Long, R. Effects of carbon content and sliding ratio on wear behavior of high-vanadium high-speed steel (HVHSS) under high-stress rolling-sliding contact. *Tribol. Int.* **2014**, *70*, 34–41. [CrossRef]
12. Xu, L.; Xing, J.; Wei, S.; Zhang, Y.; Long, R. Optimization of heat treatment technique of high-vanadium high-speed steel based on back-propagation neural networks. *Mater. Des.* **2007**, *28*, 1425–1432. [CrossRef]
13. Xu, L.; Xing, J.; Wei, S.; Zhang, Y.; Long, R. Study on relative wear resistance and wear stability of high-speed steel with high vanadium content. *Wear* **2007**, *262*, 253–261. [CrossRef]
14. Speer, J.; Matlock, D.; De Cooman, B.; Schroth, J. Carbon partitioning into austenite after martensite transformation. *Acta Mater.* **2003**, *51*, 2611–2622. [CrossRef]
15. Wu, X.; Yang, M.; Yuan, F.; Chen, L.; Zhu, Y. Combining gradient structure and TRIP effect to produce austenite stainless steel with high strength and ductility. *Acta Mater.* **2016**, *112*, 337–346. [CrossRef]
16. Shen, Y.; Qiu, L.; Sun, X.; Zuo, L.; Liaw, P.; Raabe, D. Effects of retained austenite volume fraction, morphology, and carbon content on strength and ductility of nanostructured TRIP-assisted steels. *Mater. Sci. Eng. A* **2015**, *636*, 551–564. [CrossRef]
17. Zhang, S.; Wang, P.; Li, D.; Li, Y. Investigation of the evolution of retained austenite in Fe-13%Cr-4%Ni martensitic stainless steel during intercritical tempering. *Mater. Des.* **2015**, *84*, 385–394. [CrossRef]
18. Moghaddam, P.; Hardell, J.; Vuorinen, E.; Prakash, B. The role of retained austenite in dry rolling/sliding wear of nanostructured carbide-free bainitic steels. *Wear* **2019**, *428–429*, 193–204. [CrossRef]
19. Lu, J.; Yu, H.; Kang, P.; Duan, X.; Song, C. Study of microstructure, mechanical properties and impact-abrasive wear behavior of medium-carbon steel treated by quenching and partitioning (Q&P) process. *Wear* **2018**, *414–415*, 21–30. [CrossRef]
20. Chen, P.; Li, Y.; Li, R.; Jiang, R.; Zeng, S.; Li, X. Microstructure, mechanical properties, and wear resistance of VCp-reinforced Fe-matrix composites treated by Q&P process. *Int. J. Miner. Metall. Mater.* **2018**, *25*, 1060–1069. [CrossRef]
21. Chen, P.; Zhang, Y.; Li, R.; Liu, Y.; Zeng, S. Influence of carbon-partitioning treatment on the microstructure, mechanical properties and wear resistance of in situ VCp-reinforced Fe-matrix composite. *Int. J. Miner. Metall. Mater.* **2020**, *27*, 100–111. [CrossRef]
22. Chen, P.; Zhang, Y.; Zhang, Z.; Li, R.; Zeng, S. Tuning the microstructure, mechanical properties, and tribological behavior of in-situ VCp-reinforced Fe-matrix composites via manganese-partitioning treatment. *Mater. Today Commun.* **2020**, *24*, 101135. [CrossRef]
23. Li, X.; Zhang, C.; Zhang, S.; Wu, C.; Zhang, J.; Chen, H.; Abdullah, A. Design, preparation, microstructure and properties of novel wear-resistant stainless steel-base composites using laser melting deposition. *Vacuum* **2019**, *165*, 139–147. [CrossRef]
24. Wu, Q.; Li, W.; Zhong, N.; Wang, G.; Wang, H. Microstructure and wear behavior of laser cladding VC-Cr$_7$C$_3$ ceramic coating on steel substrate. *Mater. Des.* **2013**, *49*, 10–18. [CrossRef]
25. Cheng, F.; Wang, Y.; Yang, T. Microstructure and wear properties of Fe-VC-Cr$_7$C$_3$ composite coating on surface of cast steel. *Mater. Charact.* **2008**, *59*, 488–492. [CrossRef]
26. Geng, B.; Li, Y.; Zhou, R.; Wang, Q.; Jiang, Y. Formation mechanism of stacking faults and its effect on hardness in M$_7$C$_3$ carbides. *Mater. Charact.* **2020**, *170*, 110691. [CrossRef]
27. Zhang, J.; Li, J.; Shi, C.; Huang, J. Growth and agglomeration behaviors of eutectic M$_7$C$_3$ carbide in electroslag remelted martensitic stainless steel. *J. Mater. Res. Technol.* **2021**, *11*, 1490–1505. [CrossRef]
28. Li, Z.; Ding, H.; Misra, R.; Cai, Z. Microstructure-mechanical property relationship and austenite stability in medium-Mn TRIP steels: The effect of austenite-reverted transformation and quenching-tempering treatments. *Mater. Sci. Eng. A* **2017**, *682*, 211–219. [CrossRef]
29. Moghaddam, P.; Hardell, J.; Vuorinen, E.; Prakash, B. Effect of retained austenite on adhesion-dominated wear of nanostructured carbide-free bainitic steel. *Tribol. Int.* **2020**, *150*, 106348. [CrossRef]
30. Wang, M.; Huang, M. Abnormal TRIP effect on the work hardening behavior of a quenching and partitioning steel at high strain rate. *Acta Mater.* **2020**, *188*, 551–559. [CrossRef]
31. Wei, S.; Zhu, J.; Xu, L. Research on wear resistance of high speed steel with high vanadium content. *Mater. Sci. Eng. A* **2005**, *404*, 138–145. [CrossRef]
32. Xu, L.; Wei, S.; Xiao, F.; Zhou, H.; Zhang, G.; Li, J. Effects of carbides on abrasive wear properties and failure behaviours of high speed steels with different alloy element content. *Wear* **2017**, *376–377*, 968–974. [CrossRef]
33. Reyna, S.; Bedolla-Jacuinde, A.; Guerra, F.; Mejía, I.; García, M. Effect of amount and distribution of primary TiC on the wear behavior of a 12%Cr-3%C white iron under dry sliding conditions. *Wear* **2021**, 203718. [CrossRef]
34. Liu, B.; Li, W.; Lu, X.; Jia, X.; Jin, X. The effect of retained austenite stability on impact-abrasion wear resistance in carbide-free bainitic steels. *Wear* **2019**, *428–429*, 127–136. [CrossRef]
35. Zhang, W.; Liu, Z.; Wang, G. Martensitic transformation induced by deformation and work-hardening behavior of high manganese trip steels. *Acta Metall. Sin. Engl. Lett.* **2010**, *46*, 1230–1236. [CrossRef]
36. Luo, Q.; Li, J.; Yan, Q.; Li, W.; Gao, Y.; Kitchen, M.; Bowen, L.; Farmilo, N.; Ding, Y. Sliding wear of medium-carbon bainitic/martensitic/austenitic steel treated by short-term low-temperature austempering. *Wear* **2021**, *476*, 203732. [CrossRef]
37. Wang, Y.; Song, R.; Huang, L. The effect of retained austenite on the wear mechanism of bainitic ductile iron under impact load. *J. Mater. Res. Technol.* **2021**, *11*, 1665–1671. [CrossRef]
38. Yan, X.; Hu, J.; Yu, H.; Wang, C.; Xu, W. Unraveling the significant role of retained austenite on the dry sliding wear behavior of medium manganese steel. *Wear* **2021**, *476*, 203745. [CrossRef]

 *materials*

Communication

High Temperature Oxidation Behavior of an Equimolar Cr-Mn-Fe-Co High-Entropy Alloy

Lin Wang [1,2,3], Quanqing Zeng [3], Zhibao Xie [1,2,3], Yun Zhang [1,2,3,*] and Haitao Gao [1,2,3,*]

[1] College of Mechanical and Electrical Engineering, Central South University, Changsha 410083, China; lin_wang@csu.edu.cn (L.W.); xiezhibao@csu.cn (Z.X.)

[2] State Key Laboratory of High Performance Complex Manufacturing, Central South University, Changsha 410083, China

[3] Light Alloys Research Institute, Central South University, Changsha 410083, China; 173811017@csu.edu.cn

* Correspondence: yun_zhang66@163.com (Y.Z.); 13709822710@163.com (H.G.); Tel.: +86-151-1633-7996 (Y.Z.); +86-137-0982-2710 (H.G.)

Abstract: The oxidation behavior of an equimolar Cr-Mn-Fe-Co high-entropy alloy (HEA) processed by 3D laser printing was investigated at 700 °C and 900 °C. The oxidation kinetics of the alloy followed the parabolic rate law, and the oxidation rate constant increased with the rising of the temperature. Inward diffusion of oxygen and outward diffusion of cations took place during the high-temperature oxidation process. A spinel-type oxide was formed on the surface, and the thickness of the oxide layer increased with the rising of experimental temperature or time. The exfoliation of the oxide layer took place when the test was operated at 900 °C over 12 h. During oxidation tests, the matrix was propped open by oxides and was segmented into small pieces. The formation of loose structures had great effects on the high-temperature oxidation resistance of the HEA.

Keywords: metals and alloys; 3D printing; oxidation

Citation: Wang, L.; Zeng, Q.; Xie, Z.; Zhang, Y.; Gao, H. High Temperature Oxidation Behavior of an Equimolar Cr-Mn-Fe-Co High-Entropy Alloy. *Materials* **2021**, *14*, 4259. https://doi.org/10.3390/ma14154259

Academic Editor: Jana Bidulská

Received: 2 July 2021
Accepted: 29 July 2021
Published: 30 July 2021

Publisher's Note: MDPI stays neutral with regard to jurisdictional claims in published maps and institutional affiliations.

1. Introduction

After originally proposed by Yeh et al. [1], high-entropy alloys (HEA) have attracted increasing attention due to their excellent mechanical properties [2,3], and corrosion resistance [4,5], etc. Gludovatz et al. [6] reported a deformation mechanism for Cr-Mn-Fe-Co-Ni HEA transiting from planar-slip dislocation activity to nanotwinning as the temperature decreased from room temperature to cryogenic temperature, with cryogenic temperature offering better mechanical properties. Otto et al. [7] systemically investigated the relationship between temperatures (between −196 and 800 °C), microstructures and the tensile properties of a Cr-Mn-Fe-Co-Ni HEA. The alloy showed a significant improvement in yield strength, fracture elongation and ultimate tensile strength with the decreasing temperature. What's more, some deformation-induced twins were observed as the tensile test was interrupted after more than 20% strain when the test temperature was lower from 20 °C to −196 °C. Yong et al. [8] proposed that the separation of nanoscale phase could significantly promote the yield strength of Mn-Fe-Co-Ni-Cu alloy. Keil et al. [9] systematically researched the equiatomic composition for mechanical properties optimum of (Cr-Mn-Fe-Co)x-Ni1-x and quantitatively investigated the saturation grain size, hardness and strain rate sensitivity with the variation of Ni element composition.

High-temperature resistant materials have been widely used in the nuclear industry, civil industry, military industry, etc. However, there were little researches focused on the oxidation behavior of HEA. Kai et al. [10] investigated the oxidation behavior of an equimolar Cr-Mn-Fe-Co-Ni HEA at 950 °C in various oxygen-containing atmospheres with $PO_2 = 10$, 1.0×10^3, 2.1×10^4 and 1.0×10^5 Pa, respectively. They found that the oxidation kinetics of the alloy followed a single-stage parabolic rate curve with raising oxygen pressure. While the generated triplex scales consisted of an exclusive Mn_3O_4 outer layer, a hetero phasic intermediate layer of Mn_3O_4, $(Mn, Cr)_3O_4$, and Cr_2O_3, and an exclusive Cr_2O_3 inner layer. Gorr et al. [11] studied the microstructure and high-temperature oxidation behavior of Mo-W-Al-Cr-Ti HEA which was processed by casting and heat treatment. The results revealed that the mass gain obeyed a parabolic rate law when this alloy was exposed to air at 1000 °C for 40 h, indicating that the oxide scale grew by solid-state diffusion. Pouraliakbar et al. [12] attempted to produce novel high-temperature materials by increasing the content of AlTiZr and CuFeMo in the multicomponent alloying system $(CoCrNi)_{1-x-y}(AlTiZr)_x(CuFeMo)_y$. Yong et al. [13] researched the high-temperature stability of Cu-rich filaments and discovered that the dual fcc phase alloy (e.g., Cr-Fe-Co-Cu-Ni and Cr-Fe-Co-Cu1.71-Ni HEA) owed excellent strengths even after annealing at 1000 °C. This was owing to the low internal strain energy which was resulted from continuous recrystallization during deformation processing. As mentioned in the previous literature, there are numerous high-temperature oxidation resistance applications between 700 °C to 900 °C. Thus a thorough grasp of the high-temperature oxidation in the temperature range is necessary.

HEAs have potential applications at high temperatures ranging from 700 °C to 900 °C. In this study, the high-temperature oxidation mechanism of a Cr-Mn-Fe-Co HEA produced by 3D laser printing was investigated.

2. Materials and Methods

Initially, Cr, Mn, Fe and Co metal powders were fully mixed in equal moles. In addition, the equimolar Cr-Mn-Fe-Co HEA with a size of $20 \times 60 \times 12$ mm^3 was produced by a numerical control system (3D-ZM, Dalian, China) with a laser scanning speed of 600 mm/min and laser power of 1000 W. Six cubes with the size of $6 \times 6 \times 6$ mm^3 were cut from the initial alloy utilizing wire electrical discharge machine. Each surface was polished on SiC papers up to 1600 grit, cleaned by alcohol in an ultrasonic machine and dried in high pressure air.

The oxidation experiment was operated in a non-vacuum heating furnace. It was heated in the synthetic air (79% N_2 and 21% O_2) to the temperatures of 700 °C and 900 °C up to 48 h. After 48 h, the specimens were cooled down to room temperature in the furnace under laboratory air. Analytical balance was used to accurately measure the weight of samples. An X-ray diffractometer (D8 ADVANCE, BRUKER, Leipzig, Germany) equipped with Cu kα ($\lambda = 1.5406$ Å), scan step size of 0.02 deg. and a scan rate of 1 deg.min-1 was utilized to identify phases for the physic-chemical characterization. A scanning electron microscopy (SEM) equipped with an energy dispersive spectrometer (EDS) on the platform of MIRA 3 LMU (TESCAN, Brno, The Czech Republic) was used to identify the images and the element distribution. In the EDS experiments, the accelerating voltage was 15 kV, and the distance between two measuring points was 0.35 μm.

3. Results and Discussion

3.1. Oxidation Properties of Cr-Mn-Fe-Co HEA

Figure 1 shows the EDS results of the initial composition of the alloy. Each element is evenly distributed in matrix and the content of Co, Cr, Fe, and Mn were 24.67%, 24.25%, 25.72%, and 25.37%, respectively, which has been shown in Table 1.

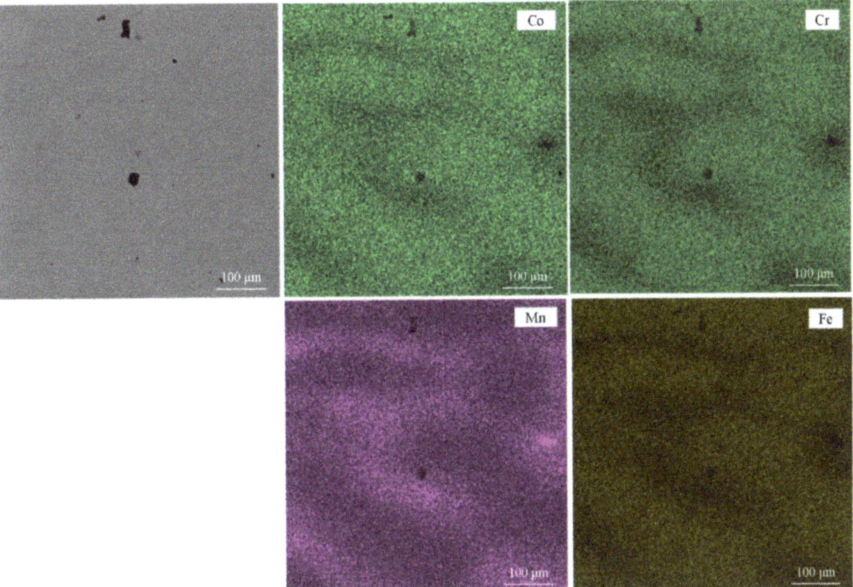

Figure 1. EDS results of the initial HEA.

Table 1. Chemical composition of the HEA.

Composition (wt.%)	Co	Cr	Fe	Mn
The initial HEA	24.67	24.25	25.72	25.37

The mass gain data were obtained from separated samples for each data point. Figure 2 shows the mass gain of the HEA at 700 °C and 900 °C over a period of 48 h. The mass gain (If there is oxide exfoliation, it was took into account.) increased at a given oxidation period as the oxidizing temperature raised. Furthermore, the mass gain rate of the alloy at a certain temperature was gradually decreased with the increasing of oxidation time. Mass gain after 48 h oxidation tests was 0.012 and 0.074 g/cm^2 for 700 °C and 900 °C, respectively. The oxidation kinetics of the alloy followed the parabolic rate law at both two temperatures as illustrated in oxidation curves (Figure 2a). The parabolic rate constant K_p can be obtained by $K_p = \frac{(\Delta m)^2}{A^2 \times t}$, where, Δm represents mass gain, A represents sample area, t represents oxidation time. The parabolic rate constant K_p of 2.79 mg^2/cm^4/h at 700 °C sharply increased to 114 mg^2/cm^4/h at 900 °C after 48 h. Compared to the oxidation behavior of equimolar Cr-Mn-Fe-Co-Ni HEA [14], its parabolic rate constant of 0.13 mg^2/cm^4/h at 900 °C verified the significant effect of Ni element, on the sluggish diffusion of alloy's high-temperature oxidation resistance. Chromium, acting as a getter for oxygen in the alloy during the initial stage, can lower the oxygen solubility in the alloy and reduce the internal oxidation speed of other elements [15]. With relative high content of chromium element, the relative low rate constants were obtained [16]. However, the rate constant increased rapidly compared to the Cr-Mn-Fe-Co-Ni alloy through the increase in the chromium element. Firstly, the oxides of Mn presented in the areas where cracks appeared and the relative high content of Mn may result in cracked oxides. In addition, Ni element can significantly increase the oxidation resistance at high temperature through reducing the volatility of Cr$_2$O$_3$ by forming a solid solution in the oxides [14]. The relative discussion has been added in the paper.

Figure 3a shows the XRD results of initial Cr-Mn-Fe-Co HEA, which is a single FCC solid solution phase [17]. Figure 3b,c present XRD analysis of Cr-Mn-Fe-Co HEA after isothermal oxidation experiments at 700 °C and 900 °C. After oxidizing at 900 °C over 12 h, the oxide layer on the surface was exfoliated. Apart from FeO and MnO identified on the surface oxidized at 700 °C for 12 h, some $MnCr_2O_4$ was also detected when the alloy was oxidized at 700 °C for 24 h, while Fe_2O_3 and $FeMn_2O_4$ appeared when the oxidation duration was prolonged to 48 h. Furthermore, $MnFe_2O_4$, $MnFeO_2$ and Cr_2O_3 were observed on the surface when the specimens were oxidized at 900 °C. The thickness of oxide layers increased with the increase in the oxidation time, indicated by the increase in the peak intensity.

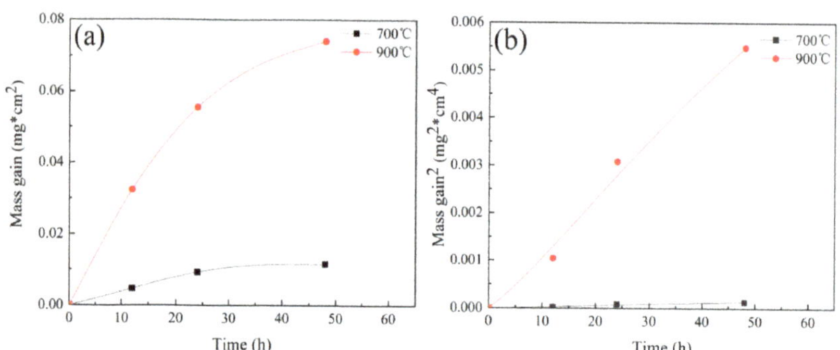

Figure 2. The mass gain of HEA: (**a**) Weight gain versus oxidation time curves; (**b**) Parabolic plot.

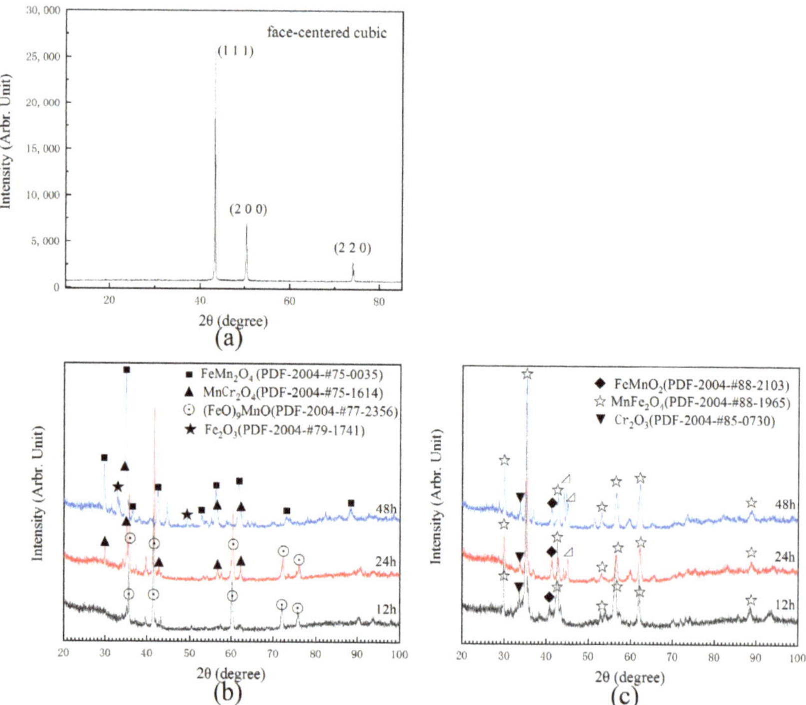

Figure 3. XRD results of HEA: (**a**) XRD spectra of the initial alloy; (**b**,**c**) XRD spectra after isothermal oxidation tests at 700 °C and 900 °C, respectively.

3.2. High-Temperature Surface Oxidation Mechanism of Cr-Mn-Fe-Co HEA

Figure 4 shows the surfaces micrographs of the HEA before and after the high-temperature oxidation test. Compared to the initial surface and the surfaces oxidized at 700 °C for 48 h, the surface after oxidation at 900 °C showed clear cracks, holes and obvious spallation of oxide layer (Figure 4a–c). The oxidation on a certain surface was obviously different on different regions, which resulted from the different speeds of oxidation penetration through grains and grain boundaries and the element diffusion. The spinel shape oxides were observed on all surfaces after oxidation tests, which was similar to the high-temperature oxidation corrosion behavior of Cr-Mn-Fe-Co-Ni HEA preferring grain boundary area [14]. Compared to the magnified images of the specimens after oxidation at 700 °C in Figure 4d,e,h,l,m, the surface after oxidized at 900 °C (Figure 4f,g,j,k,n,o) presented numerous pores and cracks and the oxidation layer peeled off. After the peeling, the surface of the specimen oxidized at 900 °C for 48 h represented numerous pores and cracks again. As a result, the maximum roughness increased from 24.4 µm to 46.8 µm with the test time increasing from 12 h to 48 h, while it was reasonable for a curtained trend of the maximum roughness measured at 900 °C.

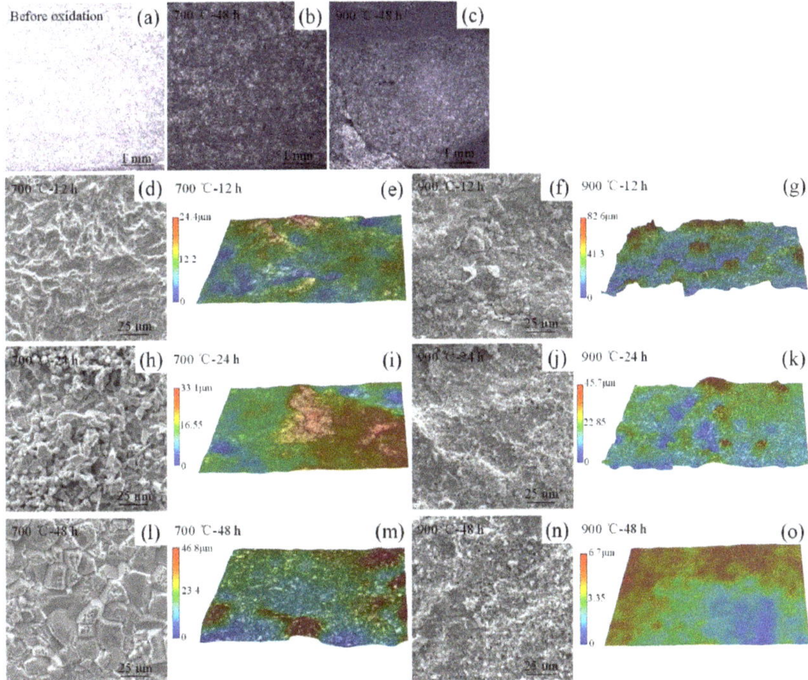

Figure 4. Surfaces micrographs of Cr-Mn-Fe-Co HEA at different temperatures and various times.

Figure 5 shows the vertical cross-sections images of the oxides. The oxide thickness increased with the oxidation time and temperatures. When tested at 700 °C, some oxides and voids were observed on the surface of HEA as well as internal oxidation. Numerous cracks were formed when the oxidation time extended to 24 h. Dense oxides formed when the tests were operated at 900 °C and spallation of the oxide scales was observed, caused by interaction between different oxides, defects and matrix. The details of specimens after oxidation at 900 °C for 24 h were magnified in Figure 5i–k. The middle and bottom parts of the oxide layer consisted of oxides (FeO and MnO) and metal matrix simultaneously. The upper part of oxidation layer ($MnFe_2O_4$, $MnFeO_2$ and Cr_2O_3) was composed of oxides and a large number of holes, which may result in easier diffusion of oxygen to the matrix and reduce the resistance of oxidation corrosion at high temperature. Furthermore, the chemical composition of the oxide in Figure 5f was analyzed and the results were shown in Figure 5l,m. An oxygen content of 57.74% (atomic) was detected at spectrum 2 (Figure 5l), verified to be oxides. While it had a small amount of oxygen element (atomic, 5.69%) at spectrum 4 (Figure 5m). The matrix was propped open by oxides and segmented into small pieces as shown in Figure 5i–k, resulting in materials failure.

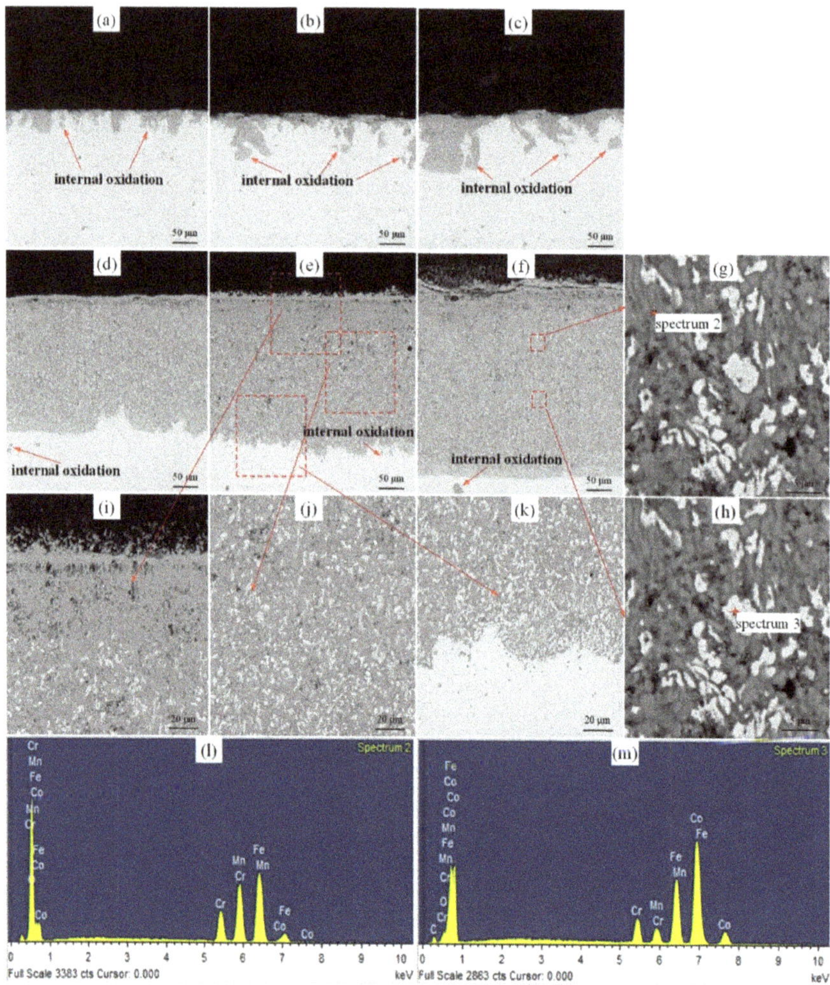

Figure 5. Cross-section images of Cr-Mn-Fe-Co HEA after oxidation test at 700 °C for (**a**) 12 h, (**b**) 24 h, (**c**) 48 h, and at 900 °C for (**d**) 12 h, (**e**) 24 h, (**f**) 48 h. (**g**,**h**) for one magnification image of (**f**), (**i**–**k**) three magnified images of (**e**), (**l**) the element distribution of spectrum 2, (**m**) the element distribution of spectrum 3.

The element distribution after oxidation is shown in Figure 6. Some voids were observed, which were Kirkendall pores, resulting from different diffusion of Cr, Mn, Fe and Co. An unstable Cr-rich, Mn-rich and Fe-rich oxide layer, respectively inferred to (Mn, Cr)$_3$O$_4$, Cr$_2$Mn$_3$O$_8$ and (Mn, Fe)$_2$O$_4$ formed during the test. They were easy to fall off from the matrix after oxidation at 900 °C. Cr, Mn and Fe had higher diffusion rates than Co, resulting in the development of Co-poor region and Co-rich region, as shown in Figure 6f. Fe and Co produced an oxidation layer in MnFe$_2$O$_4$, MnFeO$_2$ and Cr$_2$O$_3$ at high temperature, leading to a reduction in the oxidation resistance of Cr-Mn-Fe-Co HEA. After oxidation at 900 °C, few oxides pegs were observed resulting in a weakened connection between oxides and the substrate.

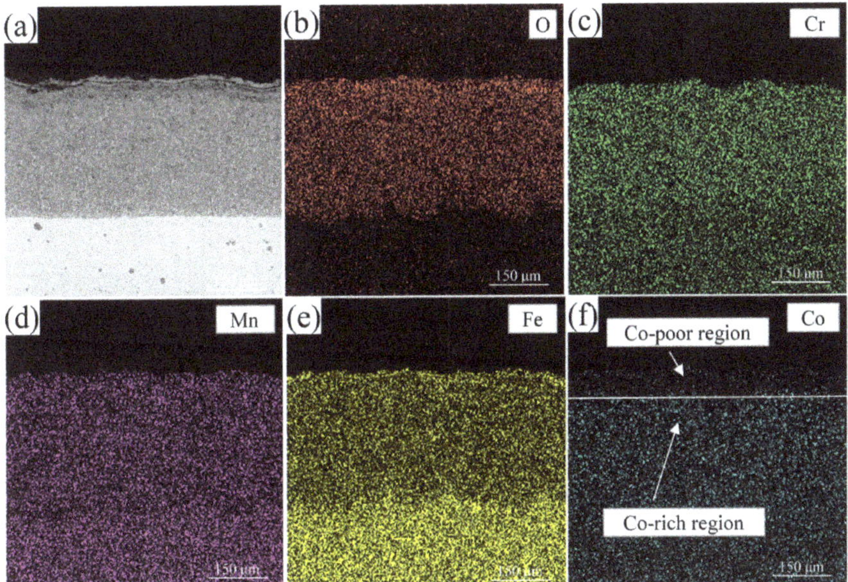

Figure 6. EDS results of Cr-Mn-Fe-Co HEA after isothermal oxidation tests at 900 °C for 48 h.

The instantaneous rate constant K_p decreased with the incremental oxidation time. Activation energy (E_a) was calculated from the K_p values and the corresponding test temperatures. It gave the value of $E_a = 212 kJ/mol$. The value ranged between 250 and 290 kJ/mol of chromia (Cr_2O_3) formation alloys during the oxidation as previously reported in the literature [18,19]. It was in good agreement with the observation in Figures 4 and 5. A chromatic scale, a diffusion barrier, appeared in the scale. However, compared to the Cr-Mn-Fe-Co-Ni HEA, the absence of Ni element greatly reduced the hindrance of cation diffusion and led to a good oxidation behavior.

4. Conclusions

The oxidation behavior of Cr-Mn-Fe-Co HEA produced by 3D laser printing was investigated at 700 °C and 900 °C up to 48 h. The main conclusions are as follows:

(1) The initial alloy was verified as FCC random solid solution. The mass gain in the oxidation process for 48 h was measured to be 0.02 and 0.15 g/cm^2 at 700 °C and 900 °C, respectively.

(2) $MnFe_2O_4$, $MnFeO_2$ and Cr_2O_3 were detected by the XRD analysis on the specimen surface. Some surface cracks were observed, especially for those oxidized at 900 °C. The spallation of oxide layer took place when the alloy was tested at 900 °C. The high-temperature oxidation resistance of the alloy sharply decreased when the test temperature increased from 700 °C to 900 °C, leading to the wide application under 700 °C.

Author Contributions: Data curation, L.W., Q.Z. and Z.X.; Funding acquisition, Y.Z.; Investigation, L.W. and Q.Z.; Project administration, Y.Z. and H.G.; Writing—original draft, L.W. and H.G.; Writing—review and editing, Y.Z. All authors have read and agreed to the published version of the manuscript.

Funding: This research was funded by "THE NATIONAL KEY RESEARCH AND DEVELOPMENT PROGRAM, grant number 2019YFB2006500", "THE SCIENCE AND TECHNOLOGY INNOVATION

PROGRAM OF HUNAN PROVINCE, grant number 2020RC2002", "INNOVATION DRIVEN PROGRAM OF CENTRAL SOUTH UNIVERSITY, grant number 2019CX006", and "THE RESEARCH FUND OF THE KEY LABORATORY OF HIGH PERFORMANCE COMPLEX MANUFACTURING".

Institutional Review Board Statement: Not applicable.

Informed Consent Statement: Not applicable.

Data Availability Statement: The data presented in this study are available on request from the corresponding author.

Conflicts of Interest: The authors declare no conflict of interest.

References

1. Yeh, J.; Chen, S.; Lin, S.; Gan, J.; Chin, T.; Shun, T.; Tsau, C.; Chang, S. Nanostructured high-entropy alloys with multiple principal elements, novel alloy design, conceptes and outcomes. *Adv. Eng. Mater.* **2004**, *6*, 299–303. [CrossRef]
2. Wu, Y.; Liu, J.; Bhatta, L.; Kong, C.; Yu, H. Study of texture analysis on asymmetric cryorolled and annealed CoCrNi medium entropy alloy. *Crystals* **2020**, *10*, 1154. [CrossRef]
3. Zhang, Z.; Wu, Y.; Li, C.; Gan, B.; Kong, C.; Wang, Y.; Yu, H. Annealing effect on mechanical properties of asymmetric cryorolled CrCoNi medium entropy alloy. *Mater. Today Comm.* **2020**. [CrossRef]
4. Vaidyal, M.; Muralikrishnal, G.; Murty, B. Multi-principal-element alloys with improved oxidation and wear resistance for thermal spray coating. *J. Mater. Res.* **2019**, *34*, 664–686. [CrossRef]
5. Tsai, K.; Tsai, M.; Yeh, J. Sluggish diffusion in Co-Cr-Fe-Mn-Ni high-entropy alloys. *Acta Mater.* **2013**, *61*, 4887–4897. [CrossRef]
6. Gludovatz, B.; Hohenwarter, A.; Catoor, D.; Chang, E.; George, E.; Ritchie, R. A fracture-resistant high-entropy alloy for cryogenic applications. *Science* **2014**, *345*, 1153–1158. [CrossRef] [PubMed]
7. Otto, F.; Dlouhy, A.; Somsen, C.; Bei, H.; Eggeler, G.; George, E. The influences of temperature and microstructure on the tensile properties of a CoCrFeMnNi high-entropy alloy. *Acta Mater.* **2013**, *61*, 5743–5755. [CrossRef]
8. Yong, K.; Sang, H.; Byung, J.; Sun, I. Correlation between mechanical properties and thermodynamic parameters of dual-fcc-phase CoCrFeCuxNi (x = 1, 1.71) and CoCu$_{1.71}$FeMnNi. *Mater. Lett.* **2020**, *272*, 127866. [CrossRef]
9. Keil, T.; Bruder, E.; Laurent-Brocq, M.; Durst, K. From diluted solid solutions to high entropy alloys: Saturation grain size and mechanical properties after high pressure torsion. *Scripta Mater.* **2021**, *192*, 43–48. [CrossRef]
10. Kai, W.; Li, C.; Cheng, F.; Chu, K.; Huang, R.; Tsay, L.; Kai, J. The oxidation behavior of an equimolar Fe-Co-Ni-Cr-Mn high-entropy alloy at 950 °C in various oxygen-containing atmospheres. *Corr. Sci.* **2016**, *108*, 209–214. [CrossRef]
11. Gorr, B.; Azim, M.; Christ, H.; Mueller, T.; Schliephake, D.; Heilmaier, M. Phase equilibria, microstructure, and high temperature oxidation resistance of novel refractory high-entropy alloys. *J. Alloy. Compd.* **2015**, *624*, 270–278. [CrossRef]
12. Pouraliakbar, H.; Sang, H.; Yong, K.; Rizi, M.; Sun, I. Microstructure evolution and mechanical properties of (CoCrNi)$_{90}$(AlTiZr)$_{5}$ (CuFeMo)$_{5}$ multicomponent alloy: A pathway through multicomponent alloys toward new superalloys. *J. Alloy. Compd.* **2021**, *860*, 158412. [CrossRef]
13. Yong, K.; Byung, J.; Soon-Ku, H.; Sun, I. Strengthening and fracture of deformation-processed dual fcc-phase CoCrFeCuNi and CoCrFeCu$_{1.71}$Ni high entropy alloys. *Mater. Sci. Eng. A* **2020**, *781*, 139241. [CrossRef]
14. Kim, Y.; Joo, Y.; Kim, H.; Lee, K. High temperature oxidation behavior of Cr-Mn-Fe-Co-Ni high entropy alloy. *Intermetallics* **2018**, *98*, 45–53. [CrossRef]
15. Wagner, C. Passivity and inhibition during the oxidation of metals at elevated temperatures. *Corr. Sci.* **1965**, *5*, 751–764. [CrossRef]
16. Niu, Y.; Zhang, X.; Wu, Y.; Gesmundo, F. The third-element effect in the oxidation of Ni-xCr-7Al (x = 0, 5, 10, 15 at.%) alloys in 1 atm O$_2$ at 900–1000 °C. *Corr. Sci.* **2006**, *48*, 4020–4036. [CrossRef]
17. Zhang, Y.; Lu, Z.; Ma, S.; Liaw, P.; Tang, Z.; Cheng, Y.; Gao, M. Guidelines in predicting phase formation of high-entropy alloys. *MRS Commun.* **2014**, *4*, 57–62. [CrossRef]
18. Chen, J.; Rogers, P.; Little, J. Oxidation behavior of several chromia-forming commercial nickel-base superalloys. *Oxid. Met.* **1997**, *47*, 381–410. [CrossRef]
19. Jacob, Y.; Haanappel, V.; Stroosnijder, M.; Buscail, H.; Fielitz, P.; Borchardt, G. The effect of gas composition on the isothermal oxidation behaviour of PM chromium. *Corros. Sci.* **2002**, *44*, 2027–2039. [CrossRef]

Article

Grain Growth Mechanism of Lamellar-Structure High-Purity Nickel via Cold Rolling and Cryorolling during Annealing

Zhide Li [1,2,†], Yuze Wu [2,3,†], Zhibao Xie [3], Charlie Kong [4] and Hailiang Yu [1,2,3,*]

[1] Light Alloys Research Institute, Central South University, Changsha 410083, China; lizhide@csu.edu.cn
[2] State Key Laboratory of High Performance Complex Manufacturing, Central South University, Changsha 410083, China; wyz2019@csu.edu.cn
[3] College of Mechanical and Electrical Engineering, Central South University, Changsha 410083, China; xiezhibao@csu.edu.cn
[4] Mark Wainwright Analytical Centre, University of New South Wales, Sydney, NSW 2052, Australia; c.kong@unsw.edu.au
* Correspondence: yuhailiang@csu.edu.cn
† Equal contribution.

Abstract: High-purity (99.999%) nickel with lamellar-structure grains (LG) was obtained by room-temperature rolling and cryorolling in this research, and then annealed at different temperatures (75 °C, 160 °C, and 245 °C). The microstructure was characterized by transmission electron microscopy. The grain growth mechanism during annealing of the LG materials obtained via different processes was studied. Results showed that the LG high-purity nickel obtained by room-temperature rolling had a static discontinuous recrystallization during annealing, whereas that obtained by cryorolling underwent static and continuous recrystallization during annealing, which was caused by the seriously inhibited dislocation recovery in the rolling process under cryogenic conditions, leading to more accumulated deformation energy storage in sheets.

Keywords: high-purity nickel; cryorolling; annealing; grain growth; lamellar-structure grains

Citation: Li, Z.; Wu, Y.; Xie, Z.; Kong, C.; Yu, H. Grain Growth Mechanism of Lamellar-Structure High-Purity Nickel via Cold Rolling and Cryorolling during Annealing. *Materials* **2021**, *14*, 4025. https://doi.org/10.3390/ma14144025

Academic Editor: Tomasz Trzepiecinski

Received: 28 June 2021
Accepted: 12 July 2021
Published: 19 July 2021

Publisher's Note: MDPI stays neutral with regard to jurisdictional claims in published maps and institutional affiliations.

1. Introduction

Nanoscale lamellar-structure grains (LG) of nickel prepared by surface mechanical grinding treatment show ultrahigh strength and excellent thermal stability [1]. The molecular dynamics simulation results showed that LG nickel with the texture of {111} <110> and staggered low-angle grain boundaries was the most stable configuration [2]. Low-angle grain boundaries of LG are normally formed through dislocation multiplication and interaction during plastic straining [3]. Various severe plastic deformation techniques to produce LG material have been developed, such as surface mechanical grinding treatment [4,5], cryorolling [6,7], cross rolling [8], and other special rolling techniques [9,10]. The study on the grain growth mechanism of LG materials prepared by different processes with annealing can support further understanding of the thermal stability of LG materials. Moreover, a preparation procedure reference of LG materials with superb properties and exceptional thermal stability can be proposed in such studies.

During annealing, the grain growth of LG material is affected by many factors. The study of Liu et al. [10] showed that the high fraction of grain boundary microstructure characteristics with small angles contributes to the improvement of thermal stability for Ni alloys. Furthermore, the 50 °C delayed recrystallization starting temperature indicates that the recovery process of LG materials is affected by grain boundary orientation. Chen et al. [11] reported that the initial cubic fraction had a significant effect on the recrystallization process of a high-purity (99.999%) nickel plate in the cold rolling process with 96% reduction, indicating that the recovery of LG material was also influenced by the original grain size. Another study showed that, under the common influence of many factors, the recovery mechanism of LG materials varies [12]. Xie et al. [13] reported the preparation of a 99.945% pure Ni sheet with LG, with an average lamellar boundary spacing of 77 nm, via dynamic plastic deformation and cold rolling. During annealing, the local faceting of lamellar boundaries leads to structural coarsening and polygonization of lamellar structure. They reported a novel mechanism via coarsening of the initial structure of nanolaminates with low angle boundaries called local faceting [14]. Cryorolling is a special rolling technique widely used to prepare LG materials [6,7,15]. LG aluminum alloy with high strength and high ductility can be obtained by cryorolling with proper reduction and a subsequent annealing process [16]. However, the grain growth mechanism of cryogenic rolled Ni with LG has not been studied.

In this paper, high-purity (99.999%) nickel with LG was prepared by cryogenic rolling and room-temperature rolling. Then, the recovery characteristics and grain growth mechanism of lamellar grain structures prepared by both processes with annealing at different temperatures were studied.

2. Materials and Methods

The material used in this study was high-purity nickel (99.999%) to minimize the additional influence of impurities on grain boundary migration rate. The initial thickness of the material was 2 mm, which was thinned to 0.2 mm by room-temperature rolling and cryorolling. The relative reduction in each rolling pass was 10%, with a total reduction of 90% and equivalent strain of $\varepsilon_{VM} = 2.7$. The roll diameter of the rolling mill used in the rolling process was 80 mm, and the rolling speed was 1 m/min. Cryorolling was carried out in liquid nitrogen ($-196\,^\circ$C). The sample was kept in the liquid nitrogen for 30 min before the first pass of cryorolling to fully cool it. Then, the sample was kept in liquid nitrogen for 8 min before each pass, maintaining a cryogenic state throughout the process. Afterward, the rolled samples were annealed in a vacuum chamber at 75 $^\circ$C, 160 $^\circ$C, and 245 $^\circ$C, with an annealing time of 1 h.

Tensile tests of the rolled and annealed samples were carried out on an AGS-X 10 kN tensile tester (Shimadzu, Kyoto, Japan) with a tensile rate of $1 \times 10^{-3}\ s^{-1}$; the gauge length of the tensile samples was 18 mm, and the width of the tensile samples was 3 mm. The microstructure of the material was characterized by a Philips CM200 transmission electron microscope (TEM) (Philips Electron Optics, Eindhoven, The Netherlands) with a working voltage of 200 kV. The transmission sample was prepared using a focusing electron beam, and the observation surface was the cross-section of the rolling direction (RD) and the normal direction (ND). In order to obtain an accurate boundary spacing, it was quantitatively measured from the TEM images using the software Image J (ImageJ 1.53c, National Institutes of Health, Bethesda, Rockville, MD, USA). The final results were obtained by analyzing statistics collected from more than 100 grains.

3. Results

3.1. Microstructure Evaluation

The microstructure of pure nickel after room-temperature rolling with 90% reduction is shown in Figure 1a, in which the characteristics of LG can be clearly observed. It can be noted from the elongated grains with light and dark distribution that a large number of dislocations accumulated inside the grains. The microstructures of samples annealed at 75 $^\circ$C, 160 $^\circ$C, and 245 $^\circ$C for 1 h are shown in Figure 1b–d. After annealing at 75 $^\circ$C and 160 $^\circ$C for 1 h, it can be observed from the TEM results in Figure 1 that the grain size did not increase significantly, indicating that the energy provided by annealing was not enough to drive the rapid grain boundary migration at the given temperature and holding time, even if dislocation recovery occurred. However, the grain was obviously coarsened after annealing at 245 $^\circ$C for 1 h, as shown in Figure 1d. Almost pure white grains can be observed in the figure, indicating the presence of recrystallized grains. For recrystallized grains, only a few dislocations were observed near grain boundaries and within the grain, as indicated by the red arrow in the figure.

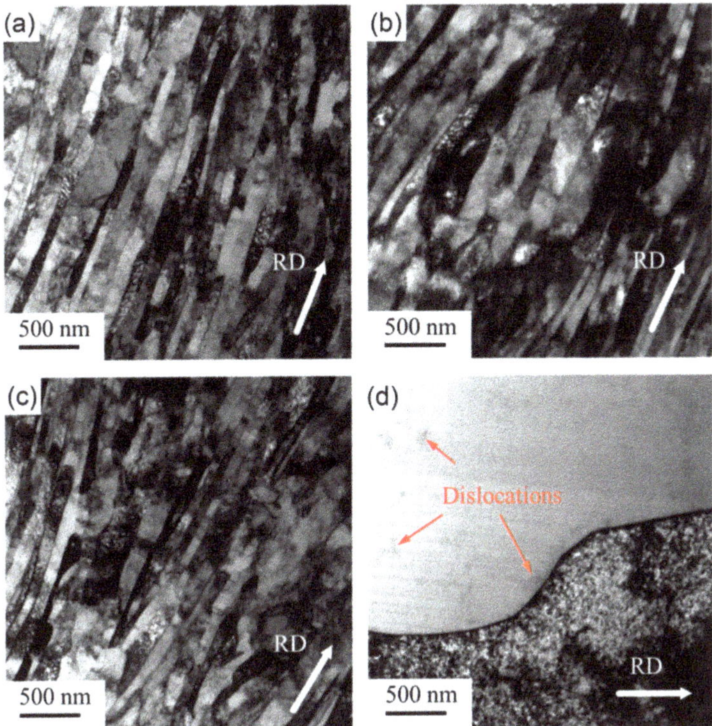

Figure 1. TEM images of Ni sheets subjected to (**a**) room-temperature rolling and subsequent annealing at (**b**) 75 °C, (**c**) 160 °C, and (**d**) 245 °C for 1 h.

The microstructure of pure nickel after cryorolling with 90% reduction is shown in Figure 2a. It can be observed that the grain size was not significantly different from that after room-temperature rolling. This shows that the grain refinement effect of cryorolling is similar to that of room-temperature rolling under the same deformation. The microstructures of samples annealed at 75 °C, 160 °C, and 245 °C for 1 h after cryorolling are shown in Figure 2b–d. The cryorolled grain size did not increase significantly after annealing at 75 °C and 160 °C for 1 h. The grains of the cryorolled samples after annealing at 245 °C for 1 h were obviously coarsened; however, unlike the recrystallization of the room-temperature rolled sample, there were a large number of intersecting subgrain boundaries in the coarsened grains. As shown in Figure 2a, clear grain boundaries could be observed after cryorolling without annealing, while there were no intersecting subgrain boundaries, as shown in Figure 2d, indicating that these subgrain boundaries were formed during annealing.

For the better understanding of the changes in the microstructure of the samples rolled and annealed, quantitative measures were carried out on the grain boundary spacing of LG as shown in Figures 1 and 2. Histograms were drawn according to the frequencies of the nine different levels, which were divided by the range of the measured statistics. Then, the mean grain boundary spacing was obtained by fitting curves of normal distribution to the frequency statistics, which are shown in Figure 3. It can be seen from Figure 3a,b that there was no obvious difference in grain boundary spacing of room-temperature rolling and cryorolling with 90% reduction. Severe plastic deformation forced the grain boundary facet to migrate, mainly by applying strong external stress as the mechanical driving force. Grains are constrained by neighboring grains during the deformation process; therefore, to achieve strain compatibility, numerous geometrically necessary dislocations (GND) need to be generated during the strain process between different grains, which separates the grains into fine grains, causing their refinement. With a 90% reduction, both room-temperature rolling and cryorolling required the materials to reduce the grain boundary spacing to adapt to macroscopic thinning, and the grain boundary spacing was similar after room-temperature rolling and cryorolling. There was a saturated grain size at a certain extent of grain refinement, depending on the deformation temperature, the initial grain size, the composition, and other factors affecting the dislocation accumulation and recovery. The size of the saturated grains could be indicated according to the Zener–Hollomon parameter, $Z = \dot{\varepsilon}\exp(Q/RT)$, where $\dot{\varepsilon}$ is the strain rate, Q is the diffusion activation energy in grain boundaries, R is the gas constant, and T is the deformation temperature [17]. The study of Pippan et al. [18] showed that the saturated grain size decreased significantly as the deformation temperature declined. This kind of dependence is stronger at the medium homologous temperature, but relatively weaker at the lower temperature. The study of Duan et al. [19] showed that, as the ARB cycles increased, the boundary spacing thickness of industrial purity (99.8%) nickel sheets decreased, reaching a saturation thickness of 75 nm after 6–8 cycles. In this paper, the grain boundary spacing of high-purity nickel reached 68 nm and 66 nm after room temperature rolling and cryorolling, respectively, which may have reached or been close to saturation thickness. In Figure 3c,e, the statistical results of grain boundary spacing after rolling at room temperature with annealing at 75 °C and 160 °C are shown, respectively. Figure 3d,f are the statistical results of grain boundary spacing of cryogenic rolling after 75 °C and 160 °C annealing. Both the statistical histogram and the fitting curve show that the grain boundary spacing after cryorolling and annealing was more evenly distributed with a smaller range fluctuation than that after room-temperature rolling and annealing, indicating that a more uniform microstructure was obtained with cryorolling compared to room-temperature rolling.

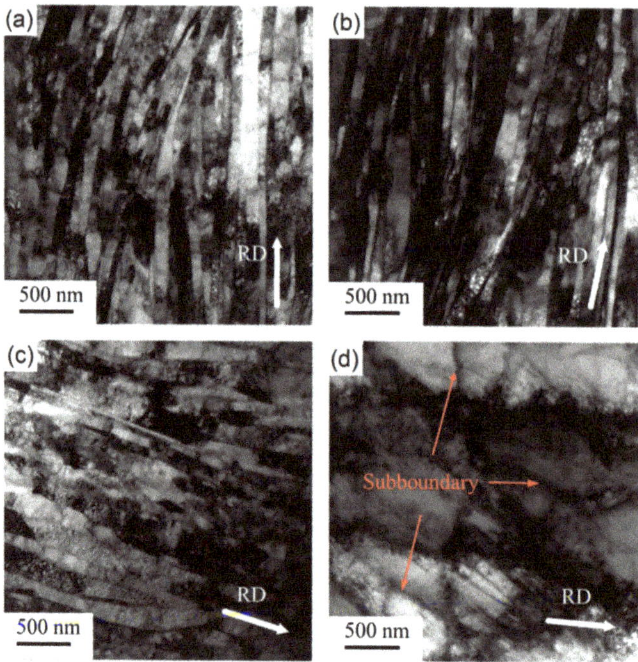

Figure 2. TEM images of Ni sheets subjected to (**a**) cryorolling and subsequent annealing at (**b**) 75 °C, (**c**) 160 °C, and (**d**) 245 °C for 1 h.

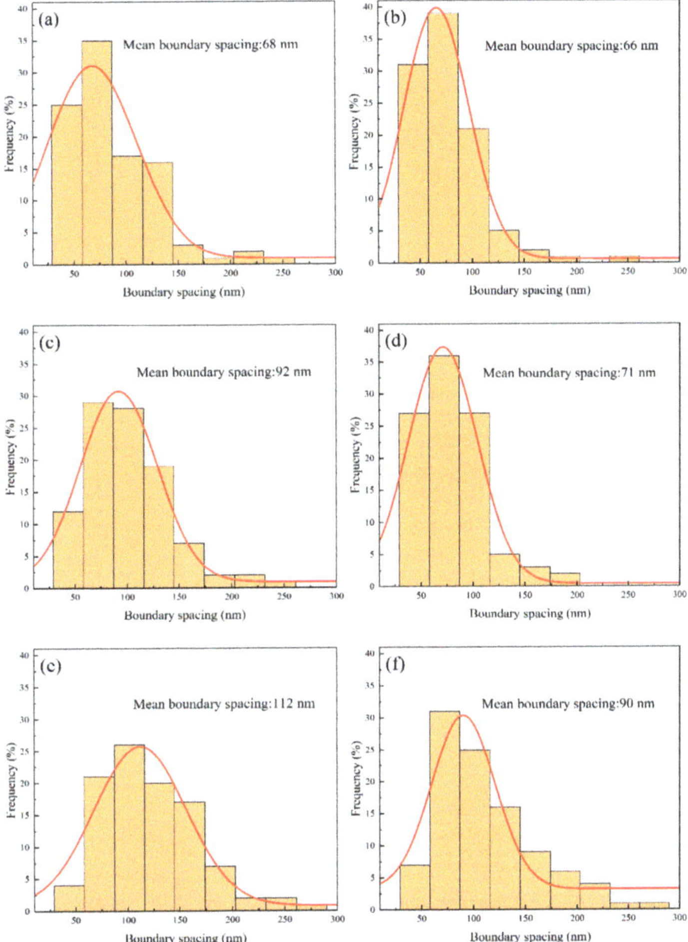

Figure 3. Histograms of boundary spacing of Ni sheets subjected to (**a**) room-temperature rolling and subsequent annealing at (**c**) 75 °C and (**e**) 160 °C for 1 h, and (**b**) cryorolling and subsequent annealing at (**d**) 75 °C and (**f**) 160 °C for 1 h.

3.2. Mechanical Properties

The tensile curves of pure nickel after room-temperature rolling, cryorolling, and subsequent annealing are shown in Figure 4. The mechanical properties of pure nickel after room-temperature rolling, cryorolling, and subsequent annealing are shown in Table 1. It can be seen that the strength of the rolled samples remained high after annealing below 160 °C. On the other hand, the strength dropped significantly after 245 °C annealing with elongation increasing, which indicates that recrystallization occurred in the 245 °C annealing samples. The difference in tensile strength and elongation between the high-purity nickel samples obtained after room-temperature rolling and cryorolling, followed by annealing at 75 °C and 160 °C or no annealing, was little, as also shown by the grain size in the TEM results. However, the elongation of the samples annealed at 245 °C after cryorolling was better than that annealed at 245 °C at room temperature, which may be related to the different ways of grain growth in the process of annealing.

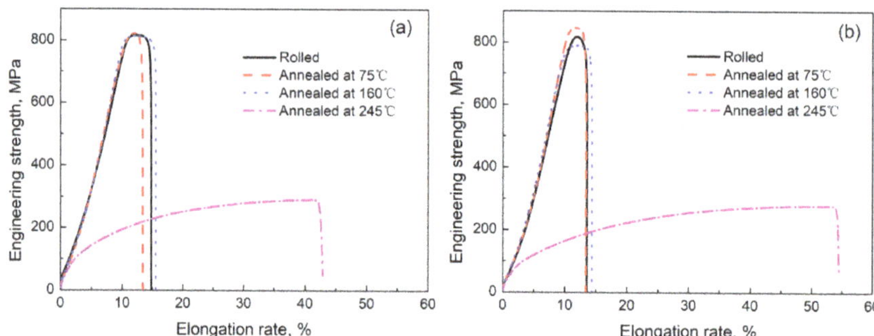

Figure 4. Tensile curves of sheets subjected to (**a**) room-temperature rolling and subsequent annealing, and (**b**) cryorolling and subsequent annealing.

Table 1. Mechanical properties of pure nickel after room-temperature rolling, cryorolling, and subsequent annealing.

Processes	Annealed Temperature	Tensile Strength (MPa)	Elongation Rate (%)
Room-temperature rolled	As rolled	811 ± 3	14 ± 1
	Annealed at 75 °C	821 ± 6	14 ± 1
	Annealed at 160 °C	768 ± 11	14 ± 1
	Annealed at 245 °C	279 ± 10	48 ± 6
Cryorolled	As rolled	820 ± 2	14 ± 1
	Annealed at 75 °C	830 ± 10	14 ± 1
	Annealed at 160 °C	779 ± 8	14 ± 1
	Annealed at 245 °C	278 ± 9	54 ± 4

4. Discussion

4.1. Recovery Characteristics of LG Nickel Annealed at Low Temperature (75–160 °C)

The grain boundary spacing of LG nickel after room-temperature rolling and cryorolling was similar. After annealing at 75 °C and 160 °C, the grain boundary spacing of samples with different rolling processes began to differ. The grain boundary spacing changes of pure nickel annealed at 75 °C and 160 °C after cryorolling and room temperature rolling are shown in Figure 5. As can be seen from the figure (black symbols), the grain boundary spacing of the sample prepared by room temperature rolling with 75 °C annealing increased at a relatively fast rate as the annealing temperature increased. On the other hand, the grain boundary spacing of the sample prepared by cryorolling with lower temperature (75 °C) annealing (red symbols) increased at a rather slow rate. At the 160 °C temperature of annealing, the grain boundary distance of samples prepared by both rolling technique increased as the annealing temperature increased, while the increase speed of the room-temperature rolled sample was still obviously higher than that of the cryorolled sample. It is worth noting that the difference between the increase rate of grain boundary distance of room-temperature rolled samples and cryorolled samples decreased. This shows that the cryorolled samples had higher initial grain-coarsening temperature in the condition where grains had similar size. When annealed at lower temperatures (75 °C and 160 °C), the nickel sheets mainly underwent a static dislocation recovery process. For the room-temperature rolled samples, even in those with a faster grain boundary spacing increase rate, only 24 nm and 44 nm increments were observed after 75 °C and 160 °C annealing. The dynamic dislocation recovery process occurs during the process of room-temperature rolling on pure nickel, while it is strongly inhibited in the low temperature of cryorolling [20]. That is to say, room-temperature rolling is a process of continuous generation and annihilation of dislocation, whereas, in the cryorolling process, there is only continuous generation of dislocation, and the annihilation process is extremely suppressed, which leads to more dislocations accumulated during cryorolling compared to room-temperature rolling. When annealed at a lower temperature (75 °C), the dislocation recovery rate was slower, such that the cryorolled samples had more dislocations to recover from in the annealing process than after room-temperature rolling. Therefore, the grain boundary spacing increase rate of the cryorolled samples was lower than that after room-temperature rolling. When annealed with a higher temperature, the thermal effect increased the dislocation recovery rate, weakening the hindering effect of dislocation on the grain boundary spacing increment, leading to a narrowed gap between the grain boundary spacing increase rate of the cryorolled and room-temperature rolled samples.

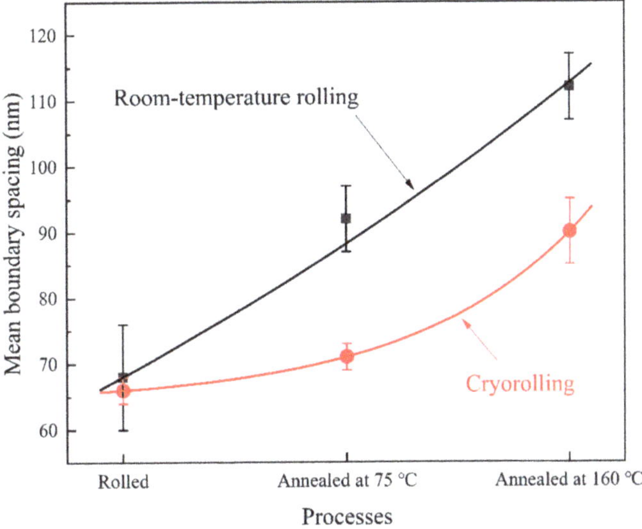

Figure 5. Mean boundary spacing of Ni sheets subjected to different processes.

4.2. Recovery Characteristics of LG Nickel Annealing at 245 °C

The microstructure shown in Figures 1d and 2d, as well as the tensile results in Figure 4, reveal that, with annealing at 245 °C, both samples recrystallized. Figure 6 further shows the recrystallization behavior of the cryorolled and room-temperature rolled samples with subsequent annealing at 160 °C and 245 °C. With annealing at 160 °C, the room-temperature rolled samples showed local facet angularization in the lamellar boundary. As shown in the area marked by the yellow box in Figure 6a, the grain boundary facet of the B1 grain migrated toward the A1 grain region, exhibiting protruding sharp angles, as pointed out by yellow arrows in the figure. A similar phenomenon also existed between the B2 and A2 grains. Different phenomena were found in the LG boundary facets with cryorolling and annealing at 160 °C. As shown in the red box in Figure 6c, the grain boundary facet of D1 grain migrated to C1 grain while maintaining a state parallel to the rolling direction as a whole, as indicated by the red arrow in the figure. Moreover, the contrast of grains showed that the orientation difference between D1 and C1 grains was small, indicating that there may be subgrain boundary rotation with small grain boundary deviation during the grain boundary facet migration. TEM results of the sample rolled at room temperature with annealing at 245 °C are shown in Figure 6b. It can be seen that the dislocation density of grain B3 was low, while that of grain A3 was high. The results show that static discontinuous recrystallization occurred in the sample, and there was a process of grain nucleation to form grains with lower dislocation density before growing in the vicinity of the area with higher dislocation density. Figure 6d shows the TEM results of cryorolled samples annealed at 245 °C. As can be seen from the figure, the number of dislocations within grains decreased in the area indicated by red arrows, and the number of subgrain boundaries increased in the red boxes in the figure. These results indicate that continuous recrystallization may occur during annealing. Migration of the subgrain boundary was caused by dislocation slip and climb, leading to a transition to the large angle grain boundary. In terms of discontinuous recrystallization and continuous recrystallization, Jazaeri et al. conducted in-depth studies [21], which showed that small initial grain size and large strains would promote continuous recrystallization. The mechanism of continuous recrystallization is thought to involve the collapse and subsequent coarsening of the LG produced when rolling to large strains, which was consistent with our research. In the HRTEM results of Li et al. [22], it was intuitively observed that the incompletely recovered dislocation density of the cryorolled material with was is significantly higher than that of the material after room-temperature rolling and annealing. The LG obtained by cryorolling had a smaller initial grain size. The main reason was that cryorolling can accumulate dislocations more efficiently than room-temperature rolling. Even with the same amount of deformation, cryorolling was equivalent to room-temperature rolling with a larger amount of deformation. The results showed that the samples after cryorolling were more prone to continuous recrystallization.

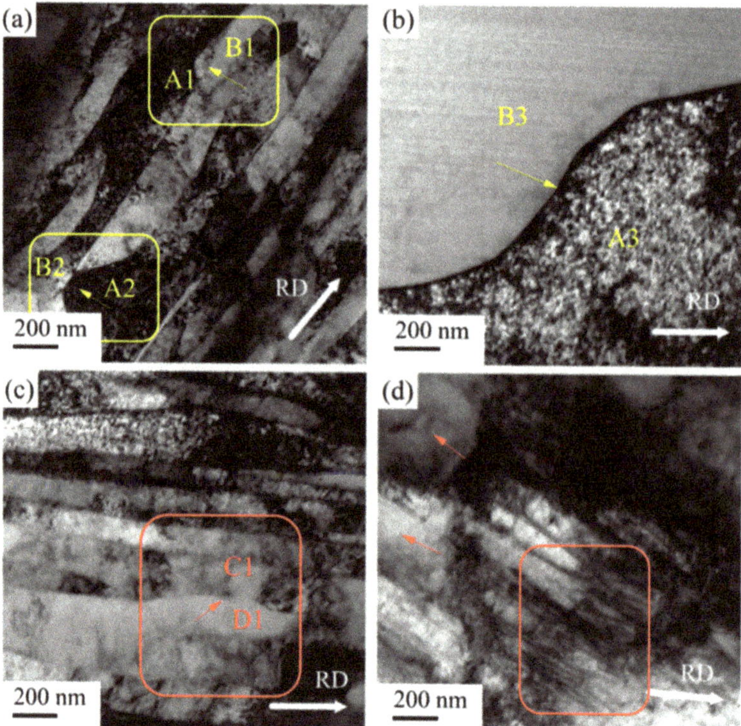

Figure 6. TEM images of the recrystallization behavior of Ni sheets subjected to room-temperature rolling with subsequent annealing at (**a**) 160 °C and (**b**) 240 °C, and cryorolling with subsequent annealing at (**c**) 160 °C and (**d**) 240 °C.

An obvious difference in grain growth phenomenon during the annealing process in cryorolled and room-temperature rolled samples can be concluded from the discussion above. Due to the inhomogeneity of grain deformation, the grain with large deformation has a high dislocation density, while the grain with a small deformation has a low dislocation density. It is a common phenomenon that the original grain boundary arched nucleus and grows under the effect of deformation energy storage during the annealing process [23–25]. The schematic diagram of this process is shown in Figure 7a,b. However, there are obvious differences in the lamellar grains obtained by cryorolling during annealing. There is dislocation recovery in crystals, diffusion and rotation of grain boundaries, and the merging of subgrain boundaries into high-angle grain boundaries, resulting in the process of grain growth [26], as shown in Figure 7c,d. Li [27] proposed that there is less boundary migration caused by two subgrains merging into a larger grain, which are involved in the process of the subgrains rotating until the adjacent subgrains have similar orientations, whereby the two subgrains involved merge into a larger grain with less boundary migration, which is the driving force generated by the reduction in boundary energy. Because the dislocation recovery was seriously inhibited during deformation, the lamellar grain structure prepared by cryorolling had a high deformation energy storage, which may have been the reason for the subgrain boundary rotation and coalescence during annealing. Wang et al. [28] observed that nanocrystalline nickel could induce grain rotation and growth during deformation using in situ transmission electron microscopy. They pointed out that the rotation of grains can cause the reduction or elimination of differences in the neighboring grains' orientation, thus leading to coalescence. Even without complete coalescence, the rotation of the entire grain may reduce the energy in the grain boundary system and effectively release stress concentration. When the grain size is small enough and the distortion of grain boundary is large enough, it is possible to release energy through grain boundary rotation during thermal activation.

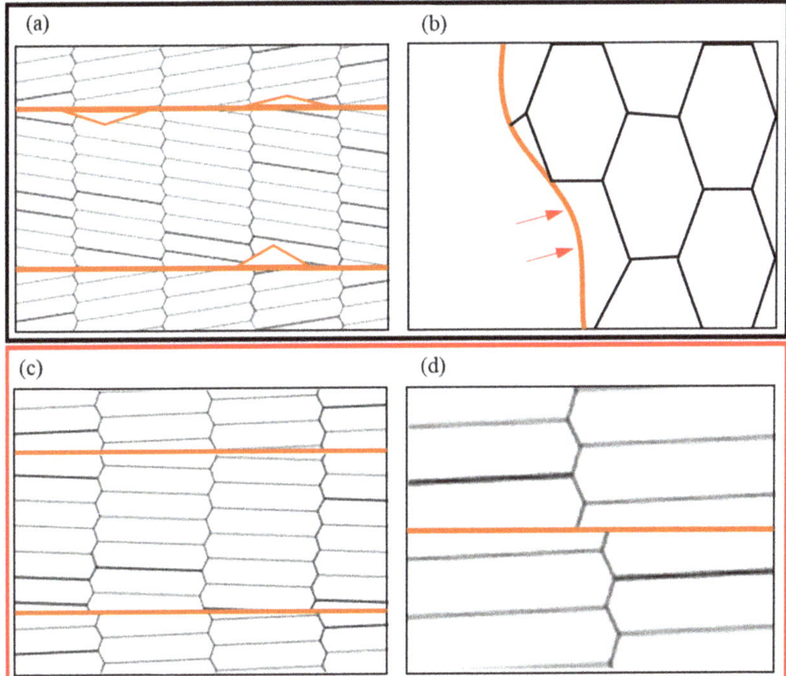

Figure 7. Schematic diagram of grain growth. (**a**,**b**) Discontinuous recrystallization processes: (**a**) serious lattice distortion and grain boundary arched nucleation; (**b**) grain growth. (**c**,**d**) Continuous recrystallization processes: (**c**) dislocations in grain recovery and subgrain boundary rotation; (**d**) slowly weakening lattice distortion with grain growth. The brown lines indicate subgrain boundaries and grain boundaries, and the black grid indicates the degree of lattice distortion.

5. Conclusions

In this paper, pure nickel was room-temperature rolled and cryorolled. The microstructures of the rolled samples and the annealed samples were observed by transmission electron microscopy (TEM). The grain growth mechanism during the annealing process of the high-purity nickel LG obtained by different rolling methods was studied, and the following conclusions were drawn:

(1)   Lamellar grain structures with a plane spacing of 68 nm and 66 nm were prepared by means of room-temperature rolling and cryorolling.

(2)   When annealed at 75 °C and 160 °C, the recovery rate of the samples after room-temperature rolling was faster than that after cryorolling.

(3)   After annealing at 245 °C, recrystallization occurred in both room-temperature rolled and cryorolled samples. Discontinuous static recrystallization occurred in the room-temperature rolled samples, whereas continuous static recrystallization occurred in the cryorolled samples; this was attributed to the greater deformation energy storage accumulated in the rolling process.

Author Contributions: Z.L.: Experiments, Investigation, Writing–original draft, review & editing. Y.W.: Experiments, Investigation, Writing—original draft, review & editing. Z.X.: Investigation, Writing—review. C.K.: Experiments, Investigation, Writing—review & editing. H.Y.: Conceptualization, Writing—review & editing, Supervision. All authors have read and agreed to the published version of the manuscript.

Funding: This research was funded by the Huxiang High-Level Talent Gathering Project of HUNAN Province (Grant No.: 2018RS3015), the Innovation-Driven Program of Central South University (Grant No.: 2019CX006), the Science and Technology Innovation Program of Hunan Province (Grant No.: 2020RC2002), and the Research Fund of the Key Laboratory of High-Performance Complex Manufacturing at Central South University.

Institutional Review Board Statement: Not applicable.

Informed Consent Statement: Not applicable.

Data Availability Statement: The data presented in this study are available on request from the corresponding author. The data are not publicly available due to ongoing studies.

Conflicts of Interest: The authors declare no conflict of interest.

References

1.   Liu, X.C.; Zhang, H.W.; Lu, K. Strain-induced ultrahard and ultrastable nanolaminated structure in nickel. *Science* **2013**, *342*, 337–340. [CrossRef]

2.   Yuan, Y.; Li, X.; Yang, W. Low-angle grain boundary structures and size effects of nickel nanolaminated structures. *J. Mech. Phys. Solids* **2019**, *130*, 280–296. [CrossRef]

3.   Lu, K. Stabilizing nanostructures in metals using grain and twin boundary architectures. *Nat. Rev. Mater.* **2016**, *1*, 16019. [CrossRef]

4.   Liu, X.; Zhang, H.; Lu, K. Formation of nano-laminated structure in nickel by means of surface mechanical grinding treatment. *Acta Mater.* **2015**, *96*, 24–36. [CrossRef]

5.   Yan, X.; Meng, X.; Luo, L.; Jing, Y.; Yi, G.; Lu, J.; Liu, Y. Mechanical behaviour of AZ31 magnesium alloy with the laminate and gradient structure. *Philos. Mag.* **2019**, *99*, 3059–3077. [CrossRef]

6.   Yu, H.; Wang, L.; Yan, M.; Gu, H.; Zhao, X.; Kong, C.; Wang, Y.; Pesin, A.; Zhilyaev, A.P.; Langdon, T.G. Microstructural evolution and mechanical properties of ultrafine-grained Ti fabricated by cryorolling and subsequent annealing. *Adv. Eng. Mater.* **2020**, *22*, 1901463. [CrossRef]

7.   Tao, J.Q.; Jiang, Y.; Liu, Y.; Liu, Y.W.; Meng, J.J.; Wang, J.T. Preparation of bulk nanolaminated aluminum alloy with high strength. *Mater. Sci. Eng.* **2020**, *770*, 138556. [CrossRef]

8.   Wang, D.-G.; Shu, X.-D.; Wang, R.; Xu, S. Mechanism of necking defect of 6082 aluminium alloy rolled by cross-wedge rolling method based on material thermal properties. *J. Cent. South Univ.* **2020**, *27*, 3721–3732. [CrossRef]

9.   Xiong, H.; Su, L.; Kong, C.; Yu, H. Development of high-performance of Al alloys via cryo-forming: A review. *Adv. Eng. Mater.* **2021**, *23*, 202001533. [CrossRef]

10.   Liu, F.; Yuan, H.; Goel, S.; Liu, Y.; Wang, J.T. Bulk nanolaminated nickel: Preparation, microstructure, mechanical property, and thermal stability. *Metall. Mater. Trans.* **2017**, *49*, 576–594. [CrossRef]

11. Chen, X.; Zhang, J.; Liu, Q. Effect of initial cube texture on the recrystallization texture of cold rolled pure nickel. *Mater. Sci. Eng.* **2013**, *585*, 66–70. [CrossRef]
12. Jin, Y.; Lin, B.; Bernacki, M.; Rohrer, G.; Rollett, A.; Bozzolo, N. Annealing twin development during recrystallization and grain growth in pure nickel. *Mater. Sci. Eng.* **2014**, *597*, 295–303. [CrossRef]
13. Xie, H.; Zhang, H.; Lu, K. Boundary faceting in goss orientated nickel with a nanolaminated structure. *Scr. Mater.* **2018**, *151*, 66–70. [CrossRef]
14. Xie, H.; Zhang, H.; Li, J.; Lu, K. Local faceting in coarsening of nanolaminates with low angle boundaries in pure nickel. *Scr. Mater.* **2016**, *122*, 110–114. [CrossRef]
15. Yu, H.; Yan, M.; Li, J.; Godbole, A.; Lu, C.; Tieu, K.; Li, H.; Kong, C. Mechanical properties and microstructure of a Ti-6Al-4V alloy subjected to cold rolling, asymmetric rolling and asymmetric cryorolling. *Mater. Sci. Eng. A* **2018**, *710*, 10–16. [CrossRef]
16. Rangaraju, N.; Raghuram, T.; Krishna, B.V.; Rao, K.P.; Venugopal, P. Effect of cryo-rolling and annealing on microstructure and properties of commercially pure aluminium. *Mater. Sci. Eng.* **2005**, *398*, 246–251. [CrossRef]
17. Ghosh, P.; Renk, O.; Pippan, R. Microtexture analysis of restoration mechanisms during high pressure torsion of pure nickel. *Mater. Sci. Eng.* **2017**, *684*, 101–109. [CrossRef]
18. Pippan, R.; Scheriau, S.; Taylor, A.; Hafok, M.; Hohenwarter, A.; Bachmaier, A. Saturation of fragmentation during severe plastic deformation. *Annu. Rev. Mater. Res.* **2010**, *40*, 319–343. [CrossRef]
19. Duan, J.Q.; Quadir, Z.; Ferry, M. An analytical framework for predicting the limit in structural refinement in accumulative roll bonded nickel. *Metall. Mater. Trans.* **2016**, *47*, 471–478. [CrossRef]
20. Lee, Y.B.; Shin, D.H.; Park, K.-T.; Nam, W.J. Effect of annealing temperature on microstructures and mechanical properties of a 5083 al alloy deformed at cryogenic temperature. *Scr. Mater.* **2004**, *51*, 355–359. [CrossRef]
21. Jazaeri, H.; Humphreys, F. The transition from discontinuous to continuous recrystallization in some aluminium alloys: II—Annealing behaviour. *Acta Mater.* **2004**, *52*, 3251–3262. [CrossRef]
22. Li, J.; Gao, H.; Kong, C.; Tandon, P.; Pesin, A.; Yu, H. Mechanical properties and thermal stability of gradient structured Zr via cyclic skin-pass cryorolling. *Mater. Lett.* **2021**, *302*, 130406. [CrossRef]
23. Nes, E. Recovery revisited. *Acta Metall. Mater.* **1995**, *43*, 2189–2207. [CrossRef]
24. Huang, Y.; Humphreys, F. Subgrain growth and low angle boundary mobility in aluminium crystals of orientation {110} <001>. *Acta Mater.* **2000**, *48*, 2017–2030. [CrossRef]
25. Ferry, M.; Humphreys, F. Discontinuous subgrain growth in deformed and annealed {110} <001> aluminium single crystals. *Acta Mater.* **1996**, *44*, 1293–1308. [CrossRef]
26. Dhal, A.; Panigrahi, S.; Shunmugam, M. Insight into the microstructural evolution during cryo-severe plastic deformation and post-deformation annealing of aluminum and its alloys. *J. Alloys Compd.* **2017**, *726*, 1205–1219. [CrossRef]
27. Li, J.C.M. Possibility of subgrain rotation during recrystallization. *J. Appl. Phys.* **1962**, *33*, 2958–2965. [CrossRef]
28. Wang, Y.B.; Li, B.Q.; Sui, M.L.; Mao, S.X. Deformation-induced grain rotation and growth in nanocrystalline Ni. *Appl. Phys. Lett.* **2008**, *92*, 011903. [CrossRef]

 *materials*

Review

Titanium in Cast Cu-Sn Alloys—A Review

Karthik Manu [1,2], Jan Jezierski [3,*], Madikkamadom Radhakrishnan Sai Ganesh [1],
Karthik Venkitaraman Shankar [1,*] and Sudarsanan Aswath Narayanan [1]

[1] Department of Mechanical Engineering, Amrita Vishwa Vidyapeetham, Amritapuri 690525, India; karthikm74125@gmail.com (K.M.); saig852000@gmail.com (M.R.S.G.); aswath610@gmail.com (S.A.N.)

[2] Department of Materials Science and Engineering, KTH Royal Institute of Technology, SE-100 44 Stockholm, Sweden

[3] Department of Foundry Engineering, Silesian University of Technology, Towarowa 7, 44-100 Gliwice, Poland

[*] Correspondence: jan.jezierski@polsl.pl (J.J.); karthikvs@am.amrita.edu (K.V.S.); Tel.: +48-691-544-485 (J.J.); +91-995-253-2461 (K.V.S.)

Abstract: The article reviews the progress made on bronze alloys processed through various casting techniques, and focuses on enhancements in the microstructural characteristics, hardness, tensile properties, and tribological behaviour of Cu-Sn and Cu-Sn-Ti alloys. Copper and its alloys have found several applications in the fields of automobiles, marine and machine tools specifically for propellers in submarines, bearings, and bushings. It has also been reported that bronze alloys are especially used as an anti-wear and friction-reducing material to make high performance bearings for roller cone cock bits and warships for defence purposes. In these applications, properties like tensile strength, yield strength, fatigue strength, elongation, hardness, impact strength, wear resistance, and corrosion resistance are very important; however, these bronze alloys possess only moderate hardness, which results in low wear resistance, thereby limiting the application of these alloys in the automobile industry. The major factor that influences the properties of bronze alloys is the microstructure. Morphological changes in these bronze alloys are achieved through different manufacturing techniques, such as casting, heat treatment, and alloy addition, which enhance the mechanical, tribological, and corrosion characteristics. Alloying of Ti to cast Cu-Sn is very effective in changing the microstructure of bronze alloys. Reinforcing the bronze matrix with several ceramic particles and surface modifications also improves the properties of bronze alloys. The present article reviews the techniques involved in changing the microstructure and enhancing the mechanical and tribological behaviours of cast Cu-Sn and Cu-Sn-Ti alloys. Moreover, this article also reviews the industrial applications and future scope of these cast alloys in the automobile and marine industries.

Keywords: creep; intermetallic; cast Cu-Sn; spinodal bronze; cast Cu-Sn-Ti; two-phase zone continuous casting

Citation: Manu, K.; Jezierski, J.; Ganesh, M.R.S.; Shankar, K.V.; Narayanan, S.A. Titanium in Cast Cu-Sn Alloys—A Review. *Materials* **2021**, *14*, 4587. https://doi.org/10.3390/ma14164587

Academic Editors: Hailiang Yu, Zhilin Liu and Xiaohui Cui

Received: 14 June 2021
Accepted: 26 July 2021
Published: 16 August 2021

1. Introduction

Understanding the different properties of an alloy is crucial if its service is to be extended under particular working conditions. Various studies on Cu-Sn alloys have been performed to expand their applications. Lead-based solders used to be very common, but there was a search for a better lead replacement due to its highly toxic nature. Cu-Sn alloys turned out to be a better replacement because of their non-toxic nature and significant abundance. Bronze alloys were found to be a viable material to make primer and detonating cords because of their high impulse transfer. The high conductivity of bronze alloys make them useful as a conducting material for winding applications, which will be described later. Likewise, many properties of Cu-Sn alloy have shown a wide range of compatibility with steel. Due to the high strength and low density of titanium, when compared to other alloying additives, the effects of titanium on various alloys have been studied by many researchers, such as in steel [1]. The paper presents changes in the impact strength and abrasive wear resistance of cast high manganese steel due to the formation of primary titanium carbides. For high chromium cast iron, another group of authors investigated the influence of titanium on its crystallization process and wear resistance [2]. Titanium also possesses the property of corrosion resistance when in contact with seawater or chlorine. This explains its use in marine industries and ships. Furthermore, the Cu-Sn-Ti alloy group has been used to make high-strength ageing materials and abrasive binders. This system showed a lower melting point and good wear properties. The outstanding wetting property of this system was attributed to the strong joints existing between abrasives and matrix. Titanium alloys have also found broad application in biomaterials because of their almost neutral effect on the human body.

The current paper reviews cast Cu-Sn-Ti alloys and presents a wide range of experiments that have been conducted and lists their applications, together with an analysis of bronze alloys and their applications. The superior nature of Cu-Sn-Ti alloys is proved along with an explanation of opportunities for future development and research.

2. Cast Bronze Alloys

Cast bronze alloys are broadly identified as C80000 and C90000 series, and are mainly produced by sand casting, and continuous or centrifugal casting process routes. The main intermetallic phases formed during processing include Cu_3Sn and Cu_6Sn_5. Deformation is based on dynamic recrystallization (DRX), where Cu_6Sn_5 particles dominate the process of particle stimulated nucleation (PSN) over those triggered by a boundary (boundary induced nucleation). Adding Sn to Cu improves various properties and allows the resulting alloy (bronze) to meet the requirements of many applications. To prove its use as a solar-reflecting substance, Alex et al. [3] researched the behaviour of the Cu-Sn intermetallic system. In this analysis, a customized replacement of aluminium or zinc in the bronze alloy was performed. A single-phase cast bulk alloy (Cu-21.2 wt%Sn) allowed in arc melting and chill casting of bronze alloys, which displayed a solar reflectance of 74.5% and a specular reflectance of 77.2%, combined with a hardness of 5 GPa. In addition, an aluminium substitution of about 4–21.2% was found to increase the solar reflectance of bulk alloy by 10% and the specular reflectance by 14%. A zinc substitution of 2–9%, on the other hand, induced a 1% decrease in bulk specular reflectance and a 1% decrease in bulk solar reflectance. Therefore, it has been shown that traditional metallurgical processing, when compared to multilayer solar-reflective coatings, can produce bulk materials with acceptable solar and specular reflectance. The next section of this analysis focuses on the effects of modifications to Sn and their effects on the microstructure.

3. Effect of Sn Additions

Because of their high corrosion resistance, as well as their outstanding thermal and electrical conductivities, Cu and its alloys are commonly used in many fields as bearings and bush sleeves, and in many other applications. An excellent way to obtain high strength and fine grains for an alloy is to use the process of severe plastic deformation (SPD). ECAP, or equal channel angular pressing, is a form of material processing that applies very high strains, leading to severe work hardening and the refining of product microstructure. Elshafey et al. [4] clarified the impact of equal channel angular pressing (ECAP) passes, a material processing method that causes very high strains in samples, on the material properties of Cu-Sn alloys. A microstructural refinement that contributes to work hardening is triggered by ECAP. The alloy was processed via ECAP for 1, 4, and 5 passes to a diameter of 12 mm at room temperature, with a 120° inner angle, and a 20° die-out arc angle. It was found that, with an increase in Sn content, or with the number of ECAP passes, the grain size decreased. It was noticed that, for 10% alloys, the proof strength, ultimate tensile strength (UTS), and Cu-2%, 5% and 10% Sn alloy hardness improved with either an improvement in the number of ECAP passes or an increase in the content of Sn. UTS reached its peak after 4 passes for 2% Sn content, and UTS only reached its limit after 3 passes for 5% Sn content. The wear resistance of bronze alloys was also found to be substantially improved by adding more Sn or by increasing the ECAP pass number. A substantial decrease in surface roughness could also be achieved by increasing the number of ECAP passes or by decreasing the content of Sn. In addition, the effects of Sn addition on the microstructure, wear properties, and surface hardness of Cu-Sn refined alloys [5] was investigated; gas tungsten arc was used in the study as the heat source to conduct surface refining. Bronze alloys with various compositions of Sn were chosen for the test. Using an optical microscope, a microstructural examination was carried out, and hardness was assessed using the Vickers hardness test. As shown in Figure 1, the Cu-10Sn alloy microstructural investigation of the as-cast condition led to confirm the occurrence of a dendritic structure. The Cu-10Sn alloy surface refining, as shown in Figure 2, resulted in a microstructure with fine grains. This shows that, by conducting surface refining, the hardness and wear properties of Cu-10Sn alloys can be improved. Homogenization of the refined Cu-10Sn alloy surface led to a microstructure without dendrites as shown in Figure 3. In addition, observations under optical microscopes showed that, even after providing alloy heat treatment, the second step did not occur. The hardness and wear properties were found to be dependent on the Sn content. The relationship between hardness and wear rate was found to comply with Archard's theory. The wear rate was found to decrease with Sn addition. In addition, it was found that the coefficient of friction was constant and independent of hardness.

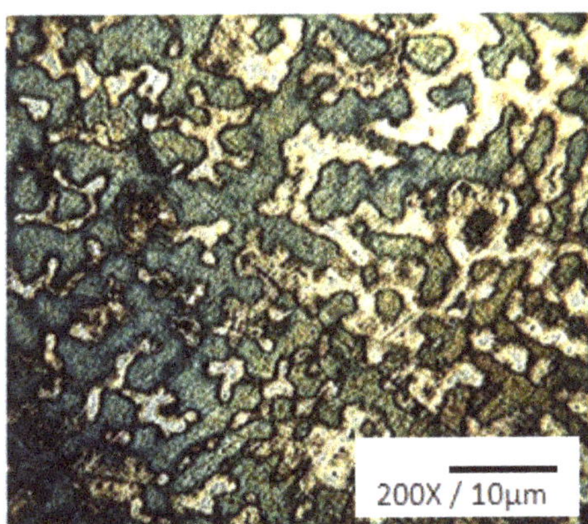

Figure 1. Microstructure of Cu-10Sn alloy in as-cast condition [5].

Figure 2. Microstructure of Cu-10Sn alloy in a refined state [5].

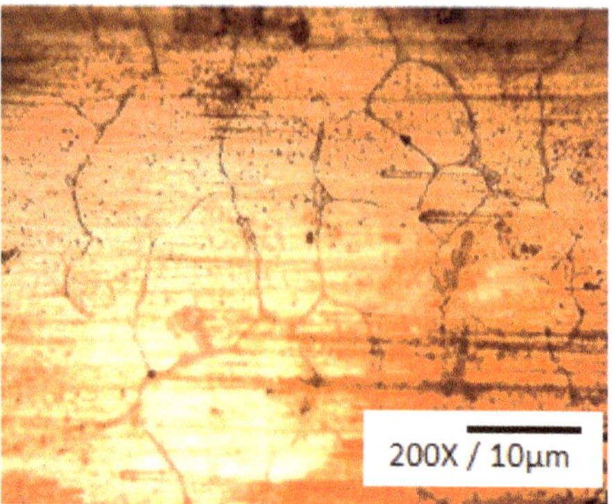

Figure 3. Microstructure of homogenized Cu-10Sn alloy [5].

To obtain the variance of hardness and wear properties for Cu-10Sn alloys, an experiment was performed. Minor additions of nickel were made to bronze alloys because bronze alloys showed a low tensile strength. This resulted in obtaining spinodal alloys with excellent corrosion resistance, fatigue property, and tensile strength. Therefore, to govern the effect of Sn addition on the microstructure and hardness of spinodal bronze alloys, an experiment was performed by Shankar et al. [6]. Three of the elements were combined by heating them in an electric furnace at a temperature of 825 °C for 10 h and then aged for 1–5 h at 250, 300, 350, 400, and 450 °C. Experimental results were obtained in the form of a plot between peak hardness and ageing temperature. Figure 4 shows that with an increase in the content of Sn from 4–12 wt%, there was an improvement in the peak hardness value. In addition, the peak ageing period was independent of the Sn content. It was found that the optimal ageing period and optimal ageing temperature were 3 h and 350 °C, respectively. When the alloy was over-aged, this analysis additionally confirmed the formation of grain boundary precipitates.

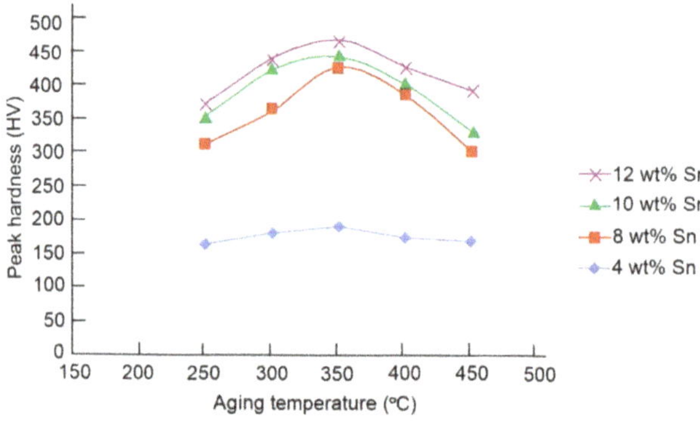

Figure 4. Peak hardness versus ageing temperature [6].

4. Properties of Bronze Alloys

Owing to their high strength, good wear resistance, good corrosion resistance, and simple casting properties, Cu-Sn alloy has been widely used in many fields. Rapid solidification processing is an alloy preparation process that, unlike ordinary processes, can result in a polished microstructure and produce a material with superior efficiency. Various properties and production methods have been studied to enhance the properties of the alloys. As presented in [7], solidification rate has a significant impact on a high-tin bronze microstructure. Other research projects presented how tin content changes the mechanical properties of bronzes [8]. Figures 5–7 show the tensile strength, hardness, and impact strength for various tin contents.

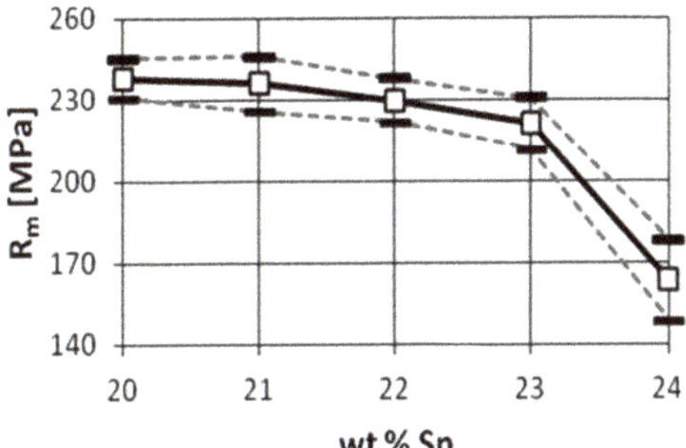

Figure 5. Tensile strength of high-tin bronzes versus tin content [8].

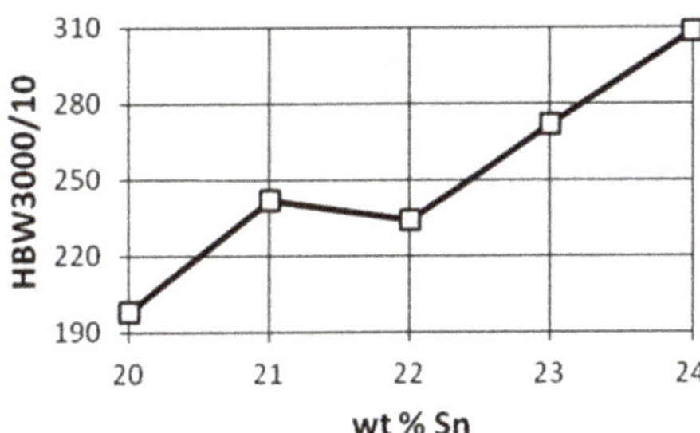

Figure 6. The hardness of high-tin bronzes versus tin content [8].

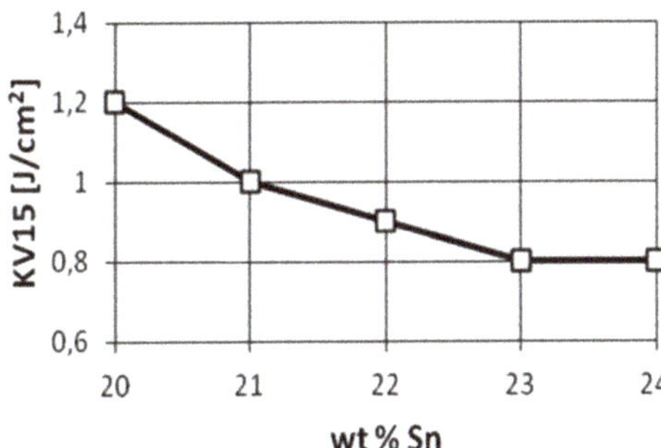

Figure 7. Impact strength of high-tin bronzes versus tin content [8].

The effects of the process parameters on surface-modified Cu-Sn alloy microstructure, hardness, and wear properties were investigated by Cherian et al. [9]. Their surface modification method used a tungsten gas arc serving as the source of heat. A pin-on-disc wear tester and a micro-hardness tester were used to determine the wear rate and layer hardness. It was found that, when compared to as-cast samples, the surface modification process enhanced the alloy hardness and wear properties. In addition, as shown in Figure 8, the presence of uniformly distributed fine grains was confirmed in the microstructure of the modified layer. Hence, a refined microstructure was provided by bronze alloy surface modification. The wear rate was found to decrease with hardness, and the coefficient of friction was found to be independent of hardness.

Figure 8. Modified structure of Cu-10Sn alloy [9].

Zhang et al. [10] investigated the resistivity and mechanical properties of various compositions of bronze alloys. Bronze (Cu-0.5 wt% Sn) alloys were isothermally annealed after cold rolling and found to have 136.95% of the International Annealed Copper Standard (IACS) electrical conductivity. With the addition of Sn, electrical conductivity and mechanical elongation decreased. On the other hand, the addition of Sn provided a substantial increase in the alloy's tensile strength. It was also concluded that the annealing procedure improved alloy conductivity and mechanical elongation. In studies conducted by Salmat et al. [11], the effect of a high-tin bronze chemical composition on the physical, mechanical, and acoustic properties of gamelan materials was investigated. Yang et al. [12] investigated the effect of the super-gravity field on Cu-Sn alloy tensile properties and grain refinement. In their analysis, two furnaces were used in the apparatus, rotating around a fixed axis. It was concluded that Cu-11 wt% Sn alloy grain refinement was promisingly higher in a super-gravity region. The average grain size was found to be 2.13 mm in the normal gravity field, while it was 0.35 mm, 0.173 mm, and 0.074 mm in the super-gravity fields of G (gravity coefficient) = 100, 300, and 600. The reduction in grain size was confirmed in Figure 9, where there is a greater strength of the super-gravity field. Cu-11 wt% Sn in the normal gravity field achieved a tensile strength value of 265 MPa while 449 MPa, 487 MPa, and 521 MPa were achieved for super-gravity fields of G = 100, 300, and 600, respectively. The morphology of failure changed from fragility to plasticity with raising the gravity coefficient. The collapse of crystal nuclei into the solidified melt was only greatly exacerbated by super-gravity at the early stage of solidification, which contributed to solid structure refinement. An increased concentration of Sn has been found to help grow grain refinement, thus increasing the rate of nucleation. The increase in tensile characteristics can be explained by the grain size reduction.

Tensile tests on tin bronzes were performed at temperatures ranging from 150 °C to 250 °C [16]. The micrographs, as shown in Figure 10, present the fractures of the samples under examination. The deformation temperature is the main factor in deciding the fractography of measured surfaces. Annealed and continuously cast bronze in the transition temperature range from ductility to brittleness, CuSn6P showed a mixed fracture in the vicinity of smooth cleavage surfaces, as shown in Figure 10a. This was a distinctive brittle inter-crystalline cracking occurrence. It was found that ductile surfaces in the fracture were less frequent with an increase in deformation temperature, as shown in Figure 10b. On the inter-crystalline surfaces, in the TE range, traces of plastic deformation and cavitation were observed. In contrast, deformation happens as there is a change in temperature which lead to fracture formation on fully inter-crystalline brittle surfaces. Inter-crystalline fissures have been discovered to nucleate at grain boundaries at micro-voids, and at the intersections between deformation bands and the grain boundary. It was also found that intercrystalline cracking found between two or three grains on the boundary surfaces ran through cavitation pores, which is typically observed in cast structures.

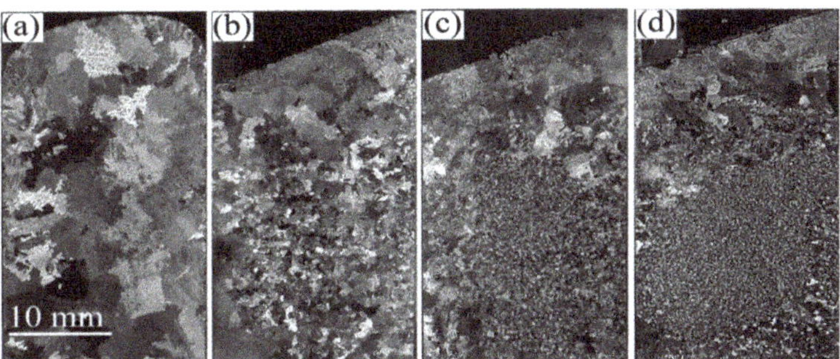

Figure 9. Cu-11 wt%Sn macrostructure at (**a**) G = 1, (**b**) G = 100, (**c**) G = 300, and (**d**) G = 600 [12].

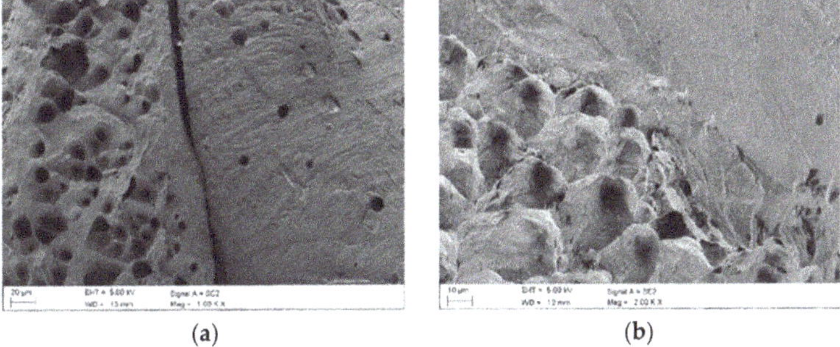

(**a**)   (**b**)

Figure 10. Fracture surface at temperature (**a**) 150 °C; (**b**) 250 °C [16].

A research project was conducted to determine the effects of thermal factors on the morphological and mechanical behaviour of directionally solidified Cu-Sn alloy [17]. Cu-20 wt% alloy temperature mapping was performed at the time of solidification by fixing thermocouples from a refrigerated base at various positions, namely, 5, 10, 15, 20, 35, 45, and 60 mm. An important improvement in the morphologic activity of the Cu-Sn alloy was observed. It was noted that, when compared to the position of 60 mm, finer dendritic structures were detected at the position of 5 mm. This is because the position of 5 mm was closer to the cooling plate and therefore had a higher cooling rate compared to the position of 60 mm, as shown in Figure 11.

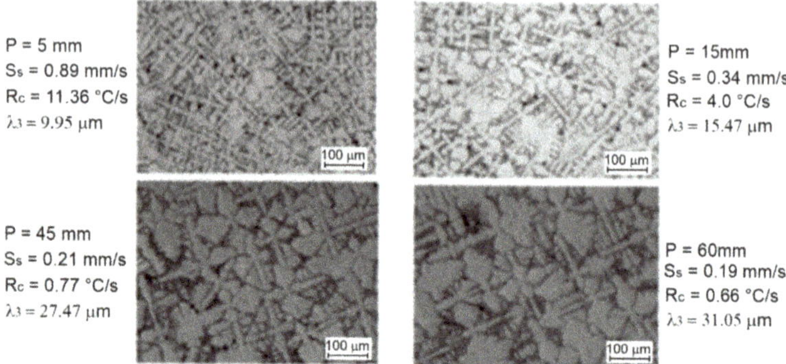

Figure 11. Morphological behaviour of Cu-Sn alloy positioned at 5, 15, 45, and 60 mm [17].

5. Analysis of Cu-Sn Cast Samples

The effects of processing factors on the microsegregation of Cu-8 % Sn alloy directionally cast samples was studied [18]. Samples of Cu-Sn alloys were prepared in this study under four experimental conditions, leading to four distinct dendritic growth varieties. By using quantitative metallographic and microprobe analyses, cylindrically formed samples were divided into transverse slices and the degree of microsegregation was observed. The three calculated microsegregation indexes included the segregation variance parameter, volume fraction of the eutectoid, and tin concentration at the centre of the primary dendrite arms. Near the surface of the cast sample, there was a decline in microsegregation. In addition, different levels of microsegregation in the columnar and equiaxial regions of the sample were observed. This was attributed to the difference in inter-dendritic distances. Moreover, this study was able to predict the microsegregation behaviour of peritectic alloys. Bayle et al. [19] experimented with flow stress and recrystallization parameters when Cu-9%Sn alloys (Cu–9Sn and Cu–9Sn–0.26Zn (wt%) were hot deformed from ambient temperature to 750 °C, with nominal (initial) strain rates ranging from 10^{-3} to 10^{-1} s. A plot between true stress and strain curves was determined. Optical metallography was used to determine the dynamic grain recrystallization (DRX). The dynamic recrystallization of grains stimulated a necklace-type structure when exposed to high temperatures. In addition, electron back scattered diffraction (EBSD) measurements were carried out and, as a result, the crucial role of twinning on these alloys was revealed. The heat strain of these alloys has led to improvements in the microstructure and stress–strain curves. Dynamic recrystallization is a slow process and is incompletely propagated when the strain magnitude is 1.2. It was also inferred that solute drag can control the behaviour of a metal alloy. The characteristics of flow curves, such as high-rate sensitivities, lack of work hardening, and traditional transient flow curves, have confirmed this. Twinning leads to recrystallized necklaces being formed along the borders of deformed grains. The electromigration phenomenon was tested by Chen et al. [20] for interfacial reactions in cast Cu-Sn joints. Samples were reacted at 170 °C and 180 °C for 24–240 h by passing an electric current with a density of 5000 A/cm^2 to perceive the effects of electromigration. At the interface, uniform layers of Cu_6Sn_5 and Cu_3Sn were formed. However, the interfacial reactions were found to be the same, even when an electric current was not supplied, i.e., the electron flows from the Sn side to Cu side. Dissolution of Cu in the Sn matrix during casting initiated a thick Cu_6Sn_5 phase. Compared with a previous study, it was found that electromigration played an important role when the direction of the electric current was from the Cu site to the Sn site. Diffusion of Cu near grain boundaries and surfaces promoted the growth of a large non-planar Cu_6Sn_5 phase in the microstructure.

Akbarifar et al. [21] compounded cast molten Al melt around a brass cylinder at 700 °C and 750 °C, with 3 and 5 as the melt–solid volume ratios. Upon increasing the temperature and melt–solid volume ratios, the rate of reaction was increased and there were three intermetallic layers. Intermetallic layers consisted of Cu-Zn, Al_4Cu_3Zn, and Al_2Cu along with eutectic α-Al/Al_2Cu and Al dendrites. The presence of lead-rich phases was confirmed in all layers. XRD and SEM analyses proved the existence of pores at the micro-level and cavities in the dendritic and eutectic layers. The presence of micropores was due to bubble entrapment. There was an indication of a solid ring around the solid core when the brass cylinder rod was cut and observed with SEM. The presence of Al_2Cu and Al_4Cu_3Zn layers prevented the saturation of the solid solution with Cu, which resulted in the absence of Cu_4Al_4. Moreover, the hardness of the intermetallic layers from the centre of the brass core to the Al outer ring was found to be 513, 477, and 650 Vickers, respectively.

6. Phase Diagram of Cu-Sn Alloys

Since the 19th century, since its introduction, several studies have been conducted to fully understand the Cu-Sn phase diagram and its phase relationships. Binary Cu-Sn alloys are very important as a primary replacement for lead-containing solder materials. By using thermal analysis, metallographic methods, and crystallographic analysis, the alloy phase diagram was determined for samples. As shown in Figure 12, Furtauer et al. [22] implemented a new phase diagram. It was noted that all Sn compositions (from 11% to 48%) were recovered and quenched in cold water, except for the β and γ phases. Since the β ($Cu_{17}Sn_3$) and γ (Cu_3Sn) phases were of a high-temperature, they were not quenched, and either transformed into a metastable phase or underwent bulk transformation to their respective low-temperature phases. This β phase can be transformed into the γ phase by a higher-order reaction. Moreover, providing an effective temperature and concentration control facilitates the random orientation of Sn atoms in the BiF_3 structure. All the results were summed up and expressed in the form of a new phase diagram.

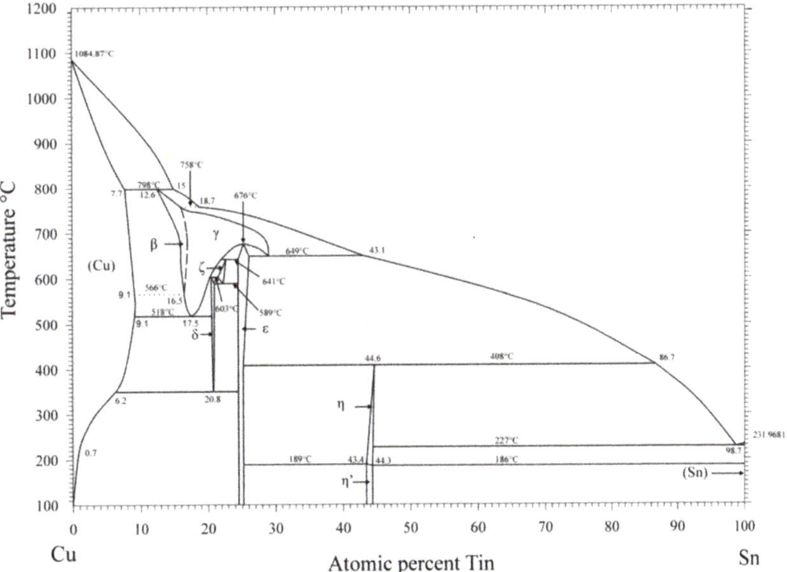

Figure 12. New Cu-Sn phase diagram, as suggested by Furtauer et al. [22].

Li et al. [23] investigated the effect of Sn content on the mechanical properties of α-phase Cu-Sn alloys. After casting, it was inferred that the Cu-Sn dendrites were prone to variations in the orientation of the α phase. This was due to a change in the Young's modulus, which was found to range from 120 to 130 GPa. By incorporating instrumented nano-indentation techniques, this experiment ascertained a linear relationship between the hardness of the α phase and successive additions of Sn.

7. Experiments on Molten/Liquid Cu-Sn Alloys

The liquid state provided a higher mobility to electrons, which enhanced the conductive property of materials. Liquid structures of bronze alloys can be broadly classified into Cux-Sn100-x where x varies from 0 to 100 (i.e., x = 0, 10, 20, 33, 40, 50, 60, 75, 80, 100). Electrical resistivity and viscosity were the two dominant guiding features in the analysis of transport properties. Jia et al. [24] arrived at three conclusions by examining the transport properties of Cu-Sn alloys. Firstly, when plotting a graph between resistivity and temperature, as shown in Figure 13, $Cu_{75}Sn_{25}$ had the lowest value for the temperature coefficient of resistivity ($-9.24\ \mu\Omega cmK^{-1}$) followed by $Cu_{80}Sn_{20}$ ($-8\ \mu\Omega cmK^{-1}$). Except for these two compositions, all other compositions of Cu-Sn alloys possessed a positive temperature coefficient of resistivity (TCR), which indicated that their resistivity will increase with increasing temperature.

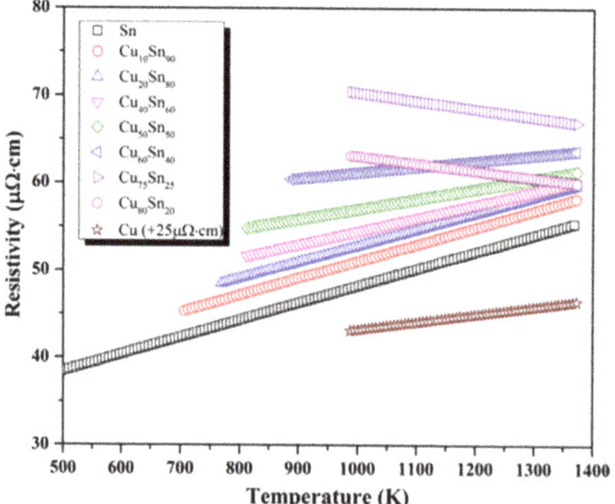

Figure 13. Temperature versus resistivity curve for liquid Cu-Sn alloys with a cooling rate of 5 K/min [24].

Secondly, the variation in resistivity with composition was also studied. Figure 14 shows that an increase in the atomic percentage of copper will result in a decrease in the TCR value until the achievement of a $Cu_{75}Sn_{25}$ composition, further increase in Cu content resulted in a lower TCR. This confirmed the presence of atomic clusters in the $Cu_{75}Sn_{25}$ alloy with a negative TCR. All the compositions that were similar to this alloy ($Cu_{75}Sn_{25}$) possessed a significant number of atomic clusters. Being more different to the composition resulted in a minimal concentration of atomic clusters in the alloy. Hence, $Cu_{33}Sn_{67}$ alloys possessed a fine structure.

Finally, as shown in Figure 15, temperature-dependent resistivity was clarified. In terms of the Arrhenius equation, all compositions of Cu-Sn alloys were found to be have a viscosity that increased with decreasing temperature. Here, $Cu_{75}Sn_{25}$ shows a maximum value for the activation energy and the volume of the flow unit. This proved the existence of a stronger bond in $Cu_{75}Sn_{25}$ alloy that confirmed the presence of atomic clusters in $Cu_{75}Sn_{25}$ alloy. An oscillating-cup viscometer was used by Tan et al. [25] to determine the dynamic viscosities of Cu-Sn alloys at the same superheating temperature. Molten Cu-Sn's viscosity increased with a decrease in temperature, as found using the Arrhenius equation. It was found that the viscosity was higher near the beta step (Cu5Sn) at a given temperature, along with the peak value occurring at Cu-25 wt% Sn. When Sn concentration was made to range from 20 to 40 wt%, the rate of the viscosity increase with temperature reduction was rapid. In addition, at the same temperature, the viscosity of pure Cu was greater than that of pure Sn alloy. Experiments to observe the effects of a magnetic field on the viscosity of molten Cu-Sn alloys were performed by Mao et al. [26]. An oscillating-cup viscometer was used along with a unique range of Gauss horizontal magnetic fields to obtain the dynamic viscosity of pure Cu and Cu-10% Sn alloy. It was found that, with a large decrease in temperature, the viscosity of Cu-10% Sn increased, while the relationship between temperature and viscosity followed the Arrhenius relationship. With an increase in the power of the magnetic field, the viscosity increased, considering the external magnetic field. When compared to Cu-10% Sn alloy, the effect of the magnetic field on the viscosity of pure Cu was greater.

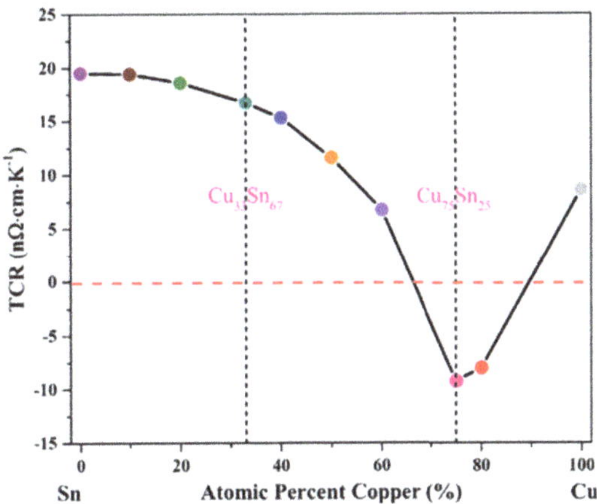

Figure 14. Composition dependence with TCR [24].

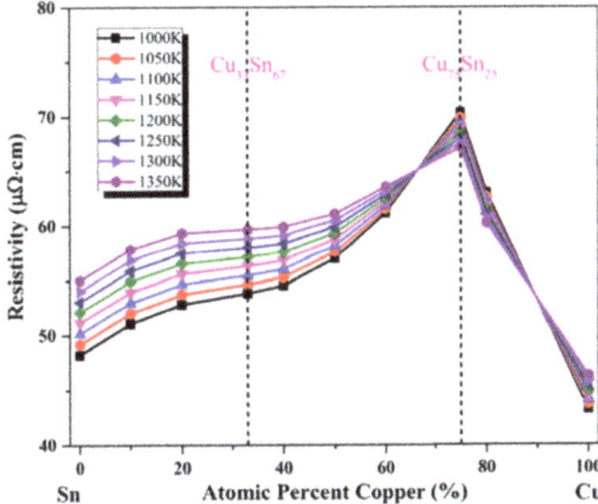

Figure 15. Composition dependent on resistivity for liquid Cu-Sn alloys [24].

8. Processes Conducted on Cu-Sn Alloy

8.1. High-Pressure Torsion

A good grain refinement normally results in SPD or extreme plastic deformation. It also contributes to different phase transitions, such as phase dissolution. To understand the grain refinement induced by Korneva et al. [27], the responses of intermetallic compounds in the Cu-Sn system to SPD were analysed, providing information on the grain refinement of intermetallic compounds when Cu-Sn alloys were subjected to high-pressure torsion (HPT). The sample of Cu-36 wt% Sn chosen for this study consisted of coarse-grained alternating plates of ζ and ε compounds with a thickness of approximately 100 μm.

At room temperature, the alloy was exposed to HPT. Increased dislocation density was caused by HPT and led to grain refinement. In the ε process, high-angle grain boundaries were more pronounced. Even after HPT, the macroscopic form of the alternating plates remained unchanged, i.e., the grain refinement took place separately within ζ and ε plates. Although the HPT created cracks in ε phase plates and increased the ε phase hardness, phase transformations and phase dissolutions were absent. Upon completion of HPT, identical conditions were found in the Cu-14 wt% Sn alloy.

8.2. Rolling

Cu-Sn alloy can be effectively used as a cladding material in explosives due to its good workability and the property of absorption of small thermal neutrons in a highly radiative environment. Liu et al. [28] incorporated electron backscatter diffraction (EBSD) technique to research the microstructure, dynamic restoration, and grain boundary, twinning, and recrystallization texture of Sn-0.5 wt% Cu alloy. After rolling at a moderate strain rate, the alloy showed a bimodal grain structure comprised of 97% fine grains (less than 30 μm in size) and 3% coarse grains (with 50–150 μm in size). Optical micrograph (OM) and back-scattered SEM micrographs showing the morphology and scale of the as-cast Sn-0.5 wt% Cu alloy in the primary β-Sn, eutectic and intermetallic processes were analysed, as shown in Figure 16. The grain size for the dendritic structure of β-Sn was found to be in the range of 100–300 μm. Both (d) and (e) in Figure 16 confirm the presence of needle-like morphology shapes, 1–3 μm in diameter and 10–20 μm long. In addition, the structure was not present in any other intermetallic phases, such as Cu_3Sn. Deformation was caused by dynamic recrystallization (DRX), where Cu_6Sn_5 particles dominated the process of particle stimulated nucleation (PSN) over those triggered by a boundary (boundary induced nucleation). {3 0 1} and {1 0 1} twins were observed as two additional mechanisms for effective deformation. In addition, the results of the paper by Liu et al. [28] proved that the deformation mechanism initiated by the PSN mechanism leads to the formation of <0 0 1>//RD oriented nuclei. Moreover, a necklace type of structure was also found along the boundary due to continuous dynamic recrystallization (CDRX). It has been observed that <1 1 0>//RD fibre structure is strengthened along with the growth of grains.

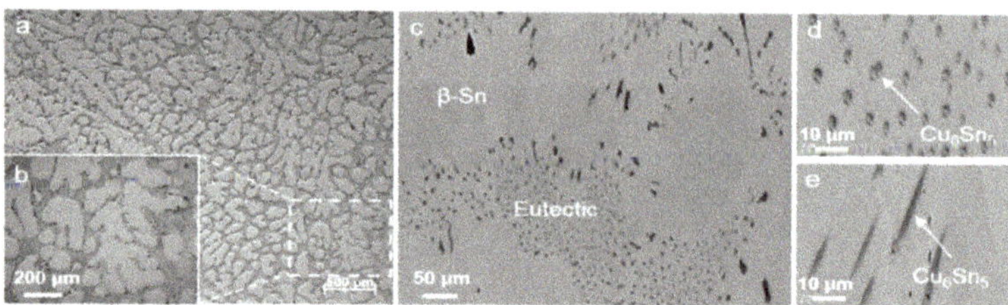

Figure 16. Micrograph and size of primary β-Sn dendrites, eutectic and intermetallic phase in F temper Cu-0.5 wt% alloy when observed under (**a**,**b**) OM, (**c–e**) SEM [28].

In the work of Wang et al. [29], spray forming was used to prepare bronze alloys with high Sn content (Cu-13.5 wt%Sn). In addition, the feasibility and characteristics of cold rolling were investigated. It was found that, as shown in Figure 17, spray-shaped Cu-13.5 wt% Sn had a fine homogeneous single-phase structure, showing excellent workability as compared to standard cast samples. With a 15% decrease in thickness for a single move, the alloy can be cold-rolled, while the overall thickness reduction can reach 80%. The spray formed Cu-13.5 wt%Sn, upon completion of proper thermomechanical treatment, acquired outstanding mechanical properties. Relationships were made between the strength, elongation, and reduction in cross-section. It was also concluded that the resulting alloy acquired a great combination of elastic modulus (88 GPa) and high flow stress (800 MPa) after cold formation. This superior Cu-13.5 wt% Sn alloy property makes them usable in the spring manufacturing industry.

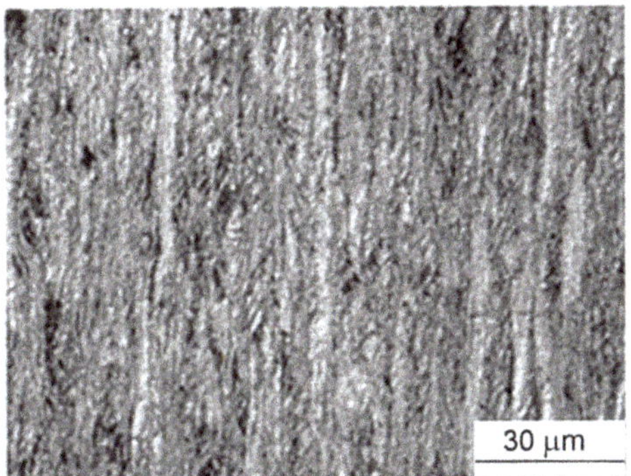

Figure 17. Optical microstructure of cold-rolled Cu-13.5 wt%Sn alloy [29].

8.3. Annealing

Four types of twins were formed at 700 °C in the initial stages of annealing. They were twins on the edge, partial twins, complete twins, and enclosed twins. In the as-cast Cu-4 wt% Sn alloy, no annealing twins were present. In the initial stages of annealing, edge twins were verified, which conformed to the Full man and Fisher model. Enclosed twins were only identified at 700 °C after 40 min of annealing. When the time approached 60 min, at the same annealing temperature, there was a loss of dendritic structures. Two mechanisms were put forth by Liu et al. [30] to explain the formation of twin grains. The first was the tin diffusion mechanism during annealing, i.e., stress fields created were attributed as the reason for recrystallization and twinning induction. Another mechanism that explained the formation of twinned grain was diffusion induced grain boundary migration (DIGM), utilizing the Kirkendall effect.

The tensile properties of bronze alloys were the least affected when annealing temperature ranged from 400 to 750 °C. Han et al. [31] fabricated bronze using a mix-alloy process, i.e., strengthened by dispersion of nano-scale refractory particles. This process seemed to be useful and stable under high-temperature applications. There was a significant improvement in tensile strength when the cold drawing ratio was increased. The increased annealing temperature for cold-drawn dispersion strengthened (DS) bronze decreased the levels of tensile strength and yield strength. This also confirmed that failure was due to the presence of splits in the surface of the DS bronze. This was the main reason for undissolved amounts of Ti and B elements, along with TiB_2 compounds. DS bronze showed high mechanical properties when compared to commercial phosphorous bronze.

9. Casting Processes Conducted on Cu-Sn Alloys

9.1. Investment Casting

One of the most used types of metal casting is investment casting. The benefits of this technique include the ability to manufacture goods with thin walls, smooth surfaces, precise shapes, and dimensions. One of the main criteria regulating the consistency of the casting is fluidity. Fluidity is a liquefied molten metal's ability to glide and fill each part of a mould. Two key parameters, composition and pouring temperature, have a major impact on this property. The main objective of Slamet et al. [32] was to examine the effect of Sn additions on Cu-(20-25) wt%Sn alloy composition and the impact of pouring temperature used in investment casting methods on microstructure and fluidity. The fluidity of Cu-Sn alloys was discovered to improve with an increasing cavity thickness and pouring temperature. A decrease in fluidity would increase the content of Sn. Therefore, the pouring temperature should be increased, and the content of Sn must be reduced to minimize casting defects in Cu-Sn alloys.

9.2. Two-Phase Zone Continuous Casting

The correlation between surface quality and process parameters was well established by Liu et al. [33]. A smooth surface of Cu-4.7Sn alloy can be obtained by controlling the speed of continuous casting and the temperature at the entrance of the two-phase zone. With a cooling water temperature of 18 °C, a smooth surface was preserved, and the cooling water flow rate was set to 400 L/h. The microstructure of the resulting alloy showed that Al formed columnar grains near the edge of the alloy plate, grain-covered grains (GCG), continuous columnar, and equiaxed grains in the middle, along with small grains with self-closed-grain boundaries. To preserve the smooth surface properties, the continuous casting speed and temperature were set to 20–30 mm/min and 1020–1040 °C, respectively, at the entrance of the two-phase zone mould.

Important mechanical properties that govern the failure mechanisms, corrosion resistance, and conductivity of Cu-Sn alloy with a wide range of solid–liquid two-phase zone interfaces were investigated by Liu et al. [34], see Figure 18. This study compared specimens processed using water-cooled mould continuous casting (WMCC) with the specimens produced using two-phase zone continuous casting (TZCC). Covered continuous columnar grains along with non-columnar small grains with self-closed grain boundaries were observed during the TZCC process. They both had the same phase microstructure. Alloys that have undergone TZCC show enhanced ductility and a 255-MPa tensile strength. The tensile strength increase was due to the existence of multiple self-closed grain boundaries that inhibited the motion of dislocation. Moreover, this reason can also be attributed to the excellent corrosion resistance of the alloy, i.e., the grain boundary corrosion was limited. An increase in ductility and a decrease in electrical resistance was due to the presence of continuous columnar grains in the microstructure of the resultant alloy. Conductivity increased to 12.2% when maintained at room temperature. In addition, elongation to failure of the resulting alloy was observed to be at the level of 49%. After TZCC processing of Cu-Sn alloy, large columnar and small grains were obtained. The term grain-covered grains (GCG) was used when small grains were found to be covered with large columnar grains. Five processing parameters (cooling distance, speed of TZCC, the temperature of the melt, mould, and cooling water) were inputted into the back propagation (BP) artificial neural network to obtain the characteristics of their microstructures. From Luo et al. [35], accurate predictions for the microstructure were made using the concept of nine quantities (size of columnar grains, size of small grains, number of columnar grains with small grains, number of columnar grains without small grains, the maximum and minimum number of small grains within columnar grains, number of small columnar grains within columnar grains and boundaries of columnar grains or at the surface of alloy).

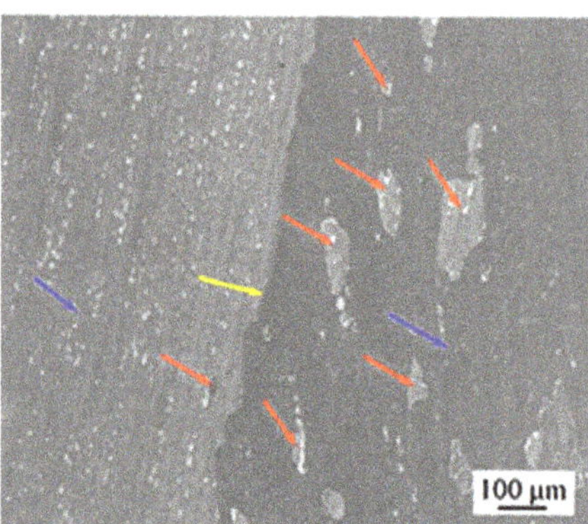

Figure 18. SEM image of GCGs in the microstructure of Cu-Sn alloy after TZCC [34].

Microstructural investigation using an optical microscope and the EBSD technique by Luo et al. [36] proved that initial nucleation of small grains resulted in the front of the solid–liquid interface region during TZCC. The angular difference between [0 0 1] columnar grains and the direction of heat flow was found to be less than 19°, which was attributed to the faster growth rate. Similarly, the slower growth rate for small grains was because of the magnitude of the angle exceeding 19° between [0 0 1] small grains and the direction of heat flow. When both columnar and small grains were in contact, the growth of columnar grains was more pronounced when compared to small grains. This led to the formation of GCGs and much smaller grains at the grain boundaries of columnar grains. Microstructure and its quantification have been well-explained by Luo et al. [36] after the TZCC process. Finally, Luo et al. [37] made use of ProCast software that simulates the TZCC process to analyse solute (Sn) distribution and surface segregation taking place along the process. Cu-4.7 wt%Sn alloy with a broader solid–liquid phase zone was under investigation. Figure 19 shows that the solidification of the alloy was first found at the centre and initiated the 'Λ' structure. Moreover, a narrow gap was formed between the solid/liquid interface and a wall in the two-phase zone mould was created. This formation of the gap was attributed to the formation of the 'Λ' structure. The alloy solidified along the direction opposite to the direction of continuous casting. SEM analysis revealed that redistribution of Sn solute occurred, which was the reason for the solid grain nucleation near the walls of the two-phase zone mould. Luo et al. [37] have also proved that the liquid enriched Sn solute caused by solid grain growth will increase the amount of Sn solute at the solid–liquid interface.

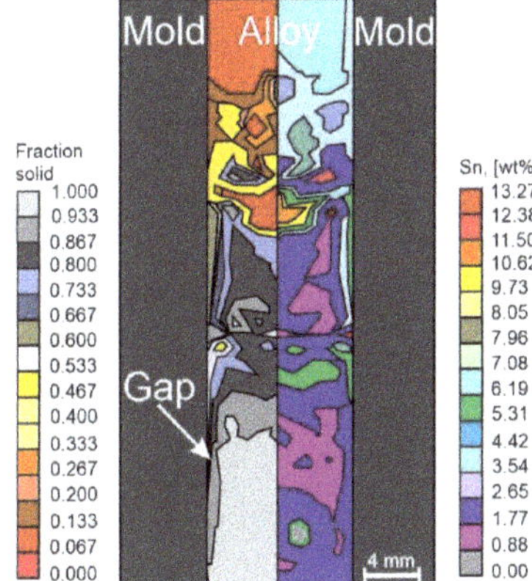

Figure 19. Schematic simulation results of solid–liquid interface morphology and Sn distribution [37].

Luo et al. [38] used Cu-4.7 wt%Sn to determine the flow of fluid and heat during the TZCC process. An alloy, 10 mm in diameter, was selected for this purpose. The speed of continuous casting, melting temperature, mould temperature, and the cooling water temperature were set to 20 mm/min, 1200 °C, 1040 °C, and 18 °C, respectively. The temperature of the alloy varied from 720 to 1081 °C. A complex circular flow caused by convection was observed in the mould. This facilitated the flow of the liquid alloy in the downward direction along the wall of the mould, followed by an upward flow in the centre. Analysing the direction of heat flow revealed that, at the centre, the flow was vertically downwards; at the upper wall, the flow was obliquely downwards, which was deflected towards the mould; and at the lower wall, the flow was deflected away from the mould.

9.3. Continuous Casting

The continuous casting of bronze alloys has an intrinsic advantage in terms of mechanical properties over other manufacturing processes because of the chilling and the sterling feeding of molten metal during solidification. Wilson [39] well-explained the continuous casting of copper-based alloys. Skoric et al. [40] provided numerical simulations and thermographic data on the structure of cast bronze alloy specimens regarding their stress status. Arsenovic et al. [41] also examined the effect of mould velocities on the fracturing of continuous cast specimens of bronze alloys [41]. Sergejevs et al. [42] suggested, a thorough description of the effect of casting velocity on the macrostructure and mechanical properties of Sn-bronzes. Due to the growing use of shape memory alloys (SMA), Kostov et al. [43] experimented on continuous cast copper-based shape memory wire with a diameter of 8 mm. Considerations of EN 1982:2008 [44] proved that macro segregation would be present in the resultant alloy after continuous casting. These macro segregations were responsible for the weakening of the alloy. Similar experiments carried out by Luo and He [45] examined the effect of process parameters on exudation thickness in continuous solidification of tin bronze. Another article by Luo [46] explained in further detail the influence of different continuous casting speeds on alloy microstructure evolution. They proved that, with a continuous casting speed increase, columnar crystals gradually replaced equiaxed crystals as the main component of the continuous unidirectional solidification of tin bronze. The diameter of the columnar crystals also begins to increase when the continuous casting speed is further increased. The effect of the distinct convection types, such as thermal buoyancy flow, solute buoyancy flow, inlet flow, and feeding flow on the formation of macrosegregation was examined by Ludwig et al. [47]. Increasing mush permeability will cause positive macrosegregation between the wall and centerline along with a negatively characterized macro segregation near the wall and in the centre. By contrast, when decreasing the permeability of the mushy zone, there will be positive macrosegregation at the wall along with a negative type of macrosegregation existing at the centreline. Grasser et al. [48] investigated the micro-macro segregation prediction based on solidification simulations for continuous casting of ternary bronze alloys. Moreover, Sugita et al. [49] utilized metal and sand moulds to experiment with solidification characteristics when casting processes were performed on Cu-20% Sn bronze alloys.

9.4. Semi-Continuous Casting and Die Casting

A significant factor that governs the material properties of bronze alloys is microstructure homogeneity. Therefore, Hao et al. [50] proposed a numerical analysis and an experimental investigation into solidification phenomenon during semi-continuous casting on phosphorous bronze alloys. To understand the solidification of these alloys, the Cu Sn-P ternary method was investigated. Data from diffusion and annealing experiments were compared with Calphad-based computational thermodynamics. For the multiphase solidification simulation, thermodynamic information was used, considering the relative velocity and microsegregation. Feeding flow-induced macro-segregation was predicted as solvent accumulation at the surface and as depletion at the centre of the casting.

Backman et al. [51] explained the thermal behaviour of die casting for high-temperature alloys as being partly solidified during machine casting. Using machine die casting, partly solidified bronze alloy 905 was prepared. Using computer simulations for both liquid and partly solid alloys, die thermal behaviour was noted. Castings were either made in a laboratory system with low pressure or in a commercial die casting system with high pressure. Using a correlation between the measured die temperatures at the casting die interfaces and computer projections, the heat transfer coefficient was determined. Various advantages were demonstrated for the use of partially solid charge materials in a die casting machine at high temperatures. By casting the bronze alloy in a partly solid state, both the surface temperature and surface temperature of the alloy could be reduced. A sudden change in the coefficient of heat transfer was also observed when the die cavity was filled. By modifying the structure of the charged material (partly solidified) itself and recognizing its thermal behaviour to enhance the transfer of heat from the casting to the die, this study led to a new approach.

10. Property Variation in Different Compositions of Cu-Sn Alloy

10.1. Eutectic Composition of Cu-Sn Alloy

The creep behaviour of eutectic C-Sn alloy, with stress ranging from 10^{-4} to 10^{-3} at 303–393 K, was investigated in [52]. Initially, the presence of pure tin (β Sn) with a dendritic structure along with finely dispersed particles of Cu_6Sn_5 intermetallic compound (IMC) dispersed in pure tin was confirmed. The creep temperature changed from 0.55 to 1.00 Tm (melting temperature). This adjustment was made to limit the transformation of tetragonal pure tin into a cubic tin (tin) at 286 K. This study has also derived that the relation for apparent activation stress exponent (Na) was found to be dependent on temperature (directly proportional) and stress was applied. The value of the apparent activation stress exponent of the alloy was also compared with pure tin. It was observed that there was an increment in the value of apparent activation stress when Sn was alloyed with Cu. The results from creep data of Cu-Sn alloy confirmed the value of the true stress exponent for creep as 7 since the slope magnitude level was 7. The experiments also showed that the six orders of magnitude for steady-state creep rate were found to range from 10^{-3} to 10^{-8} s^{-1} when the stress was within the range of 10^{-4}–10^{-3}.

10.2. Hypereutectic Composition of Cu-Sn Alloy

Sn-2.0 wt%Cu and Sn-2.8 wt%Cu alloys were prepared using a transient directional solidification system. Spinelli et al. [53] conducted the microstructural evaluation of these hyper-eutectic alloys for a wide variety of experimental tip growth rates and tip cooling rates. Cu_6Sn_5 IMC evolved into morphologies that were M or H-shaped. There were rod-like particles in the microstructure of the Sn-rich β matrix of both alloys (Sn-2.0 wt% Cu and Sn-2.8 wt% Cu). With minimum/maximum values of 16.0/37.0 mm and 2.0/12.5 mm for the respective alloys, inter-branch spacing differed significantly. For the alloy containing 2.0 wt% Cu, and 0.8 to 34.0 K/s for the alloy containing 2.8 wt% Cu, experimental cooling rates ranged from 2.7 to 33.0 K/s. In this research, Hall–Petch style equations that connected hardness, UTS, and elongation to fracture with inter-branch spacing were proposed. Similar patterns for hardness and UTS were illustrated by two alloys. For these two alloys, the ductility behaviour was found to be entirely different. The tensile strength analysis from this study has confirmed that after the process of transient directional solidification, the Sn-2.8 wt%. Cu alloy successfully meets the required mechanical strength.

10.3. Peritectic Composition of Cu-Sn Alloy

Peritectic transformations are demonstrated by several binary alloys during solidification. In these cases, the primary solid alpha phase precipitated from the liquid phase interacts at peritectic temperature with the liquid phase to form a peritectic beta phase. A few microstructures can be formed by binary alloys with a peritectic composition. Kohler et al. [54] investigated the microstructure of bronze alloys during low-speed peritectic solidification. Thereafter, Zhai et al. [55] presented different mechanical and microstructural properties of dynamically solidified peritectic Cu-70 wt%Sn alloy. These alloys were subjected to an external ultrasonic field with a frequency of 20 kHz and a 440 W power supply. The external ultrasound that is being induced promoted and completed peritectic transformation ($L + \epsilon \rightarrow \eta$). Mechanical properties, such as micro-hardness and compressive strength, were increased to a factor of 1.45 and 4.8, respectively. This was attributed to the increase in volume fraction and grain refinement. According to Zhai et al. [55], there were two methods to improve the microstructure resulting from peritectic compositions of Cu-Sn alloys. The first method was by increasing the rate of undercooling so that significant improvements in the primary phase could be achieved. If sufficiently high, undercooling stimulates the peritectic phase to directly nucleate from the metastable liquid alloy. The other method was to apply strong external fields, such as ultrasonic fields, to modify the peritectic solidification process.

The primary phase (ϵ) in the Cu-70 wt%Sn alloy was found to be Cu_3Sn IMC. The peritectic product was also an intermetallic compound Cu_6Sn_5 (η). It was found that ultrasonic waves affected peritectic transformation in two ways. Firstly, by promoting the intercooling of the primary phase (ε) and preventing the excess undercooling of the liquid alloy. It was found that the grain size of the primary phase had been refined and turned into an equiaxed shape with enhancements when exposed to ultrasound power. Secondly, refinement of peritectic grain structure resulted in a large increase in the Fourier number, which stimulates the completion of the peritectic state transformation. Looking into this, Zhai et al. [55] proved that an ultrasonic wave can be used to enhance the mechanical properties of the peritectic grain. The stress–strain curves were plotted for the top sample and bottom sample (as shown in Figure 20). This was done because there was a difference in the compressive strength for each sample, caused by the attenuation of sound waves. Figure 20 confirmed that, with an increase in the power of the ultrasound, a significant rise in stress value was observed.

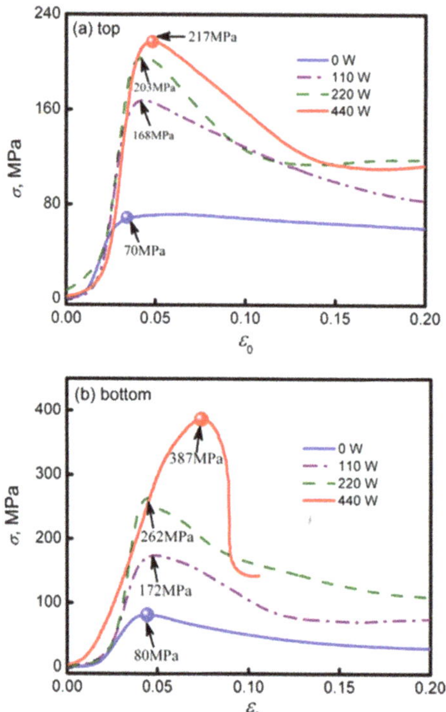

Figure 20. Compressive stress–strain curve for Cu-70 wt%Sn alloy that was solidified under static and ultrasonic conditions for the (**a**) sample top and (**b**) sample bottom [55].

On Cu-14.5 wt% Sn, Cu-21.3 wt% Sn and Cu-26.8 wt% Sn peritectic alloys, Kohler et al. [56] conducted single-path analysis. To measure the changes in the solid mass fraction, latent heat and composition during alloy solidification, a new heat flow model coupled to a Cu-Sn thermodynamic database has been developed. With the microsegregation model, including back-diffusion in the primary solid phase, a close comparison was made. In hypoperitectic Cu-Sn alloys solidified at low speed in a diffusive regime, Valloton et al. [57] performed experiments to ensure planar growth of the primary and peritectic phases. To minimize convection, samples with a reduced diameter equal to 500 μm were taken. The laminar and fibrous cooperative growth of the primary alpha and peritectic β-phases with a length of several millimetres was also found to be formed. Within a high peritectic alloy solidification interval, a quenched alpha + β front was achieved. These morphologies were observed under SEM for peritectic systems with a temperature of just a few Kelvins in the freezing range. The primary phase's high-volume fraction, differing over time, was found to be close to the quenched surface. This shows that the cooperative production front never entered a steady state. In addition, a link was formed between the primary FCC phase and the peritectic bcc phase. This was achieved to promote the peritectic process with multiple primary phase enucleations.

11. Applications of Cast Bronzes

Sn-Cu alloys are a potential substitute for soldering materials due to Pb toxicity, thanks to their good wettability, high electrical conductivity, good mechanical properties, and low cost. Electronic interconnections can be established by incorporating bronze alloys. Gain et al. [58] investigated the morphology and impact of Cu_6Sn_5 intermetallic compound on Sn-0.7Cu alloy properties at high temperatures. In the as-cast condition, there was the presence of fine Sn grains. Furthermore, hardness, electrical property, and creep behaviour were improved due to the obstruction of dislocation motion caused by the intermetallic presence at the grain boundaries (0.2–0.3 µm in diameter). After heating, the intermetallic compound grew in size (of about 915 µm) and was found to be coarsened and elongated. Finally, due to the coarsening of intermetallic compounds, Gain et al. [58] concluded that the hardness and electrical property of bronze were degraded to 18.5% and 12%, respectively, along with an increment in damping property. Hence CuSn alloys are an efficient replacement for lead alloys, that are mainly used for soldering purposes. Detailed experiments were conducted and an experiment was carried out by Rehim et al. [59]. Their research covered the microstructure and mechanical properties of five variants of bronze alloys, Sn-x Cu, where the content of Cu varied from 1–5 wt%. Heat treatment at 373, 393, 413, and 433 K for 2 h proved an increase in minimum creep rates, whenever the wt% of Cu was increased to 4 wt% (as shown in Figure 21) or on increasing the ageing temperature (as shown in Figure 22). After incorporating SEM and XRD techniques, a plot for minimum creep rates was also carried out as shown in Figure 23. These plots confirmed that Sn-4 wt% Cu alloy exhibited the best creep resistance. Furthermore, the variation in the stress component with increasing Cu content was also studied and plotted, as shown in Figure 24. This plot confirmed that stress was limited in the component whenever the annealing temperature was forced to increase. Here, a component with minimum stress resulted when the annealing temperature was set at 433 K.

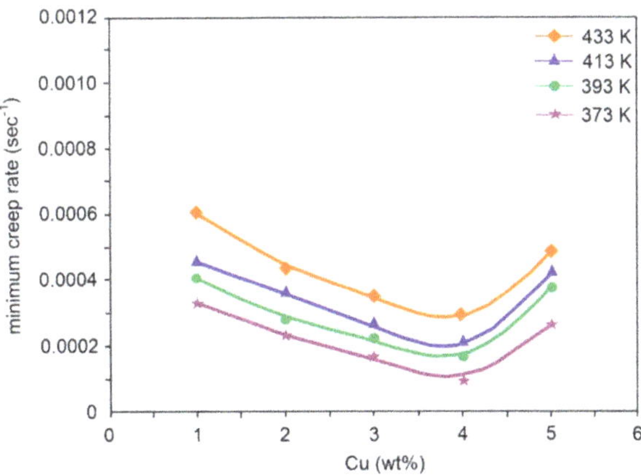

Figure 21. Variation of creep rates concerning change in wt% of Cu [59].

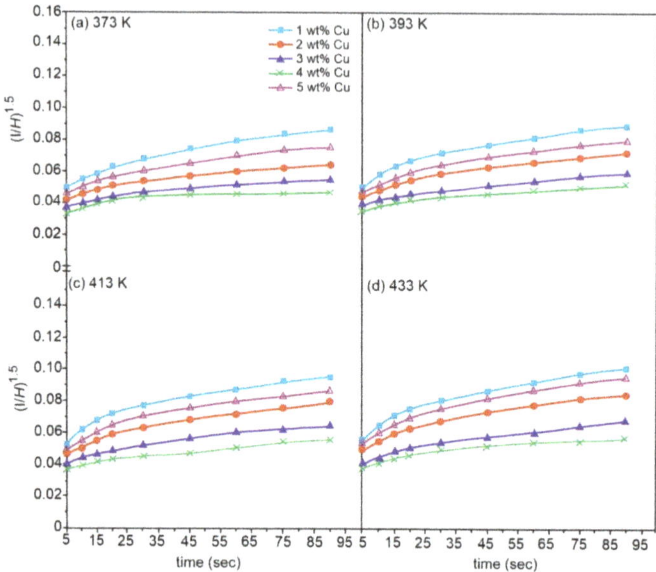

Figure 22. Variation of creep rate at increasing annealing temperature [59].

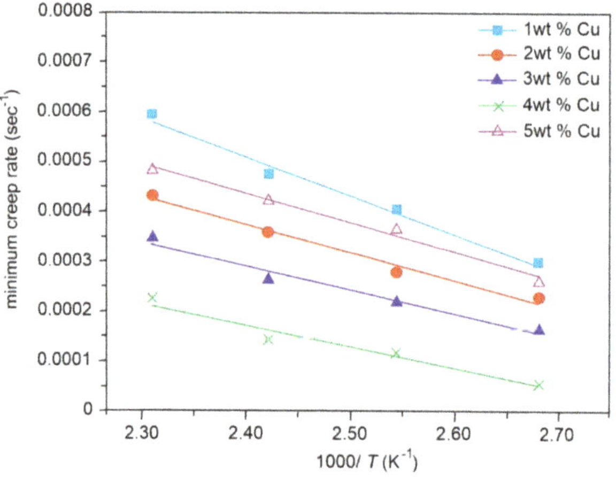

Figure 23. Variation of minimum creep rates with 1000/T, where T is the respective annealing temperature [59].

The corrosion activity of intermetallic Cu-Sn was examined at 3.5 wt%. by using galvanic corrosion and polarization processes, percent of NaCl solution, and was compared with Cu and Sn. Even though stainless steel is of superior quality, the study conducted by Tsao et al. [60] verified using of bronze alloys in the development of underwater bearings and ship propellers. The Sn, Cu, Cu_3Sn, and Cu_6Sn_5 polarization curves clearly showed that an increase in Cu material correlated with a substantial increase in corrosion current density. In the NaCl solution, corrosion values and the corrosion phase of Cu-Sn showed that Cu species diffusion occurred through the oxide layer. It was found that $Sn_3O(OH)_2Cl_2$ and CuCl, respectively, were the corrosion products occurring on the surfaces of pure Sn and Cu. In addition, the following corrosion products coexisted, Cu_3Sn, $Sn_3O(OH)_2Cl_2$ and CuCl. The formation of various compounds, including SnO_2, $Sn_3O(OH)_2Cl_2$, Cu_2O, and

$CuCl_{23}Cu(OH)_2$ was found due to the presence of the corrosion compound Cu_6Sn_5. Once the polarization tests were completed, there was an absence of SnO and CuCl compounds.

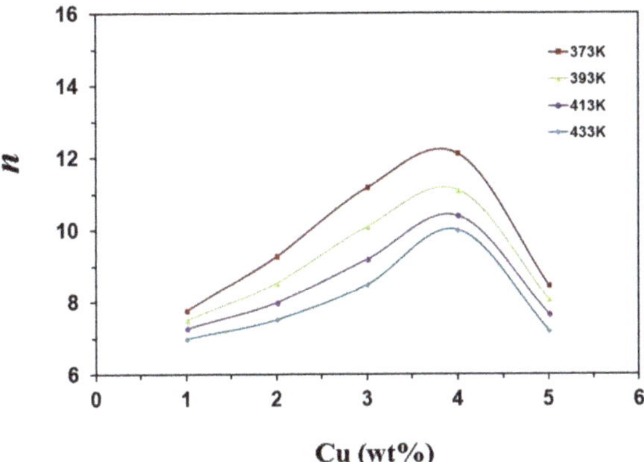

Figure 24. Plot between stress component (n) and wt% of Cu [59].

Liu et al. presented research [61] covering the microstructure and mechanical properties of Cu-Sn alloys for the development of detonating and primer cords. Due to its non-toxicity and abundance of resources, Cu-Sn alloys have been chosen for this reason. Alloys based on Cu-Sn have a high density, which was useful for supplying penetration momentum and impulse energy. Sn (0.3–1.0) wt% Cu alloy was chosen for the experiment. The phase constituent of the alloy was not altered by rolling, whereas the microstructure was significantly refined. It was observed that annealing does not affect the rolled Cu-Sn alloys. However, this increased the ductility and decreased the strength of the annealed alloy. On testing rolled Cu-Sn alloy, acceptable tensile properties for the purpose were attained. This exhibited strain-softening, which is desirable for the processing of cords.

Cu alloys with solid lubricant (mainly Pb-bronze) can be efficiently employed in anti-seizure applications. Sato et al. [62] have investigated tribological behaviour and the properties of sulphide-dispersed bronze. Friction testing under dry conditions yielded improved results for properties such as hardness and wear resistance. Sulphide-bronze prevents seizures only in certain manually set conditions, whereas sulphide was the reason for the suppression of scoring and seizure. Limited or restricted movement of Cu alloy components to the matrix steel surface by dispersed sulphide can also be attributed to the increase in hardness and wear resistance. Furthermore, adding Si to the Cu-Fe-S system and sulphide-bronze results in crystallizing the Fe-Si compound.

12. Evolution of Cast Cu-Sn-Ti Alloys

Sn content in the bronze alloy can vary from 2 wt% to 20 wt%. In Cu-Sn alloys, the amount of tin plays a major role, affecting its strength and usability. The high price of tin increases the cost of the final product. For this reason, to achieve desired properties and to be cost-effective, other allowable additives were introduced, such as zinc, lead, phosphorus, nickel, iron, titanium, aluminium, etc. Adding aluminium to bronze results in increased strength and hardness and a decrease in the level of plasticity. Research on Cu-Sn alloys with aluminium was conducted to optimize the amount of aluminium addition. Furthermore, the addition of iron in Cu-Sn alloys changes the properties of the alloy. Hence, a study based on the above-mentioned topic was put forward by Klempka et al. [63]. Both alloying elements, aluminium and iron caused grain refinement in Cu-Sn10 alloys. The major change observed due to the introduction of 0.8 wt% Al and Fe was the rise in UTS and hardness. A rise in UTS was higher when iron was added as compared to the addition of aluminium. The addition of both these alloying elements decreases the alloy plasticity. Considering plasticity, the drop due to aluminium addition was more significant with the addition of iron.

Titanium was introduced into bronze alloys because of its excellent stability and tensile properties. Therefore, the effect of Ti on bronze alloys was thoroughly investigated. An increase or decrease in reaction rate can be obtained in between solder and joining metal by adding elements to the interface layer. Moreover, this addition can alter the physical properties of phases present at the interface and can lead to the formation of new phases or new layers. Vuorinen et al. [64] performed solid-state annealing of Cu-Sn alloys and added certain amounts of Ti to investigate the behaviour of Ti on parent metal with annealing periods of up to 3000 h. Initially, the addition of Ti does not correlate with the thickness of the intermetallic compounds (Cu_6Sn_5 and Cu_3Sn) formed at the interface. Hence, the thickness of those intermetallic compounds was unaffected by the addition of Ti, even though an unevenness was developed. As per the EDS resolution limits, there was no observance of Ti in both of the intermetallic compounds. Further additions of Ti cause the heterogeneous distribution of both layers. Due to the non-reactive nature of Ti to intermetallic (Cu-Sn), this forced them to react with Sn to produce large platelets of Ti_2Sn_3 observed in the solder matrix. Hence, any addition of Ti to the matrix does not affect the Cu-Sn intermetallic formed at the interface. Rationalizing this behaviour with the phase diagram ascertains that Ti possesses very low solubility for Cu-Sn IMCs.

The results of solidifying binary Cu-Sn and ternary Cu-Sn-Ti alloys at a lower cooling rate in a differential scanning furnace were compared to fast solidification of these alloys in the experiment conducted by Li et al. [65], and properties, such as alloy composition and cooling rate, were studied. Considering binary alloys, a metastable $Cu_{5.6}Sn$ phase was observed at a high cooling rate, which was obtained from the parent β phase by diffusing less martensitic transformation. Ti was observed to form $(CuSn)_3Ti_5$ in ternary alloys in low and elevated cooling rate conditions. By having the binary system in rapid solidification conditions, the microstructure of Cu-Sn was composed of $Cu_{5.6}Sn$ and primary α dendrites. Considering the ternary system, $(CuSn_3)Ti_5$ was formed without decomposing into other ternary compounds. This was attributed to rapid cooling, which resulted in the blockage of solute diffusion in solids. When increasing the Ti content, the amount of $(CuSn_3)Ti_5$ and α phase were increased, thereby decreasing the content of $Cu_{5.6}Sn$.

13. Properties of Cu-Sn-Ti Alloys

The wettability of a solid substrate by a liquid metallic material is crucial to understanding the bonding characteristics and the interactive forces at the interface. To ensure good wetting characteristics, fluxes are used. The development of new fluxes requires studying surface tension. The wetting characteristics and mechanical properties of brazing metals, such as Cu-Sn, can be enhanced by adding a small amount of titanium. The surface and transport properties of Cu-Sn-Ti liquid alloys were studied by Novakovic et al. [66]. It was found that the addition of titanium greatly improved the wetting characteristics and mechanical properties of Cu-Sn binary alloys. Using quasi-chemical approximation and compound formation methods for binary systems, concentration and temperature dependencies of surface tension and surface composition were analysed. In the binary systems, Cu and Sn-atoms segregate in bulk compositions. All binary systems tended to form a compound. The thermodynamic property of all three binary systems exhibited a negative deviation from Raoult's law, while the surface tension isotherms exhibited a positive deviation. For the ternary system, only the surface tension has been found and the result was compared with corresponding Cu-Sn surface tension data. Moreover, alloys that were rich in copper were investigated by Lehreton et al. [67], to analyse their microstructural and mechanical properties. Three alloys, CuTi-2 wt%Sn-2.75 wt%, CuTi-3 wt%Sn-2.75 wt% and CuTi-4 wt%Sn-2.75 wt%, were selected for this purpose. $CuTi_3Sn_5$ intermetallic eutectic ternary compound was confirmed, as well as the product characteristics during the as-cast conditions. Furthermore, this compound could not be further dissolved into the matrix phase using any thermomechanical processes. A Ti content of more than 3 wt% was needed to obtain the required amount of hardness due to the decreased spinodal decomposition. Figures 25 and 26 prove that the microstructure of the dendritic structure was characterized by the formation of fine precipitates. All three samples had the same microstructure, with fine spherical precipitates that evolved from a rod-like form, with limited discontinuity in the resultant alloy. Small additions of Cr into $CuTi_4Sn_{2.75}$ resulted in enhancing hardness for the first 2 h along with the elevation of phase transformation temperature of the alloy. Through XRD patterns of CuTi-3 wt%Sn-2.75 wt%, the existence of a metastable phase limiting the coalescence was confirmed. This was the main reason for the softening of alloys rich in copper.

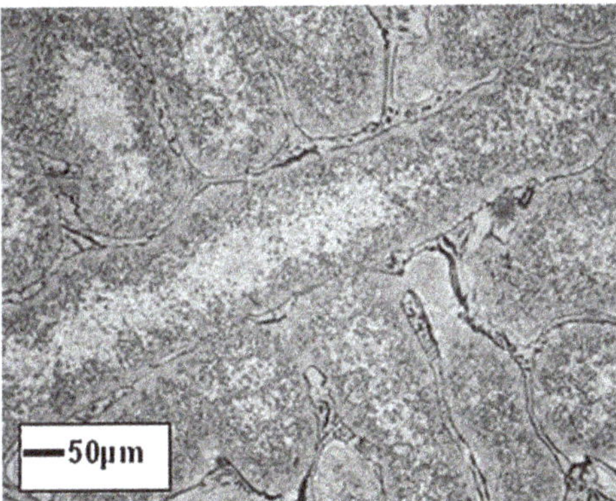

Figure 25. Microstructure of $CuTi_3Sn_{2.75}$ alloy under OM in as-cast condition [67].

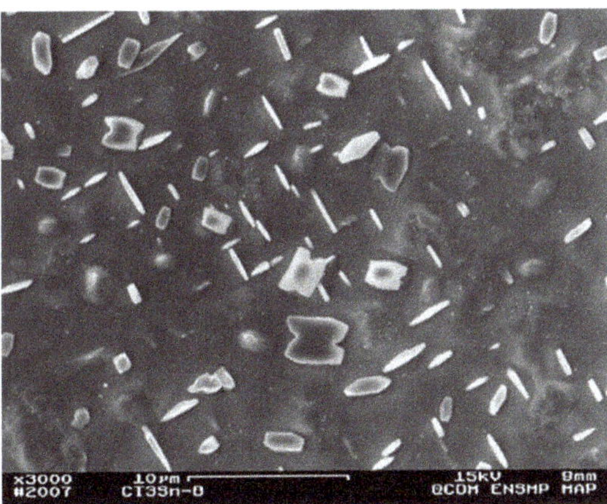

Figure 26. Observation under SEM, proving the presence of fine precipitates in the core of dendrites [67].

The experiments conducted by Lin et al. [68] showed that the minimum amount of Ti to efficiently wet alumina was found to be 6 wt%. When increasing the amount of Ti from 6 to 12 wt%, there seemed to be an increase in the volume fraction of the intermetallic phase and wettability enhancing the bond strength. Results from sessile drop tests confirmed that 70Cu-21Sn-9Ti was the best compound composition to wet alumina. Studies conducted by Lin et al. [68] provided a note to minimize the amount of Sn to below 21 wt%. At 900 °C, the best wettability was obtained when Ti content was set to 9 wt%, whereas increasing content to 12 wt% did not seem to enhance the wetting behaviour. Moreover, thermal expansion coefficients of Cu-Sn-Ti alloys were compared with Ticusil® alloy and were found to be lower. It was also observed that a combination of a lower coefficient of thermal expansion and lower wetting angle could be obtained using Cu-Sn-Ti alloy, which confirmed its dominance over commercially used Ticusil® alloy. Therefore, a minimal amount of residual stress can be expected when brazing alumina with Cu/Sn/Ti alloys.

14. Phase Diagrams and Phase Relations of Cu-Sn-Ti Ternary System

To understand the complex interface reaction during melting and casting and to get adequate mechanical properties, the knowledge of the thermodynamic properties and phase equilibria of Cu-Sn-Ti is essential. Wang et al. [69] obtained experimental data for Cu-Sn-Ti phase equilibria at 800 °C. Wang et al. [69] concluded that one single-phase point, three tie lines and four tie triangles were present, as shown in Figure 27, by integrating Euclidean distance matrix analysis (EDMA). Hence, the solid phase equilibrium calculations at 800 °C affirmed the presence of a binary compound (Sn_3Ti_5) which possess an extended solubility of Cu. Another objective of this study was to define binary systems present at 397 °C, as depicted in Figure 28, in the isothermal section and to obtain a collection of accurate and self-consistent thermodynamic parameters for this ternary Cu-Sn-Ti system. The solubility ranges of Sn-Ti alloys (Sn_3Ti_5 and Sn_5Ti_6 compounds) and the presence of two ternary compounds (CuSnTi and Cu_2SnTi) have been verified, along with the evaluation of liquid and FCC phase interaction parameters. The findings were extremely accurate on the Cu-rich side of the ternary method and can be used effectively to make decisions on Cu-based alloys high temperature (brazing) applications. The determined liquid projection of the Cu-Sn-Ti ternary system superimposed on the experimental primary phases is shown in Figure 29. Solubility ranges of the two major binary compounds have also been depicted in the figure.

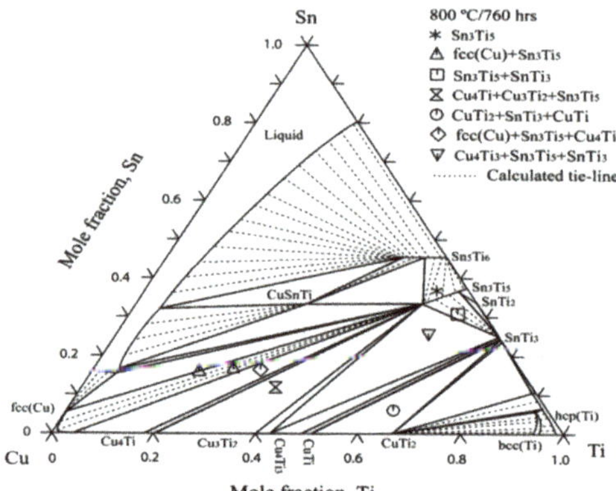

Figure 27. Ternary system of Cu-Sn-Ti ternary system at 800 °C [69].

The formation of the $CuSn_3Ti$% intermetallic phase is the main reason for the degradation of strength whenever CuSnTi filler metals are used. To predict the brazeability of CuSnTi filler metals, there is an increased need to analyse the liquidus of the CuSnTi ternary system. Naka et al. [70] came up with an experiment to sketch the liquidus of the CuSnTi ternary system. Their results lead us to the conclusion that, as shown in Figure 30, the Cu-Sn-Ti alloy method yielded a large region of primary crystallization of phases that were Cu, Ti_6Sn_5, $CuSn_3Ti_5$, and Cu-Sn-Ti. Due to the peritectic response, both ternary compounds were formed. Near the Sn-rich portion of the phase diagram, a ternary eutectic L = Sn + η (CuSn) + Ti_6Sn_5 was observed. Moreover, a pseudo-binary eutectic reaction accompanied the formation of $Cu+CuSn_3Ti_5$.

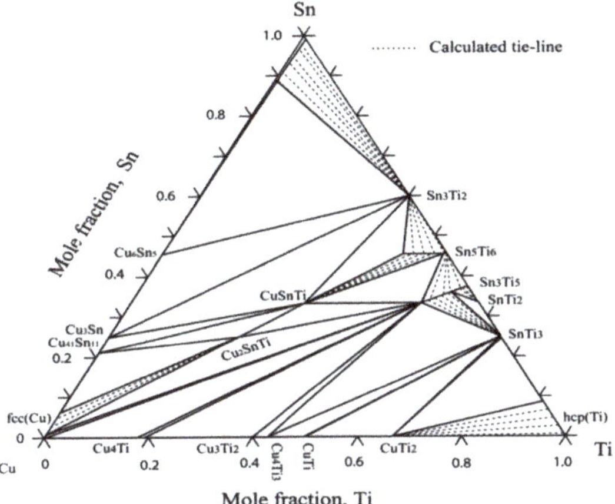

Figure 28. Isothermal section of Cu-Sn-Ti ternary system at 697 °C [69].

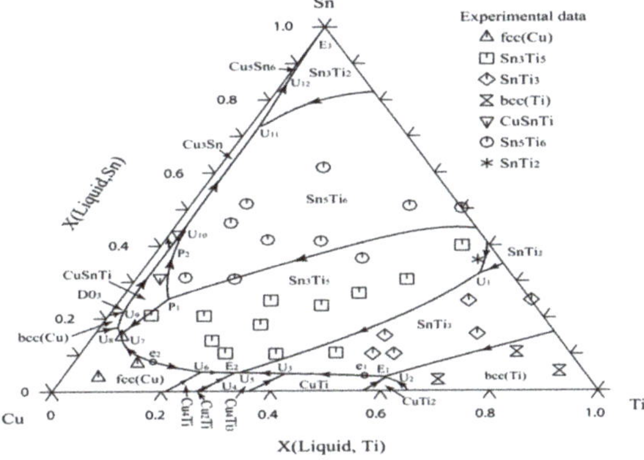

Figure 29. Liquidus projection of Cu-Sn-Ti ternary system superimposed with experimental data of primary solidification [69].

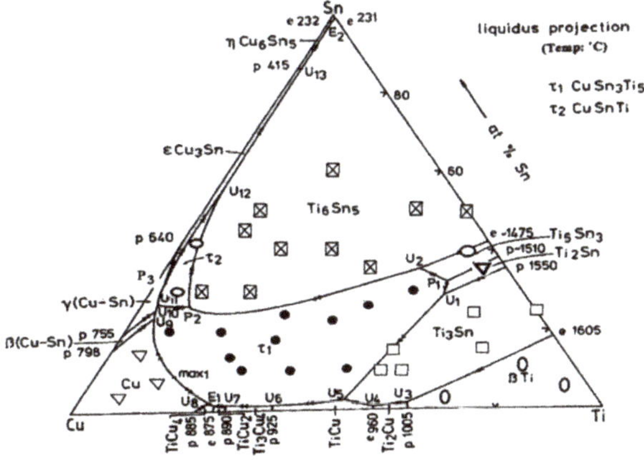

Figure 30. Liquidus surface projection in Cu-Sn-Ti ternary system [70].

Huang et al. [71] isothermally treated Cu-23at. %Ti-17 at. %Sn at 900 °C for 10 h and then quenched it in water. This study has revealed interesting facts about the intermetallic compounds formed after the commencement of the reaction. This resulted in a new ternary compound of Cu, Sn and Ti, forming $CuSnTi_3$. Figure 31 depicts the phase relationship of Cu-23at. %Ti-17at. %Sn in the Cu-Sn-Ti ternary system. Overall, alloy constituents observed after 100% commencement of reaction included Cu-14at. %Sn, $CuSnTi_3$, $CuSn_3Ti_5$ and Cu_2SnTi_3. When cooling the furnace, Cu-14at. %Sn decomposed into Cu-9 wt%Sn and $Cu_{41}Sn_{11}$. Similarly, Cu_2SnTi_3 tended to decompose into Cu and Cu-Sn-Ti when subjected to prolonged heat treatment. The only thermodynamically stable compound present at 900 °C was $CuSnTi_3$ with a hexagonal structure (a = 0.4636 nm, c = 0.5229 nm). Furthermore, it was found to be iso-structural with Ni_3Sn_2 having Cu and Sn distributed in the lattice positions of 2c (at x = 1/3; y = 2/3; z = 1/4; occ = 0.5) and Ti being distributed in 2a (at x = 0; y = 0; z = 0; occ = 1) and 2d lattice positions (at x = 1/3; y = 2/3; z = 3/4; occ = 0.5).

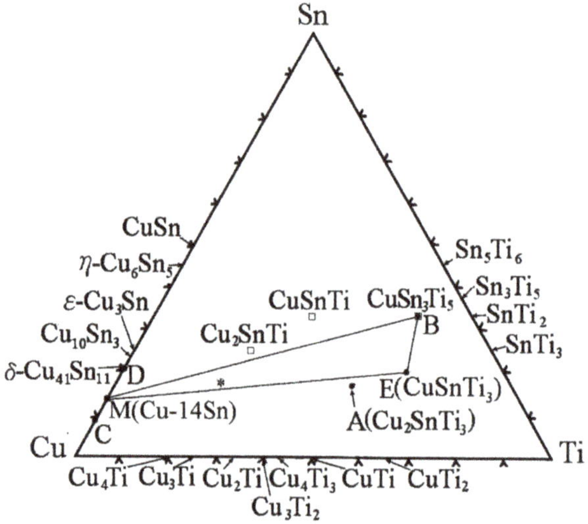

Figure 31. The phase relationship of constituent phases of Cu-23at. %Ti-17at. %Sn alloy in the ternary phase diagram of Cu-Sn-Ti when isothermally held for 10 h at 900 °C [71].

15. Effect of Elemental Additions on Cu-Sn-Ti Alloys

Cu-Sn-Ti alloy can be effectively employed to make abrasive binders. The electron microscope examination provided the image of the columnar structure of Sn and Ti near the abrasive materials (diamond or CBN grains) and the coarse particles of metal, thereby owing to an irregular microstructure of the resulting alloy. This will adversely affect the strength and ductility of the compound. The variation of properties on adding refractory metals (V, Mo, B, and Re), refractory compounds (LaB_6, TiC, TiN, Si_3Ni_4), and rare earth metals (La, Y) to the ternary system was investigated by Kizikov et al. [72]. Mechanical properties were improved when 0.01% of Si_3N_4 was added to the matrix. The micro additions of 0.01% of Si_3N_4 to Cu-20Sn-10Ti led to a structural modification, which further contributed to higher strengthening behaviour of the resultant alloy. Moreover, it was also confirmed that the micro additions of 0.01% of compounds improved strength and ductility. Tsao et al. [73] prepared four samples of Ti_7Cu_xSn (where x = 0, 1, 2.5, 5) alloy, solution treated (ST) them for 2 h at 1000 °C and quenched them in water at room temperature. Results confirmed the presence of a martensitic structure in ST Ti_7Cu alloy. Further Sn additions to Ti_7Cu_xSn refined the microstructure of $Ti_7 Cu_xSn$ alloy and resulted in a pseudo dendritic microstructure consisting of α Ti-phase. Potentiodynamic polarization curves and electrochemical impedance spectroscopy results also showed that Sn additions greatly improved the corrosion-resistant behaviour of the alloy because of the formation of a dual-layer oxide with an internal barrier and a porous external layered surface. Thus, it can be inferred that additions of Sn into ST Ti_7Cu_xSn alloys will result in an improvement of electrochemical corrosion resistance.

Another experiment on the effect of Sn content on the microstructure and corrosion behaviour of $Ti_7 Cu_xSn$ (x ranges from 0–5 wt%) was put forward by Tsao [74]. A corrosion test was conducted with 0.9 wt% NaCl solution at 25 °C. Potentiodynamic polarization curves and electrochemical impedance spectroscopy (EIS) results confirmed that the addition of up to 5% Sn improved the corrosion resistance of $Ti_7 Cu_xSn$ alloy, which also exhibited a dual-layer consisting of the porous outer covering and an inner barrier layer with oxide content. The microstructure before and after potentiodynamic testing was also obtained by the incorporation of OM (Figure 32) and SEM (Figure 33), respectively. Therefore, an increase in the volume fraction of the Ti_2Cu stage was observed by eliminating a basket-wear type of structure, thereby promoting an ultra-fine structure consisting of α-Ti and Ti_2Cu in the Ti_7Cu_5Sn alloy.

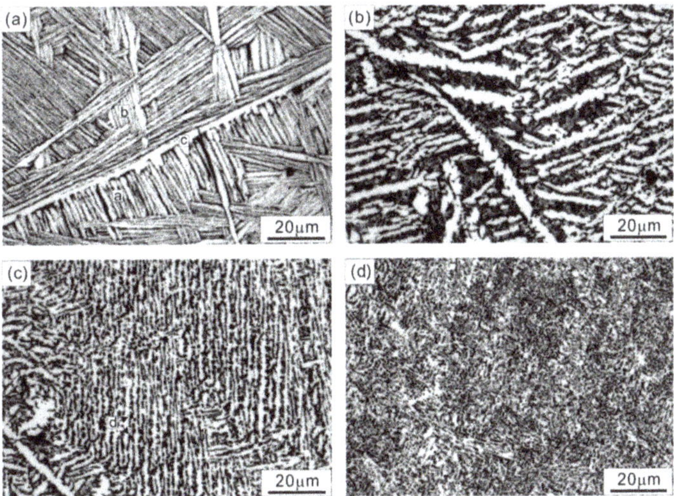

Figure 32. Microstructure of (**a**) Ti_7Cu, (**b**) Ti_7CuSn, (**c**) $Ti_7Cu_{2.5}Sn$, (**d**) Ti_7Cu_5Sn under OM [74].

The effect of Ti additions on as-cast bronze alloys was investigated by Akshay et al. [75,76]. The impact of adding small amounts of Ti on the microstructure, mechanical properties and tribological properties was recorded and compared with that of as-cast Cu-6Sn. Due to the development of blowholes, the tensile properties were found to reduce by up to 0.5 wt%. of Ti additions. Figure 34b depicts the presence of internal pores at the fracture surface of Cu-6Sn-0.5Ti alloy. Furthermore, on increasing the content of Ti addition to 1 wt%, the formation of the intermetallic compound was observed in Cu-Sn-Ti alloy, which was attributed to its high tensile strength and improved hardness (dendritic structure with fine grains). A comparison of the tensile strength and yield strength for three different as-cast Cu-Sn-Ti alloy compositions are presented in Figure 34a. By using a pin-on-disc tribometer with load variations varying from 10 to 30 N, a sliding distance of 1000 m, and sliding velocity of 1–3 m/s, tribological investigations showed that there was a rise in wear rate with increasing load or decreasing sliding distance. Hence, the wear resistance increased with Ti additions in the bronze alloy. Moreover, the mode of failure shifted from ductile to brittle as dimples were visible under microscopic observations with additions of Ti. This also confirms that Cu-Sn-Ti alloy can efficiently replace bronze alloy when working conditions are intense and high resistance to wear is required.

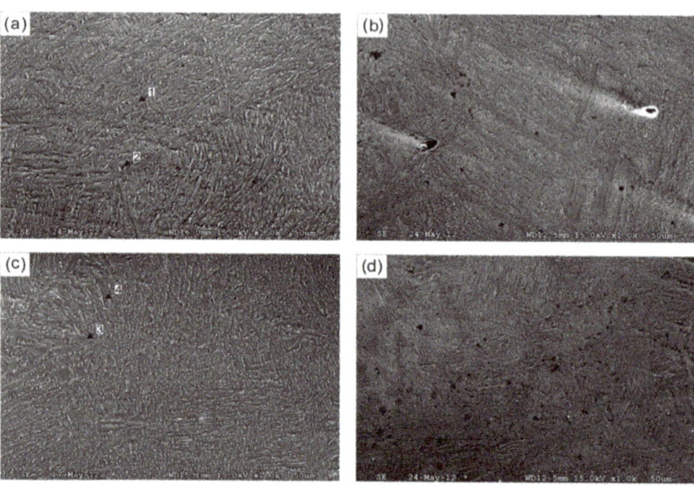

Figure 33. Microstructures under SEM after potentiodynamic testing in (**a**) Ti$_7$Cu, (**b**) Ti$_7$CuSn, (**c**) Ti$_7$Cu$_{2.5}$Sn, and (**d**) Ti$_7$Cu$_5$Sn alloy [74].

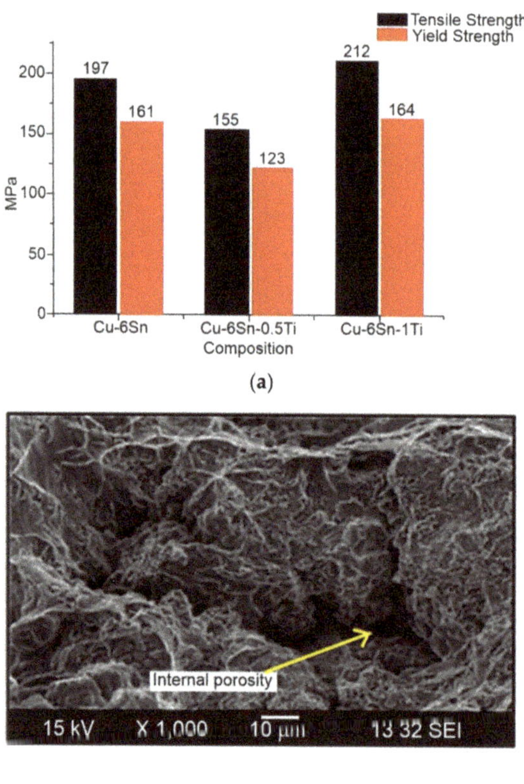

Figure 34. (**a**) Comparison of yield strength and tensile strength of as-cast alloys, (**b**) SEM image of a fracture surface of Cu-6Sn-0.5Ti [75].

16. Cu-Sn-Ti Brazing Alloys

Cu-Sn-Ti based brazing alloys are widely used for joining materials, such as diamond/steel matrix tools [77–83], metal–ceramics [84], and CBN abrasive tools [85–87]. The most common alloys previously used for brazing were alloys based on Ag-Cu-Ti. However, as pointed out by He et al. [88], these alloys had a low strength, high cost, and low wear resistance. Cu-Sn-Ti alloys were observed to have excellent mechanical properties, comparable high strength and a low cost, thus making them suitable for active braze alloys, according to many researchers [68,83,84].

Presently these alloys have been gaining importance as active braze alloys, especially in diamond/metal and polycrystalline CBN/metal tools. Studies have observed the interface microstructure of these alloys, as shown in Figures 35 and 36, and have stated that Ti plays a very important role in the brazing mechanism. Table 1 summarizes the conclusions from different research regarding Cu-Sn-Ti active braze alloys.

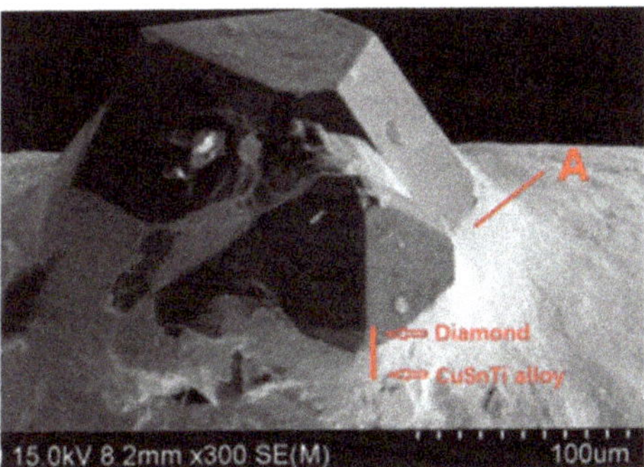

Figure 35. Interface microstructure between Cu-Sn-Ti filler alloy and diamond grits [79].

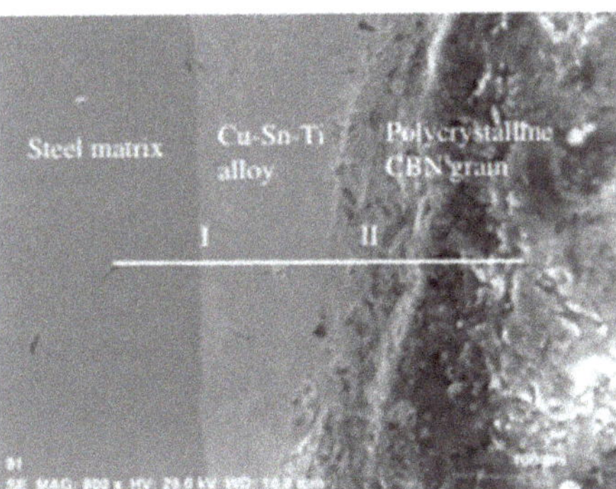

Figure 36. Interface microstructure of steel matrix/Cu–Sn–Ti alloy/polycrystalline CBN grain [84].

Table 1. Significant conclusions derived from research papers.

Sl.No.	Author and Reference	Materials/Cu-Sn-Ti braze Alloy	Conclusions Derived
1.	Lin et al. [68]	Alumina ceramics	The goal of this work was to investigate the wettability of the Cu/Sn/Ti alloy over polycrystalline alumina. The best wetting ability to alumina substrate is demonstrated by a 70Cu-21Sn-9Ti alloy with 9 wt%Ti brazed at 900 °C. Because Sn is tightly linked with Ti, Sn concentration in Cu/Sn/Ti brazes should be kept below 21% by weight. Three stages could be identified in the brazing filler metal, along with the reaction layer, according to the SEM backscattered electron image (BSE). They were (1) the Cu, Sn, Ti, oxide containing reaction layer; (2) Cu/Sn/Ti intermetallic phase, and the (3) bronze matrix solid solution containing 0.1 wt% Ti.
2.	Li et al. [78]	Between diamond grids and steel substrate	At the temperatures of 925 °C and 1050 °C, diamond grits were brazed onto a steel substrate with a Cu-10Sn-15Ti brazing alloy. In between diamond grits and the braze matrix, a TiC reaction layer was observed. The TiC layer was of grains that were around 50 nm in size, allowing it to efficiently reduce the interfacial stress caused by the diamond and TiC's lattice mismatch. An intermetallic complex comprised of Sn and Ti was also observed to nucleate and develop into a freely interwoven fine lacey structure on top of the TiC growth front.
3.	Duan et al. [79]	Diamonds, and metal matrix (Fe, Cu, Ni, and Sn)	In this study comparison between different brazing processes was made. Hence, two separate brazing equipment were used: (1) vacuum resistance furnace and (2) high-frequency induction furnace. It was observed that the surface morphology of the pre-brazed diamonds brazing in vacuum resistance furnaces was better. As with the previous studies, TiC intermediate reaction layer was formed indicating any occurrence of chemical combination between Cu–Sn–Ti alloy and the diamond grain. Ti concentration was observed to be higher in the interface of diamond/Cu-Sn-Ti alloy indicating its involvement in the joining process.
4.	Zhang et al. [80]	Diamond and KSC82 carbon steel.	Novel brazing diamond wire saws were performed by applying Cu-Sn-Ti alloy and high-temperature brazing technique to consolidate diamond grits and metallic wire matrix. The findings revealed the presence of a novel Ti2C phase at the interface, where diamond particles were brazed together via reactive wetting. When compared to KSC82 metallic wire, although tensile and yield strength decreased by 41 and 60%, the plasticity increased two-fold, which was said to meet the mechanical performance requirements as stated by the author. The failure due to the separation of diamond grits is caused by two primary reasons. To begin with, the detection of oxygen in the micro-domain of the void lip, as well as the oxidation of Ti, might result in false welding of diamond particles, resulting in early diamond particle separation. Second, the initiation and growth of fatigue crack may cause the diamond grits to lose their holding force at the interface.
5.	Yin et al. [81]	Diamond and Q460 steel substrate	Cu-Sn-Ti composite fillers reinforced with various amounts of tungsten carbide (WC) particles were used to link diamond particles to Q460 steel. The diamond grits had a superior shape and exposure height than the non-WC particle-reinforced samples. Furthermore, including WC particles caused fewer cracks to form at the interface between Cu-Sn-Ti fillers and diamond. The inclusion of 15 wt% WC particles enhanced the formation of TiC and Cu-Ti compounds while preventing the fast intermetallic reaction of Fe and Ti, decreasing the occurrence of brittle phases. As the WC particle concentration increased, the microhardness at the segment interface rose from 179 HV0.05 to 206 HV0.05. The shearing strength of brazed diamond segments reinforced with 15% WC improved by 10%.
6.	Liu et al. [82]	Diamond, c-BN, Al_2O_3 and SiC abrasive crystals on 0.45% C steel matrix	Cu-Sn-Ti active powder filler alloys were utilised to braze diamond, CBN, Al_2O_3, and SiC abrasive crystals onto a 0.45 percent Carbon steel matrix in a vacuum to construct a new superhard abrasive wheel. The Ti in the Cu-Sn-Ti filler alloy was observed to segregate primarily towards the surface of diamond, CBN, Al_2O_3 or SiC, forming a Ti intermediate reaction layer. The Ti-rich layer included phases such as (Ti-C), (Ti-N), (Ti-B), (Ti-O), (Ti-Si), and Ti-(Cu-Al) compounds thus confirming the fact that the chemical metallurgical combination was produced between the grains and the active filler. The metallurgical bonding of active-grain fillers and active filler-steel produced a strong connection between the grains and the 0.45 %C steel substrate. It has been demonstrated that the brazed grains' reliable bonding strength to the steel substrate could potentially fulfil industrial requirements.

Table 1. *Cont.*

Sl.no.	Author and Reference	Materials/Cu-Sn-Ti braze Alloy	Conclusions Derived
7.	Buhl et al. [83]	Monocrystalline block-shaped diamonds onto a stainless-steel substrate	Three distinct brazing temperatures (880, 930, and 980 °C) and two different dwell periods (10 and 30 min) were used to join monocrystalline block-shaped diamonds with a stainless-steel substrate using a Cu-Sn-based active filler alloy. At the filler-steel intermetallic layer, the following was formed: (1) intermetallic (Fe, Cr, Ni)$_2$Ti, and (2) intermetallic phases CuSn$_3$Ti, CuSnTi, Ni$_2$TiSn and NiTiSn. Residual stresses have a strong relationship with brazing parameters, i.e., either developing compressive (at temperatures of 880 °C and 930 °C) or tensile residual stresses (at temperature 980 °C). Maximum compressive residual stress of around 350 MPa was observed at a holding time of 10 min at 930 °C.
8.	Ding et al. [84]	Between Polycrystalline CBN grains and steel matrix	Cu-Sn-Ti brazing alloy was used to braze polycrystalline CBN grains with AISi 1045 steel matrix. Ti in molten Cu-Sn-Ti brazing alloy interacts with AlN binder and CBN particles of polycrystalline CBN grains forming TiN, TiB$_2$, TiB and TiAl$_3$ compounds hence proving any evidence of chemical reaction between Cu-Sn-Ti alloy and CBN grains. A strong connection between polycrystalline CBN, steel matrix and Cu-Sn-Ti filler alloy was obtained. The predominant fracture mechanism of brazed polycrystalline CBN grains was the intercrystalline fracture at the CBN-CBN particle boundary.
9.	Fan et al. [85]	CBN	When the brazing temperature is lower than 1223 K, fully uncoated and/or partly coated CBN particles with jagged edges were still observed, and the reaction layer, which is mostly made up of TiN and TiB$_2$, appears uneven and thin. When the brazing temperature reaches 1223 K, Ti diffuses completely and is enriched at the interface, resulting in a more homogeneous, continuous, and stable reaction layer composed largely of TiB, TiB$_2$ and TiN. Further raising the temperature to 1273 K is unnecessary, if not detrimental, because the reaction layer thickness stays nearly constant and some microscopic microcracks were observed in the interfacial area, reducing the grinding capacity of the final superabrasive product.
10.	Fan et al. [86]	CBN/Cu-Sn-Ti	Through FIB-TEM-EDS-SADP analysis of the interfacial reaction layer at CBN/Cu-Sn-Ti active filler metal at the temperature 1223 K, it was revealed that the interfacial reaction layer is composed mainly of continuous TiB$_2$/TiB/TiN layer and irregular TiN/TiB$_2$ layer. The reaction layer thickness was observed to be about 1.24 μm. It was proved that metallurgical interfacial bonding was observed, which according to the author can be very useful for the development of high-quality CBN grinding tools.
11.	Hsieh et al. [87]	Graphite	The wetting behaviour of Cu-Sn-Ti brazing alloys on graphite and phase formation at temperatures from 850 °C to 1000 °C was investigated in this study. To promote the wetting of the brazing alloy on graphite, a minimum brazing temperature of 1000 °C was required for Cu–Sn–Ti alloys with Ti concentrations as high as 70 wt%. High amounts of CuSn$_3$Ti$_5$ and SnTi$_3$ intermetallic compounds were observed, with an increase in Ti concentration and a reduction in Sn concentration. In a ductile Sn-rich matrix phase, however, a rise in Sn concentration and a reduction in Ti concentration might result in the precipitation of intermetallic compounds such as Sn$_3$Ti$_5$ and Sn$_3$Ti$_2$. The optimum Ti and Sn concentrations for effective wetting on graphite at low temperatures, while retaining a significant volume fraction of ductile phases, were around 10 wt% Ti and 15 wt% Sn.

17. Applications

The Cu-Sn-Ti alloy group was employed in producing high strength ageing materials and for abrasive binders. The Cu-Sn-Ti ternary system has a lower melting point and shows good wear properties along with strong interconnection between abrasives because of their outstanding wetting properties. Titanium alloys have been widely used in the manufacturing of biomaterials due to their excellent corrosion resistance, high specific strength, and non-allergenic nature. Due to the high melting point (1670 °C) and poor machinability, other metals are added to obtain combined properties, and hence Cu-Sn-Ti can efficiently replace Cu-Sn in several applications, as mentioned above in the paper by Akshay et al. [75,76].

18. Summary and Future Development

The review on cast Cu-Sn and Cu-Sn-Ti alloys was developed by evaluating up-to-date data and its main conclusions are listed.

- Firstly, cast bronze alloys were investigated in detail. The effects of Sn additions on the microstructure and morphology were precisely described.
- Important properties of bronze alloys, such as tensile strength, hardness, conductivity, corrosion, and wear resistance, were also listed above.
- Analysis of cast samples for bronze alloys along with their phase diagrams and phase relations were mentioned.
- Changes in the microstructure and morphology were obtained when different types of casting procedures, such as TZCC, investment casting and continuous casting, were performed. They were explained along with the effect of annealing and rolling.
- Eutectic, hypereutectic and peritectic compositions of bronze alloys were taken into account and their major applications were derived. They included electronic interconnections, underwater equipment (although inferior to stainless steel), detonating cords etc. They are also used in anti-seizure applications.
- Then, the additions of Ti on bronze alloys were taken into account and elaborated using research articles.
- Properties, phase diagrams, and phase relations along with major applications of cast Cu-Sn-Ti alloys were listed.

When comparing both cast alloy groups (Cu-Sn and Cu-Sn-Ti alloys), the cast Cu-Sn-Ti alloys turned out to be efficient replacement alloys for bronze because of their better wear-resistance, but they are not suitable for high load-bearing applications. The high tensile strength of Ti can promote its use as an abrasive material, and for applications where the material is exposed to high-stress levels. Future research can be made on this alloy to confirm their suitability for aerospace and marine industry applications as they show outstanding corrosion resistance, lightweight and tensile properties. Cu-Sn-Ti alloys are also used in biomedical applications.

Author Contributions: Conceptualization, K.V.S.; methodology, K.M.; writing—original draft preparation, K.M., S.A.N., M.R.S.G.; writing—review and editing, K.V.S., J.J., visualization, K.M., M.R.S.G., S.A.N.; supervision, K.V.S , J.J.; project administration, K.V.S., J.J.; funding acquisition, J.J. All authors have read and agreed to the published version of the manuscript.

Funding: Publication supported by the rector's professor's grant. Silesian University of Technology, grant number 10/080/RGP20/0061.

Institutional Review Board Statement: Not applicable.

Informed Consent Statement: Not applicable.

Data Availability Statement: The data presented in this study are available on request the corresponding author.

Acknowledgments: Not applicable.

Conflicts of Interest: The authors declare no conflict of interest.

References

1. Tęcza, G.; Zapała, R. Changes in Impact Strength and Abrasive Wear Resistance of Cast High Manganese Steel Due to the Formation of Primary Titanium Carbides. *Arch. Foundry Eng.* **2018**, *1*, 119–122.
2. Studnicki, A.; Dojka, R.; Gromczyk, M.; Kondracki, M. Influence of Titanium on Crystallization and Wear Resistance of High Chromium Cast Iron. *Arch. Foundry Eng.* **2016**, *16*, 117–123. [CrossRef]
3. Alex, S.; Chattopadhyay, K.; Basu, B. Tailored specular reflectance properties of bulk Cu based novel intermetallic alloys. *Sol. Energy Mater. Sol. Cells* **2016**, *149*, 66–74. [CrossRef]

4.	Gadallah, E.A.; Ghanem, M.A.; Abd El-Hamid, M.; El-Nikhaily, A.E. Effect of Tin Content and ECAP Passes on the Mechanical Properties of Cu/Sn Alloys. *Am. J. Sci. Technol.* **2014**, *1*, 60–68.

5.	Paul, C.; Sellamuthu, R. The Effect of Sn Content on the Properties of Surface Refined Cu-Sn Bronze Alloys. *Procedia Eng.* **2014**, *97*, 1341–1347. [CrossRef]

6.	Shankar, K.V.; Sellamuthu, R. Determination of the effect of tin content on microstructure, hardness, optimum aging temperature, and aging time for spinodal bronze alloys cast in a metal mould. *Int. J. Met.* **2017**, *11*, 189–194. [CrossRef]

7.	Bartocha, D.; Baron, C.; Suchoń, J. The Influence of Solidification Rate on High tin Bronze Microstructure. *Arch. Foundry Eng.* **2019**, *1*, 89–97.

8.	Nadolski, M. The Evaluation of Mechanical Properties of High-tin Bronzes. *Arch. Foundry Eng.* **2017**, *1*, 127–130. [CrossRef]

9.	Cherian, P.; Sellamuthu, R. An Investigation on the Effect of Process Parameters on Microstructure, Hardness and Wear Properties of Surface Modified Cu-Sn Bronze Alloy. *Appl. Mech. Mater.* **2014**, *592–594*, 58–62. [CrossRef]

10.	Zhang, J.; Cui, X.; Wang, Y.; Yang, Y.; Lin, J. Characteristics of ultrahigh electrical conductivity for Cu–Sn alloys. *Mater. Sci. Technol.* **2013**, *30*, 506–509. [CrossRef]

11.	Slamet, S.; Suyitno, S.; Kusumaningtyas, I.; Miasa, I.M. Effect of High-tin Bronze Composition on Physical, Mechanical, and Acoustic Properties of Gamelan Materials. *Arch. Foundry Eng.* **2021**, *1*, 137–145.

12.	Yang, Y.; Song, B.; Cheng, J.; Song, G.; Yang, Z.; Cai, Z. Effect of Super-gravity Field on Grain Refinement and Tensile Properties of Cu–Sn Alloys. *ISIJ Int.* **2018**, *58*, 98–106. [CrossRef]

13.	Yilmaz, O.; Turhan, H. The relationships between wear behaviour and thermal conductivity of CuSn/M7C3-M23C6 composites at ambient and elevated temperatures. *Compos. Sci. Technol.* **2001**, *61*, 2349–2359. [CrossRef]

14.	Osório, W.R.; Spinelli, J.; Afonso, C.R.M.; Peixoto, L.C.; Garcia, A. Microstructure, corrosion behaviour and microhardness of a directionally solidified Sn–Cu solder alloy. *Electrochim. Acta* **2011**, *56*, 8891–8899. [CrossRef]

15.	Martorano, M.; Capocchi, J. Heat transfer coefficient at the metal–mould interface in the unidirectional solidification of Cu–8%Sn alloys. *Int. J. Heat Mass Transf.* **2000**, *43*, 2541–2552. [CrossRef]

16.	Paradela, K.G.; Baptista, L.A.D.S.; Sales, R.C.; Junior, P.F.; Ferreira, A.F. Investigation of Thermal Parameters Effects on the Microstructure, Microhardness and Microsegregation of Cu-Sn alloy Directionally Solidified under Transient Heat Flow Conditions. *Mater. Res.* **2019**, *22*, 1–9. [CrossRef]

17.	Ozgowicz, W.; Grzegorczyk, B.; Pawelek, A.; Wajda, W.; Skuza, W.; Piatkowski, A.; Ranachowski, Z. Relation between the plastic instability and fracture of tensile tested Cu-Sn alloys investigated with the application of acoustic emission technique. *Frat. Integrità Strutt.* **2015**, *10*, 11–20. [CrossRef]

18.	Martorano, M.A.; Capocchi, J.D.T. Effects of processing variables on the microsegregation of directionally cast samples. *Met. Mater. Trans. A* **2000**, *31*, 3137–3148. [CrossRef]

19.	Bayle, B.; Bocher, P.; Jonas, J.; Montheillet, F. Flow stress and recrystallisation during hot deformation of Cu–9%Sn alloys. *Mater. Sci. Technol.* **1999**, *15*, 803–811. [CrossRef]

20.	Chen, S.-W.; Wang, C.-H. Effects of electromigration on interfacial reactions in cast Sn/Cu joints. *J. Mater. Res.* **2007**, *22*, 695–702. [CrossRef]

21.	Akbarifar, M.; Divandari, M. On the Interfacial Characteristics of Compound Cast Al/Brass Bimetals. *Int. J. Met.* **2017**, *11*, 506–512. [CrossRef]

22.	Fürtauer, S.; Li, D.; Cupid, D.; Flandorfer, H. The Cu–Sn phase diagram, Part I: New experimental results. *Intermetallics* **2013**, *34*, 142–147. [CrossRef] [PubMed]

23.	Li, Y.; He, K.; Liao, C.; Pan, C. Measurements of mechanical properties of α-phase in Cu-Sn alloys by using instru-mented nanoindentation. *J. Mater. Res.* **2012**, *27*, 192–196. [CrossRef]

24.	Jia, P.; Zhang, J.; Hu, X.; Li, C.; Zhao, D.; Teng, X.; Yang, C. Correlation between the resistivity and the atomic clusters in liquid Cu-Sn alloys. *Phys. B Condens. Matter* **2018**, *537*, 58–62. [CrossRef]

25.	Tan, M.; Xiufang, B.; Xianying, X.; Yanning, Z.; Jing, G.; Baoan, S. Correlation between viscosity of molten Cu–Sn alloys and phase diagram. *Phys. B Condens. Matter* **2007**, *387*, 1–5. [CrossRef]

26.	Mao, T.; Bian, X.; Morioka, S.; Wu, Y.; Li, X.; Lv, X. Effects of magnetic field on the viscosity of molten Cu–Sn alloys. *Phys. Lett. A* **2007**, *366*, 155–159. [CrossRef]

27.	Korneva, A.; Straumal, B.; Chulist, R.; Kilmametov, A.; Bala, P.; Cios, G.; Schcll, N.; Zięba, P. Grain refinement of in-termetallic compounds in the Cu-Sn system under high-pressure torsion. *Mater. Lett.* **2016**, *179*, 12–15. [CrossRef]

28.	Liu, G.; Ji, S. Microstructure, dynamic restoration, and recrystallization texture of Sn-Cu after rolling at room tem-perature. *Mater. Charact.* **2019**, *150*, 174–183. [CrossRef]

29.	Wang, X.; Zhao, J.; He, J.; Wang, J. Microstructural features and mechanical properties induced by the spray forming and cold rolling of the Cu-13.5 wt. pct Sn alloy. *J. Mater. Sci. Technol.* **2008**, *24*, 803–808.

30.	Liu, D.; Miller, W.A.; Aust, K.T. Annealing Twin Formation in a Cast and Annealed Cu-4% Sn Alloy. *Can. Met. Lurgical Q.* **1984**, *23*, 237–240. [CrossRef]

31.	Han, J.M.; Han, Y.S.; You, S.Y.; Kim, H.S. Mechanical behaviour of a new dispersion-strengthened bronze. *J. Mater. Sci.* **1997**, *32*, 6613–6618. [CrossRef]

32. Slamet, S.; Suyitno; Kusumaningtyas, I. Effect of Composition and Pouring Temperature of Cu-Sn Alloys on the Fluidity and Microstructure by Investment Casting. In Proceedings of the IOP Conference Series: Materials Science and Engineering, Bandung, Indonesia, 6–7 September 2019; Volume 547, p. 12010.
33. Liu, X.-F.; Luo, J.-H.; Wang, X.-C. Surface quality, microstructure and mechanical properties of Cu–Sn alloy plate prepared by two-phase zone continuous casting. *Trans. Nonferrous Met. Soc. China* **2015**, *25*, 1901–1910. [CrossRef]
34. Liu, X.; Luo, J.; Wang, X.; Wang, L.; Xie, J. Columnar grains-covered small grains Cu–Sn alloy prepared by two-phase zone continuous casting. *Prog. Nat. Sci.* **2013**, *23*, 94–101. [CrossRef]
35. Luo, J.; Liu, X.; Yang, K. Formation mechanism of microstructure in two-phase zone continuous casting Cu–Sn alloy. *Ferroelectrics* **2018**, *529*, 33–42. [CrossRef]
36. Luo, J.-H.; Liu, X.-F.; Shi, Z.-Z.; Liu, Y.-F. Microstructure quantification of Cu–4.7Sn alloys prepared by two-phase zone continuous casting and a BP artificial neural network model for microstructure prediction. *Rare Met.* **2018**, *38*, 1124–1130. [CrossRef]
37. Luo, J.; Liu, X.; Shi, L.; Cheng, C. Experimental and numerical simulation of surface segregation in two-phase zone continuous casting Cu–Sn alloy. In *Materials Science Forum*; Trans Tech Publications Ltd.: Stafa-Zurich, Switzerland, 2016; Volume 850, pp. 610–617.
38. Luo, J.; Liu, X.; Wang, X. Analysis of Temperature Field, Heat and Fluid Flow of Two-Phase Zone Continuous Casting Cu–Sn Alloy Wire. *Arch. Foundry Eng.* **2016**, *16*, 33–40. [CrossRef]
39. Wilson, R. *A Practical Approach to Continuous Casting of Copper-Based Alloys and Precious Metals*; IOM Communications Ltd.: London, UK, 2000; 266p.
40. Skoric, B.; Arsenovic, M.; Kutin, M.; Vasovic, I.; Ristic, M.; Milutinovic, Z. Thermographs and numerical simulations with respect to stress state and structure of condition cast specimens of bronze alloy. In Proceedings of the Fourth Serbian (29th Yu) Congress on Theoretical and Applied Mechanics, Vrnjacka Banja, Serbia, 4–7 June 2013; pp. 485–490.
41. Arsenovic, M.; Skoric, B.; Kutin, M.; Ristic, M. Influence of mold flow velocities on fracture of continuous cast speci-mens made of bronze alloy. In Proceedings of the 10th International Conference on Structural Integrity of Welded Structures (ISCS13), Timisoara, Romania, 11–12 July 2013.
42. Sergejevs AKromanis AOzolins, J.; Gerins, E. Influence of Casting velocity on mechanical properties and macrostructure of tin bronzes. In *Key Engineering Materials k*; Trans Tech Publications Ltd.: Stafa-Zurich, Switzerland, 2016; Volume 674, pp. 81–87.
43. Kostov, A.; Arsenovic, M.; Mitevska, N. Continuous casting of copper-based shape memory 8 mm wire. In Proceedings of the Second Balkan Conference on Development of Metallurgy in the Balkans at the beginning of 21st Century, Bucharest, Romania, 9–11 October 2000; pp. 60–63.
44. Copper and Copper Alloys. Ingots and Castings Patent EN 1982:2008, May 2008.
45. Luo, J.; He, F. Process Parameters on Exudation Thickness in Continuous Unidirectional Solidification Tin Bronze Alloy. *Arch. Foundry Eng.* **2019**, *2*, 97–100.
46. Luo, J. Preparation and Microstructure Evolution of Continuous Unidirectional Solidification Tin Bronze Alloy at Different Continuous Casting Speed. *Arch. Foundry Eng.* **2020**, *2*, 118–122.
47. Ludwig, A.; Gruber-Pretzler, M.; Wu, M.; Kuhn, A.; Riedle, J. About the Formation of Macrosegregations during Continuous Casting of Sn-Bronze. *Fluid Dyn. Mater. Process.* **2005**, *1*, 285–300.
48. Grasser, M.; Ishmurzin, A.; Mayer, F.; Wu, M.; Ludwig, A.; Hofmann, U.; Riedle, J. Micro-macrosegregation prediction based on solidification simulation for continuous casting of ternary bronze alloys. In Proceedings of the 12th International Conference on Modeling of Casting, Welding, and Advanced Solidification Processes, Vancouver, Canada, 7–14 June 2009; pp. 221–228.
49. Sugita IK, G.; Priambadi IG, N.; Kencanawati CI, P.K.; Astawa, K. Solidification characteristic of Cu-20%sn bronze alloys casting process by using sand and metal molds. In Proceedings of the IOP Conference Series: Materials Science and Engineering, Bali, Indonesia, 24–25 October 2018; IOP Publishing: Bristol, UK, 2019; Volume 539. [CrossRef]
50. Hao, J.; Grasser, M.; Wu, M.; Ludwig, A. Numerical study and experimental investigation of solidification phenomena in the semicontinuous casting of Bronze. *Adv. Mater. Res.* **2011**, *154–155*, 1401–1404.
51. Backman, D.G.; Mehrabian, R.; Flemings, M.C. Die thermal behavior in machine casting of partially solid high temperature alloys. *Met. Mater. Trans. A* **1977**, *8*, 471–477. [CrossRef]
52. Wu, C.M.L.; Huang, M.L. Creep behavior of eutectic Sn-Cu lead-free solder alloy. *J. Electron. Mater.* **2002**, *31*, 442–448. [CrossRef]
53. Spinelli, J.E.; Garcia, A. Microstructural development and mechanical properties of hypereutectic Sn-Cu solderalloys. *Mater. Sci. Eng. A* **2013**, *568*, 195–201. [CrossRef]
54. Kohler, F.; Germond, L.; Wagnière, J.D.; Rappaz, M. Peritectic solidification of Cu-Sn alloys: Microstructural competition at low speed. *Acta Mater.* **2009**, *57*, 56–68. [CrossRef]
55. Zhai, W.; Hong, Z.; Wen, X.; Geng, D.; Wei, B. Microstructural characteristics and mechanical properties of peritectic Cu Sn alloy solidified within ultrasonic field. *Mater. Des.* **2015**, *72*, 43–50. [CrossRef]
56. Kohler, F.; Campanella, T.; Nakanishi, S.; Rappaz, M. Application of single pan thermal analysis to Cu–Sn peritectic alloys. *Acta Mater.* **2008**, *56*, 1519–1528. [CrossRef]
57. Valloton, J.; Wagnière, J.-D.; Rappaz, M. Competition of the primary and peritectic phases in hypoperitectic Cu–Sn alloys solidified at low speed in a diffusive regime. *Acta Mater.* **2012**, *60*, 3840–3848. [CrossRef]
58. Gain, A.K.; Zhang, L. Growth nature of in-situ Cu_6Sn_5-phase and their influence on creep and damping charac-teristics of Sn-Cu material under high-temperature and humidity. *Microelectron. Reliab.* **2018**, *87*, 278–285. [CrossRef]

59. Abd El-Rehim, A.F.; Zahran, H.Y. Investigation of microstructure and mechanical properties of Sn-xCu solder alloys. *J. Alloy. Compd.* **2017**, *695*, 3666–3673. [CrossRef]

60. Tsao, L.; Chen, C. Corrosion characterization of Cu–Sn intermetallics in 3.5wt.% NaCl solution. *Corros. Sci.* **2012**, *63*, 393–398. [CrossRef]

61. Liu, G.; Ji, S.; Grechcini, L.; Bentley, A.; Fan, Z. Microstructure and mechanical properties of Sn-Cu alloy for detonating and explosive cords. *Mater. Sci. Technol.* **2017**, *33*, 1907–1918. [CrossRef]

62. Sato, T.; Hirai, Y.; Kobayashi, T. Development of Lead-Free Bronze with Sulfide Dispersion for Sliding Applications. *Int. J. Met.* **2017**, *11*, 148–154. [CrossRef]

63. Garbacz-Klempka, A.; Czekaj, E.; Kozana, J.; Perek-Nowak, M.; Piękoś, M. Influence of Al and Fe Additions on Structure and Properties of Cu-Sn Alloys. *Key Eng. Mater.* **2016**, *682*, 226–235. [CrossRef]

64. Vuorinen, V.; Dong, H.; Laurila, T. Effect of Ti on the interfacial reaction between Sn and Cu. *J. Mater. Sci. Mater. Electron.* **2012**, *23*, 68–74. [CrossRef]

65. Li, X.; Ivas, T.; Spierings, A.B.; Wegener, K.; Leinenbach, C. Phase and microstructure formation in rapidly solidi-fied Cu-Sn and Cu-Sn-Ti alloys. *J. Alloy. Compd.* **2018**, *735*, 1374–1382. [CrossRef]

66. Novakovic, R.; Ricci, E.; Amore, S.; Lanata, T. Surface and transport properties of Cu-Sn-Ti liquid alloys. *Rare Met.* **2006**, *25*, 457–468. [CrossRef]

67. Lebreton, V.; Pachoutinski, D.; Bienvenu, Y. An investigation of microstructure and mechanical properties in Cu–Ti–Sn alloys rich in copper. *Mater. Sci. Eng. A* **2009**, *508*, 83–92. [CrossRef]

68. Lin, C.-C.; Chen, R.-B.; Shiue, R.-K. A wettability study of Cu/Sn/Ti active braze alloys on alumina. *J. Mater. Sci.* **2001**, *36*, 2145–2150. [CrossRef]

69. Wang, J.; Liu, C.; Leinenbach, C.; Klotz, U.E.; Uggowitzer, P.J.; Löffler, J.F. Experimental investigation and ther-modynamic assessment of the CuSnTi ternary system. *Calphad Comput. Coupling Phase Diagr. Thermochem.* **2011**, *35*, 82–94. [CrossRef]

70. Naka, M.; Nakade, I.; Schuster, J.C.; Urai, S. Determination of the liquidus of the ternary system Cu-Sn-Ti. *J. Phase Equilibria* **2001**, *22*, 352–356. [CrossRef]

71. Huang, S.-F.; Tsai, H.-L.; Lin, S.-T. Crystal structure and X-ray diffraction pattern of CuSnTi3 intermetallic phase. *Intermetallics* **2005**, *13*, 87–92. [CrossRef]

72. Kizikov, E.D.; Kebko, V.P. Microadditions to alloys of the system Cu-Sn-Ti. *Met. Sci. Heat Treat.* **1987**, *29*, 68–71. [CrossRef]

73. Tsao, L.C.; Hsieh, M.; Yu, Y. Effects of Sn additions on microstructure and corrosion resistance of heat-treated Ti–Cu–Sn titanium alloys. *Corros. Eng. Sci. Technol.* **2018**, *53*, 252–258. [CrossRef]

74. Tsao, L. Effect of Sn addition on the corrosion behavior of Ti–7Cu–Sn cast alloys for biomedical applications. *Mater. Sci. Eng. C* **2015**, *46*, 246–252. [CrossRef]

75. Akshay, M.C.; Senan, V.R.A.; Surej, K.S.V.; Akhil, B.; Shankar, K.V.; Shankar, B. Determination on the Effect of Ti Addition on the Microstructural, Mechanical and Wear Behavior of Cu–6Sn Alloy in as—Cast Condition. *Trans. Indian Inst. Met.* **2019**, *73*, 309–318. [CrossRef]

76. Akshay, M.C.; Senan VR, A.; Surej KS, V.; Akhil, B.; Shankar, K.V.; Shankar, B. Determination of the effect of Ti addition on the mechanical behavior of as-cast Cu-6Sn alloy. In *AIP Conference Proceedings*; American Institute of Physics Inc.: College Park, MD, USA, 2019; Volume 2162. [CrossRef]

77. Rajendran, S.H.; Hwang, S.J.; Jung, J.P. Active Brazing of Alumina and Copper with Multicomponent Ag-Cu-Sn-Zr-Ti Filler. *Metals* **2021**, *11*, 509. [CrossRef]

78. Li, W.-C.; Lin, S.-T.; Liang, C. Interfacial segregation of Ti in the brazing of diamond grits onto a steel substrate using a Cu-Sn-Ti brazing alloy. *Met. Mater. Trans. A* **2002**, *33*, 2163–2172. [CrossRef]

79. Duan, D.-Z.; Xiao, B.; Wang, B.; Han, P.; Li, W.-J.; Xia, S.-W. Microstructure and mechanical properties of pre-brazed diamond abrasive grains using Cu–Sn–Ti alloy. *Int. J. Refract. Met. Hard Mater.* **2015**, *48*, 427–432. [CrossRef]

80. Zhang, Z.-Y.; Xiao, B.; Duan, D.-Z.; Wang, B.; Liu, S.-X. Investigation on the brazing mechanism and machining performance of diamond wire saw based on Cu-Sn-Ti alloy. *Int. J. Refract. Met. Hard Mater.* **2017**, *66*, 211–219. [CrossRef]

81. Yin, X.; Xu, F.; Min, C.; Cheng, Y.; Dong, X.; Cui, B.; Xu, D. Promoting the bonding strength and abrasion resistance of brazed diamond using Cu–Sn–Ti composite alloys reinforced with tungsten carbide. *Diam. Relat. Mater.* **2021**, *112*, 108239. [CrossRef]

82. Liu, S.; Xiao, B.; Xiao, H.; Meng, L.; Zhang, Z.; Wu, H. Characteristics of Al2O3/diamond/c-BN/SiC grain steel brazing joints using Cu–Sn–Ti active filler powder alloys. *Surf. Coat. Technol.* **2016**, *286*, 376–382. [CrossRef]

83. Buhl, S.; Leinenbach, C.; Spolenak, R.; Wegener, K. Microstructure, residual stresses and shear strength of diamond–steel-joints brazed with a Cu–Sn-based active filler alloy. *Int. J. Refract. Met. Hard Mater.* **2012**, *30*, 16–24. [CrossRef]

84. Ding, W.-F.; Xu, J.-H.; Chen, Z.-Z.; Miao, Q.; Yang, C.-Y. Interface characteristics and fracture behavior of brazed polycrystalline CBN grains using Cu-Sn-Ti alloy. *Mater. Sci. Eng. A* **2013**, *559*, 629–634. [CrossRef]

85. Fan, Y.; Fan, J.; Wang, C. Brazing temperature-dependent interfacial reaction layer features between CBN and Cu-Sn-Ti active filler metal. *J. Mater. Sci. Technol.* **2019**, *35*, 2163–2168. [CrossRef]

86. Fan, Y.; Fan, J.; Wang, C. Detailing interfacial reaction layer products between cubic boron nitride and Cu-Sn-Ti active filler metal. *J. Mater. Sci. Technol.* **2021**, *68*, 35–39. [CrossRef]

87. Hsieh, Y.-C.; Lin, S.-T. Microstructural development of Cu–Sn–Ti alloys on graphite. *J. Alloy. Compd.* **2008**, *466*, 126–132. [CrossRef]
88. He, Y.; Zhang, J.; Sun, Y.; Liu, C. Microstructure and mechanical properties of the Si3N4/42CrMo steel joints brazed with Ag–Cu–Ti+Mo composite filler. *J. Eur. Ceram. Soc.* **2010**, *30*, 3245–3251. [CrossRef]

MDPI

St. Alban-Anlage 66

4052 Basel

Switzerland

Tel. +41 61 683 77 34

Fax +41 61 302 89 18

www.mdpi.com

Materials Editorial Office

E-mail: materials@mdpi.com

www.mdpi.com/journal/materials

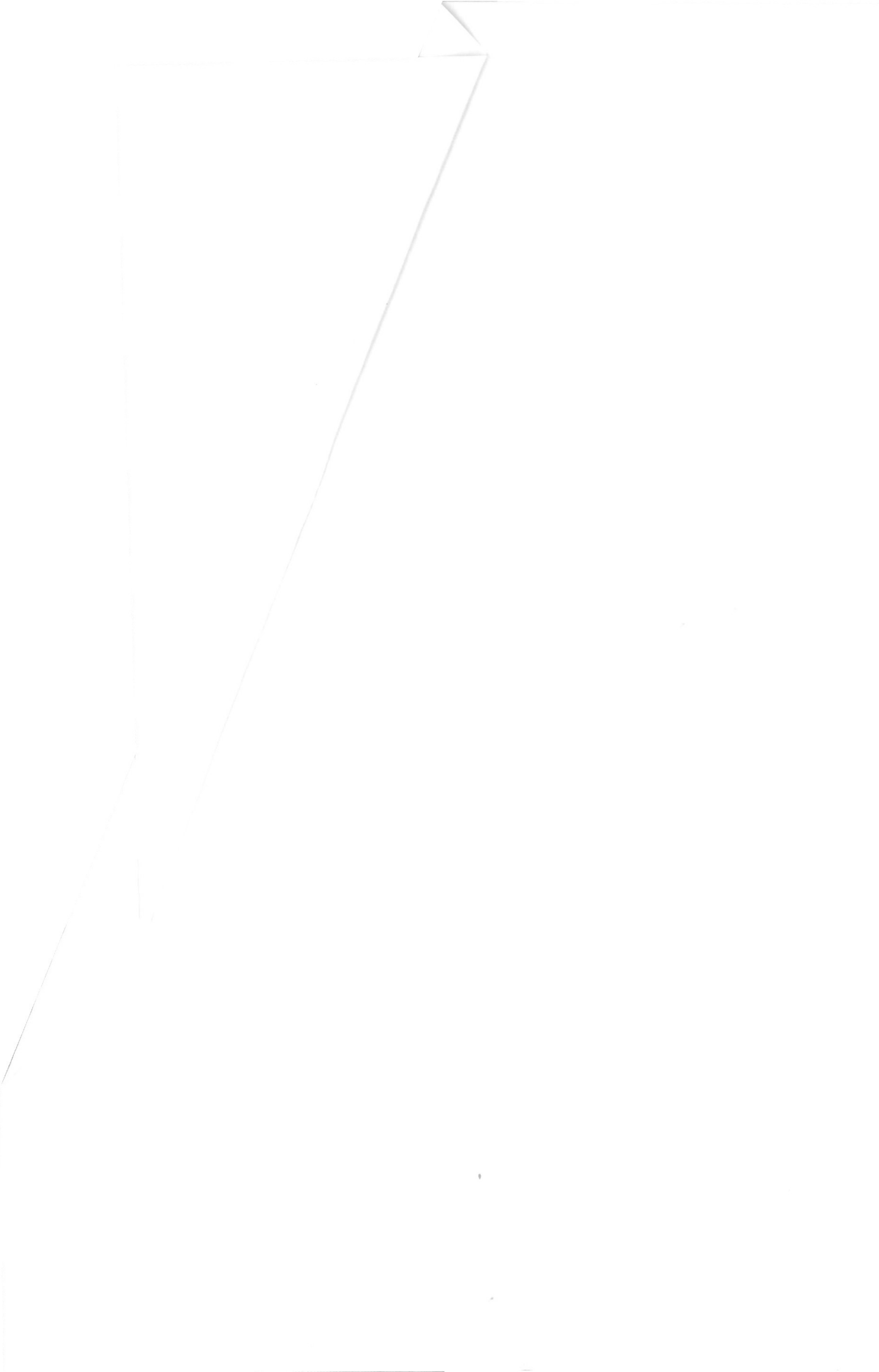

MDPI
St. Alban-Anlage 66
4052 Basel
Switzerland
Tel. +41 61 683 77 34
Fax +41 61 302 89 18
www.mdpi.com

Materials Editorial Office
E-mail: materials@mdpi.com
www.mdpi.com/journal/materials